T0280398

Lecture Notes in Mathematics

Wen Yuan · Winfried Sickel · Dachun Yang

Morrey and Campanato Meet Besov, Lizorkin and Triebel

Springer

Wen Yuan
School of Mathematical Sciences
Beijing Normal University
Laboratory of Mathematics
and Complex Systems
Ministry of Education
Beijing 100875
People's Republic of China
wyuan@mail.bnu.edu.cn

Dachun Yang
(Corresponding Author)
School of Mathematical Sciences
Beijing Normal University
Laboratory of Mathematics
and Complex Systems
Ministry of Education
Beijing 100875
People's Republic of China
dcyang@bnu.edu.cn

Winfried Sickel
Mathematisches Institut
Friedrich-Schiller-Universität Jena
Ernst-Abbe-Platz 2
Jena 07743
Germany
winfried.sickel@uni-jena.de

Corresponding author, who is supported by the National Natural Science Foundation (Grant No. 10871025) of China.

ISBN: 978-3-642-14605-3 e-ISBN: 978-3-642-14606-0
DOI: 10.1007/978-3-642-14606-0
Springer Heidelberg Dordrecht London New York

Lecture Notes in Mathematics ISSN print edition: 0075-8434
ISSN electronic edition: 1617-9692

Library of Congress Control Number: 2010935182

Mathematics Subject Classification (2010): 42B35, 46E35, 42B25, 42C40, 42B15, 47G30, 47H30

Cover design: WMXDesign GmbH, Heidelberg

Printed on acid-free paper

springer.com

Preface

This book is based on three developments in the theory of function spaces.

As *the first* we wish to mention **Besov** and **Triebel-Lizorkin spaces**. These scales

$$B^s_{p,q}(\mathbb{R}^n) \text{ and } F^s_{p,q}(\mathbb{R}^n)$$

allow a unified approach to various types of function spaces which have been known before like Hölder-Zygmund spaces, Sobolev spaces, Slobodeckij spaces and Bessel-potential spaces. Over the last 60 years these scales have proved their usefulness, there are hundreds of papers and many books using these scales in various connections. In a certain sense all these spaces are connected with the usual Lebesgue spaces $L^p(\mathbb{R}^n)$.

The *second source* we wish to mention is **Morrey** and **Campanato spaces**. Since several years there is an increasing interest in function spaces built on Morrey spaces and leading to generalizations of Campanato spaces. This interest originates, at least partly, in some applications in the field of Navier-Stokes equations.

The *third ingredient* is the so-called Q **spaces** (Q_α spaces). These spaces were originally defined as spaces of holomorphic functions on the unit disk, which are geometric in the sense that they transform naturally under conformal mappings. However, about 10 years ago, M. Essén, S. Janson, L. Peng and J. Xiao extended these spaces to the n-dimensional Euclidean space $\mathbb{R}^n$.

The aim of the book consists in giving a unified treatment of all these three types of spaces, i.e., we will define and investigate the scales

$$B^{s,\tau}_{p,q}(\mathbb{R}^n) \text{ and } F^{s,\tau}_{p,q}(\mathbb{R}^n)$$

generalizing the three types of spaces mentioned before. Such projects have been undertaken by various mathematicians during the last ten years, which have been investigating Besov-Morrey and Triebel-Lizorkin-Morrey spaces. Let us mention only the names Kozono, Yamazaki, Mazzucato, El Baraka, Sawano, Tang, Xu and two of the authors (W.Y. and D.Y.) in this connection. A more detailed history will be given in the first chapter of the book; see Sect. 1.2.

Let us further mention the approach of Hedberg and Netrusov [70] to general spaces of Besov-Triebel-Lizorkin type. There is some overlap with our treatment. Details will be given in Sect. 4.5.

The real persons Besov, Lizorkin and Triebel never met Morrey or Campanato (which we learned from personal communications with Professor Besov and Professor Triebel). However, we hope at least, the meaning of the title is clear. We shall develop a theory of spaces of Besov-Triebel-Lizorkin type built on Morrey spaces.

A second aim of the book, just a byproduct of the first, will be a completion of the theory of the Triebel-Lizorkin spaces $F^s_{\infty,q}(\mathbb{R}^n)$. By looking into the series of monographs written by Triebel over the last 30 years, these spaces play an exceptional role, in most of the cases they are even not treated. The only exception is the monograph [145], where they are introduced essentially as the dual spaces of $F^s_{1,q}(\mathbb{R}^n)$ (with some restrictions in q). Also after Jawerth and Frazier [64] have found a more appropriate definition, there have been no further contributions developing the theory of these spaces further, e. g., by establishing characterizations by differences or local oscillations (at least we do not know about).

In Chaps. 4–6 we shall prove characterizations by differences, local oscillations, and wavelets as well as assertions on the boundedness of pseudo-differential operators, nonlinear composition operators and pointwise multipliers.

In this book we only treat unweighted isotropic spaces, with other words, all directions and all points in $\mathbb{R}^n$ are of equal value. This means anisotropic and/or weighted spaces are not treated here. Further, we also do not deal with spaces of generalized smoothness or smoothness parameters depending on x (variable exponent spaces). However, some basic properties of corresponding spaces of Besov-Triebel-Lizorkin type are known in all these situations, we refer to

- Anisotropic spaces: [3, 13, 14, 148].
- Spaces of dominating mixed smoothness: [4, 128, 129, 151].
- Weighted spaces: [120, 129].
- Spaces of generalized smoothness: [57].
- Spaces of variable exponent: [47, 152].

Further investigations could be based also on a generalization of the underlying Morrey spaces, we refer to [29–31]. We believe that our methods could be applied also in these more general situations. But nothing is done at this moment.

The book contains eight chapters. Because of the generality of the spaces we use Chap. 1 for helping the reader to get an overview in various directions. First of all we summarize the contents of Chaps. 2–8. Second, we give a list of definitions of the function spaces which occur in the book. Third, we collect the various known coincidences of these spaces. Finally, we add a short history. Chapters 2–6 deal with the definition and basic properties of the spaces $B^{s,\tau}_{p,q}(\mathbb{R}^n)$ and $F^{s,\tau}_{p,q}(\mathbb{R}^n)$. Chapter 7 is devoted to the study of Besov-Hausdorff and Triebel-Lizorkin-Hausdorff spaces. Finally, in Chap. 8, parts of the theory of the *homogeneous* counterparts,

$$\dot{B}^{s,\tau}_{p,q}(\mathbb{R}^n) \text{ and } \dot{F}^{s,\tau}_{p,q}(\mathbb{R}^n),$$

of $B^{s,\tau}_{p,q}(\mathbb{R}^n)$ and $F^{s,\tau}_{p,q}(\mathbb{R}^n)$ are discussed.

The book is essentially self-contained. However, sometimes we carry over some results originally obtained for the homogeneous spaces, mainly from [163–165]. The papers [163–165] supplement the book in a certain sense. Most of the results are new in this generality and have been published never before.

Beijing and Jena
May, 2010

Wen Yuan
Winfried Sickel
Dachun Yang

Contents

Chapter 1
Introduction

The aim of this chapter is to give the reader a better orientation. For convenience of the reader we summarize the contents of the following chapters first, then we continue with some remarks to the history and finally, we collect the definitions of various function spaces and their coincidence relations.

1.1 A Short Summary of the Book

Chapter 2. For all $s, \tau \in \mathbb{R}$, all $p \in (0, \infty]$, and all $q \in (0, \infty]$, we introduce the inhomogeneous Besov-type spaces $B_{p,q}^{s,\tau}(\mathbb{R}^n)$. Triebel-Lizorkin-type spaces $F_{p,q}^{s,\tau}(\mathbb{R}^n)$ are defined for the same range of parameters except that p has to be less than infinity. Also corresponding sequence spaces, $b_{p,q}^{s,\tau}(\mathbb{R}^n)$ and $f_{p,q}^{s,\tau}(\mathbb{R}^n)$ (see Definitions 2.1 and 2.2 below), are introduced. The spaces $B_{p,q}^{s,\tau}(\mathbb{R}^n)$ and $F_{p,q}^{s,\tau}(\mathbb{R}^n)$ are the inhomogeneous counterparts of $\dot{B}_{p,q}^{s,\tau}(\mathbb{R}^n)$ and $\dot{F}_{p,q}^{s,\tau}(\mathbb{R}^n)$ introduced earlier in [164,165]. Via the Calderón reproducing formulae we establish the φ-transform characterization of these spaces in the sense of Frazier and Jawerth for all admissible values of the parameters s, τ, p, and q (see Theorem 2.1 below). On the one side this generalizes the classical results for $B_{p,q}^{s}(\mathbb{R}^n)$ and $F_{p,q}^{s}(\mathbb{R}^n)$ in [64, 65] by taking $\tau = 0$, on the other hand it also implies that $B_{p,q}^{s,\tau}(\mathbb{R}^n)$ and $F_{p,q}^{s,\tau}(\mathbb{R}^n)$ are well-defined. This method has to be traced to Frazier and Jawerth ([62,64]; see also [65]), and has been further developed by Bownik [23–25]. We continue by deriving some embedding properties for different metrics by using the φ-transform characterization; see Sect. 2.2 below. Finally, the Fatou property of $B_{p,q}^{s,\tau}(\mathbb{R}^n)$ and $F_{p,q}^{s,\tau}(\mathbb{R}^n)$ is established.

Chapter 3. To begin with, in Definition 3.1, we introduce a class of ε-almost diagonal operators on $b_{p,q}^{s,\tau}(\mathbb{R}^n)$ and $f_{p,q}^{s,\tau}(\mathbb{R}^n)$. Any ε-almost diagonal operator is an almost diagonal operator in the sense of Frazier and Jawerth [64]. The main result in the first part of this chapter is given in Theorem 3.1 and concerns the boundedness of these operators on $b_{p,q}^{s,\tau}(\mathbb{R}^n)$ and $f_{p,q}^{s,\tau}(\mathbb{R}^n)$, respectively. As an application we establish characterizations by atomic and molecular decompositions (see Theorems 3.2 and 3.3). In case $\tau = 0$, Theorems 3.1, 3.2 and 3.3 reduce to the well-known characterizations of $B_{p,q}^{s}(\mathbb{R}^n)$ and $F_{p,q}^{s}(\mathbb{R}^n)$, for which we refer to [25, 64, 65].

D. Yang et al., *Morrey and Campanato Meet Besov, Lizorkin and Triebel*,
Lecture Notes in Mathematics 2005, DOI 10.1007/978-3-642-14606-0_1,
© Springer-Verlag Berlin Heidelberg 2010

In the second section of this chapter we shall compare the spaces $B^{s,\tau}_{p,q}(\mathbb{R}^n)$ and $F^{s,\tau}_{p,q}(\mathbb{R}^n)$ with other approaches to introduce spaces of Besov-Triebel-Lizorkin type built on Morrey spaces. Let $\mathcal{N}^s_{pqu}(\mathbb{R}^n)$ denote the Besov-Morrey spaces; see (xxv) in Sect. 1.3. Then our main result consists in

$$B^{s,1/u-1/p}_{u,\infty}(\mathbb{R}^n) = \mathcal{N}^s_{p\infty u}(\mathbb{R}^n), \qquad 0 < u \le p \le \infty,$$

in the sense of equivalent quasi-norms and, if $0 < q < \infty$,

$$\mathcal{N}^s_{pqu}(\mathbb{R}^n) \subset B^{s,1/u-1/p}_{u,q}(\mathbb{R}^n), \quad \mathcal{N}^s_{pqu}(\mathbb{R}^n) \neq B^{s,1/u-1/p}_{u,q}(\mathbb{R}^n), \quad 0 < u \le p \le \infty.$$

Let $\mathcal{E}^s_{pqu}(\mathbb{R}^n)$ $(p \neq \infty)$ denote the Triebel-Lizorkin-Morrey spaces studied in [88, 126, 139]. Then we have

$$F^{s,1/u-1/p}_{u,q}(\mathbb{R}^n) = \mathcal{E}^s_{pqu}(\mathbb{R}^n), \qquad 0 < u \le p < \infty,$$

with equivalent quasi-norms. In particular, if $1 < u \le p < \infty$

$$F^{0,1/u-1/p}_{u,2}(\mathbb{R}^n) = \mathcal{E}^0_{p2u}(\mathbb{R}^n) = \mathcal{M}^p_u(\mathbb{R}^n),$$

also in the sense of with equivalent norms. Thus, these conclusions combined with Theorem 2.1 also give the φ-transform characterization of the spaces $\mathcal{N}^s_{p\infty u}(\mathbb{R}^n)$ and $\mathcal{E}^s_{pqu}(\mathbb{R}^n)$, which seems to be also new.

Chapter 4. Following a well-known but rather long and technical procedure (see, for example, [109] and [145]), we establish some equivalent characterizations of the spaces $B^{s,\tau}_{p,q}(\mathbb{R}^n)$ and $F^{s,\tau}_{p,q}(\mathbb{R}^n)$. Step by step we establish the following chain of inequalities. First we shall show that Littlewood-Paley characterizations can be dominated by characterizations by differences. The second step consists in proving that characterizations by differences can be estimated from above either by characterizations by oscillations or in terms of wavelet coefficients. The third step consists in estimating oscillations by wavelet coefficients. Finally, as an application of our atomic characterizations we can close the circle and estimate these expressions in terms of wavelet coefficients by the Littlewood-Paley characterization. Here we obtain generalizations of the well-known corresponding results for $B^s_{p,q}(\mathbb{R}^n)$ and $F^s_{p,q}(\mathbb{R}^n)$ $(p < \infty)$. They seem to be new for the classes $F^s_{\infty,q}(\mathbb{R}^n)$. A few more interesting localization properties of $B^{s,\tau}_{p,q}(\mathbb{R}^n)$ and $F^{s,\tau}_{p,q}(\mathbb{R}^n)$ will given as well. In fact, at least for small s, membership of a continuous function in $F^{s,\tau}_{p,q}(\mathbb{R}^n)$ and $B^{s,\tau}_{p,q}(\mathbb{R}^n)$ can be checked by investigating the local behavior of this function in the corresponding space with $\tau = 0$.

Chapter 5. Based on the smooth atomic and molecular decompositions, derived in Theorems 3.2 and 3.3, we shall prove here the boundedness of exotic pseudo-differential operators on $B^{s,\tau}_{p,q}(\mathbb{R}^n)$ and $F^{s,\tau}_{p,q}(\mathbb{R}^n)$ (see Theorem 5.1) under some restrictions for τ. This has several useful consequences. As applications of Theorem 5.1, we can establish mapping properties of $f \to \partial f$ as well as the so-called lifting property. Furthermore, we study the boundedness of nonlinear composition operators $T_f:\ g \to f \circ g$ on spaces $A^{s,\tau}_{p,q}(\mathbb{R}^n) \cap C(\mathbb{R}^n)$.

Chapter 6. This chapter is devoted to so-called key theorems; see [146, Chap. 4]. Assertions on pointwise multipliers (see Theorem 6.1), on diffeomorphisms (see Theorem 6.7) and traces (see Theorem 6.8) belong to this group. These theorems are basic for the definitions of Besov-Triebel-Lizorkin-type spaces on domains. We finally introduce Besov-Triebel-Lizorkin-type spaces on $\mathbb{R}^n_+$ and on bounded C^∞ domains in $\mathbb{R}^n$ and discuss a few properties.

Chapter 7. The main aim of this chapter consists in defining and investigating a class of spaces which have as duals the classes $A^{s,\tau}_{p,q}(\mathbb{R}^n)$. These spaces are introduced by using the Hausdorff capacity. For this reason we call them Besov-Hausdorff spaces $BH^{s,\tau}_{p,q}(\mathbb{R}^n)$ and Triebel-Lizorkin-Hausdorff spaces $FH^{s,\tau}_{p,q}(\mathbb{R}^n)$, respectively. They are the predual spaces of $B^{-s,\tau}_{p',q'}(\mathbb{R}^n)$ and $F^{-s,\tau}_{p',q'}(\mathbb{R}^n)$ (see Theorem 7.3 below). If $\tau = 0$, these results reduce to the classical duality assertions for Besov spaces $B^s_{p,q}(\mathbb{R}^n)$ and Triebel-Lizorkin spaces $F^s_{p,q}(\mathbb{R}^n)$. These new scales $BH^{s,\tau}_{p,q}(\mathbb{R}^n)$ and $FH^{s,\tau}_{p,q}(\mathbb{R}^n)$ have many properties in common with the classes $B^{s,\tau}_{p,q}(\mathbb{R}^n)$ and $F^{s,\tau}_{p,q}(\mathbb{R}^n)$. In particular, we establish the φ-transform characterization, characterizations by smooth atomic and molecular decompositions, boundedness of certain pseudo-differential operators, the lifting property, a pointwise multiplier and a diffeomorphism theorem and finally assertions on traces. However, the most important property is the following: let $s \in \mathbb{R}$, $p = q \in (0,\infty)$ and $\tau \in [0, \frac{1}{p}]$, then

$$({}_0B^{s,\tau}_{p,p}(\mathbb{R}^n))^* = BH^{-s,\tau}_{p',p'}(\mathbb{R}^n),$$

where ${}_0B^{s,\tau}_{p,p}(\mathbb{R}^n)$ denotes the closure of $C^\infty_c(\mathbb{R}^n) \cap B^{s,\tau}_{p,p}(\mathbb{R}^n)$ in $B^{s,\tau}_{p,p}(\mathbb{R}^n)$ (see Theorem 7.12 below). By taking $s = 0$, $p = 2$ and $\tau = 1/2$ we get back the well-known result

$$(\mathrm{cmo}\,(\mathbb{R}^n))^* = h^1(\mathbb{R}^n),$$

where $\mathrm{cmo}\,(\mathbb{R}^n)$ is the local $\mathrm{CMO}(\mathbb{R}^n)$ space and $h^1(\mathbb{R}^n)$ is the local Hardy space; see Sect. 1.3. For suitable indices, the behavior of the scales $BH^{s,\tau}_{p,q}(\mathbb{R}^n)$ and $FH^{s,\tau}_{p,q}(\mathbb{R}^n)$ under real interpolation is investigated; see Theorem 7.14 below.

Chapter 8. In the last chapter we focus on the homogeneous case. The homogeneous spaces, including homogeneous Besov-type spaces $\dot{B}^{s,\tau}_{p,q}(\mathbb{R}^n)$, Triebel-Lizorkin-type spaces $\dot{F}^{s,\tau}_{p,q}(\mathbb{R}^n)$ and their preduals, homogeneous Besov-Hausdorff spaces $B\dot{H}^{s,\tau}_{p,q}(\mathbb{R}^n)$ and Triebel-Lizorkin-Hausdorff spaces $F\dot{H}^{s,\tau}_{p,q}(\mathbb{R}^n)$, were introduced and investigated in [127,164,165,168]. We gather some corresponding results for these spaces. In particular, we establish their wavelet characterizations (see Theorem 8.2 below).

1.2 A Piece of History

Here we will give a very rough overview about the history, mentioning some pioneering work, but without having the aim to reach completeness.

1.2.1 Besov-Triebel-Lizorkin Spaces

Nikol'skij [108] introduced in 1951 the Nikol'skij-Besov spaces, nowadays denoted by $B^s_{p,\infty}(\mathbb{R}^n)$. However, he was mentioning that this was based on earlier work of Bernstein [10] ($p = \infty$) and Zygmund [170] (periodic case, $n = 1, 1 < p < \infty$). Besov [11, 12] complemented the scale by introducing the third index q in 1959. We also refer to Taibleson [136–138] for the early investigations of Besov spaces. Around 1970 Lizorkin [91, 92] and Triebel [142] started to investigate the scale $F^s_{p,q}(\mathbb{R}^n)$, nowadays named after these two mathematicians. Further, we have to mention the contributions of Peetre [111, 113, 114], who extended around 1973–1975 the range of the admissible parameters p and q to values less than one.

Of particular importance for us has been the fundamental paper [64] of Frazier and Jawerth; see also [62,63] and the monograph [65] of Frazier, Jawerth and Weiss in this connection. In these papers, the authors describe the Besov and Triebel-Lizorkin spaces in terms of a fixed countable family of functions with certain properties, namely, smooth atoms and molecules, which have been a second break-through in a certain sense (after the Fourier-analytic one in the seventieth), preparing the nowadays widely used wavelet decompositions. However, these decompositions were prepared by earlier contributions to the Calderón reproducing formula in [32, 38, 150, 155] and the studies in [41, 115]. We refer to the introduction in [64] for more details.

The theory is summarized in the monographs [14, 109, 114, 145–149]. A much more detailed history can be found in [146, 148]; see also [153].

1.2.2 Morrey-Campanato Spaces

In 1938 Morrey [102] introduced the classes $\mathcal{M}^p_u(\mathbb{R}^n)$ which are generalizations of the ordinary Lebesgue spaces. Next we would like to mention the work of John and Nirenberg, which introduced *BMO* in 1961 (see [79]). At the beginning of the sixties, in a series of papers, Campanato introduced and studied the spaces $\mathcal{L}^{p,\lambda}(\mathbb{R}^n)$, nowadays named after him; see also Meyers [101]. Peetre [110] gave a survey on the topic (to which we refer also for more detailed comments to the early history) and studied the interpolation properties of these classes. Section 2.4 in the monograph [88] of Kufner, John and Fučik is devoted to the study of Morrey and Campanato spaces and summarizes the state of the art at 1975.

Function spaces, defined by oscillations, i. e., local approximation by polynomials, were studied by Brudnij [26, 27], Il'in [13, 14], Christ [40], Bojarski [15], DeVore and Sharpley [46], Wallin [153], Seeger [130], and Triebel [146, Sect. 1.7], to mention only a few. Important for us has been also the general approach of Hedberg and Netrusov [70] to those function spaces.

1.2.3 Spaces Built on Morrey-Campanato Spaces

The Besov-Morrey spaces $\mathcal{N}^s_{pqu}(\mathbb{R}^n)$, $1 < u \le p < \infty$, $1 < q \le \infty$, were studied for the first time by Kozono and Yamazaki [88] in connection with applications to the Navier-Stokes equation. Also in connection with applications to pde the homogeneous version $\dot{\mathcal{N}}^s_{pqu}(\mathbb{R}^n)$, $1 < u \le p < \infty$, $1 < q \le \infty$, were studied by Mazzucato [97]. The next step has been done by Tang and Xu [139]. They introduced the scale $\mathcal{E}^s_{pqu}(\mathbb{R}^n)$ (the Triebel-Lizorkin counterpart of $\mathcal{N}^s_{pqu}(\mathbb{R}^n)$) and made first investigations for the extended range $0 < u \le p < \infty$, $0 < q \le \infty$, of parameters for both types of spaces. Later, Sawano and Tanaka [126] presented various decompositions including quarkonial, atomic and molecular characterizations of $\mathcal{A}^s_{pqu}(\mathbb{R}^n)$ and $\dot{\mathcal{A}}^s_{pqu}(\mathbb{R}^n)$, where $\mathcal{A} \in \{\mathcal{N}, \mathcal{E}\}$. Jia and Wang [78] investigated the Hardy-Morrey spaces, which are special cases of Triebel-Lizorkin-Morrey spaces. In [154], Wang obtained the atomic characterization and the trace theorem for Besov-Morrey and Triebel-Lizorkin-Morrey spaces independently of Sawano and Tanaka. Recently, Sawano [125] investigated the Sobolev embedding theorem for Besov-Morrey spaces. Recall that the Besov-Morrey and Triebel-Lizorkin-Morrey spaces cover many classic function spaces, such as Besov spaces, Triebel-Lizorkin spaces, Morrey spaces and Sobolev-Morrey spaces. For the Sobolev-Morrey spaces, we refer to Najafov [103–105].

The Besov-type space $B^{s,\tau}_{p,q}(\mathbb{R}^n)$ and its homogeneous version $\dot{B}^{s,\tau}_{p,q}(\mathbb{R}^n)$, restricted to the Banach space case, were first introduced by El Baraka in [49–51]. In these papers, El Baraka investigated embeddings as well as Littlewood-Paley characterizations of Campanato spaces. El Baraka showed that the spaces $B^{s,\tau}_{p,q}(\mathbb{R}^n)$ cover certain Campanato spaces (see [51]).

Triebel-Lizorkin-Morrey spaces $\dot{\mathcal{E}}^s_{pqu}(\mathbb{R}^n)$ $(p \ne \infty)$ have been studied in [88, 126, 139]. The identity

$$\dot{F}^{s,\tau}_{p,q}(\mathbb{R}^n) = \dot{\mathcal{E}}^s_{pqu}(\mathbb{R}^n)$$

has been proved in [127].

The Besov-type spaces $\dot{B}^{s,\tau}_{p,q}(\mathbb{R}^n)$ and the Triebel-Lizorkin-type spaces $\dot{F}^{s,\tau}_{p,q}(\mathbb{R}^n)$ were introduced in [164, 165].

1.2.4 Q Spaces

The history of Q_α spaces (or simply Q spaces) started in 1995 with a paper by Aulaskari, Xiao and Zhao [7]. Originally they were defined as spaces of holomorphic functions on the unit disk, which are geometric in the sense that they transform naturally under conformal mappings (see [7, 160]). Following earlier contributions of Essén and Xiao [55] and Janson [76] on the boundary values of these functions on the unit circle, Essén, Janson, Peng and Xiao [54] extended these spaces to the n-dimensional Euclidean space $\mathbb{R}^n$. There is a rapidly increasing literature devoted to this subject, we refer to [7, 44, 45, 54, 55, 76, 157–162, 169].

Most recently, in [164, 165], two of the authors (W.Y and D.Y) have introduced the scales of homogeneous Besov-Triebel-Lizorkin-type spaces $\dot{B}^{s,\tau}_{p,q}(\mathbb{R}^n)$ and $\dot{F}^{s,\tau}_{p,q}(\mathbb{R}^n)$ ($p \neq \infty$), which generalize the homogeneous Besov-Triebel-Lizorkin spaces ($\dot{B}^s_{p,q}(\mathbb{R}^n)$, $\dot{F}^s_{p,q}(\mathbb{R}^n)$) and Q spaces simultaneously, and hence answered an open question posed by Dafni and Xiao in [44] concerning the relation of these spaces. In fact, it holds

$$\dot{F}^{\alpha,\frac{1}{2}-\frac{\alpha}{n}}_{2,2}(\mathbb{R}^n) = Q_\alpha(\mathbb{R}^n)$$

if $\alpha \in (0,1)$ ($n \geq 2$).

Recently, Xiao [161], Li and Zhai [90] applied certain special cases of $\dot{B}^{s,\tau}_{p,q}(\mathbb{R}^n)$ and $\dot{F}^{s,\tau}_{p,q}(\mathbb{R}^n)$, including the Q spaces, to study the Navier-Stokes equation.

1.3 A Collection of the Function Spaces Appearing in the Book

As a service for the reader and also for having convenient references inside the book we give a list of definitions of the spaces of functions (distributions) showing up in this book. Sometimes a few comments will be added. We picked up this idea from [145, Sect. 2.2.2] and [153] and a part of our list is just a copy of the list given in [145].

As a general rule within this book we state that all spaces consist of complex-valued functions. We shall divide our collection into three groups:

- Function spaces defined by derivatives and differences.
- Function spaces defined by mean values and oscillations (local polynomial approximations).
- Function spaces defined by Fourier analytic tools.

The first item contains the classical approaches to define smoothness. In the second item we recall the definitions of spaces related to Morrey-Campanato spaces. Finally, in the third item we define spaces by Fourier analytic tools, in most of the cases by using a smooth dyadic resolution of unity.

1.3.1 Function Spaces Defined by Derivatives and Differences

(i) **Lebesgue spaces**. Let $p \in (0,\infty)$. By $L^p(\mathbb{R}^n)$ we denote the space of all measurable functions f such that

$$\|f\|_{L^p(\mathbb{R}^n)} \equiv \left(\int_{\mathbb{R}^n} |f(x)|^p\,dx\right)^{1/p} < \infty.$$

In case $p = \infty$ the space $L^\infty(\mathbb{R}^n)$ is the collection of all measurable functions f such that

$$\|f\|_{L^\infty(\mathbb{R}^n)} \equiv \operatorname*{ess\,sup}_{x\in\mathbb{R}^n} |f(x)| < \infty.$$

Of a certain importance for the book are the following modified Lebesgue-type spaces. Let $\tau \in [0,\infty)$ and $p \in (0,\infty]$. Let $L^p_\tau(\mathbb{R}^n)$ be the collection of functions $f \in L^p_{\mathrm{loc}}(\mathbb{R}^n)$ such that

$$\|f\|_{L^p_\tau(\mathbb{R}^n)} \equiv \sup \frac{1}{|P|^\tau} \left(\int_P |f(x)|^p \, dx \right)^{1/p},$$

where the supremum is taken over all dyadic cubes P with side length $l(P) \geq 1$.

(ii) The space $C(\mathbb{R}^n)$ consists of all uniformly continuous functions f such that

$$\|f\|_{C(\mathbb{R}^n)} \equiv \sup_{x\in\mathbb{R}^n} |f(x)| < \infty.$$

(iii) Let $m \in \mathbb{N}$. The space $C^m(\mathbb{R}^n)$ consists of all functions $f \in C(\mathbb{R}^n)$, having all classical derivatives $\partial^\alpha f \in C(\mathbb{R}^n)$ up to order $|\alpha| \leq m$ and such that

$$\|f\|_{C^m(\mathbb{R}^n)} \equiv \sum_{|\alpha|\leq m} \|\partial^\alpha f\|_{C(\mathbb{R}^n)} < \infty.$$

We put $C^0(\mathbb{R}^n) \equiv C(\mathbb{R}^n)$.

(iv) **Hölder spaces**. Let $m \in \mathbb{Z}_+$ and $s \in (m, m+1)$. Then $C^s(\mathbb{R}^n)$ denotes the collection of all functions $f \in C^m(\mathbb{R}^n)$ such that

$$\|f\|_{C^s(\mathbb{R}^n)} \equiv \|f\|_{C^m(\mathbb{R}^n)} + \sum_{|\alpha|=m} \sup_{x\neq y} \frac{|\partial^\alpha f(x) - \partial^\alpha f(y)|}{|x-y|^{s-m}} < \infty.$$

(v) **Lipschitz spaces**. Let $s \in (0,1]$. The Lipschitz space $\mathrm{Lip}\, s(\mathbb{R}^n)$ consists of all functions $f \in C(\mathbb{R}^n)$ such that

$$\|f\|_{\mathrm{Lip}\, s(\mathbb{R}^n)} \equiv \sup_{x\neq y} \frac{|f(x)-f(y)|}{|x-y|^s} < \infty.$$

(vi) **Zygmund spaces**. Let $m \in \mathbb{N}$. The Zygmund space $\mathscr{Z}^m(\mathbb{R}^n)$ consists of all functions $f \in C^{m-1}(\mathbb{R}^n)$ such that

$$\begin{aligned} \|f\|_{\mathscr{Z}^m(\mathbb{R}^n)} &\equiv \|f\|_{C^{m-1}(\mathbb{R}^n)} \\ &\quad + \max_{|\alpha|=m} \sup_{h\neq 0} \sup_{x\in\mathbb{R}^n} \frac{|\partial^\alpha f(x+2h) - 2\partial^\alpha f(x+h) + \partial^\alpha f(x)|}{|h|} < \infty. \end{aligned}$$

In case of $s > 0$, $s \notin \mathbb{N}$, we use the convention $\mathscr{Z}^s(\mathbb{R}^n) = C^s(\mathbb{R}^n)$.

(vii) **Sobolev spaces**. Let $p \in (1,\infty)$ and $m \in \mathbb{N}$. Then $W^m_p(\mathbb{R}^n)$ is the collection of all functions $f \in L^p(\mathbb{R}^n)$ such that the distributional derivatives $\partial^\alpha f$ are functions belonging to $L^p(\mathbb{R}^n)$ for all α, $|\alpha| \leq m$. We equip this set with the norm

$$\|f\|_{W^m_p(\mathbb{R}^n)} \equiv \sum_{|\alpha|\leq m} \|\partial^\alpha f\|_{L^p(\mathbb{R}^n)}.$$

As usual, we define $W^0_p(\mathbb{R}^n) \equiv L^p(\mathbb{R}^n)$.

(viii) **Slobodeckij spaces**. Let $p \in [1,\infty)$ and let $s \in (0,\infty)$ be not an integer. Let $m \in \mathbb{Z}_+$ such that $s \in (m,m+1)$. Then $W_p^s(\mathbb{R}^n)$ consists of all functions $f \in W_p^m(\mathbb{R}^n)$ such that

$$\|f\|_{W_p^s(\mathbb{R}^n)} \equiv \|f\|_{W_p^m(\mathbb{R}^n)} + \sum_{|\alpha|=m} \left(\int_{\mathbb{R}^n\times\mathbb{R}^n} \frac{|\partial^\alpha f(x) - \partial^\alpha f(y)|^p}{|x-y|^{n+(m+1-s)p}}\, dx\,dy \right)^{1/p} < \infty.$$

(ix) **Besov spaces** (classical variant). Let $s \in (0,\infty)$ and $p, q \in [1,\infty]$. Let $M \in \mathbb{N}$. Then, if $s \in [M-1,M)$, the space $B_{p,q}^s(\mathbb{R}^n)$ is the collection of all functions $f \in L^p(\mathbb{R}^n)$ satisfying

$$\|f\|_{B_{p,q}^s(\mathbb{R}^n)} \equiv \|f\|_{L^p(\mathbb{R}^n)} + \left(\int_{\mathbb{R}^n} |h|^{-sq} \|\Delta_h^M f(\cdot)\|_{L^p(\mathbb{R}^n)}^q \frac{dh}{|h|^n} \right)^{1/q} < \infty.$$

Besov spaces can be defined in various ways; see in particular item (xx) below. In Chaps. 2–4 we shall prove the equivalence of some of these approaches in a much more general context.

1.3.2 Function Spaces Defined by Mean Values and Oscillations

Now we turn to a group of spaces which are related to Morrey-Campanato spaces.

(x) Functions of **bounded mean oscillations**. The space $\mathrm{BMO}(\mathbb{R}^n)$ is the set of locally integrable functions f on $\mathbb{R}^n$ such that

$$\|f\|_{\mathrm{BMO}(\mathbb{R}^n)} \equiv \sup_Q \frac{1}{|Q|} \int_Q |f(x) - f_Q|\, dx < \infty,$$

where the supremum is taken on all cubes Q with sides parallel to the coordinate axes and where

$$f_Q \equiv \frac{1}{|Q|} \int_Q f(x)\, dx$$

denotes the mean value of the function f on Q.

(xi) According to Sarason [122], a function f of $\mathrm{BMO}(\mathbb{R}^n)$ which satisfies the limiting condition

$$\lim_{a\to 0} \left(\sup_{|Q|\le a} \frac{1}{|Q|} \int_Q |f(x) - f_Q|\, dx \right) = 0$$

is said to be of **vanishing mean oscillation**. The subspace of $\mathrm{BMO}(\mathbb{R}^n)$ consisting of the functions of vanishing mean oscillation is denoted by $\mathrm{VMO}(\mathbb{R}^n)$. We note that the space $\mathrm{VMO}(\mathbb{R}^n)$ considered by Coifman and Weiss [42] is different from that considered by Sarason, and it coincides with our $\mathrm{CMO}(\mathbb{R}^n)$; see the next item.

(xii) We denote by $\mathrm{CMO}(\mathbb{R}^n)$ the closure of $C_c^\infty(\mathbb{R}^n)$ in $\mathrm{BMO}(\mathbb{R}^n)$, and we endow $\mathrm{CMO}(\mathbb{R}^n)$ with the norm of $\mathrm{BMO}(\mathbb{R}^n)$.

(xiii) Functions of **local bounded mean oscillations**. The space $\mathrm{bmo}(\mathbb{R}^n)$ consists of all functions $f \in \mathrm{BMO}(\mathbb{R}^n)$ which satisfy also the following condition

$$\sup_{|Q|\geq 1} \frac{1}{|Q|} \int_Q |f(x)|\,dx < \infty.$$

We equip this space with the norm

$$\|f\|_{\mathrm{bmo}(\mathbb{R}^n)} \equiv \|f\|_{\mathrm{BMO}(\mathbb{R}^n)} + \sup_{|Q|=1} \int_Q |f(x)|\,dx.$$

(xiv) Functions of **local vanishing mean oscillations**. We set

$$\mathrm{vmo}(\mathbb{R}^n) \equiv \mathrm{VMO}(\mathbb{R}^n) \cap \mathrm{bmo}(\mathbb{R}^n),$$

and we endow the space $\mathrm{vmo}(\mathbb{R}^n)$ with the norm of $\mathrm{bmo}(\mathbb{R}^n)$.

(xv) We denote by $\mathrm{cmo}(\mathbb{R}^n)$ the closure of $C_c^\infty(\mathbb{R}^n)$ in $\mathrm{bmo}(\mathbb{R}^n)$, and we endow $\mathrm{cmo}(\mathbb{R}^n)$ with the norm of $\mathrm{bmo}(\mathbb{R}^n)$.

(xvi) **Morrey spaces**. Let $0 < u \leq p \leq \infty$. The space $\mathcal{M}_u^p(\mathbb{R}^n)$ is defined to be the set of all u-locally Lebesgue-integrable functions f on $\mathbb{R}^n$ such that

$$\|f\|_{\mathcal{M}_u^p(\mathbb{R}^n)} \equiv \sup_B |B|^{1/p-1/u} \left(\int_B |f(x)|^u\,dx \right)^{1/u} < \infty,$$

where the supremum is taken over all balls B in $\mathbb{R}^n$; see [89, Sect. 2.4].

(xvii) **Campanato spaces**. Let $\lambda \in [0,\infty)$ and $p \in [1,\infty)$. The collection of all functions $f \in L^p_{\mathrm{loc}}(\mathbb{R}^n)$ such that

$$\|f\|_{\mathcal{L}^{p,\lambda}(\mathbb{R}^n)} \equiv \sup_B \frac{1}{|B|^{\lambda/n}} \left(\int_B |f(x) - f_B|^p\,dx \right)^{1/p} < \infty,$$

where the supremum is taken over all balls B in $\mathbb{R}^n$.

This set becomes a Banach space if functions are considered modulo constants. Furthermore, $\mathcal{L}^{p,\lambda}(\mathbb{R}^n)$ consists of the constant functions only if $\lambda > n+p$; see [33–36], [110] and [89, Sect. 2.4].

(xviii) **Local approximation spaces I.** Let $p \in [1,\infty)$ and $s \in [-n/p,\infty)$. Let $B(x,t)$ be the ball with center x and radius t. Let $M \in \mathbb{Z}_+$. Denote by $\mathscr{P}_M(\mathbb{R}^n)$ the set of all polynomials of total degree less than or equal to M. For $u \in (0,\infty]$ we define the local oscillation of $f \in L^u_{\ell oc}(\mathbb{R}^n)$ by setting, for all $x \in \mathbb{R}^n$ and all $t \in (0,\infty)$,

$$\mathrm{osc}_u^M f(x,t) \equiv \inf \left(t^{-n} \int_{B(x,t)} |f(y) - P(y)|^u\,dy \right)^{1/u},$$

where the infimum is taken over all polynomials $P \in \mathscr{P}_M(\mathbb{R}^n)$ with the usual modification if $u = \infty$, i. e.,

$$\mathrm{osc}_\infty^M f(x,t) \equiv \inf \sup_{y \in B(x,t)} |f(y) - P(y)|.$$

Now we define the associated sharp maximal function

$$f_u^{M,s}(x) \equiv \sup_{0<t<1} t^{-s} \mathrm{osc}_u^M f(x,t).$$

Let $M \equiv \max\{-1, \lfloor s \rfloor\}$. Then $T_p^s(\mathbb{R}^n)$ is the collection of all functions in $L_{\mathrm{loc}}^p(\mathbb{R}^n)$ satisfying

$$\|f\|_{T_p^s(\mathbb{R}^n)} \equiv \sup_{x \in \mathbb{R}^n} \left(\int_{B(x,1)} |f(x)|^p \, dx \right)^{1/p} + \sup_{x \in \mathbb{R}^n} f_u^{M,s}(x) < \infty.$$

We followed [146, Sect. 1.7.2] (but change the notation because of item (i)); see also [153].

(xix) **Local approximation spaces II.** Let $p \in (0,\infty]$, $s \in (0,\infty)$ and $M \equiv \lfloor s \rfloor$. The local approximation space $C_s^p(\mathbb{R}^n)$ is the collection of all functions $f \in L^{\max\{p,1\}}(\mathbb{R}^n)$ such that

$$\|f\|_{C_s^p(\mathbb{R}^n)} \equiv \|f\|_{L^p(\mathbb{R}^n)} + \|f_p^{M,s}\|_{L^p(\mathbb{R}^n)} < \infty.$$

We refer to [15, 40, 46, 153] and [146, Sect. 1.7.2].

(xx) Let $\alpha \in \mathbb{R}$. The *space* $Q_\alpha(\mathbb{R}^n)$ is defined to be the collection of all $f \in L_{\mathrm{loc}}^2(\mathbb{R}^n)$ such that

$$\|f\|_{Q_\alpha(\mathbb{R}^n)} \equiv \sup_I \left\{ \frac{1}{|I|^{1-\frac{2\alpha}{n}}} \int_I \int_I \frac{|f(x) - f(y)|^2}{|x-y|^{n+2\alpha}} \, dx dy \right\}^{1/2} < \infty,$$

where I ranges over all cubes in $\mathbb{R}^n$; see, for example, [7, 44, 54].

1.3.3 *Function Spaces Defined by Fourier Analytic Tools*

Except the first two all spaces here will be defined by using a decomposition in the Fourier image induced by a smooth dyadic decomposition of unity. Let $\psi \in C_c^\infty(\mathbb{R}^n)$ be a radial function such that $\psi(x) = 1$ if $|x| \le 1$ and $\psi(x) = 0$ if $|x| \ge 3/2$. Then by means of

$$\psi^0(x) \equiv \psi(x), \qquad \psi^j(x) \equiv \psi(2^{-j}x) - \psi(2^{-j+1}x), \quad j \in \mathbb{N}, \quad x \in \mathbb{R}^n, \qquad (1.1)$$

we obtain a smooth dyadic decomposition of unity, namely,

$$\sum_{j=0}^{\infty} \psi^j(x) = 1 \qquad \text{for all} \quad x \in \mathbb{R}^n .$$

We put

$$\varphi_0 \equiv \Phi \equiv \mathscr{F}^{-1}\psi,\ \varphi(x) \equiv \mathscr{F}^{-1}[\psi(2\xi)](x) \quad \text{and} \quad \varphi_j \equiv \mathscr{F}^{-1}\psi^j, \quad j \in \mathbb{Z}_+. \tag{1.2}$$

Then $\varphi_j(x) = 2^{jn}\varphi(2^j x)$, $j \in \mathbb{Z}_+$, follows.

(xxi) **Local Hardy spaces**. Let $p \in (0,\infty)$. Let $\varphi \in \mathscr{S}(\mathbb{R}^n)$ such that $\widehat{\varphi}(0) = 1$. Then $h^p(\mathbb{R}^n)$ is the collection of all $f \in \mathscr{S}'(\mathbb{R}^n)$ such that

$$\|f\|_{h^p(\mathbb{R}^n)} \equiv \left\| \sup_{0<t<1} \mathscr{F}^{-1}[\varphi(t\xi)\mathscr{F}f(\xi)](\cdot) \right\|_{L^p(\mathbb{R}^n)} < \infty.$$

(xxii) **Bessel-potential spaces** (sometimes also called Lebesgue or Liouville spaces). Let $s \in \mathbb{R}$ and $p \in (1,\infty)$. Then $H^s_p(\mathbb{R}^n)$ is the set of all tempered distributions $f \in \mathscr{S}'(\mathbb{R}^n)$ such that $\mathscr{F}^{-1}[(1+|\xi|^2)^{s/2}\mathscr{F}f(\xi)](\cdot)$ is a regular distribution and

$$\|f\|_{H^s_p(\mathbb{R}^n)} \equiv \left\| \mathscr{F}^{-1}[(1+|\xi|^2)^{s/2}\mathscr{F}f(\xi)](\cdot) \right\|_{L^p(\mathbb{R}^n)} < \infty.$$

(xxiii) **Besov spaces** (general case). Let $p, q \in (0,\infty]$ and $s \in \mathbb{R}$. Let $\{\varphi_j\}_{j\in\mathbb{Z}_+}$ be as in (1.2). Then $B^s_{p,q}(\mathbb{R}^n)$ is the collection of all $f \in \mathscr{S}'(\mathbb{R}^n)$ such that

$$\|f\|_{B^s_{p,q}(\mathbb{R}^n)} \equiv \left\{ \sum_{j=0}^{\infty} 2^{jsq} \|\varphi_j * f\|^q_{L^p(\mathbb{R}^n)} \right\}^{1/q} < \infty.$$

(xxiv) **Triebel-Lizorkin spaces**. Let $p \in (0,\infty)$, $q \in (0,\infty]$ and $s \in \mathbb{R}$. Let $\{\varphi_j\}_{j\in\mathbb{Z}_+}$ be as in (1.2). Then $F^s_{p,q}(\mathbb{R}^n)$ is the collection of all $f \in \mathscr{S}'(\mathbb{R}^n)$ such that

$$\|f\|_{F^s_{p,q}(\mathbb{R}^n)} \equiv \left\| \left\{ \sum_{j=0}^{\infty} (2^{js}|\varphi_j * f|)^q \right\}^{1/q} \right\|_{L^p(\mathbb{R}^n)} < \infty.$$

We refer to [64, 145]. The *Triebel-Lizorkin space* $F^s_{\infty,q}(\mathbb{R}^n)$ is defined to be the set of all $f \in \mathscr{S}'(\mathbb{R}^n)$ such that

$$\|f\|_{F^s_{\infty,q}(\mathbb{R}^n)} \equiv \sup_{\substack{P \text{dyadic} \\ l(P)\le 1}} \left\{ \frac{1}{|P|} \int_P \sum_{j=j_P}^{\infty} [2^{js}|\varphi_j * f(x)|]^q \, dx \right\}^{1/q} < \infty, \tag{1.3}$$

where the supremum is taken over all dyadic cubes P with side length $l(P) \le 1$ and $j_P \equiv -\log_2 l(P)$; see [64].

(xxv) **Besov-Morrey spaces**. Let $s \in \mathbb{R}$, $q \in (0,\infty]$ and $0 < u \le p \le \infty$. Let $\{\varphi_j\}_{j\in\mathbb{Z}_+}$ be as in (1.2). Then $\mathcal{N}^s_{pqu}(\mathbb{R}^n)$ is defined to be the set of all $f \in \mathcal{S}'(\mathbb{R}^n)$ satisfying

$$\|f\|_{\mathcal{N}^s_{pqu}(\mathbb{R}^n)} \equiv \left\{\sum_{j=0}^{\infty} 2^{jsq} \sup_B |B|^{q/p-q/u} \left(\int_B |\varphi_j * f(x)|^u \, dx\right)^{q/u}\right\}^{1/q} < \infty,$$

where the supremum is taken over all balls B in $\mathbb{R}^n$; see [88, 97].

(xxvi) **Triebel-Lizorkin-Morrey spaces**. Let $s \in \mathbb{R}$, $q \in (0,\infty]$ and $0 < u \le p \le \infty$, $u \ne \infty$. Let $\{\varphi_j\}_{j\in\mathbb{Z}_+}$ be as in (1.2). The class $\mathcal{E}^s_{pqu}(\mathbb{R}^n)$ is defined to be the collection of all $f \in \mathcal{S}'(\mathbb{R}^n)$ satisfying

$$\|f\|_{\mathcal{E}^s_{pqu}(\mathbb{R}^n)} \equiv \sup_B |B|^{1/p-1/u} \left\{\int_B \left(\sum_{j=0}^{\infty} 2^{jsq} |\varphi_j * f(x)|^q\right)^{u/q} dx\right\}^{1/u} < \infty,$$

where the supremum is taken over all balls B in $\mathbb{R}^n$. We refer, e. g., to [88, 126, 139].

(xxvii) **Inhomogeneous Besov-type spaces**. Let $\tau, s \in \mathbb{R}$ and $p, q \in (0,\infty]$. Let $\{\varphi_j\}_{j\in\mathbb{Z}_+}$ be as in (1.2). The *inhomogeneous Besov-type space* $B^{s,\tau}_{p,q}(\mathbb{R}^n)$ is defined to be the set of all $f \in \mathcal{S}'(\mathbb{R}^n)$ such that

$$\|f\|_{B^{s,\tau}_{p,q}(\mathbb{R}^n)} \equiv \sup_{P\in\mathcal{Q}} \frac{1}{|P|^\tau} \left\{\sum_{j=(j_P\vee 0)}^{\infty} \left[\int_P (2^{js}|\varphi_j * f(x)|)^p \, dx\right]^{q/p}\right\}^{1/q} < \infty.$$

(xxviii) **Inhomogeneous Triebel-Lizorkin-type spaces**. Let $\tau, s \in \mathbb{R}$, $q \in (0,\infty]$ and $p \in (0,\infty)$. Let $\{\varphi_j\}_{j\in\mathbb{Z}_+}$ be as in (1.2). The *inhomogeneous Triebel-Lizorkin-type space* $F^{s,\tau}_{p,q}(\mathbb{R}^n)$ is defined to be the set of all $f \in \mathcal{S}'(\mathbb{R}^n)$ such that

$$\|f\|_{F^{s,\tau}_{p,q}(\mathbb{R}^n)} \equiv \sup_{P\in\mathcal{Q}} \frac{1}{|P|^\tau} \left\{\int_P \left[\sum_{j=(j_P\vee 0)}^{\infty} (2^{js}|\varphi_j * f(x)|)^q\right]^{p/q} dx\right\}^{1/p} < \infty.$$

A comment. The definitions of the spaces in (1.3) and (xxv)–(xxviii) are all of the same spirit. The major difference between Besov-Morrey and Triebel-Lizorkin-Morrey spaces on the one side and the spaces $B^{s,\tau}_{p,q}(\mathbb{R}^n)$ and $F^{s,\tau}_{p,q}(\mathbb{R}^n)$ on the other side consists in the starting index for the summation with respect to j. In (xxv) and (xxvi) the summation starts always with 0, whereas in the (xxvii) and (xxviii) the summation starts at $j_P \vee 0$. Comparing with (1.3) we find that this time there is a difference in the set of admissible cubes. The distribution spaces $F^{s,\tau}_{p,q}(\mathbb{R}^n)$ and $B^{s,\tau}_{p,q}(\mathbb{R}^n)$ have some overlap with all 26 different classes we have introduced above; see the next subsection.

(xxix) **Homogeneous Besov-type spaces**. Let $\tau, s \in \mathbb{R}$ and $p,q \in (0,\infty]$. Let $\varphi_j(x) \equiv 2^{jn}\varphi(2^j x)$, $j \in \mathbb{Z}$. The *Besov-type space* $\dot{B}^{s,\tau}_{p,q}(\mathbb{R}^n)$ is defined to be the set of all $f \in \mathscr{S}'_\infty(\mathbb{R}^n)$ such that $\|f\|_{\dot{B}^{s,\tau}_{p,q}(\mathbb{R}^n)} < \infty$, where

$$\|f\|_{\dot{B}^{s,\tau}_{p,q}(\mathbb{R}^n)} \equiv \sup_{P\in\mathscr{Q}} \frac{1}{|P|^\tau} \left\{ \sum_{j=j_P}^{\infty} \left[\int_P (2^{js}|\varphi_j * f(x)|)^p \, dx \right]^{q/p} \right\}^{1/q}$$

with suitable modifications made when $p = \infty$ or $q = \infty$.

(xxx) **Homogeneous Triebel-Lizorkin-type spaces**. Let $\tau, s \in \mathbb{R}$, $q \in (0,\infty]$ and $p \in (0,\infty)$. Let $\varphi_j(x) \equiv 2^{jn}\varphi(2^j x)$, $j \in \mathbb{Z}$. The *Triebel-Lizorkin-type space* $\dot{F}^{s,\tau}_{p,q}(\mathbb{R}^n)$ is defined to be the set of all $f \in \mathscr{S}'_\infty(\mathbb{R}^n)$ such that $\|f\|_{\dot{F}^{s,\tau}_{p,q}(\mathbb{R}^n)} < \infty$, where

$$\|f\|_{\dot{F}^{s,\tau}_{p,q}(\mathbb{R}^n)} \equiv \sup_{P\in\mathscr{Q}} \frac{1}{|P|^\tau} \left\{ \int_P \left[\sum_{j=j_P}^{\infty} (2^{js}|\varphi_j * f(x)|)^q \, dx \right]^{p/q} \right\}^{1/p}$$

with suitable modifications made when $q = \infty$.

(xxxi) **Besov-Hausdorff spaces and Triebel-Lizorkin-Hausdorff spaces**. The inhomogeneous classes $BH^{s,\tau}_{p,q}(\mathbb{R}^n)$ and $FH^{s,\tau}_{p,q}(\mathbb{R}^n)$ will be investigated in Chap. 7. For the homogeneous counterparts, see Sect. 8.4.
Let $s \in \mathbb{R}$, $p \in (1,\infty)$, $q \in [1,\infty)$ and $\tau \in [0, \frac{1}{(p\vee q)'}]$, $\{\varphi_j\}_{j\in\mathbb{Z}_+}$ be as in (1.2). The *Besov-Hausdorff spaces* $BH^{s,\tau}_{p,q}(\mathbb{R}^n)$ and the *Triebel-Lizorkin-Hausdorff spaces* $FH^{s,\tau}_{p,q}(\mathbb{R}^n)$ $(q \neq 1)$ are defined, respectively, to be the sets of all $f \in \mathscr{S}'(\mathbb{R}^n)$ such that

$$\|f\|_{BH^{s,\tau}_{p,q}(\mathbb{R}^n)} \equiv \inf_\omega \left\{ \sum_{j=0}^{\infty} 2^{jsq} \|\varphi_j * f[\omega(\cdot, 2^{-j})]^{-1}\|^q_{L^p(\mathbb{R}^n)} \right\}^{1/q} < \infty$$

and

$$\|f\|_{FH^{s,\tau}_{p,q}(\mathbb{R}^n)} \equiv \inf_\omega \left\| \left\{ \sum_{j=0}^{\infty} 2^{jsq} |\varphi_j * f|^q [\omega(\cdot, 2^{-j})]^{-q} \right\}^{1/q} \right\|_{L^p(\mathbb{R}^n)} < \infty,$$

the infimums here are taken over all nonnegative Borel measurable functions ω on $\mathbb{R}^n \times (0,\infty)$ with

$$\int_{\mathbb{R}^n} [N\omega(x)]^{(p\vee q)'} \, d\Lambda^{(\infty)}_{n\tau(p\vee q)'}(x) \le 1,$$

and with the restriction that $\omega(\cdot, 2^{-j})$ is allowed to vanish only where $\varphi_j * f$ vanishes, where $N\omega$ is the nontangential maximal function of ω

and $\Lambda^{(\infty)}_{n\tau(p\vee q)'}$ is the $n\tau(p\vee q)'$-dimensional Hausdorff capacity; see Sect. 7.1 below.

(xxxii) There is a number of further spaces appearing in the book. But they will be of restricted importance.

1.4 A Table of Coincidences

As mentioned above there is some overlap of these different definitions. We are collecting some of these coincidence relations in what follows.

1.4.1 Besov-Morrey Spaces

(i) It holds $B^{s,0}_{p,q}(\mathbb{R}^n) = B^s_{p,q}(\mathbb{R}^n)$ for all s, p, and q; see Lemma 2.1. This implies

$$\begin{aligned}
B^{s,0}_{\infty,\infty}(\mathbb{R}^n) &= B^s_{\infty,\infty}(\mathbb{R}^n) = \mathscr{Z}^s(\mathbb{R}^n), && s>0,\\
B^{s,0}_{\infty,\infty}(\mathbb{R}^n) &= B^s_{\infty,\infty}(\mathbb{R}^n) = C^s(\mathbb{R}^n), && s>0,\ s\notin\mathbb{N},\\
B^{s,0}_{\infty,\infty}(\mathbb{R}^n) &= B^s_{\infty,\infty}(\mathbb{R}^n), && 0<s<1,\\
B^{s,0}_{p,p}(\mathbb{R}^n) &= B^s_{p,p}(\mathbb{R}^n) = W^s_p(\mathbb{R}^n), && s>0,\ s\notin\mathbb{N},\ 1\le p<\infty
\end{aligned}$$

(all in the sense of equivalent norms); see, e. g., [145, Sect. 2.2.2] and the references given there.

(ii) Let $s\in\mathbb{R}$, $0<u\le p\le\infty$ and $q\in(0,\infty]$. On the one hand we have

$$B^{s,0}_{p,q}(\mathbb{R}^n) = \mathscr{N}^s_{pqp}(\mathbb{R}^n) = B^s_{p,q}(\mathbb{R}^n) \quad\text{and}\quad B^{s,1/u-1/p}_{u,\infty}(\mathbb{R}^n) = \mathscr{N}^s_{p\infty u}(\mathbb{R}^n)$$

(in the sense of equivalent quasi-norms), but on the other hand, it holds

$$B^{s,1/u-1/p}_{u,q}(\mathbb{R}^n) \supsetneqq \mathscr{N}^s_{pqu}(\mathbb{R}^n) \quad\text{if}\quad 0<u<p<\infty \quad\text{and}\quad 0<q<\infty;$$

see Corollary 3.3.

(iii) Let $0<p<p_0<\infty$, $k\in\mathbb{N}$ and

$$s>\frac{k}{p}+n\max\left\{0,\frac{1}{p}-1\right\}.$$

Then

$$B^{s-k/p,\frac{1}{p}\frac{n+k}{n}}_{p,q}(\mathbb{R}^n) = \mathscr{Z}^s(\mathbb{R}^n) \quad\text{if}\quad p\le q\le\infty,$$

and

$$B_{p_0,q}^{s-\frac{k+n}{p}+\frac{n}{p_0},\frac{1}{p}\frac{n+k}{n}}(\mathbb{R}^n) = \mathscr{L}^s(\mathbb{R}^n) \qquad \text{if} \qquad p \le q \le \infty,$$

in the sense of equivalent quasi-norms; see Theorem 6.9 below.

1.4.2 Triebel-Lizorkin-Morrey Spaces

(i) It holds $F_{u,q}^{s,0}(\mathbb{R}^n) = F_{p,q}^s(\mathbb{R}^n)$; see Lemma 2.1. This implies

$$\begin{aligned} F_{p,2}^{m,0}(\mathbb{R}^n) &= F_{p,2}^m(\mathbb{R}^n) = W_p^m(\mathbb{R}^n), & m \in \mathbb{N},\ 1 < p < \infty, \\ F_{p,p}^{s,0}(\mathbb{R}^n) &= F_{p,p}^s(\mathbb{R}^n) = W_p^s(\mathbb{R}^n), & s > 0,\ s \notin \mathbb{N},\ 1 \le p < \infty, \\ F_{p,2}^{0,0}(\mathbb{R}^n) &= F_{p,2}^0(\mathbb{R}^n) = h^p(\mathbb{R}^n), & 0 < p < \infty, \\ F_{\infty,2}^0(\mathbb{R}^n) &= \mathrm{bmo}\,(\mathbb{R}^n) \end{aligned}$$

(all in the sense of equivalent norms); see, e. g. [145, Sect. 2.2.2] and the references given there.

(ii) Let $p \in (0,\infty)$ and $s \in (n\max\{0, \frac{1}{p} - 1\}, \infty)$. Then

$$F_{p,\infty}^{s,0}(\mathbb{R}^n) = F_{p,\infty}^s(\mathbb{R}^n) = C_s^p(\mathbb{R}^n);$$

see [130] and [146, Theorem 1.7.2].

(iii) Let $s \in \mathbb{R}$, $p \in (0,\infty)$ and $q \in (0,\infty]$. Then

$$F_{p,q}^{s,1/p}(\mathbb{R}^n) = F_{\infty,q}^s(\mathbb{R}^n)$$

with equivalent quasi-norms; see [64] or Proposition 2.4 below. In particular,

$$F_{p,2}^{0,1/p}(\mathbb{R}^n) = F_{\infty,2}^0(\mathbb{R}^n) = \mathrm{bmo}\,(\mathbb{R}^n).$$

(iv) Let $q \in (0,\infty]$ and $0 < u \le p \le \infty$, $u \ne \infty$. Then

$$F_{u,q}^{s,1/u-1/p}(\mathbb{R}^n) = \mathscr{E}_{pqu}^s(\mathbb{R}^n).$$

For $s = 0$ and $1 < u \le p < \infty$ this yields

$$F_{u,2}^{0,1/u-1/p}(\mathbb{R}^n) = \mathscr{E}_{p2u}^0(\mathbb{R}^n) = \mathscr{M}_u^p(\mathbb{R}^n) \tag{1.4}$$

and with $1 < u = p < \infty$

$$F_{p,2}^{0,0}(\mathbb{R}^n) = \mathscr{E}_{p2p}^0(\mathbb{R}^n) = \mathscr{M}_p^p(\mathbb{R}^n) = L^p(\mathbb{R}^n),$$

all in the sense of equivalent quasi-norms; see Corollary 3.3 below.

(v) Let $\alpha \in (0,1)$ if $n \geq 2$ and $\alpha \in (0,1/2)$ if $n = 1$. Then we have

$$F_{2,2}^{\alpha,\frac{1}{2}-\frac{\alpha}{n}}(\mathbb{R}^n) = Q_\alpha(\mathbb{R}^n) \cap L^2_{\frac{1}{2}-\frac{\alpha}{n}}(\mathbb{R}^n),$$

in the sense of equivalent norms; see Corollary 4.5 and Remark 4.7.

(vi) Let $0 < p < p_0 < \infty$, $k \in \mathbb{N}$ and

$$s > \frac{k}{p} + n \max\left\{0, \frac{1}{p} - 1\right\}.$$

Then

$$F_{p,q}^{s-k/p,\frac{1}{p}\frac{n+k}{n}}(\mathbb{R}^n) = \mathscr{Z}^s(\mathbb{R}^n) \qquad \text{if} \qquad p \leq q \leq \infty,$$

and

$$F_{p_0,q}^{s-\frac{k+n}{p}+\frac{n}{p_0},\frac{1}{p}\frac{n+k}{n}}(\mathbb{R}^n) = \mathscr{Z}^s(\mathbb{R}^n) \qquad \text{if} \qquad 0 < q \leq \infty,$$

in the sense of equivalent quasi-norms; see Theorem 6.9 below.

(vii) Pointwise multipliers. For a quasi-Banach space X of functions, the space $M(X)$ denotes the associated space of all pointwise multipliers; see Sect. 6.1. Let $s \in (0,1)$. Then

$$M(F_{1,1}^s(\mathbb{R}^n)) = L^\infty(\mathbb{R}^n) \cap F_{1,1,\mathrm{unif}}^{s,\tau}(\mathbb{R}^n), \qquad \tau = 1 - s/n;$$

see Corollary 6.2 below.

1.4.3 Morrey-Campanato Spaces

(i) Let $0 < u \leq p \leq \infty$. Then

$$\mathscr{M}_u^u(\mathbb{R}^n) = L^u(\mathbb{R}^n) \qquad \text{and} \qquad \mathscr{M}_u^\infty(\mathbb{R}^n) = L^\infty(\mathbb{R}^n).$$

(ii) Let $p \in [1,\infty)$ and $\lambda \in (n, n+p)$. Then $\mathscr{L}^{p,n}(\mathbb{R}^n) = \mathrm{BMO}(\mathbb{R}^n)$,

$$\mathscr{L}^{p,n}(\mathbb{R}^n) = \mathscr{Z}^{\frac{\lambda-n}{p}}(\mathbb{R}^n) \qquad \text{and} \qquad \mathscr{L}^{p,n+p}(\mathbb{R}^n) = \mathrm{Lip}\,1(\mathbb{R}^n);$$

see [34, 36] and [89, Theorem 2.4.6.1].

(iii) Let $p \in [1,\infty)$. Then

$$T_p^{-n/p}(\mathbb{R}^n) = L^p(\mathbb{R}^n) \qquad \text{and} \qquad T_p^0(\mathbb{R}^n) = \mathrm{bmo}(\mathbb{R}^n).$$

(iv) Let $p \in [1,\infty)$ and $s \in (-n/p, 0)$. Then

$$\mathscr{L}^{p,\lambda}(\mathbb{R}^n) = \mathscr{M}_p^{-n/s}(\mathbb{R}^n) = T_p^s(\mathbb{R}^n), \qquad s = \frac{\lambda-n}{p};$$

see [89, Theorem 2.4.6.1] and [146, Sect. 1.7.2].

(v) Let $p \in [1, \infty)$ and $s \in (0, \infty)$. Then

$$T_p^s(\mathbb{R}^n) = \mathscr{X}^s(\mathbb{R}^n);$$

see [146, Sect. 1.7.2] and the references given there.

1.4.4 Homogeneous Spaces

Here we make use of the following interpretation. When comparing a class of functions, which is defined modulo polynomials of a certain order, with the spaces $\dot{A}_{p,q}^{s,\tau}(\mathbb{R}^n)$, then we always associate to an element of the first space the equivalence class

$$[f] \equiv \{g: \ g = f + p, \quad p \quad \text{is an arbitrary polynomial}\}.$$

By means of this interpretation the following relations are known.

(i) We have

$$\dot{F}_{\infty,2}^{0}(\mathbb{R}^n) = \mathrm{BMO}(\mathbb{R}^n).$$

(ii) Let $s \in \mathbb{R}$, $p \in (0, \infty)$ and $q \in (0, \infty]$. Then

$$\dot{F}_{p,q}^{s,1/p}(\mathbb{R}^n) = \dot{F}_{\infty,q}^{s}(\mathbb{R}^n)$$

with equivalent quasi-norms. In particular,

$$\dot{F}_{p,2}^{0,1/p}(\mathbb{R}^n) = \dot{F}_{\infty,2}^{0}(\mathbb{R}^n) = \mathrm{BMO}(\mathbb{R}^n).$$

(iii) Let $\alpha \in (0,1)$ if $n \geq 2$ and $\alpha \in (0, 1/2)$ if $n = 1$. Then we have

$$\dot{F}_{2,2}^{\alpha, \frac{1}{2} - \frac{\alpha}{n}}(\mathbb{R}^n) = Q_\alpha(\mathbb{R}^n)$$

in the sense of equivalent norms; see [164].

(iv) Let $\lambda \in [0, n+2)$. Then

$$\dot{F}_{2,2}^{0,2\lambda/n}(\mathbb{R}^n) = \mathscr{L}^{2,\lambda}(\mathbb{R}^n);$$

see [50].

1.5 Notation

At the end of this chapter, we make some conventions on notation.

Throughout this book, C denotes unspecified positive constants, possibly different at each occurrence; the symbol $X \lesssim Y$ means that there exists a positive constant

C such that $X \leq CY$, and $X \sim Y$ means $C^{-1}Y \leq X \leq CY$. We also use $C(\gamma, \beta, \cdots)$ to denote a positive constant depending on the indicated parameters γ, β, $\cdots$.

The real numbers are denoted by $\mathbb{R}$. Many times we shall use the abbreviations

$$a_+ \equiv \max(0, a),$$

$\lfloor a \rfloor$ for the integer part of the real number a, and $a^* \equiv a - \lfloor a \rfloor$. The symbol χ_E is used to denote the *characteristic function* of set $E \subset \mathbb{R}^n$. If $q \in [1, \infty]$ then by q' we mean its *conjugate index*, i. e., $1/q + 1/q' = 1$. Further we shall use the abbreviations

$$p \vee q \equiv \max\{p, q\}$$

and

$$p \wedge q \equiv \min\{p, q\}.$$

When dealing with the classes $A^{s,\tau}_{p,q}(\mathbb{R}^n)$, then four restrictions for the set of parameters s, p, q, τ will occur relatively often. They are connected with the quantities

$$\sigma_p \equiv \max\{n(1/p - 1), 0\} \quad \text{and} \quad \sigma_{p,q} \equiv \max\{n(1/\min\{p,q\} - 1), 0\}, \tag{1.5}$$

(restrictions for s) and

$$\tau_{s,p} \equiv \frac{1}{p} + \begin{cases} \frac{1-(\sigma_p+n-s)^*}{n} & \text{if} \quad s \leq \sigma_p, \\ \frac{s-\sigma_p}{n} & \text{if} \quad s > \sigma_p, \end{cases} \tag{1.6}$$

$$\tau_{s,p,q} \equiv \frac{1}{p} + \begin{cases} \frac{1-(\sigma_{p,q}+n-s)^*}{n} & \text{if} \quad s \leq \sigma_{p,q}, \\ \frac{s-\sigma_{p,q}}{n} & \text{if} \quad s > \sigma_{p,q} \end{cases} \tag{1.7}$$

(restrictions for τ). Also, set $\mathbb{N} \equiv \{1, 2, \cdots\}$ and $\mathbb{Z}_+ \equiv \mathbb{N} \cup \{0\}$. By $C_c^\infty(\mathbb{R}^n)$ we denote the set of all infinitely differentiable and compactly supported functions on $\mathbb{R}^n$. The symbol $\mathscr{S}(\mathbb{R}^n)$ is used in place of the set of all *Schwartz functions* φ on $\mathbb{R}^n$, i. e., φ is infinitely differentiable and

$$\|\varphi\|_{\mathscr{S}_M} \equiv \sup_{\gamma \in \mathbb{Z}^n_+, |\gamma| \leq M} \sup_{x \in \mathbb{R}^n} |\partial^\gamma \varphi(x)|(1 + |x|)^{n+M+|\gamma|} < \infty$$

for all $M \in \mathbb{N}$. The topological dual of $\mathscr{S}(\mathbb{R}^n)$, the set of *tempered distributions*, will be denoted by $\mathscr{S}'(\mathbb{R}^n)$.

For $k = (k_1, \cdots, k_n) \in \mathbb{Z}^n$ and $j \in \mathbb{Z}$, Q_{jk} denotes the dyadic cube

$$Q_{jk} \equiv \{(x_1, \cdots, x_n) : \ k_i \leq 2^j x_i < k_i + 1 \text{ for } i = 1, \cdots, n\}.$$

For the collection of all such cubes we use

$$\mathscr{Q} \equiv \{Q_{jk} : \ j \in \mathbb{Z}, k \in \mathbb{Z}^n\}.$$

Furthermore, we denote by x_Q the *lower left-corner* $2^{-j}k$ of $Q = Q_{jk}$. When the dyadic cube Q appears as an index, such as $\sum_{Q\in\mathscr{Q}}$ and $\{\cdot\}_{Q\in\mathscr{Q}}$, it is understood that Q runs over all *dyadic cubes* in $\mathbb{R}^n$. For each cube Q, we denote its *side length* by $l(Q)$, its *center* by c_Q, and for $r > 0$, we denote by rQ the *cube concentric* with Q having the side length $rl(Q)$. Further, the abbreviation $j_Q \equiv -\log_2 l(Q)$ is used.

For $j \in \mathbb{Z}$, $\varphi \in \mathscr{S}(\mathbb{R}^n)$ and $x \in \mathbb{R}^n$, we set $\widetilde{\varphi}(x) \equiv \overline{\varphi(-x)}$,

$$\widehat{\varphi}(x) \equiv \mathscr{F}\varphi(x) \equiv \int_{\mathbb{R}^n} \varphi(\xi)e^{-ix\cdot\xi}\,d\xi\,,$$

$\varphi_j(x) \equiv 2^{jn}\varphi(2^j x)$, and

$$\varphi_Q(x) \equiv |Q|^{-1/2}\varphi(2^j x - k) = |Q|^{1/2}\varphi_j(x - x_Q) \qquad \text{if} \quad Q = Q_{jk}\,.$$

For a dyadic cube Q, we shall work also with the $L^2(\mathbb{R}^n)$-normalized version

$$\widetilde{\chi}_Q(x) \equiv |Q|^{-1/2}\chi_Q(x)\,.$$

Let E denote a class of tempered distributions. Then E_{loc} denotes the collection of all $f \in \mathscr{S}'(\mathbb{R}^n)$ such that the product $\varphi \cdot f$ belongs to E for all test functions $\varphi \in C_c^\infty(\mathbb{R}^n)$. Furthermore, if E is in addition quasi-normed, then E_{unif} is the collection of all $f \in \mathscr{S}'(\mathbb{R}^n)$ such that

$$\|f\|_{E_{\mathrm{unif}}} \equiv \sup_{\lambda\in\mathbb{R}^n} \|\psi(\cdot - \lambda)f(\cdot)\|_E < \infty.$$

Here ψ is a nontrivial function in $C_c^\infty(\mathbb{R}^n)$.

Chapter 2
The Spaces $B^{s,\tau}_{p,q}(\mathbb{R}^n)$ and $F^{s,\tau}_{p,q}(\mathbb{R}^n)$

In this chapter, we introduce the spaces $B^{s,\tau}_{p,q}(\mathbb{R}^n)$ and $F^{s,\tau}_{p,q}(\mathbb{R}^n)$, establish their φ-transform characterizations, prove some embeddings and the Fatou property.

2.1 The φ-Transform for $B^{s,\tau}_{p,q}(\mathbb{R}^n)$ and $F^{s,\tau}_{p,q}(\mathbb{R}^n)$

Nowadays wavelet decompositions play an important role in the study of function spaces and their applications; see, for example, [99, 100, 149]. The φ-transform decomposition of Frazier and Jawerth [62–64] is rather close in spirit to wavelet decompositions. In this section, we establish the φ-transform characterizations of the spaces $B^{s,\tau}_{p,q}(\mathbb{R}^n)$ and $F^{s,\tau}_{p,q}(\mathbb{R}^n)$, which will play a crucial role in our considerations.

2.1.1 The Definition and Some Preliminaries

We start with the definitions of $B^{s,\tau}_{p,q}(\mathbb{R}^n)$ and $F^{s,\tau}_{p,q}(\mathbb{R}^n)$. Select a pair of Schwartz functions Φ and φ such that

$$\operatorname{supp}\widehat{\Phi}\subset\{\xi\in\mathbb{R}^n:\,|\xi|\le 2\}\quad\text{and}\quad|\widehat{\Phi}(\xi)|\ge C>0\quad\text{if }|\xi|\le\frac{5}{3}\tag{2.1}$$

and

$$\operatorname{supp}\widehat{\varphi}\subset\left\{\xi\in\mathbb{R}^n:\,\frac{1}{2}\le|\xi|\le 2\right\}\text{ and }|\widehat{\varphi}(\xi)|\ge C>0\text{ if }\frac{3}{5}\le|\xi|\le\frac{5}{3}.\tag{2.2}$$

It is easy to see that $\int_{\mathbb{R}^n} x^\gamma\varphi(x)\,dx=0$ for all multi-indices $\gamma\in\mathbb{Z}^n_+$.

Definition 2.1. Let $\tau,s\in\mathbb{R}$, $q\in(0,\infty]$, Φ and φ satisfy (2.1) and (2.2), respectively, and put $\varphi_j\equiv 2^{jn}\varphi(2^j\cdot)$.

D. Yang et al., *Morrey and Campanato Meet Besov, Lizorkin and Triebel*,
Lecture Notes in Mathematics 2005, DOI 10.1007/978-3-642-14606-0_2,

(i) Let $p \in (0,\infty]$. The *inhomogeneous Besov-type space* $B^{s,\tau}_{p,q}(\mathbb{R}^n)$ is defined to be the set of all $f \in \mathscr{S}'(\mathbb{R}^n)$ such that

$$\|f\|_{B^{s,\tau}_{p,q}(\mathbb{R}^n)} \equiv \sup_{P\in\mathscr{Q}} \frac{1}{|P|^\tau} \left\{ \sum_{j=(j_P\vee 0)}^{\infty} \left[\int_P (2^{js}|\varphi_j * f(x)|)^p \, dx \right]^{q/p} \right\}^{1/q} < \infty, \quad (2.3)$$

where φ_0 is replaced by Φ.

(ii) Let $p \in (0,\infty)$. The *inhomogeneous Triebel-Lizorkin-type space* $F^{s,\tau}_{p,q}(\mathbb{R}^n)$ is defined to be the set of all $f \in \mathscr{S}'(\mathbb{R}^n)$ such that

$$\|f\|_{F^{s,\tau}_{p,q}(\mathbb{R}^n)} \equiv \sup_{P\in\mathscr{Q}} \frac{1}{|P|^\tau} \left\{ \int_P \left[\sum_{j=(j_P\vee 0)}^{\infty} (2^{js}|\varphi_j * f(x)|)^q \right]^{p/q} dx \right\}^{1/p} < \infty, \quad (2.4)$$

where φ_0 is replaced by Φ.

Remark 2.1.

(i) When $p = q \in (0,\infty)$, $B^{s,\tau}_{p,q}(\mathbb{R}^n) = F^{s,\tau}_{p,q}(\mathbb{R}^n)$. If we replace dyadic cubes P in Definition 2.1 by arbitrary cubes P, we then obtain equivalent quasi-norms.

(ii) The definitions given here are slightly more general than those given in Sect. 1.3. The coincidence will be proved by establishing the independence of the above definitions from Φ and φ; see Corollary 2.1 below.

(iii) For $\tau > 1/p$ it is necessary to start the summation with respect to j in dependence on the size of the dyadic cube P. If the summation would start always with $j = 0$, then a Lebesgue point argument shows that only the function $f = 0$ a. e. belongs to such a space.

For simplicity, in what follows, we *use* $A^{s,\tau}_{p,q}(\mathbb{R}^n)$ *to denote either* $B^{s,\tau}_{p,q}(\mathbb{R}^n)$ *or* $F^{s,\tau}_{p,q}(\mathbb{R}^n)$. *If* $A^{s,\tau}_{p,q}(\mathbb{R}^n)$ *means* $F^{s,\tau}_{p,q}(\mathbb{R}^n)$, *then the case* $p = \infty$ *is excluded.* In the same way we shall use the abbreviation $A^s_{p,q}(\mathbb{R}^n)$ in place of $F^s_{p,q}(\mathbb{R}^n)$ and $B^s_{p,q}(\mathbb{R}^n)$, respectively.

Lemma 2.1. (i) *The classes* $A^{s,\tau}_{p,q}(\mathbb{R}^n)$ *are quasi-Banach spaces, i. e., complete quasi-normed spaces. With* $d = \min\{1,p,q\}$ *it holds*

$$\|f+g\|^d_{A^{s,\tau}_{p,q}(\mathbb{R}^n)} \le \|f\|^d_{A^{s,\tau}_{p,q}(\mathbb{R}^n)} + \|g\|^d_{A^{s,\tau}_{p,q}(\mathbb{R}^n)}$$

for all $f, g \in A^{s,\tau}_{p,q}(\mathbb{R}^n)$.

(ii) *If* $\tau = 0$, *then* $A^{s,0}_{p,q}(\mathbb{R}^n) = A^s_{p,q}(\mathbb{R}^n)$.

Proof. To prove (i) the needed arguments are standard, we refer, e. g., to [145, Sect. 2.3.3]. Details are left to the reader. The proof of (ii) is obvious. □

Sometimes it is of great service if one can restrict $\sup_{P\in\mathscr{Q}}$ in the definition of $A^{s,\tau}_{p,q}(\mathbb{R}^n)$ to a supremum taken with respect to dyadic cubes with side length ≤ 1.

Lemma 2.2. *Let $s \in \mathbb{R}$ and $q \in (0,\infty]$.*

(i) *Let $p \in (0,\infty]$ and $\tau \in [1/p,\infty)$. A tempered distribution f belongs to $B^{s,\tau}_{p,q}(\mathbb{R}^n)$ if, and only if,*

$$\|f\|^{\#}_{B^{s,\tau}_{p,q}(\mathbb{R}^n)} \equiv \sup_{\{P\in\mathcal{Q},|P|\le 1\}} \frac{1}{|P|^\tau}\left\{\sum_{j=(j_P\vee 0)}^{\infty}\left[\int_P (2^{js}|\varphi_j * f(x)|)^p\,dx\right]^{q/p}\right\}^{1/q} < \infty.$$

Furthermore, the quasi-norms $\|f\|_{B^{s,\tau}_{p,q}(\mathbb{R}^n)}$ and $\|f\|^{\#}_{B^{s,\tau}_{p,q}(\mathbb{R}^n)}$ are equivalent.

(ii) *Let $p \in (0,\infty)$ and $\tau \in [1/p,\infty)$. A tempered distribution f belongs to $F^{s,\tau}_{p,q}(\mathbb{R}^n)$ if, and only if,*

$$\|f\|^{\#}_{F^{s,\tau}_{p,q}(\mathbb{R}^n)} \equiv \sup_{\{P\in\mathcal{Q},|P|\le 1\}} \frac{1}{|P|^\tau}\left\{\int_P\left[\sum_{j=(j_P\vee 0)}^{\infty}(2^{js}|\varphi_j * f(x)|)^q\right]^{p/q} dx\right\}^{1/p} < \infty.$$

Furthermore, the quasi-norms $\|f\|_{F^{s,\tau}_{p,q}(\mathbb{R}^n)}$ and $\|f\|^{\#}_{F^{s,\tau}_{p,q}(\mathbb{R}^n)}$ are equivalent.

Proof. Let P be a dyadic cube such that $|P| = 2^{rn}$ for some $r \in \mathbb{N}$. Let $\{Q_m : m = 1,\ldots,2^{rn}\}$ be the collection of all dyadic cubes with volume 1 and such that

$$P = \bigcup_{m=1}^{2^{rn}} Q_m .$$

Then, with $g \in L^p_{\mathrm{loc}}(\mathbb{R}^n)$,

$$\begin{aligned}
\frac{1}{|P|^\tau}\left(\int_P |g(x)|^p\,dx\right)^{1/p} &= \frac{1}{|P|^\tau}\left(\sum_{m=1}^{2^{rn}}\int_{Q_m}|g(x)|^p\,dx\right)^{1/p}\\
&\le \frac{1}{|P|^\tau}\left(2^{rn}\sup_{\{Q\in\mathcal{Q},l(Q)\le 1\}}\frac{1}{|Q|^{\tau p}}\int_Q|g(x)|^p\,dx\right)^{1/p}\\
&\le \sup_{\{Q\in\mathcal{Q},l(Q)\le 1\}}\frac{1}{|Q|^\tau}\left(\int_Q|g(x)|^p\,dx\right)^{1/p}. \qquad (2.5)
\end{aligned}$$

This proves the claim for $F^{s,\tau}_{p,q}(\mathbb{R}^n)$. In case of $B^{s,\tau}_{p,q}(\mathbb{R}^n)$ one applies the inequality (2.5) either in combination with $(\sum\ldots)^{q/p} \le \sum(\ldots)^{q/p}$ if $q/p < 1$ or in combination with Minkowski's inequality if $q/p \ge 1$. □

Remark 2.2. Lemma 2.2 does not extend to values $\tau < 1/p$. A proof of this claim will be given at the end of Sect. 4.2.3 under the additional assumption $s > \sigma_p$.

2.1.2 The Calderón Reproducing Formulae and Some Consequences

The independence of $A_{p,q}^{s,\tau}(\mathbb{R}^n)$ from the choice of Φ and φ will be an immediate corollary of the φ-transform characterization of $A_{p,q}^{s,\tau}(\mathbb{R}^n)$. To establish this characterization, we need the *Calderón reproducing formulae*, which play important roles in the whole book.

Let Φ and φ satisfy, respectively, (2.1) and (2.2). By [64, pp. 130–131] or [65, Lemma (6.9)], there exist functions $\Psi \in \mathcal{S}(\mathbb{R}^n)$ satisfying (2.1) and $\psi \in \mathcal{S}(\mathbb{R}^n)$ satisfying (2.2) such that for all $\xi \in \mathbb{R}^n$,

$$\widehat{\overline{\Phi}}(\xi)\widehat{\Psi}(\xi) + \sum_{j=1}^{\infty} \widehat{\overline{\varphi}}(2^{-j}\xi)\widehat{\psi}(2^{-j}\xi) = 1. \tag{2.6}$$

Furthermore, we have the following *Calderón reproducing formula*; see [64, (12.4)].

Lemma 2.3. *Let Φ, $\Psi \in \mathcal{S}(\mathbb{R}^n)$ satisfy* (2.1) *and φ, $\psi \in \mathcal{S}(\mathbb{R}^n)$ satisfy* (2.2) *such that* (2.6) *holds. Then for all $f \in \mathcal{S}'(\mathbb{R}^n)$,*

$$\begin{aligned} f &= \Psi * \widetilde{\Phi} * f + \sum_{j=1}^{\infty} \psi_j * \widetilde{\varphi}_j * f \\ &= \sum_{k\in\mathbb{Z}^n} \widetilde{\Phi} * f(k)\,\Psi(\cdot - k) + \sum_{j=1}^{\infty} 2^{-jn} \sum_{k\in\mathbb{Z}^n} \widetilde{\varphi}_j * f(2^{-j}k)\,\psi_j(\cdot - 2^{-j}k) \\ &= \sum_{l(Q)=1} \langle f, \Phi_Q\rangle\, \Psi_Q + \sum_{j=1}^{\infty} \sum_{l(Q)=2^{-j}} \langle f, \varphi_Q\rangle\, \psi_Q \end{aligned} \tag{2.7}$$

in $\mathcal{S}'(\mathbb{R}^n)$.

The following basic estimate will be used throughout the book.

Lemma 2.4. *Let $M \in \mathbb{Z}_+$, and ψ, $\varphi \in \mathcal{S}(\mathbb{R}^n)$ with ψ satisfying $\int_{\mathbb{R}^n} x^\gamma \psi(x)\,dx = 0$ for all multi-indices $\gamma \in \mathbb{Z}_+^n$ satisfying $|\gamma| \le M$. Then there exists a positive constant $C \equiv C(M,n)$ such that for all $j \in \mathbb{Z}_+$ and $x \in \mathbb{R}^n$,*

$$\left|\psi_j * \varphi(x)\right| \le C\|\psi\|_{\mathcal{S}_{M+1}}\|\varphi\|_{\mathcal{S}_{M+1}} 2^{-jM} \frac{1}{(1+|x|)^{n+M}}. \tag{2.8}$$

Proof. Since ψ has vanishing moments of any order, we see that

$$\begin{aligned} |\psi_j * \varphi(x)| &= \left| \int_{\mathbb{R}^n} \left(\varphi(x-y) - \sum_{0\le|\alpha|\le M} \partial^\alpha \varphi(x) \frac{(-y)^\alpha}{\alpha!} \right) \psi_j(y)\,dy \right| \\ &\lesssim \int_{|y|\le(1+|x|)/2} \left| \varphi(x-y) - \sum_{0\le|\alpha|\le M} \partial^\alpha \varphi(x) \frac{(-y)^\alpha}{\alpha!} \right| |\psi_j(y)|\,dy \end{aligned}$$

$$
\begin{aligned}
&+\int_{|y|>(1+|x|)/2} |\varphi(x-y)|\,|\psi_j(y)|\,dy \\
&+\int_{|y|>(1+|x|)/2} \sum_{0\le|\alpha|\le M} |\partial^\alpha \varphi(x)|\,|y|^{|\alpha|}\,|\psi_j(y)|\,dy \\
&\equiv \mathrm{I}_1+\mathrm{I}_2+\mathrm{I}_3.
\end{aligned}
$$

Since φ, $\psi \in \mathscr{S}(\mathbb{R}^n)$, then by the mean value theorem, there exists $\theta \in [0,1]$ such that

$$
\begin{aligned}
\mathrm{I}_1 &\lesssim \int_{|y|\le(1+|x|)/2} \sup_{|\alpha|=M+1} |\partial^\alpha \varphi(x-\theta y)|\,|y|^{M+1}\frac{2^{-jM}\|\psi\|_{\mathscr{S}_M}}{(2^{-j}+|y|)^{n+M}}\,dy \\
&\lesssim \int_{|y|\le(1+|x|)/2} \frac{\|\varphi\|_{\mathscr{S}_{M+1}}|y|^{M+1}}{(1+|x-\theta y|)^{n+2M+1}}\frac{2^{-jM}\|\psi\|_{\mathscr{S}_M}}{(2^{-j}+|y|)^{n+M}}\,dy \\
&\lesssim \|\varphi\|_{\mathscr{S}_{M+1}}\|\psi\|_{\mathscr{S}_M}\frac{2^{-jM}}{(1+|x|)^{n+M}},
\end{aligned}
$$

where for the last inequality we use the fact that $1+|x-\theta y|\gtrsim 1+|x|$. Similarly,

$$
\begin{aligned}
\mathrm{I}_2 &\lesssim \int_{|y|>(1+|x|)/2} \frac{\|\varphi\|_{\mathscr{S}_M}}{(1+|x-y|)^{n+M}}\frac{2^{-jM}\|\psi\|_{\mathscr{S}_M}}{(2^{-j}+|y|)^{n+M}}\,dy \\
&\lesssim \|\varphi\|_{\mathscr{S}_M}\|\psi\|_{\mathscr{S}_M}\frac{2^{-jM}}{(1+|x|)^{n+M}},
\end{aligned}
$$

and

$$
\begin{aligned}
\mathrm{I}_3 &\lesssim \int_{|y|>(1+|x|)/2} \sum_{0\le|\alpha|\le M} \frac{\|\varphi\|_{\mathscr{S}_M}|y|^{|\alpha|}}{(1+|x|)^{n+M+|\alpha|}}\frac{2^{-j(M+1)}\|\psi\|_{\mathscr{S}_{M+1}}}{(2^{-j}+|y|)^{n+M+1}}\,dy \\
&\lesssim \|\varphi\|_{\mathscr{S}_M}\|\psi\|_{\mathscr{S}_{M+1}}\frac{2^{-jM}}{(1+|x|)^{n+M}},
\end{aligned}
$$

which completes the proof of Lemma 2.4. □

Remark 2.3. The proof of Lemma 2.4 is similar to that of [164, Lemma 2.2].

To establish the φ-transform characterization of $A^{s,\tau}_{p,q}(\mathbb{R}^n)$, we need some technical lemmas first. The following lemma is a slight variant of [65, Lemma (6.10)]. For the convenience of the reader, we give some details. Recall that a function g is called *at most polynomially increasing with order $m\in\mathbb{Z}_+$*, if there exists a positive constant C such that $|g(x)|\le C(1+|x|)^m$ for all $x\in\mathbb{R}^n$.

Lemma 2.5. *Let $h\in\mathscr{S}(\mathbb{R}^n)$ and $g\in C^\infty(\mathbb{R}^n)$ be at most polynomially increasing with order $m\in\mathbb{Z}_+$ such that* supp $\widehat{h}$, $\widehat{g}\subset\{\xi\in\mathbb{R}^n: |\xi|<2^\nu\pi\}$ *for some $\nu\in\mathbb{Z}$. Then*

$$
g*h=\sum_{k\in\mathbb{Z}^n} 2^{-\nu n}h(2^{-\nu}k)g(\cdot-2^{-\nu}k) \tag{2.9}
$$

holds pointwise as well as in $\mathscr{S}'(\mathbb{R}^n)$.

Proof. First we assume that $g \in \mathscr{S}(\mathbb{R}^n)$. Then by [65, Lemma (6.10)], (2.9) holds pointwise. We now further show that in this case, (2.9) also holds in $\mathscr{S}(\mathbb{R}^n)$.

Indeed, for any given $\alpha, \beta \in \mathbb{Z}^n_+$, let

$$\|\varphi\|_{\alpha,\beta} \equiv \sup_{x\in\mathbb{R}^n} |x^\alpha||\partial^\beta \varphi(x)|$$

denote the usual Schwartz quasi-norm, where for any $\alpha = (\alpha_1, \cdots, \alpha_n) \in \mathbb{Z}^n_+$ and $x = (x_1, \cdots, x_n) \in \mathbb{R}^n$, $x^\alpha = x_1^{\alpha_1} \cdots x_n^{\alpha_n}$ and $\partial^\alpha = (\frac{\partial}{\partial x_1})^{\alpha_1} \cdots (\frac{\partial}{\partial x_n})^{\alpha_n}$. Then

$$\begin{aligned}
\|h(2^{-\nu}k)g(\cdot - 2^{-\nu}k)\|_{\alpha,\beta} &\le |h(2^{-\nu}k)| \sup_{x\in\mathbb{R}^n} |x|^{|\alpha|} |\partial^\beta g(x - 2^{-\nu}k)| \\
&\lesssim |h(2^{-\nu}k)| \sup_{x\in\mathbb{R}^n} |x|^{|\alpha|} \frac{\|g\|_{\mathscr{S}_{|\alpha|+|\beta|}}}{(1 + |x - 2^{-\nu}k|)^{n+|\beta|+|\alpha|}} \\
&\lesssim |h(2^{-\nu}k)|(1 + |2^{-\nu}k|)^{|\alpha|} \|g\|_{\mathscr{S}_{|\alpha|+|\beta|}}.
\end{aligned}$$

Since h is a Schwartz function, then $|h(2^{-\nu}k)| \lesssim (1 + |2^{-\nu}k|)^{-n-|\alpha|-1}$. Thus,

$$\sum_{k\in\mathbb{Z}^n} \|2^{-\nu n} h(2^{-\nu}k)g(\cdot - 2^{-\nu}k)\|_{\alpha,\beta} \lesssim \sum_{k\in\mathbb{Z}^n} 2^{-\nu n} \frac{1}{(1 + |2^{-\nu}k|)^{n+1}} \|g\|_{\mathscr{S}_{|\alpha|+|\beta|}} < \infty,$$

which together with the completion of $\mathscr{S}(\mathbb{R}^n)$ implies that

$$\sum_{k\in\mathbb{Z}^n} 2^{-\nu n} h(2^{-\nu}k)g(\cdot - 2^{-\nu}k) \in \mathscr{S}(\mathbb{R}^n),$$

and hence (2.9) holds in $\mathscr{S}(\mathbb{R}^n)$ if $g \in \mathscr{S}(\mathbb{R}^n)$.

For the general case, we set $g_\delta(x) \equiv \eta(\delta x)g(x)$ for $\delta \in (0,1)$ and $x \in \mathbb{R}^n$, where $\eta \in \mathscr{S}(\mathbb{R}^n)$ satisfies $\eta(0) = 1$ and supp $\widehat{\eta} \subset \{\xi \in \mathbb{R}^n : |\xi| < 1\}$. Then $g_\delta \in \mathscr{S}(\mathbb{R}^n)$, and for sufficiently small $\delta > 0$, by the conclusion proved above, we know that

$$g_\delta * h = \sum_{k\in\mathbb{Z}^n} 2^{-\nu n} h(2^{-\nu}k) g_\delta(\cdot - 2^{-\nu}k) \tag{2.10}$$

holds in both pointwise and $\mathscr{S}(\mathbb{R}^n)$, which together with Lebesgue's dominated convergence theorem yields that (2.9) holds pointwise.

Next we show that (2.9) also holds in $\mathscr{S}'(\mathbb{R}^n)$. Notice that for all $\phi \in \mathscr{S}(\mathbb{R}^n)$,

$$\begin{aligned}
|\langle g_\delta(\cdot - 2^{-\nu}k), \phi\rangle| &\lesssim \int_{\mathbb{R}^n} |g(y - 2^{-\nu}k)||\phi(y)|\,dy \\
&\lesssim \int_{\mathbb{R}^n} \frac{(1 + |y - 2^{-\nu}k|)^m}{(1 + |y|)^{n+m+1}}\,dy \\
&\lesssim (1 + |2^{-\nu}k|)^m
\end{aligned}$$

and

$$\sum_{k\in\mathbb{Z}^n} 2^{-\nu n}|h(2^{-\nu}k)|(1+|2^{-\nu}k|)^m \lesssim \sum_{k\in\mathbb{Z}^n} \frac{2^{-\nu n}}{(1+|2^{-\nu}k|)^{n+1}} < \infty.$$

This observation together with Lebesgue's dominated convergence theorem and (2.10) implies that

$$\begin{aligned}\langle g*h,\phi\rangle &= \lim_{\delta\to 0}\langle g_\delta * h,\phi\rangle \\ &= \lim_{\delta\to 0}\sum_{k\in\mathbb{Z}^n} 2^{-\nu n}h(2^{-\nu}k)\langle g_\delta(\cdot-2^{-\nu}k),\phi\rangle \\ &= \sum_{k\in\mathbb{Z}^n} 2^{-\nu n}h(2^{-\nu}k)\langle g(\cdot-2^{-\nu}k),\phi\rangle.\end{aligned}$$

Thus (2.9) holds in $\mathscr{S}'(\mathbb{R}^n)$, which completes the proof of Lemma 2.5. □

Let γ be a fixed integer. Replacing φ_j by $\varphi_{j-\gamma}$ (φ_0 by $\Phi_{-\gamma}$) in (2.3) and (2.4), we obtain a new quasi-norm in $A^{s,\tau}_{p,q}(\mathbb{R}^n)$, denoted by $\|f\|^*_{A^{s,\tau}_{p,q}(\mathbb{R}^n)}$.

Lemma 2.6. *The quasi-norms $\|f\|^*_{A^{s,\tau}_{p,q}(\mathbb{R}^n)}$ and $\|f\|_{A^{s,\tau}_{p,q}(\mathbb{R}^n)}$ are equivalent on $\mathscr{S}'(\mathbb{R}^n)$ with equivalent constants depending on γ.*

Proof. By similarity, we only consider $B^{s,\tau}_{p,q}(\mathbb{R}^n)$ and the case $\gamma>0$.

Notice that

$$\|f\|^*_{B^{s,\tau}_{p,q}(\mathbb{R}^n)} \sim \sup_{P\in\mathscr{Q}} \frac{1}{|P|^\tau}\left\{\sum_{j=(j_P\vee 0)-\gamma}^{\infty}\left[\int_P (2^{js}|\varphi_j * f(x)|)^p\,dx\right]^{q/p}\right\}^{1/q},$$

where $\varphi_{-\gamma}$ is replaced by $\Phi_{-\gamma}$. Thus, to show $\|f\|^*_{B^{s,\tau}_{p,q}(\mathbb{R}^n)} \lesssim \|f\|_{B^{s,\tau}_{p,q}(\mathbb{R}^n)}$, it suffices to prove that for all $P\in\mathscr{Q}$ with $l(P)\ge 1$,

$$I_P \equiv \frac{1}{|P|^\tau}\left\{\sum_{j=-\gamma}^{0}\left[\int_P (2^{js}|\varphi_j * f(x)|)^p\,dx\right]^{q/p}\right\}^{1/q} \lesssim \|f\|_{B^{s,\tau}_{p,q}(\mathbb{R}^n)}$$

and that for all $P\in\mathscr{Q}$ with $l(P)<1$,

$$J_P \equiv \frac{1}{|P|^\tau}\left\{\sum_{j=j_P-\gamma}^{j_P-1}\left[\int_P (2^{js}|\varphi_j * f(x)|)^p\,dx\right]^{q/p}\right\}^{1/q} \lesssim \|f\|_{B^{s,\tau}_{p,q}(\mathbb{R}^n)}.$$

We first estimate I_P. By (2.1) and (2.2), there exist $\eta_j\in\mathscr{S}(\mathbb{R}^n)$, $j=-\gamma,\cdots,-1$, and $\zeta_1,\zeta_2\in\mathscr{S}(\mathbb{R}^n)$ such that

$$\varphi_j=\eta_j*\Phi,\quad j=-\gamma,\cdots,-1,\quad\text{and}\quad \varphi=\varphi_0=\zeta_1*\Phi+\zeta_2*\varphi_1.$$

We now consider two cases. When $p\in[1,\infty]$, by Minkowski's inequality, we see that

$$\left[\int_P|\varphi_j*f(x)|^p\,dx\right]^{1/p}=\left[\int_P|\eta_j*\Phi*f(x)|^p\,dx\right]^{1/p}$$
$$\le\int_{\mathbb{R}^n}|\eta_j(y)|\left[\int_P|(\Phi*f)(x-y)|^p\,dx\right]^{1/p}dy,$$

which further implies that

$$\frac{1}{|P|^\tau}\left\{\sum_{j=-\gamma}^{-1}\left[\int_P(2^{js}|\varphi_j*f(x)|)^p\,dx\right]^{q/p}\right\}^{1/q}$$
$$\lesssim\frac{1}{|P|^\tau}\left\{\sum_{j=-\gamma}^{-1}\left[\int_{\mathbb{R}^n}|\eta_j(y)|\left(\int_P|(\Phi*f)(x-y)|^p\,dx\right)^{1/p}dy\right]^q\right\}^{1/q}$$
$$\lesssim\|f\|_{B^{s,\tau}_{p,q}(\mathbb{R}^n)}\left\{\sum_{j=-\gamma}^{-1}\left[\int_{\mathbb{R}^n}|\eta_j(y)|\,dy\right]^q\right\}^{1/q}$$
$$\lesssim\|f\|_{B^{s,\tau}_{p,q}(\mathbb{R}^n)}.$$

The estimate for the term $j=0$ is similar.

When $p\in(0,1)$, for $j=-\gamma,\cdots,0$, by Lemma 2.3, we have

$$\varphi_j*f=\varphi_j*\Psi*\widetilde{\Phi}*f+\sum_{i=1}^\infty\varphi_j*\psi_i*\widetilde{\varphi}_i*f.$$

Notice that $\widetilde{\Phi}*f$ and $\widetilde{\varphi}_i*f$ are $C^\infty(\mathbb{R}^n)$ functions with polynomially increasing (see [134, Chap. 1, Theorem 3.13]). Then applying Lemma 2.5 with $v=0$, $h=\varphi_j*\Psi$, $g=\widetilde{\Phi}*f$ or $v=i$, $h=\varphi_j*\psi_i$ and $g=\widetilde{\varphi}_i*f$, and the monotonicity of the ℓ^q norms, in particular

$$\left(\sum_j|a_j|\right)^d\le\sum_j|a_j|^d,\qquad 0<d\le1,\quad\{a_j\}_j\subset\mathbb{C},\tag{2.11}$$

we have

$$|\varphi_j*f(x)|^p\le\sum_{k\in\mathbb{Z}^n}|\varphi_j*\Psi(k)|^p|\widetilde{\Phi}*f(x-k)|^p$$
$$+\sum_{i=1}^\infty\sum_{k\in\mathbb{Z}^n}2^{-inp}|\varphi_j*\psi_i(2^{-i}k)|^p|\widetilde{\varphi}_i*f(x-2^{-i}k)|^p.$$

Thus, using Lemma 2.4 and [164, Lemma 2.2] with $M > (n(1/p-1)\vee 0)$, we obtain

$$\begin{aligned}
&\frac{1}{|P|^\tau}\left\{\int_P (2^{js}|\varphi_j * f(x)|)^p\,dx\right\}^{1/p}\\
&\quad\lesssim \|f\|_{B^{s,\tau}_{p,q}(\mathbb{R}^n)}\left\{\sum_{k\in\mathbb{Z}^n}|\varphi_j*\Psi(k)|^p+\sum_{i=1}^\infty 2^{-inp}\sum_{k\in\mathbb{Z}^n}|\varphi_j*\psi_i(2^{-i}k)|^p\right\}^{1/p}\\
&\quad\lesssim \|f\|_{B^{s,\tau}_{p,q}(\mathbb{R}^n)}\left\{\sum_{k\in\mathbb{Z}^n}\frac{2^{-jMp}}{(1+|k|)^{(n+M)p}}+\sum_{i=1}^\infty 2^{-inp}\sum_{k\in\mathbb{Z}^n}\frac{2^{-iMp}}{(2^{-j}+|2^{-i}k|)^{(n+M)p}}\right\}\\
&\quad\lesssim \|f\|_{B^{s,\tau}_{p,q}(\mathbb{R}^n)}.
\end{aligned}$$

To prove $J_P \lesssim \|f\|_{B^{s,\tau}_{p,q}(\mathbb{R}^n)}$, denote by $P(i)$ the dyadic cube containing P with $l(P(i)) = 2^i l(P)$. We now consider two cases. If $j_P \ge \gamma+1$, by $j_{P(\gamma)} = j_P - \gamma$ and $P \subset P(\gamma)$, we have

$$J_P \lesssim \frac{1}{|P(\gamma)|^\tau}\left\{\sum_{j=j_{P(\gamma)}}^{j_P-1}\left[\int_{P(\gamma)}(2^{js}|\varphi_j * f(x)|)^p\,dx\right]^{q/p}\right\}^{1/q} \lesssim \|f\|_{B^{s,\tau}_{p,q}(\mathbb{R}^n)}.$$

If $1 \le j_P \le \gamma$, by a similar argument to the estimate for I_P, we see that

$$\frac{1}{|P|^\tau}\left\{\sum_{j=j_P-\gamma}^{j_P-1}\left[\int_P(2^{js}|\varphi_j * f(x)|)^p\,dx\right]^{q/p}\right\}^{1/q} \lesssim \|f\|_{B^{s,\tau}_{p,q}(\mathbb{R}^n)},$$

which together with the previous estimates yields $\|f\|^*_{B^{s,\tau}_{p,q}(\mathbb{R}^n)} \lesssim \|f\|_{B^{s,\tau}_{p,q}(\mathbb{R}^n)}$.

To prove the converse estimate that $\|f\|_{B^{s,\tau}_{p,q}(\mathbb{R}^n)} \lesssim \|f\|^*_{B^{s,\tau}_{p,q}(\mathbb{R}^n)}$, it suffices to show that for all $P \in \mathscr{Q}$ with $l(P) \ge 1$,

$$\frac{1}{|P|^\tau}\left\{\int_P |\Phi * f(x)|^p\,dx\right\}^{1/p} \lesssim \|f\|^*_{B^{s,\tau}_{p,q}(\mathbb{R}^n)}. \tag{2.12}$$

Indeed, similarly to the estimates for I_P, if $p\in[1,\infty]$, using the fact that there exist $\rho_j \in \mathscr{S}(\mathbb{R}^n)$, $j = -\gamma, \cdots, 1$, such that

$$\Phi * f = \rho_{-\gamma} * \Phi_{-\gamma} * f + \sum_{j=-\gamma+1}^{1} \rho_j * \varphi_j * f$$

(see, for example, [64, p. 130]), and Minkowski's inequality, we have (2.12); if $p \in (0,1)$, Lemmas 2.3, 2.4 and 2.5, and (2.11) also yield (2.12), which completes the proof of Lemma 2.6. □

2.1.3 Sequence Spaces

Now we introduce the corresponding inhomogeneous sequence spaces of $B^{s,\tau}_{p,q}(\mathbb{R}^n)$ and $F^{s,\tau}_{p,q}(\mathbb{R}^n)$, which are indexed by the set of dyadic cubes Q with $l(Q) \le 1$.

Definition 2.2. Let $\tau, s \in \mathbb{R}$ and $q \in (0,\infty]$.

(i) Let $p \in (0,\infty]$. The *inhomogeneous sequence space* $b^{s,\tau}_{p,q}(\mathbb{R}^n)$ is defined to be the set of all sequences $t \equiv \{t_Q\}_{l(Q)\le 1} \subset \mathbb{C}$ such that $\|t\|_{b^{s,\tau}_{p,q}(\mathbb{R}^n)} < \infty$, where

$$\|t\|_{b^{s,\tau}_{p,q}(\mathbb{R}^n)} \equiv \sup_{P\in\mathcal{Q}} \frac{1}{|P|^\tau} \left\{ \sum_{j=(j_P\vee 0)}^{\infty} 2^{j(s+n/2-n/p)q} \left[\sum_{\substack{l(Q)=2^{-j}\\ Q\subset P}} |t_Q|^p \right]^{\frac{q}{p}} \right\}^{\frac{1}{q}}. \quad (2.13)$$

(ii) Let $p \in (0,\infty)$. The *inhomogeneous sequence space* $f^{s,\tau}_{p,q}(\mathbb{R}^n)$ is defined to be the set of all sequences $t \equiv \{t_Q\}_{l(Q)\le 1} \subset \mathbb{C}$ such that $\|t\|_{f^{s,\tau}_{p,q}(\mathbb{R}^n)} < \infty$, where

$$\|t\|_{f^{s,\tau}_{p,q}(\mathbb{R}^n)} \equiv \sup_{P\in\mathcal{Q}} \frac{1}{|P|^\tau} \left\{ \int_P \left[\sum_{j=(j_P\vee 0)}^{\infty} \sum_{l(Q)=2^{-j}} 2^{j(s+n/2)q} |t_Q|^q \chi_Q(x) \right]^{\frac{p}{q}} dx \right\}^{\frac{1}{p}}. \quad (2.14)$$

Similarly, we use $a^{s,\tau}_{p,q}(\mathbb{R}^n)$ *to denote either* $b^{s,\tau}_{p,q}(\mathbb{R}^n)$ *or* $f^{s,\tau}_{p,q}(\mathbb{R}^n)$. *If* $a^{s,\tau}_{p,q}(\mathbb{R}^n)$ *means* $f^{s,\tau}_{p,q}(\mathbb{R}^n)$, *then the case* $p=\infty$ *is excluded.*

Under the additional restriction $p \ge q$ also the sequence spaces $\dot{F}^{s,\tau}_{p,q}(\mathbb{R}^n)$ allow a total discretization. This fact is an immediate consequence of [37, Proposition 2.2].

Remark 2.4. Let $\tau \in [0,\infty)$, $s \in \mathbb{R}$, $p \in (0,\infty)$ and $q \in (0,\infty]$. If $p \ge q$, then there exists a positive constant C, depending only on p and q, such that for all $t \in f^{s,\tau}_{p,q}(\mathbb{R}^n)$,

$$\begin{aligned} C^{-1}\|t\|_{f^{s,\tau}_{p,q}(\mathbb{R}^n)} &\le \sup_{P\in\mathcal{Q}} \frac{1}{|P|^\tau} \left\{ \sum_{j=(j_P\vee 0)}^{\infty} \sum_{\substack{l(Q)=2^{-j}\\ Q\subset P}} (|Q|^{-s/n-1/2+1/q}|t_Q|)^q \right. \\ &\quad \left. \times \left[\frac{1}{|Q|} \sum_{\substack{R\in\mathcal{Q}\\ R\subset Q}} (|R|^{-s/n-1/2+1/q}|t_R|)^q \right]^{p/q-1} \right\}^{1/p} \\ &\le C\|t\|_{f^{s,\tau}_{p,q}(\mathbb{R}^n)}. \end{aligned}$$

The homogeneous counterpart of $a^{s,\tau}_{p,q}(\mathbb{R}^n)$, denoted by $\dot{a}^{s,\tau}_{p,q}(\mathbb{R}^n)$, was already introduced in [165]. The relation between $a^{s,\tau}_{p,q}(\mathbb{R}^n)$ and $\dot{a}^{s,\tau}_{p,q}(\mathbb{R}^n)$ is trivial. In fact, define $V:\ a^{s,\tau}_{p,q}(\mathbb{R}^n) \to \dot{a}^{s,\tau}_{p,q}(\mathbb{R}^n)$ by setting

$$(Vt)_Q \equiv \begin{cases} t_Q & \text{if } \ l(Q) \le 1, \\ 0 & \text{otherwise}. \end{cases} \tag{2.15}$$

Then V is an isometric embedding of $a^{s,\tau}_{p,q}(\mathbb{R}^n)$ in $\dot{a}^{s,\tau}_{p,q}(\mathbb{R}^n)$. Define $W : \dot{a}^{s,\tau}_{p,q}(\mathbb{R}^n) \to a^{s,\tau}_{p,q}(\mathbb{R}^n)$ by setting $(Wt)_Q = t_Q$ if $l(Q) \le 1$. Then W is continuous and $W \circ V$ is the identity on $a^{s,\tau}_{p,q}(\mathbb{R}^n)$.

Next we establish the relation between $A^{s,\tau}_{p,q}(\mathbb{R}^n)$ and $a^{s,\tau}_{p,q}(\mathbb{R}^n)$. Let Φ, Ψ, φ and ψ be as in Lemma 2.3. Recall that the *φ-transform* S_φ is defined by setting $(S_\varphi f)_Q \equiv \langle f, \Phi_Q \rangle$ if $l(Q) = 1$ and $(S_\varphi f)_Q \equiv \langle f, \varphi_Q \rangle$ if $l(Q) < 1$, the *inverse φ-transform* T_ψ is defined by

$$T_\psi t \equiv \sum_{l(Q)=1} t_Q \Psi_Q + \sum_{l(Q)<1} t_Q \psi_Q;$$

see, for example, [64, p. 131].

To show that T_ψ is well defined for all $t \in a^{s,\tau}_{p,q}(\mathbb{R}^n)$, we need the following conclusion.

Lemma 2.7. *Let $s \in \mathbb{R}$, $\tau \in [0,\infty)$, $p, q \in (0,\infty]$ and Ψ, $\psi \in \mathscr{S}(\mathbb{R}^n)$ satisfy, respectively,* (2.1) *and* (2.2)*. Then for all $t \in a^{s,\tau}_{p,q}(\mathbb{R}^n)$,*

$$T_\psi t = \sum_{l(Q)=1} t_Q \Psi_Q + \sum_{l(Q)<1} t_Q \psi_Q$$

converges in $\mathscr{S}'(\mathbb{R}^n)$; moreover, $T_\psi : a^{s,\tau}_{p,q}(\mathbb{R}^n) \to \mathscr{S}'(\mathbb{R}^n)$ is continuous.

Proof. To prove Lemma 2.7, we only need to show that there exists an $M \in \mathbb{Z}$ such that for all $t \in a^{s,\tau}_{p,q}(\mathbb{R}^n)$ and $\phi \in \mathscr{S}(\mathbb{R}^n)$,

$$\sum_{l(Q)=1} |t_Q||\langle \Psi_Q, \phi \rangle| + \sum_{l(Q)<1} |t_Q||\langle \psi_Q, \phi \rangle| \lesssim \|t\|_{a^{s,\tau}_{p,q}(\mathbb{R}^n)} \|\phi\|_{\mathscr{S}_M}.$$

In fact, observe that $|t_Q| \le \|t\|_{a^{s,\tau}_{p,q}(\mathbb{R}^n)} |Q|^{s/n+1/2+\tau-1/p}$ for all dyadic cubes Q with $l(Q) \le 1$. Thus,

$$\begin{aligned} &\sum_{l(Q)=1} |t_Q||\langle \Psi_Q, \phi \rangle| + \sum_{l(Q)<1} |t_Q||\langle \psi_Q, \phi \rangle| \\ &\le \|t\|_{a^{s,\tau}_{p,q}(\mathbb{R}^n)} \left\{ \sum_{l(Q)=1} |\langle \Psi_Q, \phi \rangle| + \sum_{l(Q)<1} |Q|^{s/n+1/2+\tau-1/p} |\langle \psi_Q, \phi \rangle| \right\}. \end{aligned}$$

Let $M > (n/p - n - n\tau - s) \vee 0$. We see that

$$\begin{aligned} \sum_{l(Q)=1} |\langle \Psi_Q, \phi \rangle| &\le \sum_{k \in \mathbb{Z}^n} \int_{\mathbb{R}^n} |\Psi(x-k)||\phi(x)|\,dx \\ &\lesssim \|\Psi\|_{\mathscr{S}_M} \|\phi\|_{\mathscr{S}_M} \sum_{k \in \mathbb{Z}^n} (1+|k|)^{-n-M} \\ &\lesssim \|\Psi\|_{\mathscr{S}_M} \|\phi\|_{\mathscr{S}_M}. \end{aligned}$$

On the other hand, by Lemma 2.4, we obtain

$$\begin{aligned}
&\sum_{l(Q)<1} |Q|^{s/n+1/2+\tau-1/p}|\langle \psi_Q, \phi\rangle| \\
&\lesssim \|\psi\|_{\mathscr{S}_{M+1}}\|\phi\|_{\mathscr{S}_{M+1}} \sum_{j=1}^{\infty}\sum_{k\in\mathbb{Z}^n} 2^{-jn(s/n+1+\tau-1/p)}2^{-jM}\frac{1}{(1+|2^{-j}k|)^{n+M}} \\
&\lesssim \|\psi\|_{\mathscr{S}_{M+1}}\|\phi\|_{\mathscr{S}_{M+1}},
\end{aligned}$$

which completes the proof of Lemma 2.7. □

Now we have the following inhomogeneous analogue of [165, Theorem 3.1], which is the so-called φ-transform characterization in the sense of Frazier and Jawerth.

Theorem 2.1. *Let $s\in\mathbb{R}$, $\tau\in[0,\infty)$, $p,q\in(0,\infty]$ and Φ, Ψ, φ and ψ be as in Lemma 2.3. Then the operators $S_\varphi : A^{s,\tau}_{p,q}(\mathbb{R}^n)\to a^{s,\tau}_{p,q}(\mathbb{R}^n)$ and $T_\psi : a^{s,\tau}_{p,q}(\mathbb{R}^n)\to A^{s,\tau}_{p,q}(\mathbb{R}^n)$ are bounded. Furthermore, $T_\psi\circ S_\varphi$ is the identity on $A^{s,\tau}_{p,q}(\mathbb{R}^n)$.*

To prove this theorem, we need some technical lemmas. For a sequence $t=\{t_Q\}_{l(Q)\le 1}$, $r\in(0,\infty]$ and a fixed $\lambda\in(0,\infty)$, set

$$(t^*_{r,\lambda})_Q \equiv \left(\sum_{\{R\in\mathcal{Q}:\ l(R)=l(Q)\}}\frac{|t_R|^r}{(1+l(R)^{-1}|x_R-x_Q|)^\lambda}\right)^{1/r}, \quad Q\in\mathcal{Q},\ l(Q)\le 1$$

and $t^*_{r,\lambda}\equiv\{(t^*_{r,\lambda})_Q\}_{l(Q)\le 1}$. We have the following estimate, which is an immediate consequence of the corresponding result on homogeneous spaces $\dot a^{s,\tau}_{p,q}(\mathbb{R}^n)$ in [165, Lemma 3.4] and the fact that the operator V in (2.15) is an isometric embedding of $a^{s,\tau}_{p,q}(\mathbb{R}^n)$ in $\dot a^{s,\tau}_{p,q}(\mathbb{R}^n)$. For completeness, we give a direct proof.

Lemma 2.8. *Let $s\in\mathbb{R}$, $\tau\in[0,\infty)$, $p,q\in(0,\infty]$ and $\lambda\in(n,\infty)$. Then there exists a constant $C\in[1,\infty)$ such that for all $t\in a^{s,\tau}_{p,q}(\mathbb{R}^n)$,*

$$\|t\|_{a^{s,\tau}_{p,q}(\mathbb{R}^n)}\le \|t^*_{p\wedge q,\lambda}\|_{a^{s,\tau}_{p,q}(\mathbb{R}^n)}\le C\|t\|_{a^{s,\tau}_{p,q}(\mathbb{R}^n)}.$$

Proof. Let $t\in a^{s,\tau}_{p,q}(\mathbb{R}^n)$. Notice that $|t_Q|\le (t^*_{p\wedge q,\lambda})_Q$ holds for all dyadic cubes Q with $l(Q)\le 1$. We then obtain that $\|t\|_{a^{s,\tau}_{p,q}(\mathbb{R}^n)}\le \|t^*_{p\wedge q,\lambda}\|_{a^{s,\tau}_{p,q}(\mathbb{R}^n)}$.

To see the converses, fix a dyadic cube P. For all $l(Q)\le 1$, let $r_Q\equiv t_Q$ if $Q\subset 3P$ and $r_Q\equiv 0$ otherwise, and let $u_Q\equiv t_Q-r_Q$. Set $r\equiv\{r_Q\}_{l(Q)\le 1}$ and $u\equiv\{u_Q\}_{l(Q)\le 1}$. Then for all such Q, we have

$$(t^*_{p\wedge q,\lambda})_Q^{p\wedge q} = (r^*_{p\wedge q,\lambda})_Q^{p\wedge q} + (u^*_{p\wedge q,\lambda})_Q^{p\wedge q}. \tag{2.16}$$

Applying the fact that for all $t=\{t_Q\}_{l(Q)\le 1}$, $\|t^*_{p\wedge q,\lambda}\|_{b^s_{p,q}(\mathbb{R}^n)}\sim\|t\|_{b^s_{p,q}(\mathbb{R}^n)}$ and $\|t^*_{p\wedge q,\lambda}\|_{f^s_{p,q}(\mathbb{R}^n)}\sim\|t\|_{f^s_{p,q}(\mathbb{R}^n)}$, where $b^s_{p,q}(\mathbb{R}^n)$ and $f^s_{p,q}(\mathbb{R}^n)$ are, respectively, the corresponding sequence spaces for the inhomogeneous Besov space $B^s_{p,q}(\mathbb{R}^n)$ and the Triebel-Lizorkin space $F^s_{p,q}(\mathbb{R}^n)$ (see [64, Lemma 2.3] for its proof), we then have

$$\begin{aligned}
\mathrm{I}_P &\equiv \frac{1}{|P|^\tau}\left\{\sum_{j=(j_P\vee 0)}^{\infty}\left[\sum_{\substack{l(Q)=2^{-j}\\ Q\subset P}}[|Q|^{-s/n-1/2+1/p}(r^*_{p\wedge q,\lambda})_Q]^p\right]^{q/p}\right\}^{1/q}\\
&\le \frac{1}{|P|^\tau}\|r^*_{p\wedge q,\lambda}\|_{b^s_{p,q}(\mathbb{R}^n)}\\
&\lesssim \frac{1}{|P|^\tau}\|r\|_{b^s_{p,q}(\mathbb{R}^n)}\\
&\lesssim \|t\|_{b^{s,\tau}_{p,q}(\mathbb{R}^n)}
\end{aligned}$$

and similarly,

$$\widetilde{\mathrm{I}}_P \equiv \frac{1}{|P|^\tau}\left\{\int_P\left[\sum_{\substack{Q\subset P\\ l(Q)\le 1}}[|Q|^{-s/n-1/2}(r^*_{p\wedge q,\lambda})_Q\chi_Q(x)]^q\right]^{p/q}dx\right\}^{1/p} \lesssim \|t\|_{f^{s,\tau}_{p,q}(\mathbb{R}^n)}.$$

On the other hand, let $Q\subset P$ be a dyadic cube with side length no more then 1. Then $l(Q)=2^{-i}l(P)$ for some nonnegative integer $i\ge(-j_P)\vee 0=-(j_P\wedge 0)$. Suppose $\widetilde{Q}$ is any dyadic cube with $l(\widetilde{Q})=l(Q)=2^{-i}l(P)$ and $\widetilde{Q}\subset P+kl(P)\not\subseteq 3P$ for some $k\in\mathbb{Z}^n$, where $P+kl(P)\equiv\{x+kl(P):\ x\in P\}$. Then $|k|\ge 2$ and $1+l(\widetilde{Q})^{-1}|x_Q-x_{\widetilde{Q}}|\sim 2^i|k|$. Thus,

$$\begin{aligned}
\mathrm{J}_P &\equiv \frac{1}{|P|^\tau}\left\{\sum_{i=-(j_P\wedge 0)}^{\infty}\left[\sum_{\substack{l(Q)=2^{-i}l(P)\\ Q\subset P}}[|Q|^{-s/n-1/2+1/p}(u^*_{p\wedge q,\lambda})_Q]^p\right]^{q/p}\right\}^{1/q}\\
&\lesssim \frac{1}{|P|^\tau}\left\{\sum_{i=-(j_P\wedge 0)}^{\infty}2^{inq/p-i\lambda q/(p\wedge q)}\left[\sum_{\substack{k\in\mathbb{Z}^n\\ |k|\ge 2}}|k|^{-\lambda}\right.\right.\\
&\qquad\left.\left.\times\sum_{\substack{l(\widetilde{Q})=2^{-i}l(P)\\ \widetilde{Q}\subset P+kl(P)}}(|\widetilde{Q}|^{-s/n-1/2+1/p}|t_{\widetilde{Q}}|)^{p\wedge q}\right]^{\frac{q}{p\wedge q}}\right\}^{1/q}.
\end{aligned}$$

When $p \le q$, by $\lambda > n$, we have

$$\mathrm{J}_P \lesssim \|t\|_{b_{p,q}^{s,\tau}(\mathbb{R}^n)} \left\{ \sum_{i=-(j_P\wedge 0)}^{\infty} 2^{inq/p-i\lambda q/p} \left(\sum_{\substack{k\in\mathbb{Z}^n \\ |k|\ge 2}} |k|^{-\lambda} \right)^{q/p} \right\}^{1/q} \lesssim \|t\|_{b_{p,q}^{s,\tau}(\mathbb{R}^n)};$$

when $p > q$, by Hölder's inequality and $\lambda > n$, we obtain

$$\begin{aligned} \mathrm{J}_P &\lesssim \frac{1}{|P|^\tau} \left\{ \sum_{i=-(j_P\wedge 0)}^{\infty} 2^{inq/p-i\lambda} \left[\sum_{\substack{k\in\mathbb{Z}^n \\ |k|\ge 2}} |k|^{-\lambda} \sum_{\substack{l(\widetilde{Q})=2^{-i}l(P) \\ \widetilde{Q}\subset P+kl(P)}} (|\widetilde{Q}|^{-s/n-1/2+1/p}|t_{\widetilde{Q}}|)^q \right] \right\}^{1/q} \\ &\lesssim \|t\|_{b_{p,q}^{s,\tau}(\mathbb{R}^n)} \left\{ \sum_{i=-(j_P\wedge 0)}^{\infty} 2^{in-i\lambda} \left(\sum_{\substack{k\in\mathbb{Z}^n \\ |k|\ge 2}} |k|^{-\lambda} \right) \right\}^{1/q} \\ &\lesssim \|t\|_{b_{p,q}^{s,\tau}(\mathbb{R}^n)}. \end{aligned}$$

Therefore, by (2.16),

$$\|t^*_{\min\{p,q\},\lambda}\|_{b_{p,q}^{s,\tau}(\mathbb{R}^n)} \lesssim \sup_{P\in\mathscr{Q}} (\mathrm{I}_P + \mathrm{J}_P) \lesssim \|t\|_{b_{p,q}^{s,\tau}(\mathbb{R}^n)}.$$

To complete the proof, for any $i \in \mathbb{Z}_+$, $k \in \mathbb{Z}_+^n$ and dyadic cube P, set

$$A(i,k,P) \equiv \{\widetilde{Q} \in \mathscr{Q}:\ l(\widetilde{Q}) = 2^{-i}l(P),\ \widetilde{Q} \subset P + kl(P),\ \widetilde{Q} \cap (3P) = \emptyset\}.$$

Recall that $1 + l(\widetilde{Q})^{-1}|x_Q - x_{\widetilde{Q}}| \sim 2^i|k|$ for any $Q \subset P$ and $\widetilde{Q} \in A(i,k,P)$. Similarly to the proof of [64, Remark A.3], by (2.11), we obtain that for all $x \in P$ and $a \in (0, p\wedge q]$,

$$\begin{aligned} &\sum_{\widetilde{Q}\in A(i,k,P)} \frac{(|\widetilde{Q}|^{-s/n-1/2}|t_{\widetilde{Q}}|)^{p\wedge q}}{(1+l(\widetilde{Q})^{-1}|x_Q - x_{\widetilde{Q}}|)^\lambda} \\ &\lesssim 2^{-i(\lambda-n(p\wedge q)/a)}|k|^{-\lambda} \left[\mathrm{M}\left(\sum_{\substack{l(\widetilde{Q})=2^{-i}l(P) \\ \widetilde{Q}\subset P+kl(P)}} (|\widetilde{Q}|^{-s/n}|t_{\widetilde{Q}}|\widetilde{\chi}_{\widetilde{Q}})^a \right)(x+kl(P)) \right]^{\frac{p\wedge q}{a}}, \end{aligned}$$

where herein and in what follows, M denotes the *Hardy-Littlewood maximal function* on $\mathbb{R}^n$. Let $a \equiv \frac{2n(p\wedge q)}{n+\lambda}$. Then $a \in (0, p\wedge q)$. Applying Minkowski's inequality, Fefferman-Stein's vector-valued inequality and Hölder's inequality, we have

$$\widetilde{\mathrm{J}}_P \equiv \frac{1}{|P|^\tau}\left\{\int_P\left[\sum_{Q\subset P,l(Q)\le 1}[|Q|^{-\frac{s}{n}-\frac12}(u^*_{p\wedge q,\lambda})_Q\chi_Q(x)]^q\right]^{\frac{p}{q}}dx\right\}^{\frac1p}$$

$$\lesssim \frac{1}{|P|^\tau}\left\{\int_P\left[\sum_{i=-(j_P\wedge 0)}^{\infty}\left(\sum_{\substack{k\in\mathbb{Z}^n\\|k|\ge 2}}2^{-i(\lambda-n(p\wedge q)/a)}|k|^{-\lambda}\right.\right.\right.$$

$$\left.\left.\left.\times\left[\mathrm{M}\left(\sum_{\substack{l(\widetilde{Q})=2^{-i}l(P)\\ \widetilde{Q}\subset P+kl(P)}}(|\widetilde{Q}|^{-\frac{s}{n}}|t_{\widetilde{Q}}|\widetilde{\chi}_{\widetilde{Q}})^a\right)(x+kl(P))\right]^{\frac{p\wedge q}{a}}\right)^{\frac{q}{p\wedge q}}\right]^{\frac{p}{q}}dx\right\}^{\frac1p}$$

$$\lesssim \|t\|_{f^{s,\tau}_{p,q}(\mathbb{R}^n)}.$$

Therefore, by (2.16) again,

$$\|t^*_{\min\{p,q\},\lambda}\|_{f^{s,\tau}_{p,q}(\mathbb{R}^n)} \lesssim \sup_{P\in\mathcal{Q}}(\widetilde{I}_P+\widetilde{J}_P) \lesssim \|t\|_{f^{s,\tau}_{p,q}(\mathbb{R}^n)},$$

which completes the proof of Lemma 2.8. □

Let Φ and φ satisfy, respectively, (2.1) and (2.2). Since $\widetilde{\Phi}(x) \equiv \overline{\Phi(-x)}$ and $\widetilde{\varphi}(x) \equiv \overline{\varphi(-x)}$ also satisfy, respectively, (2.1) and (2.2), we may take $\widetilde{\Phi}$, $\widetilde{\varphi}$ in place of Φ and φ in Definition 2.1. For any $f \in \mathcal{S}'(\mathbb{R}^n)$, define the sequence $\sup(f) \equiv \{\sup_Q(f)\}_{l(Q)\le 1}$ by setting

$$\sup_Q(f) \equiv \begin{cases} |Q|^{1/2}\sup_{y\in Q}|\widetilde{\varphi}_j * f(y)| & \text{if}\quad l(Q)=2^{-j}<1,\\ \sup_{y\in Q}|\widetilde{\Phi}*f(y)| & \text{if}\quad l(Q)=1.\end{cases}$$

For any $\gamma\in\mathbb{Z}_+$, the sequence $\inf_\gamma(f) \equiv \{\inf_{Q,\gamma}(f)\}_{l(Q)\le 1}$ is defined by setting

$$\inf_{Q,\gamma}(f) \equiv |Q|^{1/2}\max\left\{\inf_{y\in\widetilde{Q}}|\widetilde{\varphi}_j * f(y)| :\ l(\widetilde{Q}) = 2^{-\gamma}l(Q),\ \widetilde{Q}\subset Q\right\}$$

if $l(Q) = 2^{-j} < 1$ and

$$\inf_{Q,\gamma}(f) \equiv \max\left\{\inf_{y\in\widetilde{Q}}|\widetilde{\Phi} * f(y)| :\ l(\widetilde{Q}) = 2^{-\gamma},\ \widetilde{Q}\subset Q\right\}$$

if $l(Q) = 1$. We then have the following lemma.

Lemma 2.9. *Let $s \in \mathbb{R}$, $\tau \in [0,\infty)$, $p,q \in (0,\infty]$ and $\gamma \in \mathbb{Z}_+$ be sufficiently large. Then there exists a constant $C \in [1,\infty)$ such that for all $f \in A^{s,\tau}_{p,q}(\mathbb{R}^n)$,*

$$C^{-1}\|\mathrm{inf}_\gamma(f)\|_{a^{s,\tau}_{p,q}(\mathbb{R}^n)} \le \|f\|_{A^{s,\tau}_{p,q}(\mathbb{R}^n)} \le \|\sup(f)\|_{a^{s,\tau}_{p,q}(\mathbb{R}^n)} \le C\|\mathrm{inf}_\gamma(f)\|_{a^{s,\tau}_{p,q}(\mathbb{R}^n)}.$$

Proof. The inequality $\|f\|_{A^{s,\tau}_{p,q}(\mathbb{R}^n)} \le \|\sup(f)\|_{a^{s,\tau}_{p,q}(\mathbb{R}^n)}$ immediately follows from the definitions of $\|f\|_{A^{s,\tau}_{p,q}(\mathbb{R}^n)}$ and $\sup(f)$.

Applying [64, Lemma A.4] to the functions $\widetilde{\Phi} * f$ and $\widetilde{\varphi}_j * f$, $j \in \mathbb{N}$, we obtain that for all $Q \in \mathcal{Q}$ with $l(Q) = 2^{-j} \le 1$,

$$(\sup(f)^*_{r,\lambda})_Q \sim (\mathrm{inf}_\gamma(f)^*_{r,\lambda})_Q,$$

where $r = p \wedge q$. Thus,

$$\|\sup(f)^*_{r,\lambda}\|_{a^{s,\tau}_{p,q}(\mathbb{R}^n)} \sim \|\mathrm{inf}_\gamma(f)^*_{r,\lambda}\|_{a^{s,\tau}_{p,q}(\mathbb{R}^n)},$$

which together with Lemma 2.8 yields that

$$\|\sup(f)\|_{a^{s,\tau}_{p,q}(\mathbb{R}^n)} \sim \|\mathrm{inf}_\gamma(f)\|_{a^{s,\tau}_{p,q}(\mathbb{R}^n)}.$$

To complete the proof of Lemma 2.9, we still need to show that

$$\|\mathrm{inf}_\gamma(f)\|_{a^{s,\tau}_{p,q}(\mathbb{R}^n)} \lesssim \|f\|_{A^{s,\tau}_{p,q}(\mathbb{R}^n)}.$$

Define a sequence $t \equiv \{t_J\}_{l(J)\le 1}$ by setting

$$t_J \equiv |J|^{1/2} \inf_{y\in J} |\widetilde{\varphi}_{i-\gamma} * f(y)| \quad \text{if} \quad l(J) = 2^{-i} < 1$$

and

$$t_J \equiv \inf_{y\in J} |\widetilde{\Phi}_{-\gamma} * f(y)| \quad \text{if} \quad l(J) = 1.$$

Then for all $r \in (0,\infty)$, dyadic cubes Q with $l(Q) = 2^{-j} \le 1$ and a fixed $\lambda > n$, we have

$$\mathrm{inf}_{Q,\gamma}(f)\widetilde{\chi}_Q \le \sum_{\substack{\widetilde{Q}\subset Q\\ l(\widetilde{Q})=2^{-\gamma}l(Q)}} \left\{ \sum_{\substack{J\subset Q\\ l(J)=2^{-\gamma}l(Q)}} t_J^r \right\}^{1/r} \widetilde{\chi}_{\widetilde{Q}} \lesssim \sum_{\substack{\widetilde{Q}\subset Q\\ l(\widetilde{Q})=2^{-\gamma}l(Q)}} (t^*_{r,\lambda})_{\widetilde{Q}}\widetilde{\chi}_{\widetilde{Q}}, \tag{2.17}$$

where $\widetilde{\chi}_Q \equiv |Q|^{-1/2}\chi_Q$. We further obtain that

$$\|\mathrm{inf}_\gamma(f)\|_{a^{s,\tau}_{p,q}(\mathbb{R}^n)} \lesssim \|t^*_{r,\lambda}\|_{a^{s,\tau}_{p,q}(\mathbb{R}^n)}.$$

In fact, for the space $f^{s,\tau}_{p,q}(\mathbb{R}^n)$, by (2.17), for each $P \in \mathcal{Q}$, we obtain that

$$
\begin{aligned}
&\frac{1}{|P|^\tau}\left\{\int_P \left[\sum_{j=(j_P\vee 0)}^{\infty}\sum_{l(Q)=2^{-j}} 2^{jsq}[\inf_{Q,\gamma}(f)\widetilde{\chi}_Q(x)]^q\right]^{p/q} dx\right\}^{1/p}\\
&\lesssim \frac{1}{|P|^\tau}\left\{\int_P \left[\sum_{j=(j_P\vee 0)}^{\infty}\sum_{l(\widetilde{Q})=2^{-j-\gamma}} 2^{jsq}[(t^*_{r,\lambda})_{\widetilde{Q}}\widetilde{\chi}_{\widetilde{Q}}(x)]^q\right]^{p/q} dx\right\}^{1/p}\\
&\sim \frac{1}{|P|^\tau}\left\{\int_P \left[\sum_{i=(j_P\vee 0)+\gamma}^{\infty}\sum_{l(\widetilde{Q})=2^{-i}} 2^{isq}[(t^*_{r,\lambda})_{\widetilde{Q}}\widetilde{\chi}_{\widetilde{Q}}(x)]^q\right]^{p/q} dx\right\}^{1/p}\\
&\equiv \mathrm{I}_P.
\end{aligned}
$$

Notice that $P = \cup_{m=1}^{2^{\gamma n}} P_m$, where $\{P_m\}_{m=1}^{2^{\gamma n}}$ are disjoint dyadic cubes with side length $l(P_m) = 2^{-\gamma} l(P) = 2^{-(j_P+\gamma)}$. This together with the facts that $\gamma \in \mathbb{Z}_+$ and

$$(j_P+\gamma)\vee 0 \le (j_P\vee 0)+\gamma$$

yields that

$$
\begin{aligned}
\mathrm{I}_P &\le \frac{1}{|P|^\tau}\left\{\int_P \left[\sum_{i=(j_P+\gamma)\vee 0}^{\infty}\sum_{l(\widetilde{Q})=2^{-i}} 2^{isq}[(t^*_{r,\lambda})_{\widetilde{Q}}\widetilde{\chi}_{\widetilde{Q}}(x)]^q\right]^{p/q} dx\right\}^{1/p}\\
&\le \frac{1}{|P|^\tau}\left\{\sum_{m=1}^{2^{\gamma n}}\int_{P_m} \left[\sum_{i=(j_{P_m}\vee 0)}^{\infty}\sum_{l(\widetilde{Q})=2^{-i}} 2^{isq}[(t^*_{r,\lambda})_{\widetilde{Q}}\widetilde{\chi}_{\widetilde{Q}}(x)]^q\right]^{p/q} dx\right\}^{1/p}\\
&\lesssim \|t^*_{r,\lambda}\|_{f^{s,\tau}_{p,q}(\mathbb{R}^n)},
\end{aligned}
$$

which further implies that

$$\|\inf_\gamma(f)\|_{f^{s,\tau}_{p,q}(\mathbb{R}^n)} \lesssim \|t^*_{r,\lambda}\|_{f^{s,\tau}_{p,q}(\mathbb{R}^n)}.$$

The proof for the space $b^{s,\tau}_{p,q}(\mathbb{R}^n)$ is similar and we leave the details to the reader.

Finally, picking $r = (p\wedge q)$, by Lemmas 2.8 and 2.6, we obtain

$$\|\inf_\gamma(f)\|_{a^{s,\tau}_{p,q}(\mathbb{R}^n)} \lesssim \|t^*_{r,\lambda}\|_{a^{s,\tau}_{p,q}(\mathbb{R}^n)} \lesssim \|t\|_{a^{s,\tau}_{p,q}(\mathbb{R}^n)} \lesssim \|f\|^*_{A^{s,\tau}_{p,q}(\mathbb{R}^n)} \lesssim \|f\|_{A^{s,\tau}_{p,q}(\mathbb{R}^n)},$$

which completes the proof of Lemma 2.9. □

Using Lemmas 2.7, 2.8 and 2.9 to replace Lemmas 3.3, 3.4 and 3.5 in [165] and repeating the proof of Theorem 3.1 in [165] then complete the proof of Theorem 2.1; see also the proof of Theorem 2.2 in [64, pp. 50–51]. For the reader's convenience, we give some details.

Proof of Theorem 2.1. By similarity, we only consider the spaces $B^{s,\tau}_{p,q}(\mathbb{R}^n)$. Let $f \in B^{s,\tau}_{p,q}(\mathbb{R}^n)$. Since for all dyadic cubes $Q = Q_{jk}$ with $l(Q) \le 1$,

$$|(S_\varphi f)_Q| = |\langle f, \varphi_Q\rangle| = |Q|^{1/2}|\widetilde{\varphi}_j * f(2^{-j}k)| \le \sup_Q(f),$$

where and in what follows, when $j = 0$, φ_0 is replaced by $\widetilde{\Phi}$. Then by Lemma 2.9, we obtain

$$\|S_\varphi f\|_{b^{s,\tau}_{p,q}(\mathbb{R}^n)} \le \|\sup_Q(f)\|_{b^{s,\tau}_{p,q}(\mathbb{R}^n)} \lesssim \|f\|_{B^{s,\tau}_{p,q}(\mathbb{R}^n)}.$$

Next we prove that the operator T_ψ is bounded from $b^{s,\tau}_{p,q}(\mathbb{R}^n)$ to $B^{s,\tau}_{p,q}(\mathbb{R}^n)$. Let $t = \{t_Q\}_{l(Q)\le 1} \in b^{s,\tau}_{p,q}(\mathbb{R}^n)$. Then by Lemma 2.7,

$$T_\psi t = \sum_{l(Q)=1} t_Q \Psi_Q + \sum_{l(Q)<1} t_Q \psi_Q$$

converges in $\mathscr{S}'(\mathbb{R}^n)$. Set

$$f \equiv T_\psi t = \sum_{l(Q)=1} t_Q \Psi_Q + \sum_{l(Q)<1} t_Q \psi_Q.$$

By (2.1) and (2.2), we obtain that

$$\widetilde{\varphi}_j * f = \sum_{i=j-1}^{j+1} \sum_{l(Q)=2^{-i}} t_Q \widetilde{\varphi}_j * \psi_Q$$

for all $j \in \mathbb{Z}_+$, where we set $\psi_Q \equiv \Psi_Q$ if $l(Q) = 1$ and $\psi_Q \equiv 0$ if $l(Q) = 2$. Notice that $\widetilde{\varphi}_j * \psi_Q \in \mathscr{S}(\mathbb{R}^n)$ for all $j \in \mathbb{Z}_+$ and dyadic cubes Q with $l(Q) = 2^{-i}$. Then for $r \in (0,\infty)$ and a fixed number $\lambda > n$, we have

$$|\widetilde{\varphi}_j * \psi_Q(x)| \lesssim |Q|^{-1/2}\left[1 + 2^i|x - x_Q|\right]^{-\lambda/\min\{1,r\}}.$$

Therefore, if $x \in Q^* \subset Q \subset Q^{**}$, where Q^*, Q and Q^{**} are respectively dyadic cubes with $l(Q^*) = 2^{-i-1}$, $l(Q) = 2^{-i}$ and $l(Q^{**}) = 2^{-i+1}$, using (2.11) when $r \in (0, 1]$, or Hölder's inequality when $r \in (1,\infty)$ together with the fact that when $r \in (1,\infty)$ and $\lambda > n$,

$$\sum_{l(Q)=2^{-i}} (1 + 2^i|x - x_Q|)^{-\lambda} \lesssim 1,$$

we obtain that

$$|\widetilde{\varphi}_j * f(x)| \lesssim \sum_{i=j-1}^{j+1} \left\{ \sum_{l(Q)=2^{-i}} |Q|^{-r/2} \frac{|t_Q|^r}{(1+2^i|x-x_Q|)^\lambda} \chi_{Q^*}(x) \right\}^{1/r}.$$

Therefore, for $x \in Q^*$, we have

$$|\widetilde{\varphi}_j * f(x)| \lesssim |Q|^{-1/2} \left\{ (t^*_{r,\lambda})_{Q^*} + (t^*_{r,\lambda})_Q + (t^*_{r,\lambda})_{Q^{**}} \right\} \chi_{Q^*}(x).$$

Taking $r = \min\{p,q\}$, then by Lemma 2.8, we obtain

$$\|T_\psi t\|_{B^{s,\tau}_{p,q}(\mathbb{R}^n)} = \|f\|_{B^{s,\tau}_{p,q}(\mathbb{R}^n)} \lesssim \|t^*_{r,\lambda}\|_{b^{s,\tau}_{p,q}(\mathbb{R}^n)} \lesssim \|t\|_{b^{s,\tau}_{p,q}(\mathbb{R}^n)},$$

which completes the proof of Theorem 2.1. □

As an immediate conclusion of Theorem 2.1, we obtain the next important property of our spaces $A^{s,\tau}_{p,q}(\mathbb{R}^n)$.

Corollary 2.1. *The definition of the spaces $A^{s,\tau}_{p,q}(\mathbb{R}^n)$ is independent of the choices of Φ and φ.*

2.2 Embeddings

From Definition 2.1, it is easy to deduce the following basic properties of the spaces $A^{s,\tau}_{p,q}(\mathbb{R}^n)$; see also Proposition 2.3.2/2 in [145]. In what follows, the symbol $\subset$ stands for continuous embedding. To begin with we collect so-called elementary embeddings.

Proposition 2.1. *Let $\tau, s \in \mathbb{R}$, $p, q \in (0,\infty]$ and $\varepsilon \in (0,\infty)$.*

(i) The scale $A^{s,\tau}_{p,q}(\mathbb{R}^n)$ is monotone with respect to q, namely, if $q_1 \le q_2$, then

$$A^{s,\tau}_{p,q_1}(\mathbb{R}^n) \subset A^{s,\tau}_{p,q_2}(\mathbb{R}^n);$$

(ii) The scale $A^{s,\tau}_{p,q}(\mathbb{R}^n)$ is monotone with respect to s, namely, for all $q_1, q_2 \in (0,\infty]$,

$$A^{s+\varepsilon,\tau}_{p,q_1}(\mathbb{R}^n) \subset A^{s,\tau}_{p,q_2}(\mathbb{R}^n);$$

(iii) For all $p \in (0,\infty)$,

$$B^{s,\tau}_{p,\min\{p,q\}}(\mathbb{R}^n) \subset F^{s,\tau}_{p,q}(\mathbb{R}^n) \subset B^{s,\tau}_{p,\max\{p,q\}}(\mathbb{R}^n);$$

(iv) If $\tau \in (-\infty,0)$, then $A^{s,\tau}_{p,q}(\mathbb{R}^n) = \{0\}$.

Proof. The properties $(i), (ii)$ are simple corollaries of both the monotonicity of the ℓ^q-norm on q and Hölder's inequality. The property (iii) follows from the (generalized) Minkowski's inequality. Property (iv) is obvious. We leave the details to the reader. □

Proposition 2.2. *Let $p, q \in (0,\infty]$.*

(i) *Let $s \in \mathbb{R}$ and $\tau \in [1/p,\infty)$. Then the scale $A^{s,\tau}_{p,q}(\mathbb{R}^n)$ is monotone with respect to τ, i. e., if $1/p \le \tau_0 \le \tau_1 < \infty$, then $A^{s,\tau_1}_{p,q}(\mathbb{R}^n) \subset A^{s,\tau_0}_{p,q}(\mathbb{R}^n)$ follows.*
(ii) *Let $s \in (\sigma_p,\infty)$, $q_1, q_2 \in (0,\infty]$ and assume $0 \le \tau_0 < \tau_1 < 1/p$. Then the spaces $A^{s,\tau_1}_{p,q_1}(\mathbb{R}^n)$ and $A^{s,\tau_0}_{p,q_2}(\mathbb{R}^n)$ are incomparable, i. e.,*

$$A^{s,\tau_1}_{p,q_1}(\mathbb{R}^n) \setminus A^{s,\tau_0}_{p,q_2}(\mathbb{R}^n) \ne \emptyset \qquad \textit{and} \qquad A^{s,\tau_0}_{p,q_1}(\mathbb{R}^n) \setminus A^{s,\tau_1}_{p,q_2}(\mathbb{R}^n) \ne \emptyset .$$

Proof. Part (i) is an immediate consequence of Lemma 2.2. Part (ii) will be proved at the end of Sect. 4.2.3. □

Remark 2.5. There is, of course, no monotonicity with respect to p (as in case of the spaces $L^p(\mathbb{R}^n)$). However, if we would restrict us to subspaces defined by means of the condition $\operatorname{supp} f \subset Q$ (where Q is a fixed cube in $\mathbb{R}^n$), then one can prove also monotonicity properties with respect to p (and with respect to $\tau \in [0,\infty)$).

Proposition 2.3. *Let $s \in \mathbb{R}$, $\tau \in [0,\infty)$ and $p, q \in (0,\infty]$. Then we have*

$$\mathscr{S}(\mathbb{R}^n) \subset A^{s,\tau}_{p,q}(\mathbb{R}^n) \subset \mathscr{S}'(\mathbb{R}^n) .$$

Proof. With $\tau = 0$ the claim has been proved by Triebel in [145, Sect. 2.3.2]. Thus, it remains to deal with $\tau > 0$.

Step 1. We shall prove $\mathscr{S}(\mathbb{R}^n) \subset A^{s,\tau}_{p,q}(\mathbb{R}^n)$. Let $f \in \mathscr{S}(\mathbb{R}^n)$ and Φ, φ be as in Definition 2.1. To prove this embedding, we need to show that there exists an $M \in \mathbb{N}$ such that $\|f\|_{A^{s,\tau}_{p,q}(\mathbb{R}^n)} \lesssim \|f\|_{\mathscr{S}_M}$ for all $f \in \mathscr{S}(\mathbb{R}^n)$.

Let P be an arbitrary dyadic cube. If $j_P > 0$, applying Lemma 2.4 with $M > \max\{0, n(1/p-1), s+n\tau\}$, we obtain

$$\begin{aligned}
&\frac{1}{|P|^\tau}\left\{\sum_{j=j_P}^{\infty}\left[\int_P (2^{js}|\varphi_j * f(x)|)^p\,dx\right]^{q/p}\right\}^{1/q}\\
&\lesssim \|\varphi\|_{\mathscr{S}_{M+1}}\|f\|_{\mathscr{S}_{M+1}}\frac{1}{|P|^\tau}\left\{\sum_{j=j_P}^{\infty}\left[\int_P \frac{2^{jsp-jMp}}{(1+|x|)^{(n+M)p}}\,dx\right]^{q/p}\right\}^{1/q}\\
&\lesssim \|\varphi\|_{\mathscr{S}_{M+1}}\|f\|_{\mathscr{S}_{M+1}}2^{j_P(s+n\tau-M)}\\
&\lesssim \|\varphi\|_{\mathscr{S}_{M+1}}\|f\|_{\mathscr{S}_{M+1}}.
\end{aligned}$$

If $j_P \le 0$, then $|P|^{-\tau} \le 1$. Notice that for all $x \in \mathbb{R}^n$,

$$|\Phi * f(x)| \lesssim \|\Phi\|_{\mathscr{S}_M}\|f\|_{\mathscr{S}_M}(1+|x|)^{-(n+M)}.$$

Applying Lemma 2.4, again with $M > \max\{0, s, n(1/p-1)\}$, we conclude

$$\begin{aligned}
&\frac{1}{|P|^\tau}\left\{\sum_{j=0}^\infty\left[\int_P(2^{js}|\varphi_j * f(x)|)^p\,dx\right]^{q/p}\right\}^{1/q}\\
&\lesssim \left(\int_P|\Phi * f(x)|^p\,dx\right)^{1/p} + \left\{\sum_{j=1}^\infty\left[\int_P(2^{js}|\varphi_j * f(x)|)^p\,dx\right]^{q/p}\right\}^{1/q}\\
&\lesssim \|\Phi\|_{\mathscr{S}_M}\|f\|_{\mathscr{S}_M} + \|\varphi\|_{\mathscr{S}_{M+1}}\|f\|_{\mathscr{S}_{M+1}}.
\end{aligned}$$

Thus, $\|f\|_{B^{s,\tau}_{p,q}(\mathbb{R}^n)} \lesssim \|f\|_{\mathscr{S}_{M+1}}$, namely, $\mathscr{S}(\mathbb{R}^n) \subset B^{s,\tau}_{p,q}(\mathbb{R}^n)$. Now we use Proposition 2.1(iii) and find

$$\mathscr{S}(\mathbb{R}^n) \subset B^{s,\tau}_{p,p\wedge q}(\mathbb{R}^n) \subset F^{s,\tau}_{p,q}(\mathbb{R}^n),$$

which completes the proof of $\mathscr{S}(\mathbb{R}^n) \subset A^{s,\tau}_{p,q}(\mathbb{R}^n)$.

Step 2. Proof of $A^{s,\tau}_{p,q}(\mathbb{R}^n) \subset \mathscr{S}'(\mathbb{R}^n)$. The proof of this assertion will be postponed and given below of Corollary 2.2. □

Proposition 2.4. *Let $s \in \mathbb{R}$, $\tau \in [0,\infty)$ and $p, q \in (0,\infty]$.*

(*i*) *If $0 < p_1 \le p_2 \le \infty$, then*

$$A^{s,\tau+\frac{1}{p_2}-\frac{1}{p_1}}_{p_2,q}(\mathbb{R}^n) \subset A^{s,\tau}_{p_1,q}(\mathbb{R}^n);$$

(*ii*) *If $\tau \in [0, 1/p]$, then*

$$A^{s}_{\frac{p}{1-\tau p},q}(\mathbb{R}^n) \subset A^{s,\tau}_{p,q}(\mathbb{R}^n);$$

(*iii*) *Let $p < \infty$. If $\tau = 1/p$, then*

$$F^{s,1/p}_{p,q}(\mathbb{R}^n) = F^{s}_{\infty,q}(\mathbb{R}^n).$$

Proof. Parts (i) and (ii) are consequences of Hölder's inequality. Concerning (iii) we apply first Lemma 2.2. From this and the inhomogeneous version of [64, Corollary 5.7] (see the comment on page 133 in [64]) it is evident that the identity $F^{s,1/p}_{p,q}(\mathbb{R}^n) = F^{s}_{\infty,q}(\mathbb{R}^n)$ holds. □

Remark 2.6.

(i) The three assertions in Proposition 2.4, in particular the last one, make clear that p and τ interact.

(ii) The homogeneous counterpart of Proposition 2.4(iii) has been proved in [165, Proposition 3.1(viii)].

Sobolev-type Embeddings

The main result of this section is the following Sobolev-type embedding, which is an immediate consequence of the corresponding result on homogeneous spaces in [165, Proposition 3.3] and the fact that the operator V in (2.15) is an isometric embedding of $a^{s,\tau}_{p,q}(\mathbb{R}^n)$ in $\dot{a}^{s,\tau}_{p,q}(\mathbb{R}^n)$. For completeness, we give a direct proof.

Proposition 2.5. *Let $\tau \in [0,\infty)$, $r, q \in (0,\infty]$ and $-\infty < s_1 < s_0 < \infty$.*

(i) If $0 < p_0 < p_1 \le \infty$ such that $s_0 - n/p_0 = s_1 - n/p_1$, then

$$b^{s_0,\tau}_{p_0,q}(\mathbb{R}^n) \subset b^{s_1,\tau}_{p_1,q}(\mathbb{R}^n).$$

(ii) If $0 < p_0 < p_1 < \infty$ such that $s_0 - n/p_0 = s_1 - n/p_1$, then

$$f^{s_0,\tau}_{p_0,r}(\mathbb{R}^n) \subset f^{s_1,\tau}_{p_1,q}(\mathbb{R}^n).$$

Proof. By Theorem 2.1, it suffices to prove the corresponding conclusions on sequence spaces $a^{s,\tau}_{p,q}(\mathbb{R}^n)$. The embedding $b^{s_0,\tau}_{p_0,q}(\mathbb{R}^n) \subset b^{s_1,\tau}_{p_1,q}(\mathbb{R}^n)$ is immediately deduced from (2.13) and (2.11).

To prove $f^{s_0,\tau}_{p_0,r}(\mathbb{R}^n) \subset f^{s_1,\tau}_{p_1,q}(\mathbb{R}^n)$, by Proposition 2.1(i), we only need to show that $f^{s_0,\tau}_{p_0,\infty}(\mathbb{R}^n) \subset f^{s_1,\tau}_{p_1,q}(\mathbb{R}^n)$. Let $t \in f^{s_0,\tau}_{p_0,\infty}(\mathbb{R}^n)$. Without loss of generality, we may assume that $\|t\|_{f^{s_0,\tau}_{p_0,\infty}(\mathbb{R}^n)} = 1$.

For any $\lambda \in (0,\infty)$ and $P \in \mathcal{Q}$, pick $N \in \mathbb{Z}$ such that

$$\left\{1 - 2^{-qn/p_1}\right\}^{1/q} \frac{\lambda}{2^{1+n/p_1}} < |P|^{\tau} 2^{nN/p_1} \le \left\{1 - 2^{-qn/p_1}\right\}^{1/q} \frac{\lambda}{2}.$$

Step 1. Let $N \ge (j_P \vee 0)$. Since

$$|Q|^{-s_0/n-1/2}|t_Q| \le |Q|^{\tau-1/p_0}\|t\|_{f^{s_0,\tau}_{p_0,\infty}(\mathbb{R}^n)} = 2^{-jn(\tau-1/p_0)}$$

for all $Q \in \mathcal{Q}$ with $l(Q) = 2^{-j} \le 1$, this together with $s_0 - n/p_0 = s_1 - n/p_1$ yields that

$$\begin{aligned}
&\left\{\sum_{j=(j_P\vee 0)}^{N} 2^{-jq(s_0-s_1)} \sup_{\substack{l(Q)=2^{-j}\\ Q\subset P}} \left(|Q|^{-s_0/n}|t_Q|\widetilde{\chi}_Q(x)\right)^q\right\}^{1/q}\\
&\le 2^{-(j_P\vee 0)n\tau} 2^{nN/p_1}\left\{1 - 2^{-qn/p_1}\right\}^{-1/q}\\
&\le |P|^{\tau} 2^{nN/p_1}\left\{1 - 2^{-qn/p_1}\right\}^{-1/q}\\
&\le \lambda/2,
\end{aligned}$$

and

$$\left\{\sum_{j=N+1}^{\infty} 2^{-jq(s_0-s_1)} \sup_{\substack{l(Q)=2^{-j}\\ Q\subset P}} \left(|Q|^{-s_0/n}|t_Q|\widetilde{\chi}_Q(x)\right)^q\right\}^{1/q}$$
$$\leq 2^{p_1(s_0-s_1)/n}\left\{1-2^{-qn/p_1}\right\}^{-p_1(s_0-s_1)/qn}\left\{1-2^{-q(s_0-s_1)}\right\}^{-1/q}$$
$$\times|P|^{\tau p_1(s_0-s_1)/n}\lambda^{-p_1(s_0-s_1)/n} \sup_{\substack{Q\subset P\\ l(Q)\leq 1}} \left(|Q|^{-s_0/n}|t_Q|\widetilde{\chi}_Q(x)\right).$$

Notice that for all dyadic cubes P,

$$\left\{\sum_{\substack{Q\subset P\\ l(Q)\leq 1}} \left(|Q|^{-s_1/n}|t_Q|\widetilde{\chi}_Q(x)\right)^q\right\}^{1/q}$$
$$=\left\{\sum_{j=(j_P\vee 0)}^{\infty} 2^{-jq(s_0-s_1)} \sup_{\substack{l(Q)=2^{-j}\\ Q\subset P}} \left(|Q|^{-s_0/n}|t_Q|\widetilde{\chi}_Q(x)\right)^q\right\}^{1/q}.$$

Then by $s_0-n/p_0=s_1-n/p_1$, we have

$$\left|\left\{x\in P:\left\{\sum_{\substack{Q\subset P\\ l(Q)\leq 1}} \left(|Q|^{-s_1/n}|t_Q|\widetilde{\chi}_Q(x)\right)^q\right\}^{1/q}>\lambda\right\}\right|$$
$$\lesssim\left|\left\{x\in P:\ \sup_{\substack{Q\subset P\\ l(Q)\leq 1}} \left(|Q|^{-s_1/n}|t_Q|\widetilde{\chi}_Q(x)\right)>2^{-1-p_1(s_0-s_1)/n}\right.\right.$$
$$\left.\left.\times\left\{1-2^{-qn/p_1}\right\}^{p_1(s_0-s_1)/qn}\left\{1-2^{-q(s_0-s_1)}\right\}^{1/q}|P|^{-\tau(p_1/p_0-1)}\lambda^{p_1/p_0}\right\}\right|,$$

and hence

$$\|t\|^{p_1}_{f^{s_1,\tau}_{p_1,q}(\mathbb{R}^n)}\lesssim\sup_{P\in\mathcal{Q}}\frac{1}{|P|^{\tau p_0}}\int_0^\infty\lambda^{p_0-1}$$
$$\times\left|\left\{x\in P:\ \sup_{\substack{Q\subset P\\ l(Q)\leq 1}} \left(|Q|^{-s_1/n}|t_Q|\widetilde{\chi}_Q(x)\right)>\lambda\right\}\right|d\lambda$$
$$\sim\|t\|^{p_0}_{f^{s_0,\tau}_{p_1,\infty}(\mathbb{R}^n)}$$
$$\sim 1.$$

Step 2. Let $N < (j_P \vee 0)$. Notice that

$$\left\{\sum_{j=(j_P\vee 0)}^{\infty} \sup_{\substack{l(Q)=2^{-j}\\ Q\subset P}} \left(|Q|^{-s_1/n}|t_Q|\widetilde{\chi}_Q(x)\right)^q\right\}^{1/q} \le \left\{\sum_{j=N+1}^{\infty} \sup_{\substack{l(Q)=2^{-j}\\ Q\subset P}} \left(|Q|^{-s_1/n}|t_Q|\widetilde{\chi}_Q(x)\right)^q\right\}^{1/q}.$$

By the same argument as above, we also obtain $\|t\|_{f_{p_1,q}^{s_1,\tau}(\mathbb{R}^n)} \lesssim 1$, which completes the proof of Proposition 2.5. □

As a corollary of Theorem 2.1 and Proposition 2.5, we have the following Sobolev-type embedding conclusions for $B_{p,q}^{s,\tau}(\mathbb{R}^n)$ and $F_{p,q}^{s,\tau}(\mathbb{R}^n)$, which generalize the classical results on Besov spaces $B_{p,q}^{s}(\mathbb{R}^n)$ and Triebel-Lizorkin spaces $F_{p,q}^{s}(\mathbb{R}^n)$ (see [145, p. 129]). In fact, if $\tau = 0$, Corollary 2.2 is just [145, Theorem 2.7.1].

Corollary 2.2. *Let* $\tau \in [0,\infty)$, $r, q \in (0,\infty]$ *and* $-\infty < s_1 < s_0 < \infty$.

(i) If $0 < p_0 < p_1 \le \infty$ *such that* $s_0 - n/p_0 = s_1 - n/p_1$, *then*

$$B_{p_0,q}^{s_0,\tau}(\mathbb{R}^n) \subset B_{p_1,q}^{s_1,\tau}(\mathbb{R}^n).$$

(ii) If $0 < p_0 < p_1 < \infty$ *such that* $s_0 - n/p_0 = s_1 - n/p_1$, *then*

$$F_{p_0,r}^{s_0,\tau}(\mathbb{R}^n) \subset F_{p_1,q}^{s_1,\tau}(\mathbb{R}^n).$$

Corollary 2.2 can be used to complete the proof of Proposition 2.3.

Proof of Proposition 2.3 (continued). To show $A_{p,q}^{s,\tau}(\mathbb{R}^n) \subset \mathscr{S}'(\mathbb{R}^n)$, we need to prove that there exists an $M \in \mathbb{N}$ such that for all $f \in A_{p,q}^{s,\tau}(\mathbb{R}^n)$ and $\phi \in \mathscr{S}(\mathbb{R}^n)$,

$$|\langle f, \phi\rangle| \lesssim \|f\|_{A_{p,q}^{s,\tau}(\mathbb{R}^n)}\|\phi\|_{\mathscr{S}_M}.$$

Let Φ, Ψ, φ and ψ be as in Lemma 2.3. Then by the Calderón reproducing formulae in Lemmas 2.3 and 2.4, we obtain

$$\begin{aligned}|\langle f, \phi\rangle| &\le \int_{\mathbb{R}^n} |\Psi * f(x)||\Phi * \phi(x)|\,dx + \sum_{j=1}^{\infty}\int_{\mathbb{R}^n} |\psi_j * f(x)||\varphi_j * \phi(x)|\,dx\\ &\lesssim \|\phi\|_{\mathscr{S}_{M+1}} \sum_{j=0}^{\infty} 2^{-jM}\int_{\mathbb{R}^n} |\psi_j * f(x)|(1+|x|)^{-(n+M)}\,dx\\ &\sim \|\phi\|_{\mathscr{S}_{M+1}} \sum_{j=0}^{\infty} 2^{-jM}\sum_{k\in\mathbb{Z}^n}\int_{Q_{0k}} |\psi_j * f(x)|(1+|x|)^{-(n+M)}\,dx,\end{aligned}$$

where we used ψ_0 to replace Ψ, and $M \in \mathbb{N}$ will be determined later.

Notice that there exists 2^n disjoint dyadic cubes $\{Q^l\}_{l=1}^{2^n}$ with $l(Q^l)=1$ such that the ball $B(0,1)\subset(\cup_{l=1}^{2^n}Q^l)$. Obviously, if $Q_{0k}\notin\{Q^l\}_{l=1}^{2^n}$ and $x\in Q_{0k}$, then $|x|\geq 1$. Moreover, if setting

$$\chi_m(k)\equiv\chi_{\{k\in\mathbb{Z}^n:\ 2^m\leq|c_{Q_{0k}}|<2^{m+1}\}}(k),$$

where $c_{Q_{0k}}$ denotes the center of Q_{0k}, we then have $\sum_{k\in\mathbb{Z}^n}\chi_m(k)\lesssim 2^{mn}$.

If $p\in[1,\infty]$, let $M>\max\{-s,0\}$. Then applying Hölder's inequality yields that

$$\begin{aligned}
|\langle f,\phi\rangle| &\lesssim \|\phi\|_{\mathscr{S}_{M+1}}\sum_{j=0}^{\infty}2^{-jM}\left\{\sum_{l=1}^{2^n}\int_{Q^l}\frac{|\psi_j*f(x)|}{(1+|x|)^{n+M}}\,dx+\sum_{m=0}^{\infty}\sum_{k\in\mathbb{Z}^n}\chi_m(k)\int_{Q_{0k}}\cdots\right\}\\
&\lesssim \|\phi\|_{\mathscr{S}_{M+1}}\sum_{j=0}^{\infty}2^{-jM}\left\{\sum_{l=1}^{2^n}2^{-js}\left(\int_{Q^l}2^{jsp}|\psi_j*f(x)|^p\,dx\right)^{1/p}\right.\\
&\qquad\left.+\sum_{m=0}^{\infty}\sum_{k\in\mathbb{Z}^n}\chi_m(k)2^{-js-m(n+M)}\left(\int_{Q_{0k}}\cdots\right)^{1/p}\right\}\\
&\lesssim \|\phi\|_{\mathscr{S}_{M+1}}\|f\|_{A^{s,\tau}_{p,q}(\mathbb{R}^n)}\sum_{j=0}^{\infty}2^{-jM}\left\{2^{-js}+\sum_{m=0}^{\infty}\sum_{k\in\mathbb{Z}^n}\chi_m(k)2^{-js-m(n+M)}\right\}\\
&\lesssim \|\phi\|_{\mathscr{S}_{M+1}}\|f\|_{A^{s,\tau}_{p,q}(\mathbb{R}^n)}.
\end{aligned}$$

The case $p\in(0,1)$ can be deduced from the embedding $A^{s,\tau}_{p,q}(\mathbb{R}^n)\subset A^{s-n/p+n,\tau}_{1,q}(\mathbb{R}^n)$ in Corollary 2.2, which completes the proof of Proposition 2.3. □

Embeddings into Hölder-Zygmund and Lebesgue-type Spaces

Later on, see Chap. 4 below, we shall need also embeddings into Hölder and Lebesgue-type spaces. Recall, that $\mathscr{Z}^s(\mathbb{R}^n)$ denotes the Hölder-Zygmund spaces; see item (vi) in Sect. 1.3. If $s>0$, then it is well known that

$$\mathscr{Z}^s(\mathbb{R}^n)=B^s_{\infty,\infty}(\mathbb{R}^n)=F^s_{\infty,\infty}(\mathbb{R}^n)$$

in the sense of equivalent norms; see [145, p. 51].

Proposition 2.6. *Let $s\in\mathbb{R}$, $\tau\in[0,\infty)$ and $p,q\in(0,\infty]$. Then*

$$A^{s,\tau}_{p,q}(\mathbb{R}^n)\subset F^{s+n\tau-n/p}_{\infty,\infty}(\mathbb{R}^n);$$

in particular,

$$A^{s,\tau}_{p,q}(\mathbb{R}^n)\subset\mathscr{Z}^{s+n\tau-n/p}(\mathbb{R}^n)$$

if $s+n\tau-n/p>0$.

Proof. Recall that

$$\|f\|_{F^s_{\infty,\infty}(\mathbb{R}^n)} \equiv \sup_{j\ge 0} 2^{js}\|\varphi_j * f\|_{L^\infty(\mathbb{R}^n)}$$

with $\varphi_0 \equiv \Phi$, where Φ and φ are as in Definition 2.1. Let $\phi \in \mathscr{S}(\mathbb{R}^n)$ such that $\widehat{\phi} \equiv 1$ on $\{\xi \in \mathbb{R}^n : |\xi| \le 2\}$. Then by (2.1) and (2.2), we obtain that

$$\varphi_j * f \equiv \phi_j * \varphi_j * f.$$

Thus, for all $j \in \mathbb{Z}_+$ and $x \in \mathbb{R}^n$, by [62, p. 782, (2.11)], we have

$$\begin{aligned}
|\varphi_j * f(x)| &\le \sum_{k\in\mathbb{Z}^n} \int_{Q_{jk}} |\phi_j(x-y)||\varphi_j * f(y)|\,dy \\
&\le \sum_{k\in\mathbb{Z}^n} \sup_{z\in Q_{jk}} |\varphi_j * f(z)| \int_{Q_{jk}} |\phi_j(x-y)|\,dy \\
&\lesssim \sum_{k\in\mathbb{Z}^n} 2^{jn/p}\left\{\sum_{l\in\mathbb{Z}^n}(1+|l|)^{-n-1}\int_{Q_{j,k+l}} |\varphi_j * f(z)|^p\,dz\right\}^{1/p} \\
&\quad \times \int_{Q_{jk}} |\phi_j(x-y)|\,dy \\
&\lesssim 2^{j(n/p-s-n\tau)}\|f\|_{A^{s,\tau}_{p,q}(\mathbb{R}^n)} \int_{\mathbb{R}^n} |\phi_j(x-y)|\,dy \\
&\lesssim 2^{j(n/p-s-n\tau)}\|f\|_{A^{s,\tau}_{p,q}(\mathbb{R}^n)},
\end{aligned}$$

which completes the proof of Proposition 2.6. □

For $\tau \in [0,\infty)$ and $p \in (0,\infty]$, set

$$\|f\|_{L^p_\tau(\mathbb{R}^n)} \equiv \sup_{\{P\in\mathscr{Q},\,|P|\ge 1\}} \frac{1}{|P|^\tau}\left(\int_P |f(x)|^p\,dx\right)^{1/p}, \tag{2.18}$$

where the supremum is taken over all dyadic cubes P with side length $l(P) \ge 1$. Denote by $L^p_\tau(\mathbb{R}^n)$ the set of all functions f satisfying $\|f\|_{L^p_\tau(\mathbb{R}^n)} < \infty$. Obviously, when $\tau = 0$, then $L^p_\tau(\mathbb{R}^n) = L^p(\mathbb{R}^n)$. The scale is monotone in τ. If $\tau_0 < \tau_1$, then $L^p_{\tau_0}(\mathbb{R}^n) \subset L^p_{\tau_1}(\mathbb{R}^n)$ follows. We also claim that $L^p_\tau(\mathbb{R}^n) \subset \mathscr{S}'(\mathbb{R}^n)$ when $p \in [1,\infty]$. In fact, by Hölder's inequality, we see that for all $\phi \in \mathscr{S}(\mathbb{R}^n)$,

$$\begin{aligned}
|\langle f, \phi\rangle| &\le \int_{\mathbb{R}^n} |f(x)||\phi(x)|\,dx \\
&\le \sum_{k\in\mathbb{Z}^n}\left\{\int_{Q_{0k}} |f(x)|^p\,dx\right\}^{1/p}\left\{\int_{Q_{0k}} |\phi(x)|^{p'}\,dx\right\}^{1/p'} \\
&\lesssim \|f\|_{L^p_\tau(\mathbb{R}^n)}\|\phi\|_{\mathscr{S}_M} \sum_{k\in\mathbb{Z}^n}\frac{1}{(1+|k|)^{n+M}} \\
&\lesssim \|f\|_{L^p_\tau(\mathbb{R}^n)}\|\phi\|_{\mathscr{S}_M},
\end{aligned}$$

which proves the above claim.

Proposition 2.7. *Let $\tau \in [0,\infty)$.*

(i) *Let $p \in [1,\infty]$. Then*

$$B^{0,\tau}_{p,1}(\mathbb{R}^n) \subset F^{0,\tau}_{p,1}(\mathbb{R}^n) \subset L^p_\tau(\mathbb{R}^n).$$

(ii) *Let $p \in (0,1)$. Then*

$$B^{\sigma_p,\tau}_{p,p}(\mathbb{R}^n) = F^{\sigma_p,\tau}_{p,p}(\mathbb{R}^n) \subset \left(L^1_\tau(\mathbb{R}^n) \cap L^p_\tau(\mathbb{R}^n)\right).$$

Proof. Step 1. Proof of (i). Let $\{\varphi_j\}_{j\in\mathbb{Z}_+}$ be the system for the definition of the spaces $A^{s,\tau}_{p,q}(\mathbb{R}^n)$; see Sect. 1.3.3. Then, for each dyadic cube P,

$$f = \sum_{j=0}^{\infty} \varphi_j * f$$

in the sense of $L^p(P)$. Thus, if $|P| \geq 1$, then

$$\frac{1}{|P|^\tau}\|f\|_{L^p(P)} \leq \frac{1}{|P|^\tau}\left[\int_P \left(\sum_{j=0}^{\infty} |\varphi_j * f(x)|\right)^p dx\right]^{1/p} \leq \|f\|_{F^{0,\tau}_{p,1}(\mathbb{R}^n)} \tag{2.19}$$

follows. The first embedding in (i) is an obvious consequence of Proposition 2.1(iii).

Step 2. Proof of (ii). We obtain

$$F^{\sigma_p,\tau}_{p,p}(\mathbb{R}^n) \subset F^{0,\tau}_{1,1}(\mathbb{R}^n) \subset L^1_\tau(\mathbb{R}^n),$$

where we used Proposition 2.1, Corollary 2.2 and Step 1. Thus, we get

$$f = \sum_{j=0}^{\infty} \varphi_j * f$$

in $L^1(P)$. Hölder's inequality yields that this identity takes place in $L^p(P)$ as well. Similarly as in (2.19) we conclude

$$\begin{aligned}\frac{1}{|P|^\tau}\|f\|_{L^p(P)} &\leq \frac{1}{|P|^\tau}\left[\int_P \left(\sum_{j=0}^{\infty} |\varphi_j * f(x)|\right)^p dx\right]^{1/p} \\ &\leq \frac{1}{|P|^\tau}\left[\int_P \sum_{j=0}^{\infty} |\varphi_j * f(x)|^p\, dx\right]^{1/p} \\ &\leq \|f\|_{F^{0,\tau}_{p,p}(\mathbb{R}^n)}\end{aligned}$$

provided that $|P| \geq 1$. This finishes the proof of Proposition 2.7. □

Remark 2.7. For $\tau = 0$ much more is known about these various types of embeddings. In particular, many times if, and only if assertions are known. We refer to [133] and the references given there.

2.3 The Fatou Property

Next we show that the spaces $A_{p,q}^{s,\tau}(\mathbb{R}^n)$ satisfy the following *Fatou property*, which is often applied in connection with nonlinear problems. We are going to use the Fatou property in Sect. 6.1.

Proposition 2.8. *Let $s \in \mathbb{R}$, $\tau \in [0,\infty)$, $p, q \in (0,\infty]$ and $\{f_m\}_{m\in\mathbb{Z}_+} \subset A_{p,q}^{s,\tau}(\mathbb{R}^n)$ be a sequence such that $f_m \rightharpoonup f$ (weak convergence in $\mathscr{S}'(\mathbb{R}^n)$) as $m \to \infty$ and*

$$\sup_{m\in\mathbb{Z}_+} \|f_m\|_{A_{p,q}^{s,\tau}(\mathbb{R}^n)} < \infty.$$

Then $f \in A_{p,q}^{s,\tau}(\mathbb{R}^n)$ and

$$\|f\|_{A_{p,q}^{s,\tau}(\mathbb{R}^n)} \leq \sup_{m\in\mathbb{Z}_+} \|f_m\|_{A_{p,q}^{s,\tau}(\mathbb{R}^n)}.$$

Proof. We follow [60]. By similarity we only consider the space $F_{p,q}^{s,\tau}(\mathbb{R}^n)$.

From assumption, it follows that for all $j \in \mathbb{Z}_+$ and $x \in \mathbb{R}^n$,

$$\varphi_j * f_m(x) = f_m(x - \cdot)(\varphi_j) \to f(x - \cdot)(\varphi_j) = \varphi_j * f(x)$$

as $m \to \infty$, where when $j = 0$, φ_0 is replaced by Φ.

Notice that for each $Q \in \mathcal{Q}$ and $N \geq (j_Q \vee 0)$, Fatou's lemma yields

$$\int_Q \left[\sum_{j=(j_Q\vee 0)}^{N} 2^{jsq}|\varphi_j * f(x)|^q\right]^{\frac{p}{q}} dx \leq \liminf_{m\to\infty} \int_Q \left[\sum_{j=(j_Q\vee 0)}^{N} 2^{jsq}|\varphi_j * f_m(x)|^q\right]^{\frac{p}{q}} dx.$$

This combined with Levi's lemma and Definition 2.1 yields the desired conclusion, which completes the proof of Proposition 2.8. □

Remark 2.8.

(i) There are many spaces which do not have the Fatou property. The most obvious examples are $L^1(\mathbb{R}^n)$ and $C(\mathbb{R}^n)$; see [59].

(ii) The Fatou property of the spaces $F_{p,q}^{s}(\mathbb{R}^n)$ $(p < \infty)$ and $B_{p,q}^{s}(\mathbb{R}^n)$ has been proved by Franke [59]; see also Franke and Runst [60]. Bourdaud and Meyer [19] gave an independent proof restricted to Besov spaces.

Chapter 3
Almost Diagonal Operators and Atomic and Molecular Decompositions

In the first part of this chapter we shall show that under certain restrictions on the parameters our spaces $A^{s,\tau}_{p,q}(\mathbb{R}^n)$ allow characterizations by smooth molecules and smooth atoms. This opens the door to the discretization of our distribution spaces $A^{s,\tau}_{p,q}(\mathbb{R}^n)$. Afterwards, based on these discretizations, we are able to compare the classes $A^{s,\tau}_{p,q}(\mathbb{R}^n)$ with the Besov-Morrey spaces $\mathcal{N}^s_{pqu}(\mathbb{R}^n)$ (see item (xxv) in Sect. 1.3) and the Triebel-Lizorkin-Morrey spaces $\mathcal{E}^s_{pqu}(\mathbb{R}^n)$ (see item (xxv) in Sect. 1.3).

3.1 Smooth Atomic and Molecular Decompositions

As an application of Theorem 2.1, we study boundedness of operators on $A^{s,\tau}_{p,q}(\mathbb{R}^n)$ by first considering their boundedness on the corresponding space $a^{s,\tau}_{p,q}(\mathbb{R}^n)$ of sequences. We first show that almost diagonal operators are bounded on $a^{s,\tau}_{p,q}(\mathbb{R}^n)$ for appropriate indices, which generalize the basic results of Frazier and Jawerth with respect to $a^s_{p,q}(\mathbb{R}^n)$; see [64, Theorem 3.3] and [65, Theorem (6.20)].

We shall say that an operator A is associated with the matrix $\{a_{QP}\}_{l(Q),l(P)\le 1}$, if for all sequences $t\equiv\{t_Q\}_{l(Q)\le 1}\subset\mathbb{C}$,

$$At\equiv\{(At)_Q\}_{l(Q)\le 1}\equiv\left\{\sum_{l(P)\le 1} a_{QP}t_P\right\}_{l(Q)\le 1}.$$

Furthermore, we shall use the abbreviation

$$J\equiv\begin{cases}\frac{n}{\min\{1,p\}} & \text{if}\quad a^{s,\tau}_{p,q}(\mathbb{R}^n)=b^{s,\tau}_{p,q}(\mathbb{R}^n),\\ \frac{n}{\min\{1,p,q\}} & \text{if}\quad a^{s,\tau}_{p,q}(\mathbb{R}^n)=f^{s,\tau}_{p,q}(\mathbb{R}^n).\end{cases}\tag{3.1}$$

Definition 3.1. Let $s\in\mathbb{R}$, $p,q\in(0,\infty]$ and $\varepsilon\in(0,\infty)$. The operator A, associated with the matrix $\{a_{QP}\}_{l(Q),l(P)\le 1}$, is called *$\varepsilon$-almost diagonal on $a^{s,\tau}_{p,q}(\mathbb{R}^n)$* if the matrix $\{a_{QP}\}_{l(Q),l(P)\le 1}$ satisfies

D. Yang et al., *Morrey and Campanato Meet Besov, Lizorkin and Triebel*, Lecture Notes in Mathematics 2005, DOI 10.1007/978-3-642-14606-0_3,

$$\sup_{l(Q),l(P)\le 1} |a_{QP}|/\omega_{QP}(\varepsilon) < \infty, \tag{3.2}$$

where

$$\omega_{QP}(\varepsilon) \equiv \left(\frac{l(Q)}{l(P)}\right)^s \left(1+\frac{|x_Q-x_P|}{\max(l(P),l(Q))}\right)^{-J-\varepsilon} \times \min\left[\left(\frac{l(Q)}{l(P)}\right)^{(n+\varepsilon)/2}, \left(\frac{l(P)}{l(Q)}\right)^{(n+\varepsilon)/2+J-n}\right].$$

We remark that each ε-almost diagonal operator is also an almost diagonal operator in the sense of Frazier and Jawerth; see [64, Sects. 3 and 12]. The following conclusion is an immediate corollary of the corresponding result on homogeneous spaces in [165, Theorem 4.1] and the fact that the operator V in (2.15) is an isometric embedding of $a^{s,\tau}_{p,q}(\mathbb{R}^n)$ in $\dot{a}^{s,\tau}_{p,q}(\mathbb{R}^n)$. For completeness, we give a direct proof.

Theorem 3.1. *Let $\varepsilon \in (0,\infty)$, $s \in \mathbb{R}$, $p,q \in (0,\infty]$ and $\tau \in [0, 1/p+\varepsilon/(2n))$. Then all ε-almost diagonal operators on $a^{s,\tau}_{p,q}(\mathbb{R}^n)$ are bounded on $a^{s,\tau}_{p,q}(\mathbb{R}^n)$.*

Proof. By similarity, we only give the proof for $b^{s,\tau}_{p,q}(\mathbb{R}^n)$.

Let

$$t = \{t_Q\}_{l(Q)\le 1} \in b^{s,\tau}_{p,q}(\mathbb{R}^n)$$

and A be a ε-almost diagonal operator associated with the matrix $\{a_{QR}\}_{l(Q),l(R)\le 1}$ and $\varepsilon \in (0,\infty)$. Without loss of generality, we may assume $s=0$. Indeed, if the conclusion holds for $s=0$, let $\widetilde{t}_R \equiv [l(R)]^{-s}t_R$ and B be the operator associated with the matrix $\{b_{QR}\}_{l(Q),l(R)\le 1}$, where

$$b_{QR} \equiv (l(R)/l(Q))^s a_{QR}$$

for all dyadic cubes Q and R with $l(Q), l(R) \le 1$. Then we have

$$\|At\|_{b^{s,\tau}_{p,q}(\mathbb{R}^n)} = \|B\widetilde{t}\|_{b^{0,\tau}_{p,q}(\mathbb{R}^n)} \lesssim \|\widetilde{t}\|_{b^{0,\tau}_{p,q}(\mathbb{R}^n)} \sim \|t\|_{b^{s,\tau}_{p,q}(\mathbb{R}^n)},$$

which yields the desired conclusion.

We now consider the case when $\min\{p,q\} > 1$. For all $Q \in \mathcal{Q}$ with $l(Q) \le 1$, we write $A \equiv A_0 + A_1$ with

$$(A_0t)_Q \equiv \sum_{\{R:\ 1\ge l(R)\ge l(Q)\}} a_{QR}t_R \quad\text{and}\quad (A_1t)_Q \equiv \sum_{\{R:\ l(R)<l(Q)\}} a_{QR}t_R.$$

By Definition 3.1, we see that for all such Q,

$$|(A_0t)_Q| \lesssim \sum_{\{R:\ 1\ge l(R)\ge l(Q)\}} \left(\frac{l(Q)}{l(R)}\right)^{\frac{n+\varepsilon}{2}} \frac{|t_R|}{(1+[l(R)]^{-1}|x_Q-x_R|)^{n+\varepsilon}},$$

and therefore

$$
\begin{aligned}
&\|A_0 t\|_{b^{0,\tau}_{p,q}(\mathbb{R}^n)}\\
&\lesssim \sup_{P\in\mathcal{Q}} \frac{1}{|P|^\tau}\left\{\sum_{j=(j_P\vee 0)}^{\infty}\left[\int_P\left(\sum_{l(Q)=2^{-j}}\sum_{\{R:\, l(Q)\le l(R)\le [l(P)\wedge 1]\}}\left(\frac{l(Q)}{l(R)}\right)^{\frac{n+\varepsilon}{2}}\right.\right.\right.\\
&\qquad\left.\left.\left.\times\frac{|t_R|\widetilde{\chi}_Q(x)}{(1+[l(R)]^{-1}|x_Q-x_R|)^{n+\varepsilon}}\right)^p dx\right]^{\frac{q}{p}}\right\}^{\frac{1}{q}}\\
&\quad+\sup_{P\in\mathcal{Q}} \frac{1}{|P|^\tau}\left\{\sum_{j=(j_P\vee 0)}^{\infty}\left[\int_P\left(\sum_{l(Q)=2^{-j}}\sum_{\{R:\, l(R)> [l(P)\wedge 1]\}}\cdots\right)^p dx\right]^{\frac{q}{p}}\right\}^{\frac{1}{q}}\\
&\equiv \mathrm{I}_1+\mathrm{I}_2.
\end{aligned}
$$

For all $i\in\mathbb{Z}_+$ and $m\in\mathbb{N}$, set

$$U_{0,i}\equiv\{R\in\mathcal{Q}:\ l(R)=2^{-i} \text{ and } |x_Q-x_R|<l(R)\}$$

and

$$U_{m,i}\equiv\{R\in\mathcal{Q}:\ l(R)=2^{-i} \text{ and } 2^{m-1}l(R)\le|x_Q-x_R|<2^m l(R)\}.$$

Then we have $\#U_{m,i}\lesssim 2^{mn}$, where $\#U_{m,i}$ denotes the cardinality of $U_{m,i}$. Notice that $|t_R|\le |R|^{1/2-1/p+\tau}\|t\|_{b^{0,\tau}_{p,q}(\mathbb{R}^n)}$ for all $R\in\mathcal{Q}$ with $l(R)\le 1$. Thus, by $0\le\tau<\frac{1}{p}+\frac{\varepsilon}{2n}$,

$$
\begin{aligned}
\mathrm{I}_2 &\lesssim \|t\|_{b^{0,\tau}_{p,q}(\mathbb{R}^n)}\sup_{P\in\mathcal{Q}}\frac{1}{|P|^\tau}\left\{\sum_{j=(j_P\vee 0)}^{\infty}\left[\int_P\left(\sum_{l(Q)=2^{-j}}\sum_{i=0}^{j_P\vee 0}\sum_{m=0}^{\infty}\sum_{R\in U_{m,i}}\left(\frac{l(Q)}{l(R)}\right)^{\frac{n+\varepsilon}{2}}\right.\right.\right.\\
&\qquad\left.\left.\left.\times\frac{|R|^{1/2-1/p+\tau}\widetilde{\chi}_Q(x)}{(1+[l(R)]^{-1}|x_Q-x_R|)^{n+\varepsilon}}\right)^p dx\right]^{q/p}\right\}^{1/q}\\
&\lesssim \|t\|_{b^{0,\tau}_{p,q}(\mathbb{R}^n)}.
\end{aligned}
$$

For I_1, let r and u be the same as in the proof of Lemma 2.8. We see that

$$
\begin{aligned}
\mathrm{I}_1 &\lesssim \sup_{P\in\mathcal{Q}}\frac{1}{|P|^\tau}\left\{\sum_{j=(j_P\vee 0)}^{\infty}\left[\int_P\left(\sum_{l(Q)=2^{-j}}\sum_{i=(j_P\vee 0)}^{j}2^{(i-j)(n+\varepsilon)/2}\right.\right.\right.\\
&\qquad\left.\left.\left.\times\sum_{l(R)=2^{-i}}\frac{|r_R|\widetilde{\chi}_Q(x)}{(1+[l(R)]^{-1}|x_Q-x_R|)^{n+\varepsilon}}\right)^p dx\right]^{q/p}\right\}^{1/q}
\end{aligned}
$$

$$+\sup_{P\in\mathcal{Q}}\frac{1}{|P|^{\tau}}\left\{\sum_{j=(j_P\vee 0)}^{\infty}\left[\int_P\left(\sum_{l(Q)=2^{-j}}\sum_{i=(j_P\vee 0)}^{j}2^{(i-j)(n+\varepsilon)/2}\right.\right.\right.$$

$$\left.\left.\left.\times\sum_{l(R)=2^{-i}}\frac{|u_R|\widetilde{\chi}_Q(x)}{(1+[l(R)]^{-1}|x_Q-x_R|)^{n+\varepsilon}}\right)^p dx\right]^{q/p}\right\}^{1/q}$$

$$\equiv \mathrm{J}_1+\mathrm{J}_2.$$

Applying [64, Remark A.3] with $a=1$, for all $x\in Q$, we have

$$\sum_{l(R)=2^{-i}}\frac{|r_R|}{(1+[l(R)]^{-1}|x_Q-x_R|)^{n+\varepsilon}}\lesssim \mathrm{M}\left(\sum_{l(R)=2^{-i}}|r_R|\chi_R\right)(x).$$

Hence Hölder's inequality and the $L^p(\mathbb{R}^n)$-boundedness for $p\in(1,\infty]$ of the Hardy-Littlewood maximal operator yield

$$\mathrm{J}_1\lesssim\sup_{P\in\mathcal{Q}}\frac{1}{|P|^{\tau}}\left\{\sum_{j=(j_P\vee 0)}^{\infty}\left[\int_P\left(\sum_{i=(j_P\vee 0)}^{j}2^{(i-j)\varepsilon/2}\right.\right.\right.$$

$$\left.\left.\left.\times\mathrm{M}\left(\sum_{l(R)=2^{-i}}|r_R|\widetilde{\chi}_R\right)(x)\right)^p dx\right]^{q/p}\right\}^{1/q}$$

$$\lesssim\sup_{P\in\mathcal{Q}}\frac{1}{|P|^{\tau}}\left\{\sum_{i=(j_P\vee 0)}^{\infty}\left[\int_{3P}\left(\sum_{l(R)=2^{-i}}|t_R|\widetilde{\chi}_R\right)^p(x)\,dx\right]^{q/p}\right\}^{1/q}$$

$$\lesssim\|t\|_{b^{0,\tau}_{p,q}(\mathbb{R}^n)},$$

where the last inequality follows from Minkowski's inequality if $q>p$ or (2.11) if $q\le p$.

To estimate J_2, we notice that if $R\cap(3P)=\emptyset$, then $R\subset P+kl(P)$ and

$$(P+kl(P))\cap(3P)=\emptyset$$

for some $k\in\mathbb{Z}^n$ with $|k|\ge 2$ and

$$1+[l(R)]^{-1}|x_Q-x_R|\sim|k|l(P)/l(R).$$

Therefore, by Hölder's inequality,

$$\mathrm{J}_2 \lesssim \sup_{P\in\mathscr{Q}} \frac{1}{|P|^{\tau+1+\frac{\varepsilon}{n}}} \left\{ \sum_{j=(j_P\vee 0)}^{\infty} 2^{-jq\varepsilon/2} \left[\int_P \left(\sum_{l(Q)=2^{-j}} \sum_{i=(j_P\vee 0)}^{j} 2^{-i(n+\varepsilon)/2} \right.\right.\right.$$

$$\left.\left.\left. \times \sum_{\substack{k\in\mathbb{Z}^n\\ |k|\geq 2}} |k|^{-n-\varepsilon} \sum_{\substack{l(R)=2^{-i}\\ R\subset P+kl(P)}} |t_R|\chi_Q(x) \right)^p dx \right]^{q/p} \right\}^{1/q}$$

$$\lesssim \|t\|_{b^{0,\tau}_{p,q}(\mathbb{R}^n)} \sup_{P\in\mathscr{Q}} \frac{1}{|P|^{\frac{\varepsilon}{n}}} \left\{ \sum_{j=(j_P\vee 0)}^{\infty} 2^{-jq\varepsilon/2} \left[\sum_{i=(j_P\vee 0)}^{j} 2^{-i\varepsilon/2} \sum_{\substack{k\in\mathbb{Z}^n\\ |k|\geq 2}} |k|^{-n-\varepsilon} \right]^q \right\}^{1/q}$$

$$\lesssim \|t\|_{b^{0,\tau}_{p,q}(\mathbb{R}^n)}.$$

Thus,

$$\|A_0 t\|_{b^{0,\tau}_{p,q}(\mathbb{R}^n)} \lesssim \|t\|_{b^{0,\tau}_{p,q}(\mathbb{R}^n)}.$$

Some similar computations to I_1 also yield that

$$\|A_1 t\|_{b^{0,\tau}_{p,q}(\mathbb{R}^n)} \lesssim \|t\|_{b^{0,\tau}_{p,q}(\mathbb{R}^n)}.$$

The case that $\min\{p,q\} \leq 1$ is a simple consequence of the case that $\min\{p,q\} > 1$. Indeed, choosing a $\delta \in (0, p\wedge q)$, then $p/\delta > 1$ and $q/\delta > 1$. Let $\widetilde{A}$ be an operator associated with the matrix

$$\{\widetilde{a}_{QP}\}_{l(Q),l(P)\leq 1} \equiv \left\{ |a_{QP}|^{\delta} (l(Q)/l(P))^{n/2-\delta n/2} \right\}_{l(Q),l(P)\leq 1}.$$

Then $\widetilde{A}$ is a $\widetilde{\varepsilon}$-almost diagonal operator with $s=0$ and $\widetilde{\varepsilon} = \delta\varepsilon$.

Define

$$\widetilde{t} \equiv \{[l(Q)]^{n/2-\delta n/2} |t_Q|^{\delta}\}_{l(Q)\leq 1}.$$

Then

$$\|\widetilde{t}\|^{1/\delta}_{b^{0,\tau\delta}_{p/\delta,q/\delta}(\mathbb{R}^n)} = \|t\|_{b^{0,\tau}_{p,q}(\mathbb{R}^n)}.$$

Since $\delta < 1$, by (2.11), we see that

$$\|At\|_{b^{0,\tau}_{p,q}(\mathbb{R}^n)} \lesssim \|\widetilde{A}\widetilde{t}\|^{1/\delta}_{b^{0,\tau\delta}_{p/\delta,q/\delta}(\mathbb{R}^n)}.$$

Applying the conclusions for $\min\{p,q\} > 1$ yields

$$\|At\|_{b^{0,\tau}_{p,q}(\mathbb{R}^n)} \lesssim \|\widetilde{A}\widetilde{t}\|^{1/\delta}_{b^{0,\tau\delta}_{p/\delta,q/\delta}(\mathbb{R}^n)} \lesssim \|\widetilde{t}\|^{1/\delta}_{b^{0,\tau\delta}_{p/\delta,q/\delta}(\mathbb{R}^n)} \sim \|t\|_{b^{0,\tau}_{p,q}(\mathbb{R}^n)},$$

which completes the proof of Theorem 3.1. □

As an application of Theorem 3.1, we establish characterizations of $A^{s,\tau}_{p,q}(\mathbb{R}^n)$ by inhomogeneous smooth atomic and molecular decompositions. For the forerunners with respect to the inhomogeneous spaces $A^s_{p,q}(\mathbb{R}^n)$ we refer to [64] and [25]. For a real number s we denote by $\lfloor s \rfloor$ the largest integer which is less than or equal to s. Let J be defined as in (3.1) if $a^{s,\tau}_{p,q}(\mathbb{R}^n)$ is replaced by $A^{s,\tau}_{p,q}(\mathbb{R}^n)$. Recall, x_Q denotes the center of the cube Q.

Definition 3.2. Let $s \in \mathbb{R}$, $\tau \in [0,\infty)$, $p,q \in (0,\infty]$, $N = \max\{\lfloor J-n-s \rfloor, -1\}$ and $s^* = s - \lfloor s \rfloor$. Let Q be a dyadic cube with $l(Q) \le 1$.

(i) A function m_Q is called an *inhomogeneous smooth synthesis molecule for* $A^{s,\tau}_{p,q}(\mathbb{R}^n)$ supported near Q if there exist a real number $\delta \in (\max\{s^*, (s+n\tau)^*\}, 1]$ and a real number $M \in (J,\infty)$ such that

$$\int_{\mathbb{R}^n} x^\gamma m_Q(x)\,dx = 0 \quad \text{if} \quad |\gamma| \le N \quad \text{and} \quad l(Q) < 1, \tag{3.3}$$

$$|m_Q(x)| \le (1+|x-x_Q|)^{-M} \quad \text{if} \quad l(Q) = 1, \tag{3.4}$$

$$|m_Q(x)| \le |Q|^{-1/2}(1+[l(Q)]^{-1}|x-x_Q|)^{-\max\{M,M-s\}} \quad \text{if} \quad l(Q) < 1, \tag{3.5}$$

$$|\partial^\gamma m_Q(x)| \le |Q|^{-1/2-|\gamma|/n}(1+[l(Q)]^{-1}|x-x_Q|)^{-M} \quad \text{if} \quad |\gamma| \le \lfloor s+n\tau \rfloor, \tag{3.6}$$

and

$$|\partial^\gamma m_Q(x) - \partial^\gamma m_Q(y)| \le |Q|^{-1/2-|\gamma|/n-\delta/n}|x-y|^\delta \times \sup_{|z|\le|x-y|} (1+[l(Q)]^{-1}|x-z-x_Q|)^{-M} \tag{3.7}$$

if $|\gamma| = \lfloor s+n\tau \rfloor$.

A collection $\{m_Q\}_{l(Q)\le 1}$ is called a family of inhomogeneous smooth synthesis molecules for $A^{s,\tau}_{p,q}(\mathbb{R}^n)$, if each m_Q is an inhomogeneous smooth synthesis molecule for $A^{s,\tau}_{p,q}(\mathbb{R}^n)$ supported near Q.

(ii) A function b_Q is called an *inhomogeneous smooth analysis molecule for* $A^{s,\tau}_{p,q}(\mathbb{R}^n)$ supported near Q if there exist a $\rho \in ((J-s)^*, 1]$ and an $M \in (J,\infty)$ such that

$$\int_{\mathbb{R}^n} x^\gamma b_Q(x)\,dx = 0 \quad \text{if} \quad |\gamma| \le \lfloor s+n\tau \rfloor \quad \text{and} \quad l(Q) < 1, \tag{3.8}$$

$$|b_Q(x)| \le (1+|x-x_Q|)^{-M} \quad \text{if} \quad l(Q) = 1, \tag{3.9}$$

$$|b_Q(x)| \le |Q|^{-1/2}(1+[l(Q)]^{-1}|x-x_Q|)^{-\max\{M,M+n+s+n\tau-J\}} \quad \text{if} \quad l(Q) < 1, \tag{3.10}$$

$$|\partial^\gamma b_Q(x)| \le |Q|^{-1/2-|\gamma|/n}(1+[l(Q)]^{-1}|x-x_Q|)^{-M} \quad \text{if} \quad |\gamma| \le N, \tag{3.11}$$

and

$$|\partial^\gamma b_Q(x) - \partial^\gamma b_Q(y)| \le |Q|^{-1/2-|\gamma|/n-\rho/n}|x-y|^\rho \times \sup_{|z|\le|x-y|} (1+[l(Q)]^{-1}|x-z-x_Q|)^{-M} \quad (3.12)$$

if $|\gamma| = N$.

A collection $\{b_Q\}_{l(Q)\le 1}$ is called a family of inhomogeneous smooth synthesis molecules for $A^{s,\tau}_{p,q}(\mathbb{R}^n)$, if each b_Q is an inhomogeneous smooth analysis molecule for $A^{s,\tau}_{p,q}(\mathbb{R}^n)$ supported near Q.

We add a remark concerning the interpretation of certain restrictions. If $s+n\tau<0$, then (3.6) and (3.7) are void. Analogously, if $J-n-s<0$, then $N=-1$ and (3.11) and (3.12) are void. Also we wish to point out that only when $l(Q)<1$, the inhomogeneous smooth synthesis and analysis molecules for $A^{s,\tau}_{p,q}(\mathbb{R}^n)$ supported near the dyadic cube Q have to fulfil the vanishing moment condition, which is the only difference between the homogeneous and the inhomogeneous case.

To establish the inhomogeneous smooth atomic and molecular decomposition characterizations, we need some elementary lemmas.

Lemma 3.1. *Let s, p, q, J, M, N and ρ be as in Definition 3.2. Assume that*

$$\tau \in \left[0, \frac{1}{p} + \left[\frac{M-J}{2n}\right] \wedge \left[\frac{\rho-(J-s)^*}{n}\right]\right)$$

if $N \ge 0$,

$$\tau \in \left[0, \frac{1}{p} + \left[\frac{M-J}{2n}\right] \wedge \left[\frac{s+n-J}{n}\right]\right)$$

if $N<0$, *and* $\delta \in (\max\{(s+n\tau)^*, s^*\}, 1]$. *Then there exist a positive real number* ε_1 *and a positive constant C such that* $\varepsilon_1 > 2n(\tau - \frac{1}{p})$ *and for all families* $\{m_Q\}_{l(Q)\le 1}$ *of inhomogeneous smooth synthesis molecules for* $A^{s,\tau}_{p,q}(\mathbb{R}^n)$ *and families* $\{b_Q\}_{l(Q)\le 1}$ *of inhomogeneous smooth analysis molecules for* $A^{s,\tau}_{p,q}(\mathbb{R}^n)$,

$$|\langle m_P, b_Q\rangle| \le C\omega_{QP}(\varepsilon_1).$$

Namely, the operator associated with the matrix $\{a_{QP}\}_{l(Q),l(P)\le 1}$ *defined by* $a_{QP} \equiv \langle m_P, b_Q\rangle$ *is* ε_1*-almost diagonal on* $a^{s,\tau}_{p,q}(\mathbb{R}^n)$.

The proof of Lemma 3.1 is a slight modification of the homogeneous analogue in [165, Lemma 4.2]; see also [64, Corollary B.3]. We give some details.

Proof. The proof for the case $l(Q)$, $l(P) < 1$ is the same as those for [165, Lemma 4.2] and [64, Corollary B.3]. In fact, if $l(Q) = 2^{-\nu} \le 2^{-\mu} = l(P)$, applying [64, Lemma B.1] when $s+n\tau \ge 0$, or [64, Lemma B.2] when $s+n\tau<0$, with $R=M$, $\theta=\delta$, $k=\nu$, $j=\mu$, $L=\lfloor s+n\tau\rfloor$, $S=M+n+s+n\tau-J$, $x_1=x_Q$, $g=m_P(x_P-\cdot)$

and $h = b_Q$, we obtain the desired estimate. If $l(Q) = 2^{-\nu} \geq 2^{-\mu} = l(P)$, applying [64, Lemma B.1] when $N \geq 0$, or [64, Lemma B.2] when $N < 0$, with $R = M$, $\theta = \rho$, $k = \mu$, $j = \nu$, $L = N$, $S = M - s$, $x_1 = x_P$, $g = \overline{b_Q(x_Q - \cdot)}$ and $h = m_P$, we also obtain the desired estimate.

Notice that in the above argument (see also the proofs of [165, Lemma 4.2] and [64, Corollary B.3]), the vanishing moment conditions for m_Q are only used when we take $h = m_P$ and apply [64, Lemma B.1], i. e., when $l(P) < l(Q)$. This never happens in the inhomogeneous case if $|P| = 1$. A similar observation holds for b_Q if $|Q| = 1$. Thus, we obtain the cases $l(Q) < l(P) = 1$ and $l(P) < l(Q) = 1$. The remaining estimate for $|\langle m_P, b_Q\rangle|$ when $|P| = |Q| = 1$ is an immediate consequence of [64, Lemma B.2]. This finishes the proof of Lemma 3.1. □

As an immediate consequence, we have the following analogues of the corresponding results on the homogeneous case in [165, Corollary 4.1].

Corollary 3.1. *Let s, p, q, τ and ε_1 be as in Lemma 3.1 and Φ, φ as in Definition 2.1. Suppose that $\{m_Q\}_{l(Q)\leq 1}$ and $\{b_Q\}_{l(Q)\leq 1}$ are families of inhomogeneous smooth synthesis and analysis molecules for $A^{s,\tau}_{p,q}(\mathbb{R}^n)$, respectively. Then the operators associated with the matrices*

$$\{a_{QP}\}_{l(Q),l(P)\leq 1} = \{\langle m_Q, \varphi_P\rangle\}_{l(Q),l(P)\leq 1}$$

and

$$\{b_{QP}\}_{l(Q),l(P)\leq 1} = \{\langle \varphi_P, b_Q\rangle\}_{l(Q),l(P)\leq 1},$$

with φ_P replaced by Φ_P when $l(P) = 1$, are both ε_1-almost diagonal on $a^{s,\tau}_{p,q}(\mathbb{R}^n)$.

To prove that $\langle f, b_Q\rangle$ is well defined for all inhomogeneous smooth analysis molecules for $A^{s,\tau}_{p,q}(\mathbb{R}^n)$, we need the following lemma. Its proof can be given as the proof of [25, Lemma 5.4].

Lemma 3.2. *Let ϕ be an inhomogeneous smooth analysis (or synthesis) molecule for $A^{s,\tau}_{p,q}(\mathbb{R}^n)$ supported near $Q \in \mathcal{Q}$. Then there exist a sequence $\{\phi_k\}_{k=1}^{\infty} \subset \mathcal{S}(\mathbb{R}^n)$ and a positive constant C such that for every $k \in \mathbb{N}$, $C\phi_k$ is also an inhomogeneous smooth analysis (or synthesis) molecule for $A^{s,\tau}_{p,q}(\mathbb{R}^n)$ supported near Q and $\phi_k(x) \to \phi(x)$ uniformly on $\mathbb{R}^n$ as $k \to \infty$.*

Now we have the following smooth molecular characterization of $A^{s,\tau}_{p,q}(\mathbb{R}^n)$. For the corresponding results of the homogeneous case, see [165, Theorem 4.2].

Theorem 3.2. *Let $s \in \mathbb{R}$, $p, q \in (0, \infty]$ and τ be as in Lemma 3.1.*

(i) Let $\{m_Q\}_{l(Q)\leq 1}$ be a family of inhomogeneous smooth synthesis molecules. Then there exists a positive constant C such that for all $t = \{t_Q\}_{l(Q)\leq 1} \in a^{s,\tau}_{p,q}(\mathbb{R}^n)$,

$$\left\| \sum_{l(Q)\leq 1} t_Q m_Q \right\|_{A^{s,\tau}_{p,q}(\mathbb{R}^n)} \leq C\|t\|_{a^{s,\tau}_{p,q}(\mathbb{R}^n)}.$$

(ii) *Let* $\{b_Q\}_{l(Q)\le 1}$ *be a family of inhomogeneous smooth analysis molecules. Then there exists a positive constant* C *such that for all* $f \in A^{s,\tau}_{p,q}(\mathbb{R}^n)$,

$$\left\|\{\langle f, b_Q\rangle\}_{l(Q)\le 1}\right\|_{a^{s,\tau}_{p,q}(\mathbb{R}^n)} \le C\|f\|_{A^{s,\tau}_{p,q}(\mathbb{R}^n)}.$$

Proof. (i) Let Φ, Ψ, φ and ψ be as in Lemma 2.3. It is easy to see that $m_P \in \mathscr{S}'(\mathbb{R}^n)$. Then by the Calderón reproducing formula in Lemma 2.3, we have

$$m_P = \sum_{l(Q)\le 1} \langle m_P, \varphi_Q\rangle \psi_Q,$$

where and in what follows, we use Φ_Q and Ψ_Q to replace, respectively, φ_Q and ψ_Q when $l(Q) = 1$. Let $\phi \in \mathscr{S}(\mathbb{R}^n)$, A and B be operators associated with, respectively, the matrices

$$\{a_{QP}\}_{l(Q),l(P)\le 1} \equiv \{\langle m_P, \varphi_Q\rangle\}_{l(Q),l(P)\le 1}$$

and

$$\{b_{QP}\}_{l(Q),l(P)\le 1} \equiv \{\langle \psi_Q, \phi_P\rangle\}_{l(Q),l(P)\le 1}.$$

Then Corollary 3.1 implies that A and B are ε_1-almost diagonal. Then, by Theorem 3.1, A and B are bounded on $a^{s,\tau}_{p,q}(\mathbb{R}^n)$. Notice that the operators $|A|$ and $|B|$ associated with, respectively, the matrices

$$\{|a_{QP}|\}_{l(Q),l(P)\le 1} \equiv \{|\langle m_P, \varphi_Q\rangle|\}_{l(Q),l(P)\le 1}$$

and

$$\{|b_{QP}|\}_{l(Q),l(P)\le 1} \equiv \{|\langle \psi_Q, \phi_P\rangle|\}_{l(Q),l(P)\le 1},$$

are also ε_1-almost diagonal on $a^{s,\tau}_{p,q}(\mathbb{R}^n)$. We then obtain

$$\begin{aligned}\sum_{l(Q)\le 1}\sum_{l(P)\le 1} |a_{QP}||t_P||\langle \psi_Q, \phi\rangle| &= \sum_{l(Q)\le 1}\sum_{l(P)\le 1} |a_{QP}||t_P||b_{QP_0}|\\ &= |((|B||A|)(|t|))_{P_0}|\\ &\le \|(|B||A|)(|t|)\|_{a^{s,\tau}_{p,q}(\mathbb{R}^n)}\\ &\lesssim \|t\|_{a^{s,\tau}_{p,q}(\mathbb{R}^n)},\end{aligned}$$

where $P_0 = [0, 1)^n$, and hence

$$T_\psi At = \sum_{l(Q)\le 1}\sum_{l(P)\le 1} a_{QP}t_P\psi_Q = \sum_{l(P)\le 1} t_P \sum_{l(Q)\le 1} a_{QP}\psi_Q = \sum_{l(P)\le 1} t_P m_P$$

holds in $\mathscr{S}'(\mathbb{R}^n)$. Then, applying Theorem 2.1, Corollary 3.1 and Theorem 3.1, we have

$$\left\|\sum_{l(P)\le 1} t_P m_P\right\|_{A^{s,\tau}_{p,q}(\mathbb{R}^n)} \le \|T_\psi At\|_{A^{s,\tau}_{p,q}(\mathbb{R}^n)} \lesssim \|At\|_{a^{s,\tau}_{p,q}(\mathbb{R}^n)} \lesssim \|t\|_{a^{s,\tau}_{p,q}(\mathbb{R}^n)}.$$

(ii) By Lemma 2.3 and Theorem 2.1, we see that

$$\langle f, b_Q\rangle = \sum_{l(P)\le 1} t_P\langle \psi_P, b_Q\rangle,$$

where

$$t \equiv \{t_P\}_{l(P)\le 1} \equiv \{\langle f, \varphi_P\rangle\}_{l(P)\le 1}$$

satisfies $\|t\|_{a^{s,\tau}_{p,q}(\mathbb{R}^n)} \lesssim \|f\|_{A^{s,\tau}_{p,q}(\mathbb{R}^n)}$. Let $a_{QP} \equiv \langle \psi_P, b_Q\rangle$ and A be the operator associated with the matrix $\{a_{QP}\}_{l(Q),l(P)\le 1}$. We see that for all $Q \in \mathcal{Q}$ with $l(Q) \le 1$,

$$\langle f, b_Q\rangle = \sum_{l(P)\le 1} a_{QP}t_P = (At)_Q.$$

Then Corollary 3.1 and Theorem 3.1 yield that

$$\|\{\langle f, b_Q\rangle\}_Q\|_{a^{s,\tau}_{p,q}(\mathbb{R}^n)} = \|At\|_{a^{s,\tau}_{p,q}(\mathbb{R}^n)} \lesssim \|t\|_{a^{s,\tau}_{p,q}(\mathbb{R}^n)},$$

which completes the proof of Theorem 3.2. □

Now we turn to the notion of a smooth atom for $A^{s,\tau}_{p,q}(\mathbb{R}^n)$; see [165, Definition 4.3] for the homogeneous case.

Definition 3.3. A function a_Q is called an *inhomogeneous smooth atom* for $A^{s,\tau}_{p,q}(\mathbb{R}^n)$ supported near a dyadic cube Q with $l(Q) \le 1$ if

$$\operatorname{supp} a_Q \subset 3Q, \tag{3.13}$$

$$\int_{\mathbb{R}^n} x^\gamma a_Q(x)\,dx = 0 \quad \text{if } |\gamma| \le \max\{\lfloor J-n-s\rfloor, -1\} \text{ and } l(Q) < 1, \tag{3.14}$$

and

$$\|\partial^\gamma a_Q\|_{L^\infty(\mathbb{R}^n)} \le |Q|^{-1/2-|\gamma|/n} \quad \text{if } |\gamma| \le \max\{\lfloor s+n\tau+1\rfloor, 0\}. \tag{3.15}$$

A collection $\{a_Q\}_{l(Q)\le 1}$ is called a family of inhomogeneous smooth atoms for $A^{s,\tau}_{p,q}(\mathbb{R}^n)$, if each a_Q is an inhomogeneous smooth atom for $A^{s,\tau}_{p,q}(\mathbb{R}^n)$ supported near Q.

Remark 3.1. We point out that in Definition 3.3, the moment condition (3.14) of smooth atoms can be strengthened into that $\int_{\mathbb{R}^n} x^\gamma a_Q(x)\,dx = 0$ if $|\gamma| \le \widetilde{N}$ and $l(Q) < 1$, the regularity condition (3.15) can be strengthened into that

$$\|\partial^\gamma a_Q\|_{L^\infty(\mathbb{R}^n)} \le |Q|^{-1/2-|\gamma|/n}$$

for all $\gamma \in \mathbb{Z}^n_+$ with $|\gamma| \le \widetilde{K}$, where $\widetilde{K}$ and $\widetilde{N}$ are arbitrary fixed integer satisfying $\widetilde{K} \ge \max\{\lfloor s+n\tau+1\rfloor, 0\}$ and $\widetilde{N} \ge \max\{\lfloor J-n-s\rfloor, -1\}$.

We also remark that, unlike the homogeneous case, an inhomogeneous smooth atom for $A^{s,\tau}_{p,q}(\mathbb{R}^n)$ supported near a dyadic cube Q has vanishing moments only when $l(Q) < 1$.

It is also clear that every inhomogeneous smooth atom for $A^{s,\tau}_{p,q}(\mathbb{R}^n)$ is a multiple of an inhomogeneous smooth synthesis molecule for $A^{s,\tau}_{p,q}(\mathbb{R}^n)$. We then have the following smooth atomic characterization of $A^{s,\tau}_{p,q}(\mathbb{R}^n)$. Recall, the quantities $\sigma_p, \sigma_{p,q}, \tau_{s,p}$ and $\tau_{s,p,q}$ have been defined in (1.5)–(1.7).

Theorem 3.3. *Let $s \in \mathbb{R}$ and $p, q \in (0,\infty]$ ($p < \infty$ if $A^{s,\tau}_{p,q}(\mathbb{R}^n) = F^{s,\tau}_{p,q}(\mathbb{R}^n)$). Let τ satisfy the inequality*

$$0 \le \tau < \begin{cases} \tau_{s,p} & \text{if} \quad A^{s,\tau}_{p,q}(\mathbb{R}^n) = B^{s,\tau}_{p,q}(\mathbb{R}^n), \\ \tau_{s,p,q} & \text{if} \quad A^{s,\tau}_{p,q}(\mathbb{R}^n) = F^{s,\tau}_{p,q}(\mathbb{R}^n). \end{cases} \tag{3.16}$$

Then for each $f \in A^{s,\tau}_{p,q}(\mathbb{R}^n)$, there exist a family $\{a_Q\}_{l(Q)\le 1}$ of inhomogeneous smooth atoms for $A^{s,\tau}_{p,q}(\mathbb{R}^n)$, a sequence $t \equiv \{t_Q\}_{Q\le 1} \subset \mathbb{C}$ of coefficients such that $f = \sum_{l(Q)\le 1} t_Q a_Q$ in $\mathscr{S}'(\mathbb{R}^n)$, $t \in a^{s,\tau}_{p,q}(\mathbb{R}^n)$ and

$$\|t\|_{a^{s,\tau}_{p,q}(\mathbb{R}^n)} \le C\|f\|_{A^{s,\tau}_{p,q}(\mathbb{R}^n)},$$

where C is a positive constant independent of t and f.

Conversely, there exists a positive constant C such that for all families $\{a_Q\}_{l(Q)\le 1}$ of inhomogeneous smooth atoms for $A^{s,\tau}_{p,q}(\mathbb{R}^n)$ and $t \equiv \{t_Q\}_{Q\le 1} \in a^{s,\tau}_{p,q}(\mathbb{R}^n)$,

$$\left\| \sum_{l(Q)\le 1} t_Q a_Q \right\|_{A^{s,\tau}_{p,q}(\mathbb{R}^n)} \le C\|t\|_{a^{s,\tau}_{p,q}(\mathbb{R}^n)}.$$

The proof of Theorem 3.3 is a simple modification of [165, Theorem 4.3]; see also [64, Theorem 4.1]. For completeness, we give the details.

Proof. The second claim follows immediately from Theorem 3.2. We only need to prove the first one.

Let Φ, φ, Ψ and ψ be as in (2.6). For $f \in A^{s,\tau}_{p,q}(\mathbb{R}^n)$, by Lemma 2.3, we write $f = \sum_{l(Q)\le 1} t_Q \psi_Q$ in $\mathscr{S}'(\mathbb{R}^n)$, where

$$t \equiv \{t_Q\}_{l(Q)\le 1} \equiv \{\langle f, \varphi_Q\rangle\}_{l(Q)\le 1}$$

satisfies $\|t\|_{a^{s,\tau}_{p,q}(\mathbb{R}^n)} \lesssim \|f\|_{A^{s,\tau}_{p,q}(\mathbb{R}^n)}$. Choose $\theta \in \mathscr{S}(\mathbb{R}^n)$ such that $\operatorname{supp}\theta \subset \{x \in \mathbb{R}^n : |x| \le 1\}$, $|\widehat{\theta}(\xi)| \ge C > 0$ if $1/2 \le |\xi| \le 2$ and $\int_{\mathbb{R}^n} x^\gamma \theta(x)\,dx = 0$ if $|\gamma| \le \widetilde{N}$. By (2.1) and (2.2), there exists a $\eta \in \mathscr{S}(\mathbb{R}^n)$ such that $\psi = \theta * \eta$. Set

$$g_k \equiv \int_{Q_{0k}} \theta(\cdot - y)\eta(y)\,dy$$

for all $k \in \mathbb{Z}^n$. Then $\operatorname{supp} g_k \subset 3Q_{0k}$, $\int_{\mathbb{R}^n} x^\gamma g_k(x)\,dx = 0$ if $|\gamma| \le \widetilde{N}$,

$$|\partial^\gamma g_k(x)| \le C(M,\gamma)(1+|k|)^{-M}$$

for any $M \in \mathbb{N}$ and $\psi = \sum_{k\in\mathbb{Z}^n} g_k$. Thus, for all $j \in \mathbb{Z}_+$ and $l \in \mathbb{Z}^n$,

$$\psi_{Q_{jl}} \equiv |Q_{jl}|^{-1/2} \sum_{k\in\mathbb{Z}^n} g_k(2^j x - l).$$

For $Q = Q_{jl}$ with $l(Q) \le 1$, set $r_Q \equiv C(t^*_{\min\{p,q\},\lambda})$ and

$$a_Q \equiv |Q|^{-1/2} \sum_{l\in\mathbb{Z}^n} t_{Q_{jl}} g_{k-l}(2^j x - l)/r_Q,$$

where C is a positive constant and will be determined later. From the above representation of f, we deduce that $f = \sum_Q r_Q a_Q$ in $\mathscr{S}'(\mathbb{R}^n)$. Let $\lambda > n$, M and C be large enough. It is easy to check that each a_Q is a smooth atom for $A^{s,\tau}_{p,q}(\mathbb{R}^n)$. Moreover, by Lemma 2.8, we see that

$$\|r\|_{a^{s,\tau}_{p,q}(\mathbb{R}^n)} \lesssim \|t\|_{a^{s,\tau}_{p,q}(\mathbb{R}^n)} \lesssim \|f\|_{A^{s,\tau}_{p,q}(\mathbb{R}^n)},$$

which completes the proof of Theorem 3.3. □

Remark 3.2.
(i) Observe that in case $s \le \sigma_p$ the interval $[0, 1/p]$ is always admissible. Furthermore, if s tends to infinity, also τ can become arbitrarily large.
(ii) If $\tau = 0$, Theorem 3.3 reduces to the known results on $B^s_{p,q}(\mathbb{R}^n)$ and $F^s_{p,q}(\mathbb{R}^n)$; see [23, 25, 64, 146].

Atomic decompositions have the essential advantage that they are localized, whereas the Fourier-analytic characterizations, used in the definition of the classes $A^{s,\tau}_{p,q}(\mathbb{R}^n)$, are not. We formulate a first simple consequence. By $A^s_{p,q}(P)$ we denote the corresponding space on the cube P, defined by restrictions. We refer to [145, 148] and Definition 6.2 below.

Corollary 3.2. *Let s, p, q and τ be as in Theorem 3.3. Then there exists a positive constant C such that for all $f \in A^{s,\tau}_{p,q}(\mathbb{R}^n)$,*

$$\sup_{\{P\in\mathscr{Q},\,|P|\ge 1\}} \frac{1}{|P|^\tau} \|f\|_{A^s_{p,q}(P)} \le C \|f\|_{A^{s,\tau}_{p,q}(\mathbb{R}^n)}.$$

Proof. For given f let $\sum_{l(Q)\le 1} t_Q a_Q$ be an optimal atomic decomposition of f in the sense of Theorem 3.3, i. e.,

$$f = \sum_{l(Q)\le 1} t_Q a_Q$$

and

$$\|f\|_{A^{s,\tau}_{p,q}(\mathbb{R}^n)} \sim \|t\|_{a^{s,\tau}_{p,q}(\mathbb{R}^n)}$$

and $\{a_Q\}_{l(Q)\le 1}$ is a family of inhomogeneous smooth atoms for $A^{s,\tau}_{p,q}(\mathbb{R}^n)$. Now we fix a dyadic cube P such that $|P| \ge 1$. Then

$$f = f^P \equiv \sum_{\substack{l(Q)\le 1\\ 3Q\cap 3P\neq\emptyset}} t_Q a_Q$$

coincide on P. Define a sequence $t^P = \{t^P_Q\}_{l(Q)\le 1}$ by

$$t^P_Q \equiv \begin{cases} t_Q & \text{if } \ 3Q\cap 3P \neq \emptyset, \\ 0 & \text{otherwise.} \end{cases}$$

By applying the definition of $A^s_{p,q}(P)$ and Theorem 3.3 with $\tau = 0$ (observe that $\{a_Q\}_{l(Q)\le 1}$ is a family of inhomogeneous smooth atoms for $A^{s,0}_{p,q}(\mathbb{R}^n)$ as well) we obtain

$$\|f\|_{A^s_{p,q}(P)} = \|f^P\|_{A^s_{p,q}(P)} \le \|f^P\|_{A^s_{p,q}(\mathbb{R}^n)} = \|f^P\|_{A^{s,0}_{p,q}(\mathbb{R}^n)} \lesssim \|t^P\|_{a^{s,0}_{p,q}(\mathbb{R}^n)}.$$

Since $(j_P \vee 0) = 0$ we find that

$$\frac{1}{|4P|^\tau}\|t^P\|_{a^{s,0}_{p,q}(\mathbb{R}^n)} \le \|t\|_{a^{s,\tau}_{p,q}(\mathbb{R}^n)}.$$

This proves the claim. □

Remark 3.3. Under the given restrictions in Corollary 3.2 we observe the following: any $f \in A^{s,\tau}_{p,q}(\mathbb{R}^n)$ belongs locally to $A^s_{p,q}(\mathbb{R}^n)$ and the quasi-norm $\|f\|_{A^s_{p,q}(P)}$ may only grow in dependence of $|P|$ and τ.

3.2 The Relation of $A^{s,\tau}_{p,q}(\mathbb{R}^n)$ to Besov-Triebel-Lizorkin-Morrey Spaces

We need another set of sequence spaces. Let p, q, s, τ be as in Definition 2.2. Define the spaces $\widetilde{a}^{s,\tau}_{p,q}(\mathbb{R}^n)$ to be set of all $\{t_Q\}_{l(Q)\le 1} \subset \mathbb{C}$ satisfying $\|t\|_{\widetilde{a}^{s,\tau}_{p,q}(\mathbb{R}^n)} < \infty$, where when $\widetilde{a}^{s,\tau}_{p,q}(\mathbb{R}^n) = \widetilde{b}^{s,\tau}_{p,q}(\mathbb{R}^n)$,

$$\|t\|_{\widetilde{b}^{s,\tau}_{p,q}(\mathbb{R}^n)} \equiv \left\{\sum_{j=0}^{\infty} 2^{j(s+n/2)q} \sup_{P\in\mathscr{Q}} \frac{1}{|P|^{\tau q}} \left[\int_P \left(\sum_{l(Q)=2^{-j}} |t_Q|\chi_Q(x)\right)^p dx\right]^{q/p}\right\}^{1/q},$$

and when $\widetilde{a}^{s,\tau}_{p,q}(\mathbb{R}^n) = \widetilde{f}^{s,\tau}_{p,q}(\mathbb{R}^n)$,

$$\|t\|_{\widetilde{f}^{s,\tau}_{p,q}(\mathbb{R}^n)} \equiv \sup_{P\in\mathscr{Q}} \frac{1}{|P|^\tau} \left\{ \int_P \left[\sum_{j=0}^{\infty} \sum_{l(Q)=2^{-j}} 2^{j(s+n/2)q} |t_Q|^q \chi_Q(x) \right]^{p/q} dx \right\}^{1/p}.$$

There is an essential difference in the definitions of $\|\cdot\|_{\widetilde{b}^{s,\tau}_{p,q}(\mathbb{R}^n)}$ and $\|\cdot\|_{b^{s,\tau}_{p,q}(\mathbb{R}^n)}$. The latter one does not contain an integration, it is totally discrete. The situation for the f-classes is different.

Proposition 3.1. *Let $s \in \mathbb{R}$.*

(i) Let $p \in (0,\infty)$. Let either $\tau \in [0, 1/p)$ and $q \in (0,\infty]$ or $\tau = 1/p$ and $q = \infty$. Then $\|t\|_{\widetilde{f}^{s,\tau}_{p,q}(\mathbb{R}^n)}$ and $\|t\|_{f^{s,\tau}_{p,q}(\mathbb{R}^n)}$ are equivalent quasi-norms in $f^{s,\tau}_{p,q}(\mathbb{R}^n)$.
(ii) If $p \in (0,\infty]$ and $\tau \in [0, 1/p]$, then $\|t\|_{\widetilde{b}^{s,\tau}_{p,\infty}(\mathbb{R}^n)}$ and $\|t\|_{b^{s,\tau}_{p,\infty}(\mathbb{R}^n)}$ are equivalent quasi-norms in $b^{s,\tau}_{p,\infty}(\mathbb{R}^n)$; however, if $q \in (0,\infty)$, then $\|t\|_{\widetilde{b}^{s,\tau}_{p,q}(\mathbb{R}^n)}$ and $\|t\|_{b^{s,\tau}_{p,q}(\mathbb{R}^n)}$ are not equivalent quasi-norms in $b^{s,\tau}_{p,q}(\mathbb{R}^n)$.

Proof. Step 1. Without restrictions concerning τ we have

$$\|t\|_{f^{s,\tau}_{p,q}(\mathbb{R}^n)} \le \|t\|_{\widetilde{f}^{s,\tau}_{p,q}(\mathbb{R}^n)} \qquad \text{and} \qquad \|t\|_{b^{s,\tau}_{p,q}(\mathbb{R}^n)} \le \|t\|_{\widetilde{b}^{s,\tau}_{p,q}(\mathbb{R}^n)}.$$

Step 2. To prove $\|t\|_{\widetilde{f}^{s,\tau}_{p,q}(\mathbb{R}^n)}$ and $\|t\|_{f^{s,\tau}_{p,q}(\mathbb{R}^n)}$ are equivalent, it suffices to show that for all dyadic cubes P with $j_P \ge 1$,

$$\mathrm{I}_P \equiv \frac{1}{|P|^\tau} \left\{ \int_P \left[\sum_{j=0}^{j_P-1} 2^{j(s+n/2)q} \sum_{l(Q)=2^{-j}} |t_Q|^q \chi_Q(x) \right]^{p/q} dx \right\}^{1/p} \lesssim \|t\|_{f^{s,\tau}_{p,q}(\mathbb{R}^n)}.$$

Indeed, for each $j = 0, \cdots, j_P - 1$, there exists a unique $Q_j \in \mathscr{Q}$ such that $l(Q_j) = 2^{-j}$ and $P \subset Q_j$. Then, by geometric properties of dyadic cubes, $\tau < 1/p$ and

$$|t_{Q_j}| \le \|t\|_{f^{s,\tau}_{p,q}(\mathbb{R}^n)} |Q_j|^{s/n+1/2+\tau-1/p},$$

we have

$$\begin{aligned}
\mathrm{I}_P &= \frac{1}{|P|^\tau} \left\{ \sum_{j=0}^{j_P-1} 2^{j(s+n/2)q} |P|^{q/p} |t_{Q_j}|^q \right\}^{1/q} \\
&\le \|t\|_{f^{s,\tau}_{p,q}(\mathbb{R}^n)} \frac{1}{|P|^\tau} \left\{ \sum_{j=0}^{j_P-1} 2^{-jn(\tau-1/p)q} |P|^{q/p} \right\}^{1/q} \\
&\lesssim \|t\|_{f^{s,\tau}_{p,q}(\mathbb{R}^n)},
\end{aligned}$$

which yields the claim.

Step 3. Next we consider the Besov-type spaces. Using the same type of arguments as in Step 2 it is easy to see that the converse inequality

$$\|t\|_{\widetilde{b}^{s,\tau}_{p,\infty}(\mathbb{R}^n)} \lesssim \|t\|_{b^{s,\tau}_{p,\infty}(\mathbb{R}^n)}$$

holds as long as $\tau \le 1/p$. However, if $q \in (0,\infty)$, the converse inequality does not hold. In fact, for all $j \in \mathbb{Z}_+$, set $R_j \equiv [2^{-j}, 2^{-j+1})^n$. Define a sequence $\{t_Q\}_{l(Q)\le 1}$ by setting

$$t_{R_j} \equiv 2^{-j(s+n/2+n\tau-n/p)}$$

for all $j \in \mathbb{Z}_+$ and $t_Q \equiv 0$ otherwise. Then, similarly to the arguments in [127, Theorem 1.1], it is easy to check that $\|t\|_{b^{s,\tau}_{p,q}(\mathbb{R}^n)} < \infty$ but $\|t\|_{\widetilde{b}^{s,\tau}_{p,q}(\mathbb{R}^n)} = \infty$, which completes the proof of Proposition 3.1. □

From the proof of Proposition 3.1 we also deduce a further equivalent quasi-norm of $\|f\|_{b^{s,\tau}_{p,q}(\mathbb{R}^n)}$.

Lemma 3.3. Let $s \in \mathbb{R}$, $p, q \in (0,\infty]$ and $\tau \in [0, 1/p)$. Then

$$\|t\|_{b^{s,\tau}_{p,q}(\mathbb{R}^n)} \sim \sup_{P\in\mathcal{Q}} \frac{1}{|P|^\tau} \left\{ \sum_{j=0}^{\infty} 2^{j(s+n/2)q} \left[\int_P \left(\sum_{l(Q)=2^{-j}} |t_Q|\chi_Q(x) \right)^p dx \right]^{q/p} \right\}^{1/q}.$$

In Sect. 1.3 we recalled the definition of Morrey spaces $\mathcal{M}^p_u(\mathbb{R}^n)$ (see item (xvi)), Besov-Morrey spaces $\mathcal{N}^s_{pqu}(\mathbb{R}^n)$ (see item (xxv)) and Triebel-Lizorkin-Morrey spaces $\mathcal{E}^s_{pqu}(\mathbb{R}^n)$ (see item (xxvi)). Let $\mathcal{A}^s_{pqu}(\mathbb{R}^n)$ denote either the Besov-Morrey space $\mathcal{N}^s_{pqu}(\mathbb{R}^n)$ or the Triebel-Lizorkin-Morrey space $\mathcal{E}^s_{pqu}(\mathbb{R}^n)$ $(p \ne \infty)$. Sawano and Tanaka [126] presented various decompositions including quarkonial, atomic and molecular characterizations of $\mathcal{A}^s_{pqu}(\mathbb{R}^n)$. Now it is easy to compare the spaces $A^{s,\tau}_{p,q}(\mathbb{R}^n)$ and $\mathcal{A}^s_{pqu}(\mathbb{R}^n)$. We only need to compare the associated sequence spaces. Notice that $\widetilde{a}^{0,1/u-1/p}_{u,q}(\mathbb{R}^n)$ is just the sequence space $a_{pqu}(\mathbb{R}^n)$ defined in [126, Definition 4.2]. There it has been shown that $\widetilde{a}^{0,1/u-1/p}_{u,q}(\mathbb{R}^n)$ is the sequence space of all admissible coefficient sequences of atomic decompositions of functions $f \in \mathcal{A}^s_{pqu}(\mathbb{R}^n)$; see [126, Theorem 4.9]. Arguing as in [127, Theorem 1.1], taking into account also the wavelet characterizations of Besov-Morrey and Triebel-Lizorkin-Morrey spaces (see [123] and [126, Theorem 3.9]), Proposition 3.1 and Theorem 3.3 yield the following.

Corollary 3.3. *Let* $s \in \mathbb{R}$.

(i) If $q \in (0,\infty]$ *and* $0 < u \le p \le \infty$, $u \ne \infty$, *then*

$$F^{s,1/u-1/p}_{u,q}(\mathbb{R}^n) = \mathcal{E}^s_{pqu}(\mathbb{R}^n)$$

with equivalent quasi-norms. In particular, if $1 < u \le p < \infty$*, then*

$$F^{0,1/u-1/p}_{u,2}(\mathbb{R}^n) = \mathscr{E}^0_{p2u}(\mathbb{R}^n) = \mathscr{M}^p_u(\mathbb{R}^n)$$

holds with equivalent norms.

(ii) If $0 < u = p \le \infty$ *and* $q \in (0, \infty]$*, then*

$$B^{s,1/u-1/p}_{u,q}(\mathbb{R}^n) = \mathscr{N}^s_{pqu}(\mathbb{R}^n) = B^s_{p,q}(\mathbb{R}^n)$$

with equivalent quasi-norms. If $0 < u \le p \le \infty$*, then*

$$B^{s,1/u-1/p}_{u,\infty}(\mathbb{R}^n) = \mathscr{N}^s_{p\infty u}(\mathbb{R}^n)$$

with equivalent quasi-norms. If $0 < u < p < \infty$ *and* $q \in (0, \infty)$*, then*

$$\mathscr{N}^s_{pqu}(\mathbb{R}^n) \subset B^{s,1/u-1/p}_{u,q}(\mathbb{R}^n)$$

and the embedding is proper.

Chapter 4
Several Equivalent Characterizations

Using the characterizations of $A_{p,q}^{s,\tau}(\mathbb{R}^n)$ by atoms and molecules obtained in Chap. 3, in this chapter, for certain p, q, s, τ, we establish characterizations of $A_{p,q}^{s,\tau}(\mathbb{R}^n)$ by wavelets, differences and oscillations (local approximation by polynomials). In addition we consider a localization property of $A_{p,q}^{s,\tau}(\mathbb{R}^n)$ by using $A_{p,q}^{s}(\mathbb{R}^n)$.

4.1 Preparations

Here we collect some assertions which will be of some use for us when establishing characterizations of $A_{p,q}^{s,\tau}(\mathbb{R}^n)$ by wavelets, differences and oscillations.

4.1.1 An Equivalent Definition

In this section, we give an equivalent definition of $A_{p,q}^{s,\tau}(\mathbb{R}^n)$ with the conditions on Φ and φ different from those in Definition 2.1. The following Definition 4.1 is more convenient than Definition 2.1 in applications.

Definition 4.1. Let $L \in (0, 1/2]$ and $\Theta(\mathbb{R}^n, L)$ be the set of all sequences $\{\phi_j\}_{j=0}^{\infty} \subset \mathscr{S}(\mathbb{R}^n)$ such that for all $j \in \mathbb{N}$,

$$\operatorname{supp} \widehat{\phi_0} \subset \{\xi \in \mathbb{R}^n : |\xi| \le 2\} \text{ and } \operatorname{supp} \widehat{\phi_j} \subset \{\xi \in \mathbb{R}^n : 2^j L \le |\xi| \le 2^{j+1}\}, \quad (4.1)$$

and that for every $M \in \mathbb{N}$, there exists a positive constant $C(n,M)$ such that for all $j \in \mathbb{Z}_+$ and $\xi \in \mathbb{R}^n$,

$$|\phi_j(\xi)| \le C(n,M) \frac{2^{jn}}{(1+2^j|\xi|)^{n+M}} \quad (4.2)$$

D. Yang et al., *Morrey and Campanato Meet Besov, Lizorkin and Triebel*, Lecture Notes in Mathematics 2005, DOI 10.1007/978-3-642-14606-0_4,

and

$$\sum_{j=0}^{\infty}\widehat{\phi}_j(\xi)=1 \quad \text{for all} \quad \xi\in\mathbb{R}^n. \tag{4.3}$$

It was proved in [145, p. 45] that $\Theta(\mathbb{R}^n,L)$ is not empty. Replacing $\{\Phi,\varphi_j: j\in\mathbb{N}\}$ in Definition 2.1 by $\{\phi_j\}_{j=0}^\infty\in\Theta(\mathbb{R}^n,L)$, we then obtain *two classes of spaces*, denoted respectively by $B^{s,\tau}_{p,q}(\Theta(\mathbb{R}^n,L))$ and $F^{s,\tau}_{p,q}(\Theta(\mathbb{R}^n,L))$. Similarly, we use $A^{s,\tau}_{p,q}(\Theta(\mathbb{R}^n,L))$ to denote either $B^{s,\tau}_{p,q}(\Theta(\mathbb{R}^n,L))$ or $F^{s,\tau}_{p,q}(\Theta(\mathbb{R}^n,L))$, and use $\|\cdot\|_{A^{s,\tau}_{p,q}(\Theta(\mathbb{R}^n,L))}$ to denote the quasi-norm of $A^{s,\tau}_{p,q}(\Theta(\mathbb{R}^n,L))$.

Lemma 4.1. *The space $A^{s,\tau}_{p,q}(\Theta(\mathbb{R}^n,L))$ is independent of the choices of $L\in(0,1/2]$ and $\{\phi_j\}_{j=0}^\infty\in\Theta(\mathbb{R}^n,L)$.*

Proof. Let $L_1,L_2\in(0,1/2]$, $\{\phi_j\}_{j=0}^\infty\in\Theta(\mathbb{R}^n,L_1)$ and $\{\psi_j\}_{j=0}^\infty\in\Theta(\mathbb{R}^n,L_2)$. By symmetry, we only need to show that for all $f\in A^{s,\tau}_{p,q}(\Theta(\mathbb{R}^n,L_2))$,

$$\|f\|_{A^{s,\tau}_{p,q}(\Theta(\mathbb{R}^n,L_1))}\lesssim\|f\|_{A^{s,\tau}_{p,q}(\Theta(\mathbb{R}^n,L_2))}. \tag{4.4}$$

Set $\psi_{-r}\equiv 0$ for all $r\in\mathbb{N}$. From (4.1), it follows that for all $j\in\mathbb{Z}_+$,

$$\phi_j=\sum_{r=\lfloor\log_2\frac{L_1}{2}\rfloor}^{\lceil\log_2\frac{2}{L_2}\rceil}\phi_j*\psi_{j+r}, \tag{4.5}$$

where $\lfloor\log_2\frac{L_1}{2}\rfloor$ and $\lceil\log_2\frac{2}{L_2}\rceil$ respectively denote the maximal integer *no more than* $\log_2\frac{L_1}{2}$ and the minimal integer *no less than* $\log_2\frac{2}{L_2}$. Thus,

$$\begin{aligned}&\|f\|_{B^{s,\tau}_{p,q}(\Theta(\mathbb{R}^n,L_1))}\\ &\quad\lesssim\sum_{r=\lfloor\log_2\frac{L_1}{2}\rfloor}^{\lceil\log_2\frac{2}{L_2}\rceil}\sup_{P\in\mathcal{Q}}\frac{1}{|P|^\tau}\left\{\sum_{j=(j_P\vee 0)}^{\infty}2^{jsq}\left[\int_P|\phi_j*\psi_{j+r}*f(x)|^p\,dx\right]^{q/p}\right\}^{1/q}.\end{aligned}$$

When $r=\lfloor\log_2\frac{L_1}{2}\rfloor,\cdots,0$, then $j+r\le j$. Applying [65, Lemma (6.10)], for all $x\in\mathbb{R}^n$, we have

$$\phi_j*\psi_{j+r}*f(x)=\sum_{k\in\mathbb{Z}^n}2^{-jn}\phi_j(x-2^{-j}k)\,\psi_{j+r}*f(2^{-j}k), \tag{4.6}$$

which together with the fact $\phi\in\mathcal{S}(\mathbb{R}^n)$ yields that

$$\mathrm{I}_P\equiv\frac{1}{|P|^\tau}\left\{\sum_{j=(j_P\vee 0)}^{\infty}2^{jsq}\left[\int_P|\phi_j*\psi_{j+r}*f(x)|^p\,dx\right]^{\frac{q}{p}}\right\}^{\frac{1}{q}}$$

$$\lesssim \frac{1}{|P|^\tau}\left\{\sum_{j=(j_P\vee 0)}^{\infty} 2^{jsq}\left[\sum_{\substack{l\in\mathbb{Z}^n\\ Q_{jl}\subset P}}\int_{Q_{jl}}\left(\sum_{k\in\mathbb{Z}^n}\frac{|\psi_{j+r}*f(2^{-j}k)|}{(1+2^j|x-2^{-j}k|)^{n+M}}\right)^p dx\right]^{\frac{q}{p}}\right\}^{\frac{1}{q}},$$

where $M\in\mathbb{N}$ will be determined later.

Let $\delta\in(0,\min\{1,p,q\})$. Since

$$1+2^j|x-2^{-j}k|\sim 1+2^j|2^{-j}l-2^{-j}k|$$

for all $l,k\in\mathbb{Z}^n$, $j\in\mathbb{Z}_+$ and $x\in Q_{jl}$, then for each j, applying [64, Lemma A. 4] to the function $\psi_{j+r}*f(2^{-j}y)$ and (2.11), we find that there exists a natural number γ such that for all $x\in Q_{jl}$,

$$\begin{aligned}
&\sum_{k\in\mathbb{Z}^n}\frac{|\psi_{j+r}*f(2^{-j}k)|}{(1+2^j|x-2^{-j}k|)^{n+M}}\\
&\quad\lesssim\left\{\sum_{k\in\mathbb{Z}^n}\frac{1}{(1+2^j|2^{-j}l-2^{-j}k|)^{(n+M)\delta}}\sup_{z\in Q_{jk}}|\psi_{j+r}*f(z)|^\delta\right\}^{1/\delta}\\
&\quad\lesssim\left\{\sum_{k\in\mathbb{Z}^n}\frac{\max\left\{\inf_{z\in\widetilde{Q}}|\psi_{j+r}*f(z)|^\delta:\ \widetilde{Q}\subset Q_{jk},\ l(\widetilde{Q})=2^{-\gamma}l(Q_{jk})\right\}}{(1+|l-k|)^{(n+M)\delta}}\right\}^{1/\delta}\\
&\quad\lesssim\left\{\sum_{k\in\mathbb{Z}^n}\frac{2^{jn}}{(1+|l-k|)^{(n+M)\delta}}\int_{Q_{jk}}|\psi_{j+r}*f(z)|^\delta\,dz\right\}^{1/\delta}.
\end{aligned}\tag{4.7}$$

Noticing that $3P$ is a union of dyadic cubes, by (4.7), we have

$$\begin{aligned}
\mathrm{I}_P&\lesssim\frac{1}{|P|^\tau}\left\{\sum_{j=(j_P\vee 0)}^{\infty}2^{jsq}\left[\sum_{\substack{l\in\mathbb{Z}^n\\ Q_{jl}\subset P}}\int_{Q_{jl}}\left(\sum_{k\in\mathbb{Z}^n}\chi_{\{k\in\mathbb{Z}^n:\,Q_{jk}\subset 3P\}}(k)\right.\right.\right.\\
&\qquad\left.\left.\left.\times\frac{2^{jn}}{(1+|l-k|)^{(n+M)\delta}}\int_{Q_{jk}}|\psi_{j+r}*f(z)|^\delta\,dz\right)^{\frac{p}{\delta}}dx\right]^{\frac{q}{p}}\right\}^{\frac{1}{q}}\\
&\quad+\frac{1}{|P|^\tau}\left\{\sum_{j=(j_P\vee 0)}^{\infty}2^{jsq}\left[\sum_{\substack{l\in\mathbb{Z}^n\\ Q_{jl}\subset P}}\int_{Q_{jl}}\left(\sum_{k\in\mathbb{Z}^n}\chi_{\{k\in\mathbb{Z}^n:\,Q_{jk}\cap 3P=\emptyset\}}(k)\cdots\right)^{\frac{p}{\delta}}dx\right]^{\frac{q}{p}}\right\}^{\frac{1}{q}}\\
&\equiv\mathrm{I}_1+\mathrm{I}_2.
\end{aligned}$$

Since $3P \subset \cup_{\{i\in\mathbb{Z}^n:\,|i|\le\sqrt{n}\}}(P+il(P))$, where

$$P+il(P)=\{x+il(P):x\in P\},$$

if we choose $M>\max\{(1/\delta-1)n, n(p/(\delta q)-1)\}$, by $p/\delta>1$, Hölder's inequality and the $L^{\frac{p}{\delta}}(\mathbb{R}^n)$-boundedness of the Hardy-Littlewood operator M, we have

$$\begin{aligned}
\mathrm{I}_1 &\lesssim \sum_{\substack{i\in\mathbb{Z}^n\\|i|\le\sqrt{n}}}\frac{1}{|P|^\tau}\left\{\sum_{j=(j_P\vee 0)}^{\infty}2^{jsq}\left[\sum_{\substack{l\in\mathbb{Z}^n\\Q_{jl}\subset P}}\int_{Q_{jl}}\left(\sum_{k\in\mathbb{Z}^n}\chi_{\{k\in\mathbb{Z}^n:Q_{jk}\subset(P+il(P))\}}(k)\right.\right.\right.\\
&\quad\left.\left.\left.\times\frac{2^{jn}}{(1+|l-k|)^{(n+M)\delta}}\int_{Q_{jk}}|\psi_{j+r}*f(z)|^\delta\chi_{P+il(P)}(z)\,dz\right)^{p/\delta}dx\right]^{q/p}\right\}^{1/q}\\
&\lesssim \sum_{\substack{i\in\mathbb{Z}^n\\|i|\le\sqrt{n}}}\frac{1}{|P|^\tau}\left\{\sum_{j=(j_P\vee 0)}^{\infty}2^{jsq}\left[\sum_{\substack{l\in\mathbb{Z}^n\\Q_{jl}\subset P}}\int_{Q_{jl}}\left(\sum_{k\in\mathbb{Z}^n}\frac{\chi_{\{k\in\mathbb{Z}^n:Q_{jk}\subset(P+il(P))\}}(k)}{(1+|l-k|)^{(n+M)\delta}}\right.\right.\right.\\
&\quad\left.\left.\left.\times\mathrm{M}(|\psi_{j+r}*f|^\delta\chi_{P+il(P)})(x+(k-l)2^{-j})\right)^{p/\delta}dx\right]^{q/p}\right\}^{1/q}\\
&\lesssim \sum_{\substack{i\in\mathbb{Z}^n\\|i|\le\sqrt{n}}}\frac{1}{|P|^\tau}\left\{\sum_{j=(j_P\vee 0)}^{\infty}2^{jsq}\left[\sum_{\substack{l\in\mathbb{Z}^n\\Q_{jl}\subset P}}\int_{Q_{jl}}\sum_{k\in\mathbb{Z}^n}\frac{\chi_{\{k\in\mathbb{Z}^n:Q_{jk}\subset(P+il(P))\}}(k)}{(1+|l-k|)^{(n+M)\delta}}\right.\right.\\
&\quad\left.\left.\times\left(\mathrm{M}(|\psi_{j+r}*f|^\delta\chi_{P+il(P)})(x+(k-l)2^{-j})\right)^{p/\delta}dx\right]^{q/p}\right\}^{1/q}\\
&\lesssim \sum_{\substack{i\in\mathbb{Z}^n\\|i|\le\sqrt{n}}}\frac{1}{|P|^\tau}\left\{\sum_{j=(j_P\vee 0)}^{\infty}2^{jsq}\left[\int_{P+il(P)}\right.\right.\\
&\quad\left.\left.\times\left(\mathrm{M}(|\psi_{j+r}*f|^\delta\chi_{P+il(P)})(x)\right)^{p/\delta}dx\right]^{q/p}\right\}^{1/q}\\
&\lesssim \sum_{\substack{i\in\mathbb{Z}^n\\|i|\le\sqrt{n}}}\frac{1}{|P|^\tau}\left\{\sum_{j=(j_P\vee 0)}^{\infty}2^{jsq}\left[\int_{P+il(P)}|\psi_{j+r}*f(x)|^p\,dx\right]^{q/p}\right\}^{1/q}\\
&\lesssim \|f\|_{B^{s,\tau}_{p,q}(\Theta(\mathbb{R}^n,L_2))}.
\end{aligned}$$

For I_2, we remark that if $Q_{jk}\cap 3P=\emptyset$, then $Q_{jk}\subset P+il(P)$ for a unique $i\in\mathbb{Z}^n$ with $|i|\ge 2$; moreover, $1+|l-k|\sim 2^j|i|l(P)$ for any $l\in\mathbb{Z}^n$ satisfying $Q_{jl}\subset P$. Similarly to the estimate of I_1, by Hölder's inequality, we obtain

$$
\begin{aligned}
\mathrm{I}_2 \lesssim \frac{1}{|P|^\tau}\Bigg\{ \sum_{j=(j_P\vee 0)}^{\infty} 2^{jsq} \Bigg[\sum_{\substack{l\in\mathbb{Z}^n\\ Q_{jl}\subset P}} \int_{Q_{jl}} \Bigg(\sum_{\{i\in\mathbb{Z}^n:\,|i|\geq 2\}} \sum_{k\in\mathbb{Z}^n} \chi_{\{k\in\mathbb{Z}^n:\,Q_{jk}\subset(P+il(P))\}}(k) 2^{jn} \\
\times \frac{2^{(j_P-j)(n+M)\delta}}{|i|^{(n+M)\delta}} \int_{Q_{jk}} |\psi_{j+r}*f(z)|^\delta \chi_{P+il(P)}(z)\,dz \Bigg)^{p/\delta} dx \Bigg]^{q/p} \Bigg\}^{1/q} \\
\lesssim \frac{1}{|P|^\tau}\Bigg\{ \sum_{j=(j_P\vee 0)}^{\infty} 2^{jsq} \Bigg[\sum_{\substack{l\in\mathbb{Z}^n\\ Q_{jl}\subset P}} \int_{Q_{jl}} \Bigg(\sum_{\{i\in\mathbb{Z}^n:\,|i|\geq 2\}} \sum_{k\in\mathbb{Z}^n} \chi_{\{k\in\mathbb{Z}^n:\,Q_{jk}\subset(P+il(P))\}}(k) \\
\times \frac{2^{(j_P-j)(n+M)\delta}}{|i|^{(n+M)\delta}} \mathrm{M}(|\psi_{j+r}*f|^\delta \chi_{P+il(P)})(x+(k-l)2^{-j}) \Bigg)^{p/\delta} dx \Bigg]^{q/p} \Bigg\}^{1/q} \\
\lesssim \frac{1}{|P|^\tau}\Bigg\{ \sum_{j=(j_P\vee 0)}^{\infty} 2^{jsq} \Bigg[\sum_{\substack{l\in\mathbb{Z}^n\\ Q_{jl}\subset P}} \int_{Q_{jl}} \sum_{\{i\in\mathbb{Z}^n:\,|i|\geq 2\}} \sum_{k\in\mathbb{Z}^n} \chi_{\{k\in\mathbb{Z}^n:\,Q_{jk}\subset(P+il(P))\}}(k) \\
\times \frac{2^{(j_P-j)(n+M)\delta}}{|i|^{(n+M)\delta}} \Big(\mathrm{M}(|\psi_{j+r}*f|^\delta \chi_{P+il(P)})(x+(k-l)2^{-j}) \Big)^{p/\delta} dx \Bigg]^{q/p} \Bigg\}^{1/q} \\
\lesssim \frac{1}{|P|^\tau}\Bigg\{ \sum_{j=(j_P\vee 0)}^{\infty} 2^{jsq} \Bigg[\sum_{\{i\in\mathbb{Z}^n:\,|i|\geq 2\}} |i|^{-(n+M)\delta} \\
\times \int_{P+il(P)} \Big(\mathrm{M}(|\psi_{j+r}*f|^\delta \chi_{P+il(P)})(x) \Big)^{p/\delta} dx \Bigg]^{q/p} \Bigg\}^{1/q},
\end{aligned}
$$

which by (2.11) or Hölder's inequality, $p/\delta > 1$ and the $L^{\frac{p}{\delta}}(\mathbb{R}^n)$-boundedness of the Hardy-Littlewood operator M, we have

$$
\begin{aligned}
\mathrm{I}_2 &\lesssim \frac{1}{|P|^\tau}\Bigg\{ \sum_{\{i\in\mathbb{Z}^n:\,|i|\geq 2\}} |i|^{-(n+M)\delta\min\{q/p,1\}} \sum_{j=(j_P\vee 0)}^{\infty} 2^{jsq} \\
&\quad \times \Bigg[\int_{P+il(P)} \Big(\mathrm{M}(|\psi_{j+r}*f|^\delta \chi_{P+il(P)})(x) \Big)^{p/\delta} dx \Bigg]^{q/p} \Bigg\}^{1/q} \\
&\lesssim \|f\|_{B^{s,\tau}_{p,q}(\Theta(\mathbb{R}^n,L_2))}.
\end{aligned}
$$

Thus,

$$
I_P \lesssim \|f\|_{B^{s,\tau}_{p,q}(\Theta(\mathbb{R}^n,L_2))}
$$

for all $r = \lfloor \log_2 \frac{L_1}{2} \rfloor, \cdots, 0$.

When $r = 1, \cdots, \lceil \log_2 \frac{2}{L_2} \rceil$, the proof is similar. In fact, in this case, we use [65, Lemma (6.10)] again to obtain that for all $x \in \mathbb{R}^n$,

$$\phi_j * \psi_{j+r} * f(x) = \sum_{k \in \mathbb{Z}^n} 2^{-(j+r)n} \phi_j(x - 2^{-j-r}k)\, \psi_{j+r} * f(2^{-j-r}k). \tag{4.8}$$

With (4.6) replaced by (4.8), following the previous arguments, we also obtain

$$I_P \lesssim \|f\|_{B^{s,\tau}_{p,q}(\Theta(\mathbb{R}^n, L_2))},$$

which further yields

$$\|f\|_{B^{s,\tau}_{p,q}(\Theta(\mathbb{R}^n, L_1))} \lesssim \|f\|_{B^{s,\tau}_{p,q}(\Theta(\mathbb{R}^n, L_2))}.$$

For the Triebel-Lizorkin-type spaces, applying (4.5) again, yields

$$\begin{aligned}
&\|f\|_{F^{s,\tau}_{p,q}(\Theta(\mathbb{R}^n, L_1))} \\
&\quad \lesssim \sum_{r=\lfloor \log_2 \frac{L_1}{2} \rfloor}^{\lceil \log_2 \frac{2}{L_2} \rceil} \sup_{P \in \mathscr{Q}} \frac{1}{|P|^\tau} \left\{ \int_P \left[\sum_{j=(j_P \vee 0)}^{\infty} 2^{jsq} |\phi_j * \psi_{j+r} * f(x)|^q \right]^{\frac{p}{q}} dx \right\}^{\frac{1}{p}}.
\end{aligned}$$

Similarly, when $r = \lfloor \log_2 \frac{L_1}{2} \rfloor, \cdots, 0$, by (4.6) and (4.7), we see that

$$\begin{aligned}
\mathrm{J}_P &\equiv \frac{1}{|P|^\tau} \left\{ \int_P \left[\sum_{j=(j_P \vee 0)}^{\infty} 2^{jsq} |\phi_j * \psi_{j+r} * f(x)|^q \right]^{p/q} dx \right\}^{1/p} \\
&\lesssim \frac{1}{|P|^\tau} \left\{ \int_{\mathbb{R}^n} \left[\sum_{j=(j_P \vee 0)}^{\infty} 2^{jsq} \sum_{\substack{l \in \mathbb{Z}^n \\ Q_{jl} \subset P}} \chi_{Q_{jl}}(x) \left(\sum_{k \in \mathbb{Z}^n} \frac{\chi_{\{k \in \mathbb{Z}^n : Q_{jk} \subset 3P\}}(k)}{(1+|l-k|)^{(n+M)\delta}} \right. \right. \right. \\
&\quad \left. \left. \left. \times \max\left\{ \inf_{z \in \widetilde{Q}} |\psi_{j+r} * f(z)|^\delta : \widetilde{Q} \subset Q_{jk},\ l(\widetilde{Q}) = 2^{-\gamma} l(Q_{jk}) \right\} \right)^{q/\delta} \right]^{p/q} dx \right\}^{1/p} \\
&\quad + \frac{1}{|P|^\tau} \left\{ \int_{\mathbb{R}^n} \left[\sum_{j=(j_P \vee 0)}^{\infty} 2^{jsq} \left(\sum_{k \in \mathbb{Z}^n} \chi_{\{k \in \mathbb{Z}^n : Q_{jk} \cap 3P = \emptyset\}}(k) \sum_{\substack{l \in \mathbb{Z}^n \\ Q_{jl} \subset P}} \chi_{Q_{jl}}(x) \right. \right. \right. \\
&\quad \left. \left. \left. \times \frac{2^{jn}}{(1+|l-k|)^{(n+M)\delta}} \int_{Q_{jk}} |\psi_{j+r} * f(z)|^\delta \, dz \right)^{q/\delta} \right]^{p/q} dx \right\}^{1/p} \\
&\equiv \mathrm{J}_1 + \mathrm{J}_2.
\end{aligned}$$

Define

$$t_{Q_{jk}} \equiv \max\left\{ \inf_{z\in\widetilde{Q}} |\psi_{j+r} * f(z)|^{\delta} : \widetilde{Q} \subset Q_{jk},\ l(\widetilde{Q}) = 2^{-\gamma} l(Q_{jk}) \right\}$$

if $Q_{jk} \subset 3P$ and $t_{Q_{jk}} \equiv 0$ otherwise. Then by [64, Lemma A. 2], if $M > n(1/\delta - 1)$, we obtain that for each $x \in Q_{jl}$,

$$\sum_{k\in\mathbb{Z}^n} \frac{\chi_{\{k\in\mathbb{Z}^n:\, Q_{jk}\subset 3P\}}(k)}{(1+|l-k|)^{(n+M)\delta}} t_{Q_{jk}} \lesssim \mathrm{M}\left(\sum_{k\in\mathbb{Z}^n} t_{Q_{jk}} \chi_{Q_{jk}}\right)(x).$$

Thus, choosing

$$M > \max\{(1/\delta - 1)n, n(q/(\delta p) - 1)\}$$

and applying the Fefferman-Stein vector-valued maximal inequality, we have

$$\begin{aligned}
\mathrm{J}_1 &\lesssim \frac{1}{|P|^{\tau}} \left\{ \int_{\mathbb{R}^n} \left[\sum_{j=(j_P\vee 0)}^{\infty} 2^{jsq} \left(\mathrm{M}\left(\sum_{k\in\mathbb{Z}^n} t_{Q_{jk}} \chi_{Q_{jk}} \right)(x) \right)^{q/\delta} \right]^{p/q} dx \right\}^{1/p} \\
&\lesssim \|f\|_{F^{s,\tau}_{p,q}(\Theta(\mathbb{R}^n, L_2))}.
\end{aligned}$$

Similarly to the estimate of I_2, by (2.11) or Hölder's inequality, $p/\delta > 1, q/\delta > 1$ and the Fefferman-Stein vector-valued maximal inequality, we see that

$$\begin{aligned}
\mathrm{J}_2 &\lesssim \frac{1}{|P|^{\tau}} \left\{ \int_{\mathbb{R}^n} \left[\sum_{j=(j_P\vee 0)}^{\infty} 2^{jsq} \left(\sum_{\{i\in\mathbb{Z}^n:\, |i|\geq 2\}} \sum_{k\in\mathbb{Z}^n} \chi_{\{k\in\mathbb{Z}^n:\, Q_{jk}\subset(P+il(P))\}}(k) \right.\right.\right. \\
&\qquad \left.\left.\left. \times \sum_{\substack{l\in\mathbb{Z}^n \\ Q_{jl}\subset P}} \chi_{Q_{jl}}(x) \frac{2^{(j_P-j)(n+M)\delta}}{|i|^{(n+M)\delta}} 2^{jn} \int_{Q_{jk}} |\psi_{j+r} * f(z)|^{\delta}\, dz \right)^{q/\delta} \right]^{p/q} dx \right\}^{1/p} \\
&\lesssim \frac{1}{|P|^{\tau}} \left\{ \int_{\mathbb{R}^n} \left[\sum_{j=(j_P\vee 0)}^{\infty} 2^{jsq} \left(\sum_{\{i\in\mathbb{Z}^n:\, |i|\geq 2\}} |i|^{-(n+M)\delta} \right.\right.\right. \\
&\qquad \left.\left.\left. \times \mathrm{M}(|\psi_{j+r} * f|^{\delta} \chi_{P+il(P)})(x + il(P)) \right)^{q/\delta} \right]^{p/q} dx \right\}^{1/p} \\
&\lesssim \|f\|_{F^{s,\tau}_{p,q}(\Theta(\mathbb{R}^n, L_2))}.
\end{aligned}$$

A similar argument also holds for $r = 1, \cdots, \lceil \log_2 \frac{2}{L_2} \rceil$ if we replace (4.6) by (4.8). Thus, we obtain

$$\|f\|_{F^{s,\tau}_{p,q}(\Theta(\mathbb{R}^n, L_1))} \lesssim \|f\|_{F^{s,\tau}_{p,q}(\Theta(\mathbb{R}^n, L_2))},$$

which completes the proof of Lemma 4.1. □

From Lemma 4.1, we immediately deduce that the spaces $A^{s,\tau}_{p,q}(\Theta(\mathbb{R}^n,L))$ and $A^{s,\tau}_{p,q}(\mathbb{R}^n)$ coincide.

Corollary 4.1. *Let $L\in(0,1/2]$ and p, q, s, τ be as in Definition 2.1. Then we have the coincidence of the spaces $A^{s,\tau}_{p,q}(\mathbb{R}^n)$ and $A^{s,\tau}_{p,q}(\Theta(\mathbb{R}^n,L))$ in the sense of equivalent quasi-norms.*

Proof. It suffices to construct a pair of Schwartz functions Φ and φ such that $\{\Phi,\varphi_j : j\in\mathbb{N}\}\in\Theta(\mathbb{R}^n,1/2)$ and satisfy, respectively, (2.1) and (2.2).

Indeed, let $\psi\in\mathscr{S}(\mathbb{R}^n)$ such that $\widehat{\psi}\ge 0$, supp $\widehat{\psi}\subset\{\xi\in\mathbb{R}^n : 1/2\le|\xi|\le 2\}$ and $\widehat{\psi}(\xi)\ge C>0$ if $3/5\le|\xi|\le 5/3$, where C is a positive constant independent of ξ. Define φ and Φ by setting

$$\widehat{\varphi}(\xi)\equiv\widehat{\psi}(\xi)\left(\sum_{i\in\mathbb{Z}}\widehat{\psi}(2^{-i}\xi)\right)^{-1}\quad\text{and}\quad\widehat{\Phi}(\xi)\equiv 1-\sum_{j=1}^{\infty}\widehat{\varphi}_j(\xi).$$

It is easy to check that φ and Φ are desired functions, which completes the proof of Corollary 4.1. □

4.1.2 Several Technical Lemmas on Differences

To determine the relation between the Littlewood-Paley characterization and characterization of $A^{s,\tau}_{p,q}(\mathbb{R}^n)$ by differences, we establish some technical lemmas first.

Let $L\in(0,1/2]$ and $\Phi\in\mathscr{S}(\mathbb{R}^n)$ such that $\widehat{\Phi}\equiv 1$ on $\{\xi\in\mathbb{R}^n : |\xi|\le 2L\}$ and

$$\operatorname{supp}\widehat{\Phi}\subset\{\xi\in\mathbb{R}^n : |\xi|\le 2\}.$$

Define φ_j by setting

$$\widehat{\varphi}_j(\xi)=\widehat{\Phi}(2^{-j}\xi)-\widehat{\Phi}(2^{-j+1}\xi)$$

for all $j\in\mathbb{N}$ and $\xi\in\mathbb{R}^n$. Then $\{\Phi,\varphi_j : j\in\mathbb{N}\}\in\Theta(\mathbb{R}^n,L)$. We then let

$$\|f\|^{\square}_{B^{s,\tau}_{p,q}(\mathbb{R}^n)}\equiv\sup_{P\in\mathscr{Q}}\frac{1}{|P|^{\tau}}\left\{\sum_{j=(j_P\vee 0)}^{\infty}2^{jsq}\left[\int_P|f(x)-\Phi_j*f(x)|^p\,dx\right]^{q/p}\right\}^{1/q}$$

and

$$\|f\|^{\square}_{F^{s,\tau}_{p,q}(\mathbb{R}^n)}\equiv\sup_{P\in\mathscr{Q}}\frac{1}{|P|^{\tau}}\left\{\int_P\left[\sum_{j=(j_P\vee 0)}^{\infty}2^{jsq}|f(x)-\Phi_j*f(x)|^q\right]^{p/q}dx\right\}^{1/p}.$$

Notice that $\varphi_j=\Phi_j-\Phi_{j-1}$ for all $j\ge 1$. From this, it is easy to deduce the following lemma. We omit the details.

Lemma 4.2. *Let $s \in \mathbb{R}$, $\tau \in [0,\infty)$ and $p, q \in (0,\infty]$. Then there exists a positive constant C such that*

$$\|f\|_{A^{s,\tau}_{p,q}(\mathbb{R}^n)} \le C\left\{\|f\|^{\square}_{A^{s,\tau}_{p,q}(\mathbb{R}^n)} + \sup_{\{P\in\mathscr{Q}: l(P)\ge 1\}} \frac{1}{|P|^{\tau}}\left(\int_P |\Phi * f(x)|^p\,dx\right)^{1/p}\right\}.$$

Next, using a trick of Nikol'skij [109, Sect. 5.2.1], we obtain a representation of $f - \Phi_j * f$ as an integral mean of differences of f. Recall that for all $x, h \in \mathbb{R}^n$,

$$\Delta_h^M f(x) \equiv \sum_{j=0}^{M} (-1)^j \binom{M}{j} f(x + (M-j)h). \tag{4.9}$$

Let $\psi \in \mathscr{S}(\mathbb{R}^n)$ such that

$$\widehat{\psi}(x) \ge 0,\ \widehat{\psi}(x) = 1 \quad \text{if} \quad |x| \le 1 \quad \text{and} \quad \widehat{\psi}(x) = 0 \quad \text{if} \quad |x| \ge 2.$$

For a fixed $M \in \mathbb{N}$, we define Φ by setting, for all $\xi \in \mathbb{R}^n$,

$$\widehat{\Phi}(\xi) \equiv (-1)^{M+1} \sum_{j=0}^{M-1} \binom{M}{j} (-1)^j \widehat{\psi}((M-j)\xi). \tag{4.10}$$

Then the function Φ satisfies

$$\widehat{\Phi}(\xi) = 1 \quad \text{if} \quad |\xi| \le 1/M \quad \text{and} \quad \operatorname{supp} \widehat{\Phi} \subset \{\xi \in \mathbb{R}^n : |\xi| \le 2\}.$$

Define φ_j by setting

$$\widehat{\varphi}_j(\xi) \equiv \widehat{\Phi}(2^{-j}\xi) - \widehat{\Phi}(2^{-j+1}\xi)$$

for all $j \in \mathbb{N}$ and $\xi \in \mathbb{R}^n$. Then

$$\{\Phi, \varphi_j : j \in \mathbb{N}\} \in \Theta(\mathbb{R}^n, 1/(2M))$$

and for all $f \in L^p_\tau(\mathbb{R}^n)$ with $p \in [1,\infty]$ and $j \in \mathbb{Z}_+$,

$$f(x) - \Phi_j * f(x) = (-1)^M \int_{\mathbb{R}^n} \left(\Delta^M_{-2^{-j}y} f(x)\right) \psi(y)\,dy. \tag{4.11}$$

Denote by $L^1_{\mathrm{loc}}(\mathbb{R}^n)$ the set of all locally integrable functions on $\mathbb{R}^n$. For $f \in L^1_{\mathrm{loc}}(\mathbb{R}^n)$, $P \in \mathscr{Q}$ and $N \in \mathbb{N}$, we define

$$a_t(x) \equiv t^{-n} \int_{t/2 \le |h| < t} |\Delta_h^M f(x)|\,dh \tag{4.12}$$

and

$$T_{N,P} \equiv \left(\sum_{m=(j_P \vee 0)}^{\infty} \frac{2^{(n+s-N)md}}{|P|^{\tau d}} \left\{ \int_1^{2^{m-(j_P\vee 0)+2}} t^{-sq} \left(\int_P [a_t(x)]^p \, dx \right)^{\frac{q}{p}} \frac{dt}{t} \right\}^{\frac{d}{q}} \right)^{\frac{1}{d}};$$

(4.13)

$$\widetilde{T}_{N,P} \equiv \left(\sum_{m=(j_P \vee 0)}^{\infty} \frac{2^{(n+s-N)md}}{|P|^{\tau d}} \left\{ \int_P \left(\int_1^{2^{m-(j_P\vee 0)+2}} t^{-sq} [a_t(x)]^q \frac{dt}{t} \right)^{\frac{p}{q}} dx \right\}^{\frac{d}{p}} \right)^{\frac{1}{d}}.$$

Here s, p, q and d will be fixed later on. Recall, $\overline{p} = \max\{1, p\}$.

Lemma 4.3. *Let $q \in (0,\infty]$, $s \in [0,\infty)$, $\tau \in [0,\infty)$ and P be a dyadic cube. Then there exist a positive constant C and $N \in \mathbb{N}$, independent of P, such that for all $p \in (0,\infty]$ and $f \in L_\tau^{\overline{p}}(\mathbb{R}^n)$,*

$$\begin{aligned} \mathrm{I}_P &\equiv \frac{1}{|P|^\tau} \left\{ \sum_{j=(j_P\vee 0)}^{\infty} 2^{jsq} \left(\int_P |f(x) - \Phi_j * f(x)|^p \, dx \right)^{q/p} \right\}^{1/q} \\ &\le C \sup_{\{Q \in \mathcal{Q}: P \subset Q\}} \frac{1}{|Q|^\tau} \left\{ \int_0^{2(l(Q)\wedge 1)} t^{-sq} \left(\int_Q [a_t(x)]^p \, dx \right)^{q/p} \frac{dt}{t} \right\}^{1/q} + C T_{N,P}, \end{aligned}$$

and that for all $p \in (0,\infty)$ and $f \in L_\tau^{\overline{p}}(\mathbb{R}^n)$,

$$\begin{aligned} \mathrm{J}_P &\equiv \frac{1}{|P|^\tau} \left\{ \int_P \left(\sum_{j=(j_P\vee 0)}^{\infty} 2^{jsq} |f(x) - \Phi_j * f(x)|^q \right)^{p/q} dx \right\}^{1/p} \\ &\le C \sup_{\{Q \in \mathcal{Q}: P \subset Q\}} \frac{1}{|Q|^\tau} \left\{ \int_Q \left(\int_0^{2(l(Q)\wedge 1)} t^{-sq} [a_t(x)]^q \frac{dt}{t} \right)^{p/q} dx \right\}^{1/p} + C \widetilde{T}_{N,P}. \end{aligned}$$

Proof. For a dyadic cube P we define

$$A_P \equiv \sup_{\{Q \in \mathcal{Q}: P \subset Q\}} \frac{1}{|Q|^\tau} \left\{ \int_0^{2(l(Q)\wedge 1)} t^{-sq} \left(\int_Q [a_t(x)]^p \, dx \right)^{q/p} \frac{dt}{t} \right\}^{1/q},$$

and

$$\widetilde{A}_P \equiv \sup_{\{Q \in \mathcal{Q}: P \subset Q\}} \frac{1}{|Q|^\tau} \left\{ \int_Q \left(\int_0^{2(l(Q)\wedge 1)} t^{-sq} [a_t(x)]^q \frac{dt}{t} \right)^{p/q} dx \right\}^{1/p}.$$

Step 1. Estimate of I_P. Using (4.11) and then splitting the integral into the regions $|y| \geq 1$ and $|y| < 1$, we see that

$$\begin{aligned}\mathrm{I}_P &\lesssim \frac{1}{|P|^\tau}\left\{\sum_{j=(j_P\vee 0)}^{\infty} 2^{jsq}\left[\int_P\left(\int_{|y|<1}|\Delta^M_{2^{-j}y}f(x)||\psi(y)|\,dy\right)^p dx\right]^{q/p}\right\}^{1/q} \\ &\quad+\frac{1}{|P|^\tau}\left\{\sum_{j=(j_P\vee 0)}^{\infty} 2^{jsq}\left[\int_P\left(\int_{|y|\geq 1}\cdots\right)^p dx\right]^{q/p}\right\}^{1/q} \\ &\equiv \mathrm{I}_1+\mathrm{I}_2.\end{aligned}$$

Since $\psi \in \mathscr{S}(\mathbb{R}^n)$, we have

$$|\psi(y)| \leq C(N)(1+|y|)^{-N}$$

for all $y \in \mathbb{R}^n$, where $N \in \mathbb{N}$ is at our disposal and will be chosen later on and $C(N)$ is a positive constant depending on N. Hence

$$\mathrm{I}_1 \lesssim \frac{1}{|P|^\tau}\left\{\sum_{j=(j_P\vee 0)}^{\infty} 2^{jsq}\left[\int_P\left(\sum_{m=0}^{\infty} 2^{-mn}a_{2^{-m-j}}(x)\right)^p dx\right]^{q/p}\right\}^{1/q}$$

and

$$\mathrm{I}_2 \lesssim \frac{1}{|P|^\tau}\left\{\sum_{j=(j_P\vee 0)}^{\infty} 2^{jsq}\left[\int_P\left(\sum_{m=0}^{\infty} 2^{m(n-N)}a_{2^{m-j+1}}(x)\right)^p dx\right]^{q/p}\right\}^{1/q}.$$

Step 1.1. Estimate of I_1. Observe that for all $x \in \mathbb{R}^n$, $l \in \mathbb{Z}$ and $2^{l-j} \leq t \leq 2^{l-j+1}$,

$$a_{2^{l-j}}(x) \leq 2^n(a_t(x)+a_{t/2}(x)). \tag{4.14}$$

Furthermore, recall the elementary inequality

$$\left(\sum_{j=0}^{\infty}\left\|\sum_{m=0}^{\infty} f_{m,j}\right\|^q_{L^p(\Omega)}\right)^{d/q} \leq \sum_{m=0}^{\infty}\left(\sum_{j=0}^{\infty}\|f_{m,j}\|^q_{L^p(\Omega)}\right)^{d/q}, \tag{4.15}$$

valid for all sequences $\{f_{m,j}\}_{m,j\in\mathbb{N}}$ of locally integrable functions, Ω a measurable subset of $\mathbb{R}^n$ and with $d = \min\{1, p, q\}$. This yields

$$\begin{aligned}(\mathrm{I}_1)^d &\lesssim \frac{1}{|P|^{\tau d}}\sum_{m=0}^{\infty} 2^{-mnd}\left\{\sum_{j=(j_P\vee 0)}^{\infty} 2^{jsq}\left(\int_P [a_{2^{-m-j}}(x)]^p\,dx\right)^{q/p}\right\}^{d/q} \\ &\lesssim \frac{1}{|P|^{\tau d}}\sum_{m=0}^{\infty} 2^{-mnd}\end{aligned}$$

$$\times \left\{ \sum_{j=(j_P\vee 0)}^{\infty} 2^{jsq} \int_{2^{-m-j}}^{2^{-m-j+1}} \left(\int_P [a_t(x)+a_{t/2}(x)]^p \, dx \right)^{q/p} \frac{dt}{t} \right\}^{d/q}$$

$$\lesssim \frac{1}{|P|^{\tau d}} \sum_{m=0}^{\infty} 2^{-m(n+s)d}$$

$$\times \left\{ \int_0^{2^{-m-(j_P\vee 0)+1}} t^{-sq} \left(\int_P [a_t(x)+a_{t/2}(x)]^p \, dx \right)^{q/p} \frac{dt}{t} \right\}^{d/q}$$

$$\lesssim \frac{1}{|P|^{\tau d}} \left\{ \int_0^{2^{-(j_P\vee 0)+1}} t^{-sq} \left(\int_P [a_t(x)]^p \, dx \right)^{q/p} \frac{dt}{t} \right\}^{d/q}$$

$$\lesssim (A_P)^d .$$

Step 1.2. Estimate of I_2. Applying again (4.14) and (4.15) we find

$$(\mathrm{I}_2)^d \lesssim \frac{1}{|P|^{\tau d}} \sum_{m=0}^{\infty} 2^{(n-N)md} \left\{ \sum_{j=(j_P\vee 0)}^{\infty} 2^{jsq} \int_{2^{m-j+1}}^{2^{m-j+2}} \right.$$

$$\left. \times \left(\int_P [a_t(x)+a_{t/2}(x)]^p \, dx \right)^{q/p} \frac{dt}{t} \right\}^{d/q}$$

$$\lesssim \frac{1}{|P|^{\tau d}} \sum_{m=0}^{\infty} 2^{(n-N+s)md} \left\{ \int_0^{2^{m-(j_P\vee 0)+2}} t^{-sq} \left(\int_P [a_t(x)]^p \, dx \right)^{q/p} \frac{dt}{t} \right\}^{d/q} .$$

Therefore, for the case $j_P \ge 1$, it holds $(\mathrm{I}_2)^d \le \mathrm{I}_2^1 + \mathrm{I}_2^2 + \mathrm{I}_2^3$, where

$$\mathrm{I}_2^1 \equiv \frac{1}{|P|^{\tau d}} \sum_{m=0}^{\infty} 2^{(n-N+s)md} \left\{ \int_0^{2l(P)} t^{-sq} \left(\int_P [a_t(x)]^p \, dx \right)^{q/p} \frac{dt}{t} \right\}^{d/q} ,$$

$$\mathrm{I}_2^2 \equiv \frac{1}{|P|^{\tau d}} \sum_{m=0}^{j_P-1} 2^{(n-N+s)md} \left\{ \int_{2l(P)}^{2^{m-j_P+2}} \cdots \frac{dt}{t} \right\}^{d/q} ,$$

$$\mathrm{I}_3^2 \equiv \frac{1}{|P|^{\tau d}} \sum_{m=j_P}^{\infty} 2^{(n-N+s)md} \left\{ \int_{2l(P)}^{2^{m-j_P+2}} \cdots \frac{dt}{t} \right\}^{d/q} .$$

Choose $N > n+s+n\tau$. The fact $\mathrm{I}_2^1 \lesssim (A_P)^d$ is trivial. To estimate the others, let P_m be the dyadic cube containing P and having side length $2^m l(P)$. Then, under the above condition on N, we obtain

$$\mathrm{I}_2^2 \lesssim \frac{1}{|P|^{\tau d}} \sum_{m=0}^{j_P-1} 2^{(n+s-N)md} \left\{ \sum_{i=1}^{m+1} \int_{l(P_i)}^{2l(P_i)} t^{-sq} \left(\int_{P_i} [a_t(x)]^p \, dx \right)^{q/p} \frac{dt}{t} \right\}^{d/q}$$

$$\lesssim (A_P)^d \sum_{m=0}^{j_P-1} 2^{(n+s-N)md} \left\{ \sum_{i=0}^{m+1} 2^{in\tau q} \right\}^{d/q}$$

$$\lesssim (A_P)^d .$$

Finally, we split I_2^3 once again. Using our abbreviation $T_{N,P}$ we get

$$\mathrm{I}_2^3 \lesssim \frac{1}{|P|^{\tau d}} \sum_{m=j_P}^{\infty} 2^{(n+s-N)md} \left\{ \int_{2l(P)}^{1} \left(\int_P [a_t(x)]^p \, dx \right)^{q/p} \frac{dt}{t} \right\}^{d/q}$$

$$+ \frac{1}{|P|^{\tau d}} \sum_{m=j_P}^{\infty} 2^{(n+s-N)md} \left\{ \int_1^{2^{m-j_P+2}} \cdots \frac{dt}{t} \right\}^{d/q}$$

$$\lesssim (A_P)^d \sum_{m=j_P}^{\infty} 2^{(n+s-N)md} \left\{ \sum_{i=1}^{j_P-1} 2^{in\tau q} \right\}^{d/q} + (T_{N,P})^d$$

$$\lesssim (A_P)^d + (T_{N,P})^d .$$

In the case $j_P \le 0$ we argue in a similar way and find

$$(\mathrm{I}_2)^d \lesssim \frac{1}{|P|^{\tau d}} \sum_{m=0}^{\infty} 2^{(n+s-N)md} \left\{ \int_0^2 t^{-sq} \left(\int_P [a_t(x)]^p \, dx \right)^{q/p} \frac{dt}{t} \right\}^{d/q}$$

$$+ \frac{1}{|P|^{\tau d}} \sum_{m=0}^{\infty} 2^{(n+s-N)md} \left\{ \int_2^{2^{m+2}} \cdots \frac{dt}{t} \right\}^{d/q}$$

$$\lesssim (A_P)^d + (T_{N,P})^d .$$

Thus, we have proved the claim in case of I_P.

Step 2. Estimate of J_P. As in Step 1 we conclude

$$\mathrm{J}_P \lesssim \frac{1}{|P|^{\tau}} \left\{ \int_P \left[\sum_{j=(j_P \vee 0)}^{\infty} 2^{jsq} \left(\sum_{m=0}^{\infty} 2^{-mn} a_{2^{-m-j}}(x) \right)^q \right]^{p/q} dx \right\}^{1/p}$$

$$+ \frac{1}{|P|^{\tau}} \left\{ \int_P \left[\sum_{j=(j_P \vee 0)}^{\infty} 2^{jsq} \left(\sum_{m=0}^{\infty} 2^{m(n-N)} a_{2^{m-j+1}}(x) \right)^q \right]^{p/q} dx \right\}^{1/p}$$

$$\equiv \mathrm{J}_1 + \mathrm{J}_2 .$$

This time we continue by using

$$\left\| \left(\sum_{j=0}^{\infty} \left| \sum_{m=0}^{\infty} f_{m,j} \right|^q \right)^{1/q} \right\|_{L^p(\Omega)}^d \le \sum_{m=0}^{\infty} \left\| \left(\sum_{j=0}^{\infty} |f_{m,j}|^q \right)^{1/q} \right\|_{L^p(\Omega)}^d , \tag{4.16}$$

valid for all sequences $\{f_{m,j}\}_{m,j\in\mathbb{N}}$ of locally integrable functions, Ω a measurable subset of $\mathbb{R}^n$ and with $d = \min\{1, p, q\}$.

Step 2.1. Estimate of J_1. Applying (4.14) in combination with (4.16), we see that

$$\begin{aligned}
(\mathrm{J}_1)^d &\lesssim \frac{1}{|P|^{\tau d}} \sum_{m=0}^{\infty} 2^{-mnd} \left\{ \int_P \left[\sum_{j=(j_P\vee 0)}^{\infty} 2^{jsq} \right.\right. \\
&\quad \times \left.\left. \int_{2^{-m-j}}^{2^{-m-j+1}} (a_t(x) + a_{t/2}(x))^q \frac{dt}{t} \right]^{p/q} dx \right\}^{d/p} \\
&\lesssim \frac{1}{|P|^{\tau d}} \sum_{m=0}^{\infty} 2^{-m(n+s)d} \left\{ \int_P \left[\int_0^{2^{-m-(j_P\vee 0)+1}} t^{-sq} [a_t(x)]^q \frac{dt}{t} \right]^{p/q} dx \right\}^{d/p} \\
&\lesssim (\widetilde{A}_P)^d .
\end{aligned}$$

Step 2.2. Estimate of J_2. Again we use (4.16) and find

$$(\mathrm{J}_2)^d \lesssim \frac{1}{|P|^{\tau d}} \sum_{m=0}^{\infty} 2^{(n-N)md} \left\{ \int_P \left[\sum_{j=(j_P\vee 0)}^{\infty} 2^{jsq} (a_{2^{m-j+1}}(x))^q \right]^{p/q} dx \right\}^{d/p} .$$

We split the right-hand side into three terms, i. e., $(\mathrm{J}_2)^d \lesssim \mathrm{J}_2^1 + \mathrm{J}_2^2 + \mathrm{J}_2^3$, where

$$\mathrm{J}_2^1 \equiv \frac{1}{|P|^{\tau d}} \sum_{m=0}^{\infty} 2^{(n-N)md} \left\{ \int_P \left[\sum_{j=(j_P\vee 0)+m+1}^{\infty} 2^{jsq} (a_{2^{m-j+1}}(x))^q \right]^{p/q} dx \right\}^{d/p}$$

$$\mathrm{J}_2^2 \equiv \frac{1}{|P|^{\tau d}} \sum_{m=0}^{(j_P\vee 0)} 2^{(n-N)md} \left\{ \int_P \left[\sum_{j=(j_P\vee 0)}^{(j_P\vee 0)+m} \cdots \right]^{p/q} dx \right\}^{d/p}$$

$$\mathrm{J}_2^3 \equiv \frac{1}{|P|^{\tau d}} \sum_{m=(j_P\vee 0)+1}^{\infty} 2^{(n-N)md} \left\{ \int_P \left[\sum_{j=(j_P\vee 0)}^{(j_P\vee 0)+m} \cdots \right]^{p/q} dx \right\}^{d/p} .$$

Similarly to the estimate of I_2, we obtain $\mathrm{J}_2^1, \mathrm{J}_2^2 \lesssim (\widetilde{A}_P)^d$. As above the estimate of the third term J_2^3 is more complicated. We split our considerations into the cases $j_P < 0$ and $j_P \ge 0$. First, let $j_P < 0$. Then (4.14) leads to

$$\mathrm{J}_2^3 \lesssim \frac{1}{|P|^{\tau d}} \sum_{m=1}^{\infty} 2^{(n+s-N)md} \left\{ \int_P \left[\int_2^{2^{m+2}} t^{-sq} [a_t(x)]^q \frac{dt}{t} \right]^{p/q} dx \right\}^{d/p}.$$

Hence $\mathrm{J}_2^3 \lesssim (\widetilde{T}_{N,P})^d$. In case $j_P \geq 0$, similarly,

$$\begin{aligned} \mathrm{J}_2^3 &\lesssim \frac{1}{|P|^{\tau d}} \sum_{m=j_P}^{\infty} 2^{(n-N)md} \left\{ \int_P \left[\sum_{j=j_P}^{m} 2^{jsq} (a_{2^{m-j+1}}(x))^q \right]^{p/q} dx \right\}^{d/p} \\ &\quad + \frac{1}{|P|^{\tau d}} \sum_{m=j_P}^{\infty} 2^{(n-N)md} \left\{ \int_P \left[\sum_{j=m+1}^{m+j_P} 2^{jsq} (a_{2^{m-j+1}}(x))^q \right]^{p/q} dx \right\}^{d/p} \\ &\lesssim (\widetilde{T}_{N,P})^d + (\widetilde{A}_P)^d, \end{aligned}$$

which completes the proof of Lemma 4.3. □

Lemma 4.4. *Let $s \in (0,\infty)$ and $P \in \mathcal{Q}$.*

(i) *Let $p \in [1,\infty]$. Then, for sufficiently large N, there exists a positive constant C, independent of P, such that*

$$T_{N,P} + \widetilde{T}_{N,P} \leq C \|f\|_{L^p_\tau(\mathbb{R}^n)}$$

holds for all f with finite right-hand side.

(ii) *Let $p \in (0,1)$ and $\tau \in [1/p,\infty)$. Then, for sufficiently large N, there exists a positive constant C, independent of P, such that*

$$T_{N,P} + \widetilde{T}_{N,P} \leq C (\|f\|_{L^1_\tau(\mathbb{R}^n)} + \|f\|_{L^p_\tau(\mathbb{R}^n)})$$

holds for all f with finite right-hand side.

(iii) *Let $p \in (0,1)$ and $\sigma_p < s_0 < s$. Then, for sufficiently large N, there exists a positive constant C, independent of P, such that*

$$T_{N,P} + \widetilde{T}_{N,P} \leq C \sup_{\{P \in \mathcal{Q}, |P| \geq 1\}} \frac{\|f\|_{B^{s_0}_{p,\infty}(2P)}}{|P|^\tau}$$

holds for all f with finite right-hand side.

Proof. Step 1. Proof of (i). Let $p \in [1,\infty]$. We choose

$$N > \max\{2n+s, 2n+s+nd(\tau-1)\}.$$

Minkowski's inequality then yields

$$\begin{aligned}
(T_{N,P})^d &\le \sum_{m=(j_P\vee 0)}^{\infty} 2^{(n+s-N)md} \frac{1}{|P|^{\tau d}} \left\{ \int_1^{2^{m-(j_P\vee 0)+2}} t^{-sq-nq} \right. \\
&\quad \times \left. \left[\int_P \left(\int_{1/2\le|h|<2^{m-(j_P\vee 0)+2}} |\Delta_h^M f(x)|\,dh \right)^p dx \right]^{q/p} \frac{dt}{t} \right\}^{d/q} \\
&\le \sum_{m=(j_P\vee 0)}^{\infty} 2^{(n+s-N)md} \frac{1}{|P|^{\tau d}} \left\{ \int_1^{2^{m-(j_P\vee 0)+2}} t^{-sq-nq} \right. \\
&\quad \times \left. \left[\int_{1/2\le|h|<2^{m-(j_P\vee 0)+2}} \left(\int_P |\Delta_h^M f(x)|^p\,dx \right)^{1/p} dh \right]^{q} \frac{dt}{t} \right\}^{d/q} \\
&\lesssim \|f\|^d_{L^p_\tau(\mathbb{R}^n)} 2^{(j_P\vee 0)n\tau d} \sum_{m=(j_P\vee 0)}^{\infty} 2^{(n+s-N)md} 2^{(m-(j_P\vee 0)+2)nd} \\
&\lesssim \|f\|^d_{L^p_\tau(\mathbb{R}^n)},
\end{aligned}$$

since $s>0$. This proves $T_{N,P} \lesssim \|f\|_{L^p_\tau(\mathbb{R}^n)}$. We describe the modifications needed in case of $\widetilde{T}_{N,P}$. Since $s>0$ we observe

$$\begin{aligned}
&\left\{ \int_P \left(\int_1^{2^{m-(j_P\vee 0)+2}} t^{-sq} [a_t(x)]^q \frac{dt}{t} \right)^{\frac{p}{q}} dx \right\}^{1/p} \\
&\quad\lesssim \left\{ \int_P \left[\sup_{1<t<2^{m-(j_P\vee 0)+2}} a_t(x) \right]^p dx \right\}^{1/p} \\
&\quad\lesssim \left\{ \int_P \left(\int_{|h|<2^{m-(j_P\vee 0)+2}} |\Delta_h^M f(x)|\,dh \right)^p dx \right\}^{1/p} \\
&\quad\lesssim \int_{|h|<2^{m-(j_P\vee 0)+2}} \left(\int_P |\Delta_h^M f(x)|^p dx \right)^{1/p} dh.
\end{aligned}$$

This implies

$$(\widetilde{T}_{N,P})^d \lesssim \|f\|^d_{L^p_\tau(\mathbb{R}^n)} 2^{(j_P\vee 0)n\tau} \sum_{m=(j_P\vee 0)}^{\infty} 2^{(n+s-N)md} 2^{(m-(j_P\vee 0)+2)nd} \lesssim \|f\|^d_{L^p_\tau(\mathbb{R}^n)}$$

if

$$N > \max\{2n+s,\, 2n+s+nd(\tau-1)\}.$$

Thus, (i) is proved.

Step 2. Let $p\in(0,1)$. For $t>1$ we have

$$a_t(x) \lesssim |f(x)| + t^{n\tau-n} \|f\|_{L^1_\tau(\mathbb{R}^n)}. \tag{4.17}$$

Substep 2.1. Small cubes. Let $|P| < 1$. By using (4.17) the estimate of $T_{N,P}$ is split into two parts. According to the first one we have

$$\sum_{m=(j_P\vee 0)}^{\infty} \frac{2^{(n+s-N)md}}{|P|^{\tau d}} \left\{ \int_1^{2^{m-(j_P\vee 0)+2}} t^{-sq} \left(\int_P |f(x)|^p\, dx \right)^{\frac{q}{p}} \frac{dt}{t} \right\}^{\frac{d}{q}}$$

$$\leq \|f\|^d_{L^p_\tau(\mathbb{R}^n)} \sum_{m=j_P}^{\infty} \frac{2^{(n+s-N)md}}{|P|^{\tau d}} \left\{ \int_1^{2^{m-j_P+2}} t^{-sq} \frac{dt}{t} \right\}^{\frac{d}{q}}$$

$$\lesssim \|f\|^d_{L^p_\tau(\mathbb{R}^n)}$$

if $N > s+n+n\tau$ and $s > 0$. Now we turn to the estimate of the second part. Let $\varepsilon > 0$. We obtain

$$\sum_{m=(j_P\vee 0)}^{\infty} \frac{2^{(n+s-N)md}}{|P|^{\tau d}} \left\{ \int_1^{2^{m-(j_P\vee 0)+2}} t^{-sq} \left(\int_P [t^{n\tau-n} \|f\|_{L^1_\tau(\mathbb{R}^n)}]^p\, dx \right)^{\frac{q}{p}} \frac{dt}{t} \right\}^{\frac{d}{q}}$$

$$\lesssim \|f\|^d_{L^1_\tau(\mathbb{R}^n)} \sum_{m=j_P}^{\infty} \frac{2^{(n+s-N)md}}{|P|^{\tau d}} |P|^{d/p} \left\{ \int_1^{2^{m-j_P+2}} t^{(-s+n\tau-n)q} \frac{dt}{t} \right\}^{\frac{d}{q}}$$

$$\lesssim \|f\|^d_{L^1_\tau(\mathbb{R}^n)} \frac{|P|^{d/p}}{|P|^{\tau d}} \sum_{m=j_P}^{\infty} 2^{(n+s-N)md}\, 2^{d(m-j_P)[(n\tau-s-n)_+ +\varepsilon]}$$

$$\lesssim \|f\|^d_{L^1_\tau(\mathbb{R}^n)} 2^{-dj_P(\frac{1}{p}-\tau)n} 2^{-dj_P[(n\tau-s-n)_+ +\varepsilon]} 2^{j_P[n+s-N+(n\tau-s-n)_+ +\varepsilon]d}$$

$$\lesssim \|f\|^d_{L^1_\tau(\mathbb{R}^n)}$$

if

$$N > \max\left\{ n+s+(n\tau-s-n)_+ +\varepsilon,\ n+s-\frac{n}{p}+\frac{n}{\tau} \right\}.$$

Substep 2.2. Large cubes. Let $|P| = 2^{rn}$ for some $r \in \mathbb{N}$. If $N > s+n > n$, then the first part (related to $|f(x)|$, see (4.17)) can be estimated from above by $\lesssim \|f\|^d_{L^p_\tau(\mathbb{R}^n)}$. For the estimate of the second part we argue as in Substep 2.1 and find

$$\sum_{m=r}^{\infty} \frac{2^{(n+s-N)md}}{|P|^{\tau d}} \left\{ \int_1^{2^{m+2}} t^{-sq} \left(\int_P [t^{n\tau-n} \|f\|_{L^1_\tau(\mathbb{R}^n)}]^p\, dx \right)^{\frac{q}{p}} \frac{dt}{t} \right\}^{\frac{d}{q}}$$

$$\lesssim \|f\|^d_{L^1_\tau(\mathbb{R}^n)} \frac{|P|^{d/p}}{|P|^{\tau d}} \sum_{m=r}^{\infty} 2^{(n+s-N)md} \left\{ \int_1^{2^{m+2}} t^{(-s+n\tau-n)q} \frac{dt}{t} \right\}^{\frac{d}{q}}$$

$$\lesssim \|f\|^d_{L^1_\tau(\mathbb{R}^n)} 2^{rnd(\frac{1}{p}-\tau)} \sum_{m=r}^{\infty} 2^{(n+s-N)md}\, 2^{md[(n\tau-s-n)_+ +\varepsilon]}$$

$$\lesssim \|f\|^d_{L^1_\tau(\mathbb{R}^n)},$$

if

$$N > \max\left\{n+s+(n\tau-s-n)_{+}+\varepsilon, n+s+(n\tau-s-n)_{+}+\varepsilon+\frac{n}{p}-n\tau\right\}.$$

It remains to estimate $\sum_{m=0}^{r}$. Here we need the assumption $\tau \geq 1/p$. Then, arguing as before, we find

$$\begin{aligned}
&\sum_{m=0}^{r} \frac{2^{(n+s-N)md}}{|P|^{\tau d}} \left\{\int_1^{2^{m+2}} t^{-sq} \left(\int_P [t^{n\tau-n} \|f\|_{L_\tau^1(\mathbb{R}^n)}]^p \, dx\right)^{\frac{q}{p}} \frac{dt}{t}\right\}^{\frac{d}{q}} \\
&\lesssim \|f\|_{L_\tau^1(\mathbb{R}^n)}^d \sum_{m=0}^{r} 2^{(n+s-N)md} \left\{\int_1^{2^{m+2}} t^{(-s+n\tau-n)q} \frac{dt}{t}\right\}^{\frac{d}{q}} \\
&\lesssim \|f\|_{L_\tau^1(\mathbb{R}^n)}^d \sum_{m=0}^{r} 2^{(n+s-N)md} 2^{md[(n\tau-s-n)_{+}+\varepsilon]} \\
&\lesssim \|f\|_{L_\tau^1(\mathbb{R}^n)}^d, \qquad (4.18)
\end{aligned}$$

if

$$N > n+s+(n\tau-s-n)_{+}+\varepsilon.$$

The estimate of $\widetilde{T}_{N,P}$ can be done in the same way. This finishes the proof of (ii).

Step 3. Let $p \in (0,1)$ and assume $\tau \in [0,\infty)$. We only need to modify the estimate (4.18). To all dyadic cubes P of sidelength $l(P) \geq 1$ we associate to f an extension $\mathscr{E}_P f$ such that $\mathscr{E}_P f$ denotes an extension of the restriction of f to $2P$ and

$$\|\mathscr{E}_P f\|_{B_{p,\infty}^{s_0}(\mathbb{R}^n)} \leq 2\,\|\mathscr{E}_P f\|_{B_{p,\infty}^{s_0}(2P)}.$$

For the definition of $B_{p,\infty}^{s_0}(2P)$ we refer to Sect. 6.4. Then

$$\begin{aligned}
&\sum_{m=0}^{r} \frac{2^{(n+s-N)md}}{|P|^{\tau d}} \left\{\int_1^{2^{m+2}} t^{-sq} \left(\int_P [a_t(x)]^p \, dx\right)^{\frac{q}{p}} \frac{dt}{t}\right\}^{\frac{d}{q}} \\
&= \sum_{m=0}^{r} \frac{2^{(n+s-N)md}}{|P|^{\tau d}} \left\{\int_1^{2^{m+2}} t^{-sq}\right. \\
&\qquad \times \left.\left(\int_P \left[t^{-n} \int_{t/2<|h|<t} |\Delta_h^M(\mathscr{E}_P f)(x)|\,dh\right]^p dx\right)^{\frac{q}{p}} \frac{dt}{t}\right\}^{\frac{d}{q}}.
\end{aligned}$$

Next we apply $t \geq 1$ and the known characterizations of $B_{p,q}^{s_0}(\mathbb{R}^n)$ by differences; see [146, Theorem 3.5.3]. It follows that

$$\left\{\int_1^{2^{m+2}} t^{-sq} \left(\int_P \left[t^{-n} \int_{t/2<|h|<t} |\Delta_h^M(\mathscr{E}_P f)(x)|\,dh\right]^p dx\right)^{\frac{q}{p}} \frac{dt}{t}\right\}^{\frac{1}{q}}$$

$$\lesssim \sup_{1<t<2^{m+2}} t^{-s_0} \left(\int_P \left[t^{-n} \int_{t/2<|h|<t} |\Delta_h^M(\mathscr{E}_P f)(x)|\, dh \right]^p dx \right)^{\frac{1}{p}}$$
$$\lesssim \|\mathscr{E}_P f\|_{B^{s_0}_{p,\infty}(\mathbb{R}^n)}$$
$$\lesssim \|f\|_{B^{s_0}_{p,\infty}(2P)}.$$

This proves the claim. □

Finally, we deal with a supplement of Lemma 4.3.

Lemma 4.5. *Let* $p, q \in (0,\infty]$ *and* $\tau \in [1/p,\infty)$. *Then*

$$\sup_{\{Q\in\mathscr{Q}:P\subset Q\}} \frac{1}{|Q|^\tau} \left\{ \int_0^{2(l(Q)\wedge 1)} t^{-sq} \left(\int_Q [a_t(x)]^p\, dx \right)^{q/p} \frac{dt}{t} \right\}^{1/q}$$
$$\lesssim \sup_{\{Q\in\mathscr{Q}:|Q|\le 1\}} \frac{1}{|Q|^\tau} \left\{ \int_0^{2l(Q)} t^{-sq} \left(\int_Q [a_t(x)]^p\, dx \right)^{q/p} \frac{dt}{t} \right\}^{1/q}$$

and

$$\sup_{\{Q\in\mathscr{Q}:P\subset Q\}} \frac{1}{|Q|^\tau} \left\{ \int_Q \left(\int_0^{2(l(Q)\wedge 1)} t^{-sq} [a_t(x)]^q \frac{dt}{t} \right)^{p/q} dx \right\}^{1/p}$$
$$\lesssim \sup_{\{Q\in\mathscr{Q}:|Q|\le 1\}} \frac{1}{|Q|^\tau} \left\{ \int_Q \left(\int_0^{2l(Q)} t^{-sq} [a_t(x)]^q \frac{dt}{t} \right)^{p/q} dx \right\}^{1/p}$$

hold for all $f \in L^{\overline{p}}_\tau(\mathbb{R}^n)$ *and all* $P \in \mathscr{Q}$.

Proof. Let Q be a dyadic cube such that $|Q| = 2^{rn}$ for some $r \in \mathbb{N}$. Then the integral with respect to t extends over the interval $(0,2)$, i.e., is independent of Q itself. In such a situation we may argue as in proof of Lemma 2.2. □

4.1.3 Means of Differences

We consider different types of means of differences. We set

$$\|f\|^{\clubsuit}_{B^{s,\tau}_{p,q}(\mathbb{R}^n)} \equiv \sup_{P\in\mathscr{Q}} \frac{1}{|P|^\tau} \left\{ \int_0^{2(l(P)\wedge 1)} t^{-sq} \left(\int_P [a_t(x)]^p\, dx \right)^{q/p} \frac{dt}{t} \right\}^{1/q},$$

$$\|f\|^{\clubsuit}_{F^{s,\tau}_{p,q}(\mathbb{R}^n)} \equiv \sup_{P\in\mathscr{Q}} \frac{1}{|P|^\tau} \left\{ \int_P \left(\int_0^{2(l(P)\wedge 1)} t^{-sq} [a_t(x)]^q \frac{dt}{t} \right)^{p/q} dx \right\}^{1/p},$$

$$\|f\|^{\heartsuit}_{B^{s,\tau}_{p,q}(\mathbb{R}^n)} \equiv \sup_{P\in\mathcal{Q}} \frac{1}{|P|^\tau} \left\{ \int_0^{2(l(P)\wedge 1)} t^{-sq} \right.$$
$$\left. \times \left(\int_P t^{-n} \int_{t/2\le|h|<t} |\Delta_h^M f(x)|^p \, dh \, dx \right)^{q/p} \frac{dt}{t} \right\}^{1/q},$$
$$\|f\|^{\heartsuit}_{F^{s,\tau}_{p,q}(\mathbb{R}^n)} \equiv \sup_{P\in\mathcal{Q}} \frac{1}{|P|^\tau} \left\{ \int_P \left(\int_0^{2(l(P)\wedge 1)} t^{-sq} t^{-n} \right.\right.$$
$$\left.\left. \times \int_{t/2\le|h|<t} |\Delta_h^M f(x)|^q \, dh \frac{dt}{t} \right)^{p/q} dx \right\}^{1/p},$$

and

$$\|f\|^{\spadesuit}_{B^{s,\tau}_{p,q}(\mathbb{R}^n)} \equiv \sup_{P\in\mathcal{Q}} \frac{1}{|P|^\tau} \left\{ \int_0^{2(l(P)\wedge 1)} t^{-sq} \sup_{t/2\le|h|<t} \left(\int_P |\Delta_h^M f(x)|^p \, dx \right)^{q/p} \frac{dt}{t} \right\}^{1/q},$$
$$\|f\|^{\spadesuit}_{F^{s,\tau}_{p,q}(\mathbb{R}^n)} \equiv \sup_{P\in\mathcal{Q}} \frac{1}{|P|^\tau} \left\{ \int_P \left(\int_0^{2(l(P)\wedge 1)} t^{-sq} \sup_{t/2\le|h|<t} |\Delta_h^M f(x)|^q \frac{dt}{t} \right)^{p/q} dx \right\}^{1/p}.$$

The following conclusion is an immediate corollary of Hölder's inequality.

Lemma 4.6. *Let $s \in [0,\infty)$ and $\tau \in [0,\infty)$.*

(i) Let $p \in [1,\infty]$ and $q \in (0,\infty]$. Then

$$\|f\|^{\clubsuit}_{B^{s,\tau}_{p,q}(\mathbb{R}^n)} \le \|f\|^{\heartsuit}_{B^{s,\tau}_{p,q}(\mathbb{R}^n)} \le \|f\|^{\spadesuit}_{B^{s,\tau}_{p,q}(\mathbb{R}^n)}.$$

(ii) Let $p \in (0,\infty)$ and $q \in [1,\infty]$. Then

$$\|f\|^{\clubsuit}_{F^{s,\tau}_{p,q}(\mathbb{R}^n)} \le \|f\|^{\heartsuit}_{F^{s,\tau}_{p,q}(\mathbb{R}^n)} \le \|f\|^{\spadesuit}_{F^{s,\tau}_{p,q}(\mathbb{R}^n)}.$$

4.2 Characterizations by Wavelets

The main aim of this section consists in proving characterizations of our spaces $A^{s,\tau}_{p,q}(\mathbb{R}^n)$ in terms of wavelet coefficients. In addition we prepare the characterization of $A^{s,\tau}_{p,q}(\mathbb{R}^n)$ via differences. To begin with we recall some basics of wavelet theory.

4.2.1 Wavelets and Besov-Triebel-Lizorkin Spaces

Wavelet bases in Besov and Triebel-Lizorkin spaces are a well-developed concept. We refer to the monographs of Meyer [99], Wojtasczyk [156] and Triebel [148, 149] for the general n-dimensional case (for the one-dimensional case we refer to the books of Hernandez and Weiss [72], Kahane and Lemarie-Rieuseut [82] and the article of Bourdaud [18]). Let $\widetilde{\phi}$ be an *orthonormal scaling function* on $\mathbb{R}$ with compact support and of sufficiently high regularity. Let $\widetilde{\psi}$ be the *corresponding orthonormal wavelet*. Then the *tensor product ansatz* yields a scaling function ϕ and associated wavelets $\psi_1, \cdots, \psi_{2^n-1}$, all defined now on $\mathbb{R}^n$; see, e.g., [156, Proposition 5.2]. We suppose

$$\phi \in C^{N_1}(\mathbb{R}^n) \quad \text{and} \quad \operatorname{supp} \phi \subset [-N_2, N_2]^n \tag{4.19}$$

for certain natural numbers N_1 and N_2. This implies

$$\psi_i \in C^{N_1}(\mathbb{R}^n) \quad \text{and} \quad \operatorname{supp} \psi_i \subset [-N_3, N_3]^n, \quad i = 1, \cdots, 2^n - 1 \tag{4.20}$$

for some $N_3 \in \mathbb{N}$. For $k \in \mathbb{Z}^n$, $j \in \mathbb{Z}_+$ and $i = 1, \cdots, 2^n - 1$, we shall use the standard abbreviations in this context:

$$\phi_{j,k}(x) \equiv 2^{jn/2}\phi(2^j x - k) \quad \text{and} \quad \psi_{i,j,k}(x) \equiv 2^{jn/2}\psi_i(2^j x - k), \quad x \in \mathbb{R}^n.$$

Furthermore, it is well known that

$$\int_{\mathbb{R}^n} \psi_{i,j,k}(x) x^\gamma \, dx = 0 \qquad \text{if} \qquad |\gamma| \le N_1$$

(see [156, Proposition 3.1]) and

$$\{\phi_{0,k} : k \in \mathbb{Z}^n\} \cup \{\psi_{i,j,k} : k \in \mathbb{Z}^n, \ j \in \mathbb{Z}_+, \ i = 1, \cdots, 2^n - 1\} \tag{4.21}$$

yields an *orthonormal basis* of $L^2(\mathbb{R}^n)$; see [99, Sect. 3.9] or [148, Sect. 3.1].

Recall that for $p, q \in (0, \infty]$, any function $f \in B^s_{p,q}(\mathbb{R}^n)$ when $s \in (\sigma_p, N_1)$ or $f \in F^s_{p,q}(\mathbb{R}^n)$ when $s \in (\sigma_{p,q}, N_1)$ admits a representation

$$f = \sum_{k \in \mathbb{Z}^n} a_k \phi_{0,k} + \sum_{i=1}^{2^n-1} \sum_{j \in \mathbb{Z}_+} \sum_{k \in \mathbb{Z}^n} a_{i,j,k} \psi_{i,j,k} \tag{4.22}$$

in $\mathscr{S}'(\mathbb{R}^n)$, where $a_k \equiv \langle f, \phi_{0,k} \rangle$ and $a_{i,j,k} \equiv \langle f, \psi_{i,j,k} \rangle$ (here it will be sufficient to interpret $\langle \cdot, \cdot \rangle$ as scalar product in $L^2(\mathbb{R}^n)$, since the functions $\phi_{0,k}$ and $\psi_{i,j,k}$ are compactly supported continuous functions and $f \in L^1_{\text{loc}}(\mathbb{R}^n)$). Moreover,

$$\|f\|_{B^s_{p,q}(\mathbb{R}^n)} \sim \left(\sum_{k \in \mathbb{Z}^n} |a_k|^p \right)^{1/p} + \left[\sum_{i=1}^{2^n-1} \sum_{j \in \mathbb{Z}_+} 2^{j(s+n/2)q} \left(\sum_{k \in \mathbb{Z}^n} 2^{-jn} |a_{i,j,k}|^p \right)^{q/p} \right]^{1/q},$$

and

$$\|f\|_{F^s_{p,q}(\mathbb{R}^n)} \sim \left\| \left(\sum_{k\in\mathbb{Z}^n} |a_k \widetilde{\chi}_{Q_{0k}}|^q \right)^{1/q} + \left[\sum_{i=1}^{2^n-1} \sum_{j\in\mathbb{Z}_+} 2^{jsq} \left(\sum_{k\in\mathbb{Z}^n} |a_{i,j,k} \widetilde{\chi}_{Q_{jk}}| \right)^q \right]^{1/q} \right\|_{L^p(\mathbb{R}^n)}$$

in the sense of equivalent quasi-norms; see, e. g., [149, Theorem 1.20]. In fact, even more is true. If, for a function $f \in L^{\max\{1,p\}}(\mathbb{R}^n)$, the right-hand side is finite, then this function belongs to $A^s_{p,q}(\mathbb{R}^n)$.

For each $j \in \mathbb{Z}$, define a projection W_j by setting

$$W_j f \equiv \sum_{k\in\mathbb{Z}^n} \langle f, \phi_{j,k} \rangle \phi_{j,k} \text{ if } j > 1 \quad \text{and} \quad W_j f \equiv \sum_{k\in\mathbb{Z}^n} \langle f, \phi_{0,k} \rangle \phi_{0,k} \text{ if } j \le 0. \tag{4.23}$$

Then these functions have a second representation

$$W_j f = \sum_{k\in\mathbb{Z}^n} \langle f, \phi_{0,k} \rangle \phi_{0,k} + \sum_{i=1}^{2^n-1} \sum_{t=0}^{j-1} \sum_{k\in\mathbb{Z}^n} \langle f, \psi_{i,t,k} \rangle \psi_{i,t,k}, \tag{4.24}$$

where, when $j < 1$, the second summation in the right part of (4.24) is void.

For any functions $f \in L^{\overline{p}}_\tau(\mathbb{R}^n)$ with $p \in (0,\infty)$, we have the convergence of the wavelet expansions in the following sense:

$$\lim_{j\to\infty} \|f - W_j f\|_{L^p(Q)} = 0$$

for any dyadic cube Q; see [156, Theorem 8.4].

4.2.2 *Estimates of Mean-Values of Differences by Wavelet Coefficients*

To estimate $\|f\|^{\clubsuit}_{A^{s,\tau}_{p,q}(\mathbb{R}^n)}$, $\|f\|^{\heartsuit}_{A^{s,\tau}_{p,q}(\mathbb{R}^n)}$ and $\|f\|^{\spadesuit}_{A^{s,\tau}_{p,q}(\mathbb{R}^n)}$ in terms of wavelet coefficients, we need some more abbreviations. For any $P \in \mathcal{Q}$ we set

$$\|f\|_{B^{\clubsuit}(P)} \equiv \frac{1}{|P|^\tau} \left\{ \int_0^{2(l(P)\wedge 1)} t^{-sq} \times \left[\int_P \left(t^{-n} \int_{t/2\le|h|<t} |\Delta_h^M f(x)|\, dh \right)^p dx \right]^{q/p} \frac{dt}{t} \right\}^{1/q},$$

$$\|f\|_{F^{\clubsuit}(P)} \equiv \frac{1}{|P|^{\tau}} \left\{ \int_P \left[\int_0^{2(l(P)\wedge 1)} t^{-sq} \right.\right.$$
$$\left.\left. \times \left(t^{-n} \int_{t/2 \le |h| < t} |\Delta_h^M f(x)|\, dh \right)^q \frac{dt}{t} \right]^{p/q} dx \right\}^{1/p},$$

$$\|f\|_{B^{\spadesuit}(P)} \equiv \frac{1}{|P|^{\tau}} \left\{ \int_0^{2(l(P)\wedge 1)} t^{-sq} \sup_{t/2 \le |h| < t} \left(\int_P |\Delta_h^M f(x)|^p \, dx \right)^{q/p} \frac{dt}{t} \right\}^{1/q}$$

and

$$\|f\|_{F^{\spadesuit}(P)} \equiv \frac{1}{|P|^{\tau}} \left\{ \int_P \left(\int_0^{2(l(P)\wedge 1)} t^{-sq} \sup_{t/2 \le |h| < t} |\Delta_h^M f(x)|^q \frac{dt}{t} \right)^{p/q} dx \right\}^{1/p}.$$

Similarly, $\|f\|_{B^{\heartsuit}(P)}$ and $\|f\|_{F^{\heartsuit}(P)}$ are defined. We also use the abbreviations that for $Q = Q_{jk} \in \mathscr{Q}$ and $m \in \mathbb{Z}_+$,

$$J_Q \equiv \{r \in \mathbb{Z}^n : |\operatorname{supp} \phi_{0,r} \cap Q| > 0\},$$
$$I_{Q,m} \equiv \{r \in \mathbb{Z}^n : \text{there exists}\, i \in \{1, \cdots, 2^n - 1\} \text{ such that } |\operatorname{supp} \psi_{i,m,r} \cap Q| > 0\},$$

where $|\cdot|$ denotes the Lebesgue measure in $\mathbb{R}^n$. Let $|J_Q|$ and $|I_{Q,m}|$ denote the *cardinalities* of these sets. It is easy to check that there exists a positive constant $C \equiv C(N_2, N_3)$ such that

$$|J_Q| \le C \max(1, |Q|) \qquad \text{and} \qquad |I_{Q,m}| \le C \max(1, 2^{mn}|Q|). \tag{4.25}$$

For $Q = Q_{jk}$ and $m \in \mathbb{Z}_+$, we set

$$\mathscr{I}_{Q,m} \equiv \bigcup_{|l-k| \le M} I_{Q_{jl},m} \quad \text{and} \quad \mathscr{J}_Q \equiv \bigcup_{|l-k| \le M} J_{Q_{jl}}.$$

The natural number M will be fixed later on.

By taking into account the wavelet characterizations of the preceding subsection it is not difficult to obtain local estimates of the difference $f - W_j f$.

Lemma 4.7. *Let $\tau \in [0, \infty)$, $p \in (0, \infty)$, $q \in (0, \infty]$ and $Q \equiv 2^{-j_Q}([0,1)^n + k_Q)^n$.*

(i) If $\sigma_p < s < M \le N_1$, then there exists a positive constant C such that for all $f \in L_{\tau}^{\overline{p}}(\mathbb{R}^n)$,

$$\|f - W_{j_Q} f\|_{B^{\spadesuit}(Q)}$$
$$\le C \frac{1}{|Q|^{\tau}} \left\{ \sum_{\omega = (j_Q \vee 0)}^{\infty} 2^{\omega(s+n/2)q} \sum_{i=1}^{2^n - 1} \left(2^{-\omega n} \sum_{r \in \mathscr{I}_{Q,\omega}} |a_{i,\omega,r}|^p \right)^{q/p} \right\}^{1/q}. \tag{4.26}$$

The inequality (4.26) *remains to hold if* $\|f - W_{j_Q}f\|_{B^{\spadesuit}(Q)}$ *is replaced by either*

$$\|f - W_{j_Q}f\|_{B^{\clubsuit}(Q)} \quad or \quad \|f - W_{j_Q}f\|_{B^{\heartsuit}(Q)}.$$

(ii) If $\sigma_{p,q} < s < M \le N_1$*, then there exists a positive constant* C *such that for all* $f \in L^{\overline{p}}_{\tau}(\mathbb{R}^n)$,

$$\|f - W_{j_Q}f\|_{F^{\clubsuit}(Q)} \le C\frac{1}{|Q|^{\tau}}\left\|\left[\sum_{\omega=(j_Q\vee 0)}^{\infty}\sum_{i=1}^{2^n-1}\sum_{r\in\mathscr{I}_{Q,\omega}} 2^{wsq}|a_{i,\omega,r}\widetilde{\chi}_{Q_{\omega r}}|^q\right]^{1/q}\right\|_{L^p(Q)}. \quad (4.27)$$

The inequality (4.27) *remains to hold if* $\|f - W_{j_Q}f\|_{F^{\clubsuit}(Q)}$ *is replaced by*

$$\|f - W_{j_Q}f\|_{F^{\heartsuit}(Q)}.$$

(iii) If $n/\min\{p,q\} < s < M \le N_1$*, then the inequality* (4.27) *remains to hold if* $\|f - W_{j_Q}f\|_{F^{\clubsuit}(Q)}$ *is replaced by* $\|f - W_{j_Q}f\|_{F^{\spadesuit}(Q)}$.

Proof. Let $f \in L^{\overline{p}}_{\tau}(\mathbb{R}^n)$. For convenience we put $a_{i,\omega,r} \equiv 0$ if $\omega < 0$. Then, by (4.22) and (4.24), we have

$$f(x) - W_{j_Q}f(x) = \sum_{i=1}^{2^n-1}\sum_{w=j_Q}^{\infty}\sum_{r\in\mathscr{I}_{Q,\omega}} a_{i,\omega,r}\psi_{i,\omega,r}(x),$$

valid in $L^p(\Omega)$, where $\Omega \equiv \cup_{|l-k_Q|\le M} Q_{j_Q l}$. For simplicity we denote the function on the right-hand side of the above formula by $g(x)$.

Step 1. Proof of (i). By making use of the characterization of $B^s_{p,q}(\mathbb{R}^n)$ by differences (see [146, Theorem 2.6.1]), the wavelet characterization of $B^s_{p,q}(\mathbb{R}^n)$ (see Sect. 4.2.1), and the fact that (4.21) is an orthonormal basis of $L^2(\mathbb{R}^n)$, we find that

$$\begin{aligned}
&\|f - W_{j_Q}f\|_{B^{\clubsuit}(Q)}\\
&= \frac{1}{|Q|^{\tau}}\left\{\int_0^{2(l(Q)\wedge 1)} t^{-sq}\sup_{t/2\le|h|<t}\left(\int_P |\Delta_h^M g(x)|^p\,dx\right)^{q/p}\frac{dt}{t}\right\}^{1/q}\\
&\le \frac{1}{|Q|^{\tau}}\|g\|_{B^s_{p,q}(\mathbb{R}^n)}\\
&\sim \frac{1}{|Q|^{\tau}}\left(\sum_{k\in\mathbb{Z}^n}|\langle g, \phi_{0k}\rangle|^p\right)^{1/p}
\end{aligned}$$

$$+ \frac{1}{|Q|^\tau}\left[\sum_{\omega\in\mathbb{Z}_+} 2^{w(s+n/2)q}\sum_{i=1}^{2^n-1}\left(\sum_{k\in\mathbb{Z}^n} 2^{-\omega n}|\langle g, \psi_{i,\omega,k}\rangle|^p\right)^{q/p}\right]^{1/q}$$

$$\sim \frac{1}{|Q|^\tau}\left[\sum_{\omega=(j_Q\vee 0)}^{\infty} 2^{w(s+n/2)q}\sum_{i=1}^{2^n-1}\left(\sum_{r\in\mathscr{I}_{Q,\omega}} 2^{-\omega n}|a_{i,\omega,r}|^p\right)^{q/p}\right]^{1/q}.$$

For $\|f-W_{j_Q}f\|_{B^\clubsuit(Q)}$ and for $\|f-W_{j_Q}f\|_{B^\heartsuit(Q)}$, by [146, Theorems 2.6.1/3.5.3], the above argument is also feasible.

Step 2. Proof of (ii). Similarly, by the same group of arguments (for the characterization of $F^s_{p,q}(\mathbb{R}^n)$ by differences see [146, Theorem 3.5.3]; for the wavelet characterization of $F^s_{p,q}(\mathbb{R}^n)$ see [149, Theorem 1.20]), we have

$$\begin{aligned}
&\|f-W_{j_Q}f\|_{F^\clubsuit(Q)}\\
&= \frac{1}{|Q|^\tau}\left\{\int_Q\left[\int_0^{2(l(Q)\wedge 1)} t^{-sq}\left(t^{-n}\int_{t/2\le|h|<t}|\Delta_h^M g(x)|\,dh\right)^q\frac{dt}{t}\right]^{p/q}dx\right\}^{1/p}\\
&\le \frac{1}{|Q|^\tau}\|g\|_{F^s_{p,q}(\mathbb{R}^n)}\\
&\sim \frac{1}{|Q|^\tau}\left\{\left\|\left(\sum_{k\in\mathbb{Z}^n}|\langle g, \phi_{0k}\rangle\widetilde{\chi}_{Q_{0k}}|^q\right)^{1/q}\right.\right.\\
&\qquad+\left.\left.\left[\sum_{\omega\in\mathbb{Z}_+}\sum_{i=1}^{2^n-1}\sum_{k\in\mathbb{Z}^n}2^{wsq}|\langle g, \psi_{i,\omega,k}\rangle\widetilde{\chi}_{Q_{\omega k}}|^q\right]^{1/q}\right\|_{L^p(\mathbb{R}^n)}\right\}\\
&\sim \frac{1}{|Q|^\tau}\left\|\left[\sum_{\omega=(j_Q\vee 0)}^{\infty}\sum_{i=1}^{2^n-1}\sum_{r\in\mathscr{I}_{Q,\omega}}2^{wsq}|a_{i,\omega,r}\widetilde{\chi}_{Q_{\omega r}}|^q\right]^{1/q}\right\|_{L^p(Q)}.
\end{aligned}$$

By [146, Theorem 2.6.2], we know that the above argument is also feasible for $\|f-W_{j_Q}f\|_{F^\heartsuit(Q)}$, which completes the proof of Lemma 4.7.

Step 3. Proof of (iii). We can argue as before but this time the restriction for the characterizations by differences is different; see [146, Theorems 2.6.2/3.5.3].

Step 4. A technical remark. Of course, in all preceding steps we have to argue by starting with the finiteness of the right-hand side for a function f belonging to $L^{\bar{p}}_{\tau}(\mathbb{R}^n)$. This always implies that $g\in A^s_{p,q}(\mathbb{R}^n)$ as a consequence of the wavelet characterization recalled in Sect. 4.2.1. □

It remains to estimate the projections $W_j f$. For having a more compact notation we set

$$\|f\|_{B^{s,\tau}_{p,q}(\mathbb{R}^n)}^{\blacktriangle} \equiv \sup_{\{P\in\mathcal{Q}:\,|P|\geq 1\}} \frac{1}{|P|^\tau}\left(\sum_{k\in\mathcal{J}_P} |\langle f,\phi_{0,k}\rangle|^p\right)^{\frac{1}{p}}$$
$$+\sup_{P\in\mathcal{Q}} \frac{1}{|P|^\tau}\left\{\sum_{j=(j_P\vee 0)}^{\infty} 2^{j(s+n/2)q}\sum_{i=1}^{2^n-1}\left[\sum_{k\in\mathcal{I}_{P,j}} 2^{-jn}|\langle f,\psi_{i,j,k}\rangle|^p\right]^{\frac{q}{p}}\right\}^{\frac{1}{q}}$$

and

$$\|f\|_{F^{s,\tau}_{p,q}(\mathbb{R}^n)}^{\blacktriangle} \equiv \sup_{\{P\in\mathcal{Q}:\,|P|\geq 1\}} \frac{1}{|P|^\tau}\left(\sum_{k\in\mathcal{J}_P} |\langle f,\phi_{0,k}\rangle|^p\right)^{\frac{1}{p}}$$
$$+\sup_{P\in\mathcal{Q}} \frac{1}{|P|^\tau}\left\|\left[\sum_{j=(j_P\vee 0)}^{\infty}\sum_{i=1}^{2^n-1}\sum_{k\in\mathcal{I}_{P,j}} 2^{jsq}|\langle f,\psi_{i,j,k}\rangle\,\widetilde{\chi}_{Q_{jk}}|^q\right]^{\frac{1}{q}}\right\|_{L^p(P)}.$$

We then have the following conclusions.

Lemma 4.8. *Let* $\tau\in[0,\infty)$, $p\in(0,\infty)$, $q\in(0,\infty]$, $0<M\leq N_1$ *and* $Q\in\mathcal{Q}$.

(i) If $\sigma_p<s<\{M\wedge(M+n(1/p-\tau))\}$, *then there exists a positive constant C such that for all* $f\in L^{\overline{p}}_\tau(\mathbb{R}^n)$,

$$\|W_{j_Q}f\|_{B^\clubsuit(Q)}+\|W_{j_Q}f\|_{B^\heartsuit(Q)}+\|W_{j_Q}f\|_{B^\spadesuit(Q)}\leq C\|f\|_{B^{s,\tau}_{p,q}(\mathbb{R}^n)}^{\blacktriangle}.$$

(ii) If $\sigma_{p,q}<s<\{M\wedge(M+n(1/p-\tau))\}$, *then there exists a positive constant C such that for all* $f\in L^{\overline{p}}_\tau(\mathbb{R}^n)$,

$$\|W_{j_Q}f\|_{F^\clubsuit(Q)}+\|W_{j_Q}f\|_{F^\heartsuit(Q)}\leq C\|f\|_{F^{s,\tau}_{p,q}(\mathbb{R}^n)}^{\blacktriangle}.$$

(iii) If $n/\min\{p,q\}<s<\{M\wedge M+n(1/p-\tau)\}$, *then there exists a positive constant C such that for all* $f\in L^{\overline{p}}_\tau(\mathbb{R}^n)$,

$$\|W_{j_Q}f\|_{F^\spadesuit(Q)}\leq C\|f\|_{F^{s,\tau}_{p,q}(\mathbb{R}^n)}^{\blacktriangle}.$$

Proof. We shall concentrate on the terms $\|W_{j_Q}f(x)\|_{A^\spadesuit(Q)}$. By (4.24), for all $f\in L^{\overline{p}}_\tau(\mathbb{R}^n)$, we have

$$W_{j_Q}f(x)=\sum_{k\in\mathbb{Z}^n} a_k\phi_{0,k}(x)+\sum_{i=1}^{2^n-1}\sum_{j=0}^{j_Q-1}\sum_{k\in\mathcal{I}_{Q,j}} a_{i,j,k}\psi_{i,j,k}(x),$$

valid in $L^p(\cup_{|l-k_Q|\le M} Q_{j_Q l})$. Recall, if $j_Q < 1$, then the second summation on the right-hand side of above equality is void. Our main tool will be the elementary inequality

$$|\Delta_h^M g(x)| \lesssim |h|^M \sup_{|\alpha|=M} \sup_{|x-y|\le M|h|} |\partial^\alpha g(y)|, \tag{4.28}$$

valid for all functions $g \in C^M(\mathbb{R}^n)$. We shall apply this inequality with respect to the elements of our wavelet basis which belong to $C^{N_1}(\mathbb{R}^n)$ by assumption.

Step 1. Estimate of $\sum_{k\in\mathbb{Z}^n} a_k \phi_{0,k}$. Notice that $\Delta_h^M \phi_{0,k}(x) = 0$ for $x \in Q$ and $k \notin \mathscr{J}_Q$. We put

$$g(x) \equiv \sum_{k\in\mathscr{J}_Q} a_k \phi_{0,k}.$$

In analogy to the proof of Lemma 4.7 it follows that

$$\begin{aligned}
&\left\| \sum_{k\in\mathbb{Z}^n} a_k \phi_{0,k} \right\|_{B^\spadesuit(Q)} \\
&\le \frac{1}{|Q|^\tau} \left\{ \int_0^{2(l(Q)\wedge 1)} t^{-sq} \sup_{t/2\le|h|<t} \left(\int_Q |\Delta_h^M g(x)|^p \, dx \right)^{q/p} \frac{dt}{t} \right\}^{1/q} \\
&\le \frac{1}{|Q|^\tau} \|g\|_{B^s_{p,q}(\mathbb{R}^n)} \\
&\lesssim \frac{1}{|Q|^\tau} \left(\sum_{k\in\mathscr{J}_Q} |a_k|^p \right)^{1/p} .
\end{aligned}$$

This estimate will be applied for large cubes Q, i. e., $|Q| \ge 1$. If $|Q| < 1$, we argue as follows. The cardinality of $\mathscr{J}_Q$ is uniformly bounded, for simplicity say 1. Then

$$\begin{aligned}
\|g\|_{B^\spadesuit(Q)} &\le \frac{1}{|Q|^\tau} \left\{ \int_0^{2(l(Q)\wedge 1)} t^{-sq} \sup_{t/2\le|h|<t} \left(\int_Q |\Delta_h^M a_{k_Q} \phi_{0,k_Q}(x)|^p \, dx \right)^{q/p} \frac{dt}{t} \right\}^{1/q} \\
&\lesssim \frac{|Q|^{1/p}}{|Q|^\tau} |a_{k_Q}| \|\phi\|_{C^M(\mathbb{R}^n)} [l(Q)]^{M-s} \\
&\lesssim |a_{k_Q}|,
\end{aligned}$$

if $M + n/p - n\tau \ge s$. Summarizing we get

$$\left\| \sum_{k\in\mathbb{Z}^n} a_k \phi_{0,k} \right\|_{B^\spadesuit(Q)} \lesssim \sup_{\{P\in\mathscr{Q}:\, |P|\ge 1\}} \frac{1}{|P|^\tau} \left(\sum_{k\in\mathscr{J}_P} |a_k|^p \right)^{1/p} . \tag{4.29}$$

Similarly, for the Triebel-Lizorkin-type spaces, we also have

$$\left\| \sum_{k\in\mathbb{Z}^n} a_k \phi_{0,k} \right\|_{F^\spadesuit(Q)} \lesssim \sup_{\{P\in\mathscr{Q}:\, |P|\ge 1\}} \frac{1}{|P|^\tau} \left(\sum_{k\in\mathscr{J}_P} |a_k|^p \right)^{1/p} ,$$

if $n/\min\{p,q\} < s < M + n/p - n\tau$. Replacing $\|\sum_{k\in\mathbb{Z}^n} a_k\phi_{0,k}\|_{F^\spadesuit(Q)}$ either by $\|\sum_{k\in\mathbb{Z}^n} a_k\phi_{0,k}\|_{F^\clubsuit(Q)}$ or by $\|\sum_{k\in\mathbb{Z}^n} a_k\phi_{0,k}\|_{F^\heartsuit(Q)}$, the restrictions

$$\sigma_{p,q} < s < M + n/p - n\tau$$

are sufficient.

Step 2. Estimate of $\sum_{j=0}^{j_Q-1}\sum_{k\in\mathscr{I}_{Q,j}} a_{i,j,k}\psi_{i,j,k}$. Let $j_Q \geq 1$. For all $j \in \{0,\cdots, j_Q-1\}$, $i \in \{1,\cdots,2^n-1\}$ and $x \in \mathbb{R}^n$ we define

$$g_{ij}(x) \equiv \sum_{k\in\mathscr{I}_{Q,j}} a_{i,j,k}\psi_{i,j,k}(x).$$

Recall that $|\mathscr{I}_{Q,j}| \lesssim 2^{-[(j_Q-j)\wedge 0]n}$; see (4.25). By using (4.28) we obtain

$$\begin{aligned}
\left\|\sum_{j=0}^{j_Q-1} g_{ij}\right\|_{B^\spadesuit(Q)} &\lesssim \frac{|Q|^{1/p}}{|Q|^\tau}\left\{\int_0^{2l(Q)} t^{-sq}t^{Mq}\left(\sum_{j=0}^{j_Q-1}\sum_{k\in\mathscr{I}_{Q,j}} |a_{i,j,k}|2^{jn/2}2^{jM}\right)^q \frac{dt}{t}\right\}^{1/q}\\
&\lesssim |Q|^{(M-s)/n+1/p-\tau}\sum_{j=0}^{j_Q-1}\sum_{k\in\mathscr{I}_{Q,j}} 2^{j(M+n/2)}|a_{i,j,k}|\\
&\lesssim |Q|^{(M-s)/n+1/p-\tau}\sum_{j=0}^{j_Q-1} 2^{j(M-s+n/p-n\tau)}\\
&\quad\times\left(\sup_{j\in\{0,\cdots,j_Q-1\}}\sup_{k\in\mathscr{I}_{Q,j}} 2^{j(s+n/2+n\tau-n/p)}|a_{i,j,k}|\right)\\
&\lesssim \sup_{j\in\{0,\cdots,j_Q-1\}}\sup_{k\in\mathscr{I}_{Q,j}} 2^{j(s+n/2+n\tau-n/p)}|a_{i,j,k}|,
\end{aligned}$$

because of $M > s + n\tau - n/p$. Similarly,

$$\left\|\sum_{j=0}^{j_Q-1} g_{ij}\right\|_{F^\clubsuit(Q)} \lesssim \sup_{i\in\{1,\cdots,2^n-1\}}\sup_{j\in\{0,\cdots,j_Q-1\}}\sup_{k\in\mathscr{I}_{Q,j}} 2^{j(s+n/2+n\tau-n/p)}|a_{i,j,k}|.$$

Notice that $\mathscr{I}_{Q,j} \subset \mathscr{I}_{P,j}$ if $Q \subset P$. It is easy to check that

$$\begin{aligned}
&\sup_{i\in\{1,\cdots,2^n-1\}}\sup_{j\in\{0,\cdots,j_Q-1\}}\sup_{k\in\mathscr{I}_{Q,j}} 2^{j(s+n/2+n\tau-n/p)}|a_{i,j,k}|\\
&\lesssim \sup_{P\in\mathscr{Q}}\frac{1}{|P|^\tau}\left\{\sum_{j=(j_P\vee 0)}^{\infty} 2^{j(s+n/2)q}\sum_{i=1}^{2^n-1}\left[\sum_{k\in\mathscr{I}_{P,j}} 2^{-jn}|a_{i,j,k}|^p\right]^{q/p}\right\}^{1/q}
\end{aligned}$$

and

$$\sup_{i\in\{1,\cdots,2^n-1\}}\sup_{j\in\{0,\cdots,j_Q-1\}}\sup_{k\in\mathscr{I}_{Q,j}} 2^{j(s+n/2+n\tau-n/p)}|a_{i,j,k}|$$
$$\lesssim \sup_{P\in\mathscr{Q}}\frac{1}{|P|^\tau}\left\{\int_P\left[\sum_{j=(j_P\vee 0)}^{\infty}\sum_{i=1}^{2^n-1}\sum_{k\in\mathscr{I}_{P,j}}(2^{js}|a_{i,j,k}|\widetilde{\chi}_{Q_{jk}}(x))^q\right]^{p/q}dx\right\}^{1/q},$$

which completes the proof of Lemma 4.8. □

Lemma 4.9. *Let $p,q\in(0,\infty]$, $\tau\in[0,\infty)$ and $s\in(0,\infty)$. Then there exists a positive constant C such that for all $f\in L^{\overline{p}}_{\tau}(\mathbb{R}^n)$,*

$$\Phi_{\tau,p}(f)\equiv\sup_{\{P\in\mathscr{Q}:\,|P|\ge 1\}}\frac{1}{|P|^\tau}\left(\int_P|\Phi*f(x)|^p\,dx\right)^{1/p}\le C\|f\|_{\overset{\blacktriangle}{A}{}^{s,\tau}_{p,q}(\mathbb{R}^n)}.$$

Proof. By the wavelet expansion (4.22) of f, we see that

$$\frac{1}{|P|^\tau}\left(\int_P|\Phi*f(x)|^p\,dx\right)^{1/p}\lesssim\frac{1}{|P|^\tau}\left[\int_P\left(\sum_{k\in\mathbb{Z}^n}|a_k||\Phi*\phi_{0,k}(x)|\right.\right.$$
$$\left.\left.+\sum_{i=1}^{2^n-1}\sum_{j=0}^{\infty}\sum_{k\in\mathbb{Z}^n}|a_{i,j,k}||\Phi*\psi_{i,j,k}(x)|\right)^p dx\right]^{1/p}.$$

It is easy to check that for all $x\in P$ and $k\in\mathbb{Z}^n$,

$$|\Phi*\phi_{0,k}(x)|\lesssim\frac{1}{(1+|x-k|)^{n+\delta}},$$

where we choose

$$\delta>\max\{n[\tau\vee(1/p)-1],0\}.$$

This estimate combined with (2.11) when $p\le 1$ or Hölder's inequality when $p>1$ yields

$$\mathrm{A}_P\equiv\frac{1}{|P|^\tau}\left[\int_P\left(\sum_{k\in\mathbb{Z}^n}|a_k||\Phi*\phi_{0,k}(x)|\right)^p dx\right]^{1/p}$$
$$\lesssim\frac{1}{|P|^\tau}\left[\int_P\left(\sum_{k\in\mathbb{Z}^n}|a_k|\frac{1}{(1+|x-k|)^{n+\delta}}\right)^p dx\right]^{1/p}$$

$$\lesssim \frac{1}{|P|^\tau}\left[\sum_{k\in\mathbb{Z}^n}|a_k|^p\int_P\frac{1}{(1+|x-k|)^{(n+\delta)p}}\,dx\right]^{1/p}$$

$$\lesssim \frac{1}{|P|^\tau}\left[\sum_{m=0}^{\infty}\sum_{k\in J_{2^mP\setminus 2^{m-1}P}}|a_k|^p\int_P\frac{1}{(1+|x-k|)^{(n+\delta)p}}\,dx\right]^{1/p},$$

where when $m=0$, $J_{P\setminus 2^{-1}P}$ is replaced by J_P.

Notice that for all $x\in P$ and $k\in J_{2^mP\setminus 2^{m-1}P}$, $|x-k|\sim 2^m l(P)$, which implies that

$$\begin{aligned}
\mathrm{A}_P &\lesssim \frac{1}{|P|^\tau}\left[\sum_{m=0}^{\infty}2^{-m(n+\delta)p}\sum_{k\in J_{2^mP}}|a_k|^p|P|^{1-(n+\delta)p/n}\right]^{1/p}\\
&\lesssim \left(\sum_{m=0}^{\infty}2^{-m(n+\delta)p+mn\tau p}|P|^{1-(n+\delta)p/n}\right)^{1/p}\sup_{\{P\in\mathcal{Q}:\,|P|\ge 1\}}\frac{1}{|P|^\tau}\left(\sum_{k\in\mathcal{J}_P}|a_k|^p\right)^{1/p}\\
&\lesssim \sup_{\{P\in\mathcal{Q}:\,|P|\ge 1\}}\frac{1}{|P|^\tau}\left(\sum_{k\in\mathcal{J}_P}|a_k|^p\right)^{1/p}.
\end{aligned}$$

Similarly, we have that for all $i\in\{1,\cdots,2^n-1\}$, $j\in\mathbb{Z}_+$, $k\in\mathbb{Z}^n$ and $x\in\mathbb{R}^n$,

$$|\Phi*\psi_{i,j,k}(x)|\lesssim\frac{2^{-jn/2}}{(1+|x-2^{-j}k|)^{n+\delta}},$$

where we choose

$$\delta>\max\{0,n(1/p-1),n(\tau-1),n(\tau p-1)\}.$$

This estimate yields

$$\begin{aligned}
\mathrm{B}_P &\equiv \frac{1}{|P|^\tau}\left[\int_P\left(\sum_{i=1}^{2^n-1}\sum_{j=0}^{\infty}\sum_{k\in\mathbb{Z}^n}|a_{i,j,k}||\Phi*\psi_{i,j,k}(x)|\right)^p dx\right]^{1/p}\\
&\lesssim \frac{1}{|P|^\tau}\left[\int_P\left(\sum_{i=1}^{2^n-1}\sum_{j=0}^{\infty}\sum_{k\in\mathbb{Z}^n}|a_{i,j,k}|\frac{2^{-jn/2}}{(1+|x-2^{-j}k|)^{n+\delta}}\right)^p dx\right]^{1/p}.
\end{aligned}$$

We remark that for all $j\ge 0$, $k\in I_{2^mP\setminus 2^{m-1}P,j}$ and $x\in P$, $1+|x-2^{-j}k|\gtrsim 2^m l(P)$. Then for the case $p\le 1$, applying (2.11), we obtain

$$\mathrm{B}_P\lesssim\frac{1}{|P|^\tau}\left[\sum_{m=0}^{\infty}2^{-m(n+\delta)p}\sum_{i=1}^{2^n-1}\sum_{j=0}^{\infty}\sum_{k\in I_{2^mP\setminus 2^{m-1}P,j}}|a_{i,j,k}|^p2^{-jnp/2}|P|^{1-(n+\delta)p/n}\right]^{1/p}.$$

If $q \le p$, using (2.11) again yields

$$\mathrm{B}_P \lesssim \left[\sum_{m=0}^{\infty} 2^{-m(n+\delta)q+mn\tau q} \sup_{j\in\mathbb{Z}_+} 2^{-j[s-n(1/p-1)]q}\right]^{1/q}$$
$$\times \sup_{\{P\in\mathcal{Q}:|P|\ge1\}} \frac{1}{|P|^\tau}\left\{\sum_{j=0}^{\infty} 2^{j(s+n/2)q}\sum_{i=1}^{2^n-1}\left[\sum_{k\in\mathcal{I}_{P,j}} 2^{-jn}|a_{i,j,k}|^p\right]^{q/p}\right\}^{1/q}$$
$$\lesssim \sup_{\{P\in\mathcal{Q}:|P|\ge1\}} \frac{1}{|P|^\tau}\left\{\sum_{j=0}^{\infty} 2^{j(s+n/2)q}\sum_{i=1}^{2^n-1}\left[\sum_{k\in\mathcal{I}_{P,j}} 2^{-jn}|a_{i,j,k}|^p\right]^{q/p}\right\}^{1/q};$$

if $q < p$, using Hölder's inequality instead of (2.11) yields the same estimate.

For the case $p > 1$, let $\varepsilon \in (2/p-1-2s/n, 2/p-1)$. Then by Hölder's inequality, we see that

$$\mathrm{B}_P \lesssim \frac{1}{|P|^\tau}\left[\sum_{i=1}^{2^n-1}\sum_{j=0}^{\infty}\sum_{k\in\mathbb{Z}^n}|a_{i,j,k}|^p\int_P \frac{2^{-jn\varepsilon p/2}}{(1+|x-2^{-j}k|)^{n+\delta}}\,dx\right]^{1/p}$$
$$\lesssim \frac{1}{|P|^\tau}\left[\sum_{m=0}^{\infty} 2^{-m(n+\delta)}\sum_{i=1}^{2^n-1}\sum_{j=0}^{\infty}\sum_{k\in I_{2^mP\setminus 2^{m-1}P,j}} |a_{i,j,k}|^p 2^{-jn\varepsilon p/2}|P|^{-\delta/n}\right]^{1/p}.$$

Similarly, applying (2.11) when $q \le p$ or Hölder's inequality when $q > p$, we also obtain

$$\mathrm{B}_P \lesssim \sup_{\{P\in\mathcal{Q}:|P|\ge1\}} \frac{1}{|P|^\tau}\left\{\sum_{j=0}^{\infty} 2^{j(s+n/2)q}\sum_{i=1}^{2^n-1}\left[\sum_{k\in\mathcal{I}_{P,j}} 2^{-jn}|a_{i,j,k}|^p\right]^{q/p}\right\}^{1/q},$$

which together with the estimate of A_P yields the desired inequality for $B^{s,\tau}_{p,q}(\mathbb{R}^n)$.

To prove the desired inequality for $F^{s,\tau}_{p,q}(\mathbb{R}^n)$, notice that when $p \le q$, the desired inequality for $F^{s,\tau}_{p,q}(\mathbb{R}^n)$ is an immediate corollary of the above inequality and Minkowski's inequality. We only consider the remaining case $p > q$.

Notice that

$$\mathrm{B}_P \lesssim \frac{1}{|P|^\tau}\left[\int_P\left(\sum_{m=0}^{\infty}\sum_{i=1}^{2^n-1}\sum_{j=0}^{\infty}\sum_{k\in I_{2^mP\setminus 2^{m-1}P,j}}|a_{i,j,k}|\frac{2^{-jn/2}}{(1+|x-2^{-j}k|)^{n+\delta}}\right)^p dx\right]^{1/p}.$$

Then when $p \le 1$, applying (2.11), we have

$$\mathrm{B}_P \lesssim \frac{1}{|P|^\tau}\left[\sum_{m=0}^{\infty}\int_P\left(\sum_{i=1}^{2^n-1}\sum_{j=0}^{\infty}\sum_{k\in I_{2^mP\setminus 2^{m-1}P,j}}|a_{i,j,k}|\frac{2^{-jn/2}}{(1+|x-2^{-j}k|)^{n+\delta}}\right)^p dx\right]^{1/p},$$

and when $p > 1$, applying Minkowski's inequality, we obtain

$$\mathrm{B}_P \lesssim \frac{1}{|P|^\tau} \sum_{m=0}^{\infty} \left[\int_P \left(\sum_{i=1}^{2^n-1} \sum_{j=0}^{\infty} \sum_{k \in I_{2^m P \setminus 2^{m-1} P, j}} |a_{i,j,k}| \frac{2^{-jn/2}}{(1+|x-2^{-j}k|)^{n+\delta}} \right)^p dx \right]^{1/p}.$$

Recall that $1+|x-2^{-j}k| \gtrsim 2^m l(P)$ holds for all $j \geq 0$, $k \in I_{2^m P \setminus 2^{m-1} P, j}$ and $x \in P$. Let $a \in (0, \min\{1, p, q\})$. Then for all $x \in P$,

$$\begin{aligned}
&\sum_{k \in I_{2^m P \setminus 2^{m-1} P, j}} \frac{|a_{i,j,k}|^q 2^{-jnq/2}}{(1+|x-2^{-j}k|)^{(n+\delta)q}} \\
&\lesssim 2^{-m(n+\delta)q} [l(P)]^{-(n+\delta)q} 2^{-jnq/2} \left(\sum_{k \in I_{2^m P, j}} |a_{i,j,k}|^a \right)^{\frac{q}{a}} \\
&\lesssim 2^{-m(n+\delta)q} [l(P)]^{-(n+\delta)q} 2^{-jnq} (2^m |P|)^{q/a} \\
&\quad \times \left(\frac{1}{|2^m P|} \int_{2^m P} \sum_{k \in I_{2^m P, j}} |a_{i,j,k}|^a \widetilde{\chi}_{Q_{jk}}^a(y)\, dy \right)^{\frac{q}{a}} \\
&\lesssim 2^{-m(n+\delta)q} [l(P)]^{-(n+\delta)q} 2^{-jnq} (2^m |P|)^{q/a} \left[\mathrm{M} \left(\sum_{k \in I_{2^m P, j}} |a_{i,j,k}|^a \widetilde{\chi}_{Q_{jk}}^a \right)(x) \right].
\end{aligned}$$

Since $q < p$, using (2.11) and the Fefferman-Stein vector valued maximal inequality, we obtain the desired conclusion, and then complete the proof of Lemma 4.9. □

4.2.3 The Wavelet Characterization of $A_{p,q}^{s,\tau}(\mathbb{R}^n)$

After these preparations we are now in a position to turn to the wavelet characterization of our classes $A_{p,q}^{s,\tau}(\mathbb{R}^n)$. Let N_1 be as in (4.19). If $N_1 \geq \lfloor s+n\tau \rfloor$, then for all $k \in \mathbb{Z}^n$, $i = 1, \cdots, 2^n-1$ and $j \in \mathbb{Z}_+$, ϕ_{0k} and ψ_{ijk} are multiples of inhomogeneous smooth analysis $A_{p,q}^{s,\tau}(\mathbb{R}^n)$-molecules, supported respectively near Q_{0k} and Q_{jk}. Then Theorem 3.2(ii) tells us that for all $f \in A_{p,q}^{s,\tau}(\mathbb{R}^n)$,

$$\sup_{\{P \in \mathcal{Q}: |P| \geq 1\}} \frac{1}{|P|^\tau} \left\{ \sum_{\substack{k \in \mathbb{Z}^n \\ Q_{0k} \subset P}} |\langle f, \phi_{0,k} \rangle|^p \right\}^{1/p} \lesssim \|f\|_{A_{p,q}^{s,\tau}(\mathbb{R}^n)},$$

$$\|\{|\langle f, \psi_{i,j,k} \rangle|\}_{l(Q_{jk}) \leq 1}\|_{a_{p,q}^{s,\tau}(\mathbb{R}^n)} \lesssim \|f\|_{A_{p,q}^{s,\tau}(\mathbb{R}^n)}, \qquad i = 1, \ldots, 2^n-1,$$

which further implies that $\|f\|^{\blacktriangle}_{A^{s,\tau}_{p,q}(\mathbb{R}^n)} \lesssim \|f\|_{A^{s,\tau}_{p,q}(\mathbb{R}^n)}$. These estimates allow to close the circle which we started with Lemma 4.2. This yields the following wavelet characterization of $A^{s,\tau}_{p,q}(\mathbb{R}^n)$.

Theorem 4.1. *Let the generators ϕ and ψ of the wavelet system satisfy the conditions in (4.19), (4.20) with respect to $N_1, N_2, N_3 \in \mathbb{N}$. Let $s \in \mathbb{R}$ and $p, q \in (0,\infty]$.*

(i) *We further suppose $\sigma_p < s$, $\max\{s, s+n\tau-n/p\} < N_1$ and $0 \le \tau < \tau_{s,p}$. Then $f \in B^{s,\tau}_{p,q}(\mathbb{R}^n)$ if, and only if f is locally integrable and $\|f\|^{\blacktriangle}_{B^{s,\tau}_{p,q}(\mathbb{R}^n)} < \infty$. Further $\|f\|^{\blacktriangle}_{B^{s,\tau}_{p,q}(\mathbb{R}^n)}$ and $\|f\|_{B^{s,\tau}_{p,q}(\mathbb{R}^n)}$ are equivalent.*

(ii) *This time we suppose $p \in (0,\infty)$, $\sigma_{p,q} < s$, $\max\{s, s+n\tau-n/p\} < N_1$ and $0 \le \tau < \tau_{s,p,q}$. Then $f \in F^{s,\tau}_{p,q}(\mathbb{R}^n)$ if, and only if f is locally integrable and $\|f\|^{\blacktriangle}_{F^{s,\tau}_{p,q}(\mathbb{R}^n)} < \infty$. Further $\|f\|^{\blacktriangle}_{F^{s,\tau}_{p,q}(\mathbb{R}^n)}$ and $\|f\|_{F^{s,\tau}_{p,q}(\mathbb{R}^n)}$ are equivalent.*

Proof. To finish the proof, it remains to show $\|f\|_{A^{s,\tau}_{p,q}(\mathbb{R}^n)} \lesssim \|f\|^{\blacktriangle}_{A^{s,\tau}_{p,q}(\mathbb{R}^n)}$. In fact, by Lemmas 4.2–4.9, we have

$$\begin{aligned}\|f\|_{B^{s,\tau}_{p,q}(\mathbb{R}^n)} &\lesssim \|f\|^{\square}_{B^{s,\tau}_{p,q}(\mathbb{R}^n)} + \sup_{\{P\in\mathcal{Q}: l(P)\ge 1\}} \frac{1}{|P|^\tau}\left(\int_P |\Phi * f(x)|^p\,dx\right)^{1/p}\\ &\lesssim \|f\|^{\blacktriangle}_{B^{s,\tau}_{p,q}(\mathbb{R}^n)} + \sup_{P\in\mathcal{Q}} T_{N,P}\end{aligned}$$

and

$$\|f\|_{F^{s,\tau}_{p,q}(\mathbb{R}^n)} \lesssim \|f\|^{\blacktriangle}_{F^{s,\tau}_{p,q}(\mathbb{R}^n)} + \sup_{P\in\mathcal{Q}} \widetilde{T}_{N,P}\,.$$

Thus, we need to estimate $T_{N,P}$ ($\widetilde{T}_{N,P}$ in the F-case). The first step consists in applying Lemma 4.4. We have to give estimates of $\|f\|_{L^p_\tau(\mathbb{R}^n)}$, $\|f\|_{L^1_\tau(\mathbb{R}^n)}$, and

$$\sup_{\{P\in\mathcal{Q}, |P|\ge 1\}} \frac{\|f\|_{B^{s_0}_{p,\infty}(2P)}}{|P|^\tau}.$$

Step 1. Estimate of $\|f\|_{L^p_\tau(\mathbb{R}^n)}$. Let P be a dyadic cube such that $|P| \ge 1$. Using the compact support of the generators ϕ and ψ we immediately obtain the so-called L^p-stability of the corresponding shifts, i. e.,

$$\left(\int_P \left|\sum_{k\in J_P} a_k \phi_{0,k}\right|^p dx\right)^{1/p} \sim \left(\sum_{k\in J_P} |a_k|^p\right)^{1/p},$$

$$\left(\int_P \left|\sum_{k\in I_{P,m}} a_{i,m,k}\psi_{i,m,k}\right|^p dx\right)^{1/p} \sim 2^{mn(\frac{1}{2}-\frac{1}{p})}\left(\sum_{k\in I_{P,m}} |a_{i,m,k}|^p\right)^{1/p}$$

and the constants behind $\sim$ do neither depend on the dyadic cube P nor on $m \in \mathbb{Z}_+$. Since $s > 0$ we conclude

$$\|f\|_{L^p_\tau(\mathbb{R}^n)} \lesssim \|f\|^{\blacktriangle}_{A^{s,\tau}_{p,q}(\mathbb{R}^n)}\,. \tag{4.30}$$

Let $0 < p < 1$. Since $s > \sigma_p$ we have $A^{s,\tau}_{p,q}(\mathbb{R}^n) \subset B^{s-\frac{n}{p}+n,\tau}_{1,\infty}(\mathbb{R}^n)$. Using (4.30) with $p = 1$ we have proved

$$\|f\|_{L^1_\tau(\mathbb{R}^n)} \lesssim \|f\|^{\blacktriangle}_{A^{s,\tau}_{p,q}(\mathbb{R}^n)}\,.$$

Step 2. Estimate of $\sup_{P\in\mathcal{Q},\,|P|\geq 1} \|f\|_{B^{s_0}_{p,\infty}(2P)}/|P|^\tau$. Starting point is the identity

$$f(x) = \sum_{k\in J_P} a_k \phi_{0,k}(x) + \sum_{i=1}^{2^n-1} \sum_{j=0}^{\infty} \sum_{k\in\mathcal{I}_{P,j}} a_{i,j,k}\psi_{i,j,k}(x)\,,$$

which is valid on P. Taking the right-hand side of this formula we obtain an extension of the restriction of f to P to $\mathbb{R}^n$. Hence

$$\begin{aligned}
\|f\|_{B^{s_0}_{p,\infty}(P)} &\leq \left\| \sum_{k\in J_P} a_k \phi_{0,k}(x) + \sum_{i=1}^{2^n-1} \sum_{j=0}^{\infty} \sum_{k\in\mathcal{I}_{P,j}} a_{i,j,k}\psi_{i,j,k}(x) \right\|_{B^{s_0}_{p,\infty}(\mathbb{R}^n)} \\
&\lesssim \left(\sum_{k\in J_P} |a_k|^p\right)^{1/p} + \sum_{i=1}^{2^n-1} \sup_{j\in\mathbb{Z}_+} 2^{j(s_0+n/2)} \left(\sum_{k\in\mathcal{I}_{P,j}} 2^{-jn}|a_{i,j,k}|^p\right)^{1/p};
\end{aligned}$$

see Sect. 4.2.1. With the obvious modifications, needed for the replacement of P by $2P$, this implies

$$\sup_{P\in\mathcal{Q},\,|P|\geq 1} \|f\|_{B^{s_0}_{p,\infty}(2P)}/|P|^\tau \lesssim \|f\|^{\blacktriangle}_{B^{s_0,\tau}_{p,\infty}(\mathbb{R}^n)}\,.$$

Since $\|f\|^{\blacktriangle}_{B^{s_0,\tau}_{p,\infty}(\mathbb{R}^n)} \lesssim \|f\|^{\blacktriangle}_{A^{s,\tau}_{p,q}(\mathbb{R}^n)}$, the proof is complete. □

Remark 4.1. When $\tau = 0$, Theorem 4.1 is just the classical result for $A^s_{p,q}(\mathbb{R}^n)$; see, e. g., [99, Sect. 2.9], [82, Sect. 6.5], [65], [72, Remark 6.8/8], [149, Theorem 1.20] and [156, Corollary 9.10].

For $\tau \in [1/p,\infty)$ we can simplify the characterization. Therefore we need a further abbreviation. Let

$$\begin{aligned}
|||f|||^{\blacktriangle}_{B^{s,\tau}_{p,q}(\mathbb{R}^n)} \equiv \sup_{k\in\mathbb{Z}^n} |\langle f,\phi_{0,k}\rangle| + \sup_{\{P\in\mathcal{Q}:\,|P|\leq 1\}} \frac{1}{|P|^\tau} \\
\times \left\{ \sum_{j=j_P}^{\infty} 2^{j(s+n/2)q} \sum_{i=1}^{2^n-1} \left[\sum_{k\in\mathcal{I}_{P,j}} 2^{-jn}|\langle f,\psi_{i,j,k}\rangle|^p\right]^{q/p} \right\}^{1/q}
\end{aligned}$$

and

$$\||f\||_{F^{s,\tau}_{p,q}(\mathbb{R}^n)}^{\blacktriangle} \equiv \sup_{k\in\mathbb{Z}^n} |\langle f,\phi_{0,k}\rangle| + \sup_{\{P\in\mathcal{Q}:\,|P|\le 1\}} \frac{1}{|P|^\tau}$$
$$\times \left\| \left[\sum_{j=j_P}^{\infty} \sum_{i=1}^{2^n-1} \sum_{k\in\mathcal{I}_{P,j}} 2^{jsq} |\langle f,\psi_{i,j,k}\rangle \widetilde{\chi}_{Q_{jk}}|^q \right]^{1/q} \right\|_{L^p(P)} .$$

Theorem 4.2. *Under the same conditions as in Theorem 4.1, but assuming additionally $\tau \in [1/p,\infty)$, we have the following: all assertions in Theorem 4.1 remain true if $\|f\|^{\blacktriangle}_{A^{s,\tau}_{p,q}(\mathbb{R}^n)}$ is replaced by $\||f\||^{\blacktriangle}_{A^{s,\tau}_{p,q}(\mathbb{R}^n)}$.*

Proof. Instead of Lemma 4.3 we employ Lemma 4.5. The equivalence of

$$\sup_{\{P\in\mathcal{Q}:\,|P|\ge 1\}} \frac{1}{|P|^\tau} \left(\sum_{k\in\mathcal{J}_P} |\langle f,\phi_{0,k}\rangle|^p \right)^{1/p}$$

and $\sup_{k\in\mathbb{Z}^n} |\langle f,\phi_{0,k}\rangle|$ is obvious if $\tau \in [1/p,\infty)$. □

We finish this subsection by completing the proof of Proposition 2.2.

Proof of Proposition 2.2 (Continued). We have to prove that the spaces $A^{s,\tau_1}_{p,q_1}(\mathbb{R}^n)$ and $A^{s,\tau_0}_{p,q_2}(\mathbb{R}^n)$ are incomparable, if $s\in(\sigma_p,\infty)$ and $0\le\tau_0<\tau_1<1/p$.

Step 1. We shall prove $A^{s,\tau_1}_{p,q}(\mathbb{R}^n) \not\subset A^{s,\tau_0}_{p,q}(\mathbb{R}^n)$. Let $\ell > n\tau$. For any dyadic cube Q with $|Q|\ge 1$ we define

$$f_Q \equiv \sum_{k\in J_Q} |k|^\ell \phi_{0,k}.$$

Theorem 4.1 implies that in case $\tau < 1/p$,

$$\|f_Q\|_{B^{s,\tau}_{p,q}(\mathbb{R}^n)} \sim |Q|^{\frac{\ell}{n}+\frac{1}{p}-\tau}.$$

By Q_m we denote the cube with center in $x_m \equiv (2^{m+2}+2^m,0,\ldots,0)$ and side length 2^m. For all $\alpha\in\mathbb{R}$, let

$$g_\alpha \equiv \sum_{m=1}^{\infty} m^{-\alpha} |Q_m|^{-\frac{\ell}{n}-\frac{1}{p}+\tau_1} f_{Q_m}.$$

Lemma 2.1 in combination with Proposition 2.1(iii) implies that

$$g_\alpha \in A^{s,\tau_1}_{p,q}(\mathbb{R}^n) \qquad \text{if} \qquad \alpha > \frac{1}{\min\{1,p,q\}}.$$

On the other hand, by the disjointness of the cubes Q_m, we obtain

$$\frac{1}{|Q_m|^{\tau_0}}\left(\sum_{k\in\mathscr{J}_{Q_m}}|\langle g_\alpha,\phi_{0,k}\rangle|^p\right)^{1/p}\sim m^{-\alpha}\,|Q_m|^{\tau_1-\tau_0}.$$

The right-hand side is unbounded, hence $g_\alpha\notin A^{s,\tau_0}_{p,q}(\mathbb{R}^n)$ (applying again Proposition 2.1(iii) to avoid the restriction $s>\sigma_{p,q}$).

Step 2. It remains to prove $A^{s,\tau_0}_{p,q}(\mathbb{R}^n)\not\subset A^{s,\tau_1}_{p,q}(\mathbb{R}^n)$. We consider the sequence $\{\psi_{1,j,0}\}_{j=0}^\infty$ of functions. Obviously

$$\|\psi_{1,j,0}\|_{\overset{\blacktriangle}{A}{}^{s,\tau}_{p,q}(\mathbb{R}^n)}\sim 2^{j(n\tau+s+\frac{n}{2}-\frac{n}{p})}.$$

Let

$$h_\alpha\equiv\sum_{j=1}^\infty j^{-\alpha}\,2^{-j(n\tau_0+s+\frac{n}{2}-\frac{n}{p})}\,\psi_{1,j,0}.$$

As above it follows that

$$h_\alpha\in A^{s,\tau_0}_{p,q}(\mathbb{R}^n)\setminus A^{s,\tau_1}_{p,q}(\mathbb{R}^n).$$

This finishes the proof of Proposition 2.2. □

4.2.4 The Wavelet Characterization of $F^s_{\infty,q}(\mathbb{R}^n)$

A case of particular importance is given by $F^s_{\infty,q}(\mathbb{R}^n)$. By means of Proposition 2.4(iii) and Theorem 4.2 we obtain in case $s>n\max\{0,\frac{1}{q}-1\}$: a locally integrable function f belongs to $F^s_{\infty,q}(\mathbb{R}^n)$ if, and only if f belongs to $F^{s,1/p}_{p,q}(\mathbb{R}^n)$ for some $p<\infty$, if, and only if

$$|||f|||_{\overset{\blacktriangle}{F}{}^s_{\infty,q}(\mathbb{R}^n)}\equiv\sup_{k\in\mathbb{Z}^n}|\langle f,\phi_{0,k}\rangle|+\sup_{\{P\in\mathscr{Q}:\,|P|\le 1\}}\frac{1}{|P|^{1/p}}\times\left\|\left[\sum_{j=j_P}^\infty\sum_{i=1}^{2^n-1}\sum_{k\in\mathscr{I}_{P,j}}2^{jsq}|\langle f,\psi_{i,j,k}\rangle\widetilde{\chi}_{Q_{jk}}|^q\right]^{1/q}\right\|_{L^p(P)}<\infty$$

for some $p\in(\frac{n}{n+s},\infty)$. Let $q\in(0,\infty)$. Then we can choose $p=q$. Interchanging integration and summation (or use $F^{s,1/q}_{q,q}(\mathbb{R}^n)=B^{s,1/q}_{q,q}(\mathbb{R}^n)$) we can carry out the integration and obtain the following corollary.

Corollary 4.2. *Let $q\in(0,\infty)$ and $s\in(n\max\{0,\frac{1}{q}-1\},\infty)$. A tempered distribution $f\in\mathscr{S}'(\mathbb{R}^n)$ belongs to $F^s_{\infty,q}(\mathbb{R}^n)$ if, and only if*

$$|||f|||^{\blacktriangle}_{F^s_{\infty,q}(\mathbb{R}^n)} \equiv \sup_{k\in\mathbb{Z}^n} |\langle f,\phi_{0,k}\rangle| + \sup_{\{P\in\mathcal{Q}:\,|P|\le 1\}} \frac{1}{|P|^{1/q}} \times \left[\sum_{j=j_P}^{\infty}\sum_{i=1}^{2^n-1}\sum_{k\in\mathcal{I}_{P,j}} 2^{j(s+n(\frac{1}{2}-\frac{1}{q}))q}|\langle f,\psi_{i,j,k}\rangle|^q\right]^{1/q} < \infty. \quad (4.31)$$

Furthermore, $|||f|||^{\blacktriangle}_{F^s_{\infty,q}(\mathbb{R}^n)}$ *and* $\|f\|_{F^s_{\infty,q}(\mathbb{R}^n)}$ *are equivalent.*

Remark 4.2.

(i) Using Proposition 5.1 below, we obtain

$$F^s_{\infty,2}(\mathbb{R}^n) = I_{-s}(F^0_{\infty,2}(\mathbb{R}^n)) = I_{-s}(\mathrm{bmo}\,(\mathbb{R}^n)).$$

Wavelet characterizations of the homogeneous counterparts of these spaces have been obtained in [5].

(ii) Interesting limiting cases are $\mathrm{bmo}\,(\mathbb{R}^n)$ and $\mathrm{vmo}\,(\mathbb{R}^n)$; see items (xiii) and (xiv) in Sect. 1.3. Also $\mathrm{BMO}\,(\mathbb{R}^n)$ is of interest, of course. Let $\widetilde{\psi}$ be a compactly supported, continuously differentiable wavelet on $\mathbb{R}$ and let $\psi_1,\ldots,\psi_{2^n-1}$ be the associated generators for a wavelet basis of $L^2(\mathbb{R}^n)$. Only here we shall use the convention $\psi_{i,j,k}(x) \equiv 2^{jn/2}\,\psi_i(2^jx-k)$ also for $j<0$. Then a locally integrable function f belongs to $\mathrm{BMO}\,(\mathbb{R}^n)$ if, and only if

$$\sup_{P\in\mathcal{Q}} \frac{1}{|P|^{1/2}}\left[\sum_{j=j_P}^{\infty}\sum_{i=1}^{2^n-1}\sum_{k\in\mathcal{I}_{P,j}} |\langle f,\psi_{i,j,k}\rangle|^2\right]^{1/2} < \infty;$$

see [99, Sect. 5.6] and [156, Example 8.8]. In the literature sometimes the convention

$$|\langle f,\psi_Q\rangle| \equiv \left(\sum_{i=1}^{n} |\langle f,\psi_{i,j,k}\rangle|^2\right)^{1/2}$$

is used with $Q=Q_{j,k}$. In this language we obtain that a locally integrable function f belongs to $\mathrm{BMO}\,(\mathbb{R}^n)$ if, and only if

$$\sup_{P\in\mathcal{Q}}\left[\frac{1}{|P|}\sum_{Q\subset P}|\langle f,\psi_Q\rangle|^2\right]^{1/2} < \infty,$$

The formula (4.31) remains to be true for $\mathrm{bmo}\,(\mathbb{R}^n)$, i. e., if $s=0$ and $q=2$. For this result we refer to [5]. Moreover, there one can also find a wavelet characterization of $\mathrm{vmo}\,(\mathbb{R}^n)$. Defining for $\varepsilon>0$

$$N_\varepsilon(f) \equiv \sup_{l(P)\le\varepsilon}\left(\frac{1}{|P|}\sum_{Q\subset P}|\langle f,\psi_Q\rangle|^2\right)^{1/2}$$

and

$$N_0(f) \equiv \lim_{\varepsilon \to 0} N_\varepsilon(f),$$

then a function $f \in \mathrm{bmo}(\mathbb{R}^n)$ belongs to $\mathrm{vmo}(\mathbb{R}^n)$ if, and only if $N_0(f) = 0$.

4.3 Characterizations of $A^{s,\tau}_{p,q}(\mathbb{R}^n)$ by Differences

Since the restrictions for Besov spaces differ from those for Triebel-Lizorkin spaces, we split the description of our results. Recall, the three expressions $\|f\|^{\clubsuit}_{A^{s,\tau}_{p,q}(\mathbb{R}^n)}$, $\|f\|^{\heartsuit}_{A^{s,\tau}_{p,q}(\mathbb{R}^n)}$ and $\|f\|^{\spadesuit}_{A^{s,\tau}_{p,q}(\mathbb{R}^n)}$ have been defined in Sect. 4.1.3. In addition we need the following local versions. By $|||f|||^{\clubsuit}_{A^{s,\tau}_{p,q}(\mathbb{R}^n)}$ we denote the quantity obtained by replacing, in the definition of $\|f\|^{\clubsuit}_{A^{s,\tau}_{p,q}(\mathbb{R}^n)}$,

$$\sup_{P \in \mathcal{Q}} \quad \text{by} \quad \sup_{\{P \in \mathcal{Q}:\, |P| \le 1\}}.$$

Similar are the definitions of $|||f|||^{\heartsuit}_{A^{s,\tau}_{p,q}(\mathbb{R}^n)}$ and $|||f|||^{\spadesuit}_{A^{s,\tau}_{p,q}(\mathbb{R}^n)}$. Observe that in all local cases the integral (supremum) with respect to t extends over the interval $(0, 2l(P))$ instead of $2(l(P) \wedge 1)$.

The following general rule applies to all results within this section. If we can prove a characterization of $A^{s,\tau}_{p,q}(\mathbb{R}^n)$ by using $|||f|||^{\clubsuit}_{A^{s,\tau}_{p,q}(\mathbb{R}^n)}$, then we get a characterization as well by using $\|f\|^{\clubsuit}_{A^{s,\tau}_{p,q}(\mathbb{R}^n)}$; the same is true for $|||f|||^{\heartsuit}_{A^{s,\tau}_{p,q}(\mathbb{R}^n)}$ and $|||f|||^{\spadesuit}_{A^{s,\tau}_{p,q}(\mathbb{R}^n)}$. However, in such a situation we concentrate on the first possibility. For later applications we also mention that we can always switch in our characterizations from the used annulus $\{h \in \mathbb{R}^n : t/2 \le |h| < t\}$ to the associated ball $\{h \in \mathbb{R}^n : |h| < t\}$ without changing the assertion.

4.3.1 Characterizations of $F^{s,\tau}_{p,q}(\mathbb{R}^n)$ by Differences

We shall distinguish three cases by following the different estimates we got for the remainder $\widetilde{T}_{N,P}$ in Lemma 4.4. Most transparent is the situation if $p \in [1,\infty)$.

Theorem 4.3. *Let* $p \in [1,\infty)$, $q \in (0,\infty]$,

$$\sigma_{1,q} < s \le \max\{s, s + n\tau - n/p\} < M$$

with $M \in \mathbb{N}$ *and* $\tau \in [0, \tau_{s,p,q})$.

(i) *A function* $f \in F^{s,\tau}_{p,q}(\mathbb{R}^n)$ *if, and only if* $f \in L^p_\tau(\mathbb{R}^n)$ *and* $\|f\|^{\clubsuit}_{F^{s,\tau}_{p,q}(\mathbb{R}^n)} < \infty$. *Further,* $\|f\|_{L^p_\tau(\mathbb{R}^n)} + \|f\|^{\clubsuit}_{F^{s,\tau}_{p,q}(\mathbb{R}^n)}$ *and* $\|f\|_{F^{s,\tau}_{p,q}(\mathbb{R}^n)}$ *are equivalent. Moreover, when* $q \in [1,\infty]$, *then these assertions remain true if* $\|f\|^{\clubsuit}_{F^{s,\tau}_{p,q}(\mathbb{R}^n)}$ *is replaced by* $\|f\|^{\heartsuit}_{F^{s,\tau}_{p,q}(\mathbb{R}^n)}$.

(ii) *Let in addition* $\tau \in [1/p,\infty)$. *A function* $f \in F^{s,\tau}_{p,q}(\mathbb{R}^n)$ *if, and only if* $N_p(f) < \infty$ *and* $|||f|||^{\clubsuit}_{F^{s,\tau}_{p,q}(\mathbb{R}^n)} < \infty$. *Further,* $N_p(f) + |||f|||^{\clubsuit}_{F^{s,\tau}_{p,q}(\mathbb{R}^n)}$ *and* $\|f\|_{F^{s,\tau}_{p,q}(\mathbb{R}^n)}$ *are equivalent. Moreover, when* $q \in [1,\infty]$, *then these assertions remain true if* $|||f|||^{\clubsuit}_{F^{s,\tau}_{p,q}(\mathbb{R}^n)}$ *is replaced by* $|||f|||^{\heartsuit}_{F^{s,\tau}_{p,q}(\mathbb{R}^n)}$.

Proof. Step 1. We deal with $\|f\|^{\clubsuit}_{F^{s,\tau}_{p,q}(\mathbb{R}^n)}$. By Minkowski's inequality, we find

$$\frac{1}{|P|^\tau}\left(\int_P |\Phi * f(x)|^p\,dx\right)^{1/p} \lesssim \|f\|_{L^p_\tau(\mathbb{R}^n)}$$

as long as $P \in \mathcal{Q}$ and $|P| \geq 1$. Moreover, we know $F^{s,\tau}_{p,q}(\mathbb{R}^n) \subset L^p_\tau(\mathbb{R}^n)$; see Propositions 2.1 and 2.7. Now we make use of Lemmas 4.2, 4.3 and 4.4(i) to obtain

$$\begin{aligned}
\|f\|_{F^{s,\tau}_{p,q}(\mathbb{R}^n)} &\lesssim \left\{\|f\|^{\Box}_{A^{s,\tau}_{p,q}(\mathbb{R}^n)} + \sup_{\{P\in\mathcal{Q}:l(P)\geq 1\}} \frac{1}{|P|^\tau}\left(\int_P |\Phi * f(x)|^p\,dx\right)^{1/p}\right\}\\
&\lesssim \|f\|^{\clubsuit}_{F^{s,\tau}_{p,q}(\mathbb{R}^n)} + \widetilde{T}_{N,P} + \|f\|_{L^p_\tau(\mathbb{R}^n)}.\\
&\lesssim \|f\|^{\clubsuit}_{F^{s,\tau}_{p,q}(\mathbb{R}^n)} + \|f\|_{L^p_\tau(\mathbb{R}^n)}.
\end{aligned}$$

Next we employ Lemmas 4.7, 4.8, combine it with Theorem 4.1(ii), and obtain

$$\|f\|^{\clubsuit}_{F^{s,\tau}_{p,q}(\mathbb{R}^n)} + \|f\|_{L^p_\tau(\mathbb{R}^n)} \lesssim \|f\|^{\blacktriangle}_{F^{s,\tau}_{p,q}(\mathbb{R}^n)}.$$

This proves the claim for $\|f\|^{\clubsuit}_{F^{s,\tau}_{p,q}(\mathbb{R}^n)}$.

Step 2. We deal with $\|f\|^{\heartsuit}_{F^{s,\tau}_{p,q}(\mathbb{R}^n)}$. In comparison with Step 1 we have to modify the estimate from below. For this, under the restriction $1 \leq q \leq \infty$, we refer to Lemma 4.6.

Step 3. Proof of (ii). We employ Lemma 4.5 instead of Lemma 4.3, $\|f\|^{\clubsuit}_{F^{s,\tau}_{p,q}(\mathbb{R}^n)}$ (resp. $\|f\|^{\heartsuit}_{F^{s,\tau}_{p,q}(\mathbb{R}^n)}$) replaced by $|||f|||^{\clubsuit}_{F^{s,\tau}_{p,q}(\mathbb{R}^n)}$ (resp. $|||f|||^{\heartsuit}_{F^{s,\tau}_{p,q}(\mathbb{R}^n)}$). For the replacement of $\|f\|_{L^p_\tau(\mathbb{R}^n)}$ by $N_p(f)$, see (2.5). □

Remark 4.3. For $\tau = 0$ we are back in the classical situation; see [83, 85, 91] and [145, Sect. 2.5.11].

Next we consider the case $p \in (0,1)$ and $\tau \in [1/p,\infty)$.

Theorem 4.4. *Let* $p \in (0,1)$, $q \in (0,\infty]$,

$$\sigma_{p,q} < s \le \max\{s, s+n\tau - n/p\} < M$$

with $M \in \mathbb{N}$ *and* $\tau \in [1/p, \tau_{s,p,q})$. *Then* $f \in F^{s,\tau}_{p,q}(\mathbb{R}^n)$ *if, and only if* $N_1(f) < \infty$ *and* $|||f|||^{\clubsuit}_{F^{s,\tau}_{p,q}(\mathbb{R}^n)} < \infty$. *Further,* $N_1(f) + |||f|||^{\clubsuit}_{F^{s,\tau}_{p,q}(\mathbb{R}^n)}$ *and* $\|f\|_{F^{s,\tau}_{p,q}(\mathbb{R}^n)}$ *are equivalent. Moreover, when* $q \in [1,\infty]$, *then these assertions remain true if* $|||f|||^{\clubsuit}_{F^{s,\tau}_{p,q}(\mathbb{R}^n)}$ *is replaced by* $|||f|||^{\heartsuit}_{F^{s,\tau}_{p,q}(\mathbb{R}^n)}$.

Proof. The used arguments are the same as in proof of Theorem 4.3 except the estimate of

$$\Phi_{\tau,p}(f) = \sup_{\{P\in\mathscr{Q}:\,|P|\ge 1\}} \frac{1}{|P|^{\tau}} \left(\int_P |\Phi * f(x)|^p\,dx\right)^{1/p}.$$

For the estimate of $\Phi_{\tau,p}(f)$ we shall use Lemma 4.9. This proves that $\|f\|_{F^{s,\tau}_{p,q}(\mathbb{R}^n)}$ and

$$\Phi_{\tau,p}(f) + \|f\|_{L^p_\tau(\mathbb{R}^n)} + \|f\|_{L^1_\tau(\mathbb{R}^n)} + \|f\|^{\clubsuit}_{F^{s,\tau}_{p,q}(\mathbb{R}^n)}$$

are equivalent. Making use of the argument in (2.5) we obtain for any dyadic cube P with $|P| \ge 1$,

$$\begin{aligned}\frac{1}{|P|^{\tau}} \left(\int_P |\Phi * f(x)|^p\,dx\right)^{1/p} &\le \sup_{\{Q\in\mathscr{Q}:\,|Q|=1\}} \left(\int_Q |\Phi * f(x)|^p\,dx\right)^{1/p}\\ &\le \sup_{\{Q\in\mathscr{Q}:\,|Q|=1\}} \int_Q |\Phi * f(x)|\,dx \lesssim N_1(f),\end{aligned}$$

where we used in the last step an convolution inequality. Similarly,

$$\|f\|_{L^1_\tau(\mathbb{R}^n)} + \|f\|_{L^p_\tau(\mathbb{R}^n)} \lesssim N_1(f)$$

can be proved. This shows

$$\Phi_{\tau,p}(f) + \|f\|_{L^p_\tau(\mathbb{R}^n)} + \|f\|_{L^1_\tau(\mathbb{R}^n)} + \|f\|^{\clubsuit}_{F^{s,\tau}_{p,q}(\mathbb{R}^n)} \lesssim N_1(f) + |||f|||^{\clubsuit}_{F^{s,\tau}_{p,q}(\mathbb{R}^n)};$$

see Lemma 4.5. □

Finally we study the case $p \in (0,1)$ and $\tau \in [0,1/p)$.

Theorem 4.5. *Let* $p \in (0,1)$, $q \in (0,\infty]$,

$$\sigma_{p,q} < s \le \max\{s, s+n\tau - n/p\} < M$$

with $M \in \mathbb{N}$ and $\tau \in [0, \min\{\tau_{s,p,q}, 1/p\})$. Let $\sigma_p < s_0 < s$. Then $f \in F_{p,q}^{s,\tau}(\mathbb{R}^n)$ if, and only if $f \in L_\tau^p(\mathbb{R}^n)$,

$$\sup_{\{P \in \mathcal{Q}, |P| \geq 1\}} \frac{\|f\|_{B_{p,\infty}^{s_0}(2P)}}{|P|^\tau} < \infty,$$

and $\|f\|_{F_{p,q}^{s,\tau}(\mathbb{R}^n)}^{\clubsuit} < \infty$. Further,

$$\sup_{\{P \in \mathcal{Q}, |P| \geq 1\}} \frac{\|f\|_{B_{p,\infty}^{s_0}(2P)}}{|P|^\tau} + \|f\|_{L_\tau^p(\mathbb{R}^n)} + \|f\|_{F_{p,q}^{s,\tau}(\mathbb{R}^n)}^{\clubsuit}$$

and $\|f\|_{F_{p,q}^{s,\tau}(\mathbb{R}^n)}$ are equivalent. Moreover, when $q \in [1, \infty]$, then these assertions remain true if $\|f\|_{F_{p,q}^{s,\tau}(\mathbb{R}^n)}^{\clubsuit}$ is replaced by $\|f\|_{F_{p,q}^{s,\tau}(\mathbb{R}^n)}^{\heartsuit}$.

Proof. We only comment on the modifications. The estimate of

$$\sup_{\{P \in \mathcal{Q}, |P| \geq 1\}} \frac{\|f\|_{B_{p,\infty}^{s_0}(2P)}}{|P|^\tau}$$

from above follows from an argument as in Step 2 of the proof for Theorem 4.1. Further, to estimate of $\Phi_{\tau,p}(f)$ we shall have a closer look to the proof of Lemma 4.9 and combine it with Theorem 4.1 (applied with s_0). Then we obtain

$$\Phi_{\tau,p}(f) \lesssim \sup_{\{P \in \mathcal{Q}, |P| \geq 1\}} \frac{\|f\|_{B_{p,\infty}^{s_0}(2P)}}{|P|^\tau}.$$

The proof is complete. □

Nothing has been said about $\|f\|_{F_{p,q}^{s,\tau}(\mathbb{R}^n)}^{\spadesuit}$ up to now. We summarize our findings also with respect to this quantity as follows.

Theorem 4.6. *We may replace $\|f\|_{F_{p,q}^{s,\tau}(\mathbb{R}^n)}^{\clubsuit}$ by $\|f\|_{F_{p,q}^{s,\tau}(\mathbb{R}^n)}^{\spadesuit}$ in Theorems 4.3–4.5 if additionally $s \in (n/\min\{p,q\}, \infty)$ is satisfied.*

Remark 4.4.

(i) Further characterizations by differences for $F_{p,q}^{s,\tau}(\mathbb{R}^n)$ can be obtained from the identity $F_{u,q}^{s,1/u-1/p}(\mathbb{R}^n) = \mathcal{E}_{pqu}^s(\mathbb{R}^n)$ (see Corollary 3.3(i)) in combination with Corollary 4.11 below.

(ii) For $\tau = 0$ we refer to Seeger [130] and Triebel [146, Sect. 3.5.3].

4.3.2 Characterizations of $B^{s,\tau}_{p,q}(\mathbb{R}^n)$ by Differences

We keep the same structure as in Sect. 4.3.1.

Theorem 4.7. *Let* $p \in [1,\infty]$, $q \in (0,\infty]$,

$$0 < s \le \max\{s, s+n\tau-n/p\} < M$$

with $M \in \mathbb{N}$ *and*

$$0 \le \tau < \tau_{s,p} = \frac{s}{n} + \frac{1}{p}.$$

(i) *Then* $f \in B^{s,\tau}_{p,q}(\mathbb{R}^n)$ *if, and only if* $f \in L^p_\tau(\mathbb{R}^n)$ *and* $\|f\|^{\clubsuit}_{B^{s,\tau}_{p,q}(\mathbb{R}^n)} < \infty$. *Furthermore,* $\|f\|_{L^p_\tau(\mathbb{R}^n)} + \|f\|^{\clubsuit}_{B^{s,\tau}_{p,q}(\mathbb{R}^n)}$ *and* $\|f\|_{B^{s,\tau}_{p,q}(\mathbb{R}^n)}$ *are equivalent. Moreover, these assertions remain true if* $\|f\|^{\clubsuit}_{B^{s,\tau}_{p,q}(\mathbb{R}^n)}$ *is replaced either by* $\|f\|^{\heartsuit}_{B^{s,\tau}_{p,q}(\mathbb{R}^n)}$ *or by* $\|f\|^{\spadesuit}_{B^{s,\tau}_{p,q}(\mathbb{R}^n)}$.

(ii) *Let in addition* $\tau \in [1/p,\infty)$. *Then* $f \in B^{s,\tau}_{p,q}(\mathbb{R}^n)$ *if, and only if* $N_p(f) < \infty$ *and* $|||f|||^{\clubsuit}_{B^{s,\tau}_{p,q}(\mathbb{R}^n)} < \infty$ *(or* $|||f|||^{\heartsuit}_{B^{s,\tau}_{p,q}(\mathbb{R}^n)} < \infty$ *or* $|||f|||^{\spadesuit}_{B^{s,\tau}_{p,q}(\mathbb{R}^n)} < \infty$*). Furthermore,* $N_p(f) + |||f|||^{\clubsuit}_{B^{s,\tau}_{p,q}(\mathbb{R}^n)}$, $N_p(f) + |||f|||^{\heartsuit}_{B^{s,\tau}_{p,q}(\mathbb{R}^n)}$, $N_p(f) + |||f|||^{\spadesuit}_{B^{s,\tau}_{p,q}(\mathbb{R}^n)}$ *and* $\|f\|_{B^{s,\tau}_{p,q}(\mathbb{R}^n)}$ *are equivalent.*

Proof. The arguments are essentially the same as in the case of Theorem 4.3. Changes are caused by Lemmas 4.7, 4.8 (σ_p instead of $\sigma_{p,q}$) and Lemma 4.6. □

Most interesting is the case $p = q = \infty$.

Corollary 4.3. *Let* $0 < s \le s + n\tau < M$ *with* $M \in \mathbb{N}$ *and* $\tau \in [0, s/n)$. *Then* $f \in B^{s,\tau}_{\infty,\infty}(\mathbb{R}^n)$ *if, and only if* $f \in L^\infty(\mathbb{R}^n)$ *and*

$$|||f|||^{\spadesuit}_{B^{s,\tau}_{\infty,\infty}(\mathbb{R}^n)} \equiv \sup_{\{P\in\mathcal{Q}:\,|P|\le 1\}} \frac{1}{|P|^\tau} \sup_{0<t<2l(P)} t^{-s} \sup_{t/2\le|h|<t} \sup_{x\in P} |\Delta^M_h f(x)| < \infty.$$

Further, $\|f\|_{L^\infty(\mathbb{R}^n)} + |||f|||^{\spadesuit}_{B^{s,\tau}_{\infty,\infty}(\mathbb{R}^n)}$ *and* $\|f\|_{B^{s,\tau}_{\infty,\infty}(\mathbb{R}^n)}$ *are equivalent norms. Moreover, these assertions remain true if* $|||f|||^{\spadesuit}_{B^{s,\tau}_{\infty,\infty}(\mathbb{R}^n)}$ *is replaced either by* $|||f|||^{\heartsuit}_{B^{s,\tau}_{\infty,\infty}(\mathbb{R}^n)}$ *or by* $|||f|||^{\clubsuit}_{B^{s,\tau}_{\infty,\infty}(\mathbb{R}^n)}$.

Remark 4.5. There are many references for the case $\tau = 0$. We refer to [14, 109], [9, Theorem 6.2.5], and [145, Sect. 2.5.12].

Next we consider the case $p \in (0,1)$ and $\tau \in [1/p,\infty)$.

Theorem 4.8. *Let $p \in (0,1)$, $q \in (0,\infty]$,*

$$\sigma_p < s \le \max\{s, s+n\tau-n/p\} < M$$

with $M \in \mathbb{N}$ and

$$\frac{1}{p} \le \tau < \tau_{s,p} = 1+\frac{s}{n}.$$

Then $f \in B^{s,\tau}_{p,q}(\mathbb{R}^n)$ if, and only if $N_1(f) < \infty$ and $|||f|||^{\clubsuit}_{B^{s,\tau}_{p,q}(\mathbb{R}^n)} < \infty$. Further, $N_1(f) + \|f\|^{\clubsuit}_{B^{s,\tau}_{p,q}(\mathbb{R}^n)}$ and $\|f\|_{B^{s,\tau}_{p,q}(\mathbb{R}^n)}$ are equivalent. Moreover, these assertions remain true if $|||f|||^{\clubsuit}_{B^{s,\tau}_{p,q}(\mathbb{R}^n)}$ is replaced by $|||f|||^{\spadesuit}_{B^{s,\tau}_{p,q}(\mathbb{R}^n)}$.

Finally we study the case $p \in (0,1)$ and $\tau \in [0,1/p)$.

Theorem 4.9. *Let $p \in (0,1)$, $q \in (0,\infty]$,*

$$\sigma_p < s \le \max\{s, s+n\tau-n/p\} < M$$

with $M \in \mathbb{N}$ and

$$0 \le \tau < \min\left\{\tau_{s,p},\, 1/p\right\} = \min\left\{1+\frac{s}{n}, \frac{1}{p}\right\}.$$

Let $\sigma_p < s_0 < s$. Then $f \in B^{s,\tau}_{p,q}(\mathbb{R}^n)$ if, and only if $f \in L^p_\tau(\mathbb{R}^n)$,

$$\sup_{\{P\in\mathcal{Q},\,|P|\ge 1\}} \frac{\|f\|_{B^{s_0}_{p,\infty}(2P)}}{|P|^\tau} < \infty,$$

and $\|f\|^{\clubsuit}_{B^{s,\tau}_{p,q}(\mathbb{R}^n)} < \infty$. Further,

$$\sup_{\{P\in\mathcal{Q},\,|P|\ge 1\}} \frac{\|f\|_{B^{s_0}_{p,\infty}(2P)}}{|P|^\tau} + \|f\|_{L^p_\tau(\mathbb{R}^n)} + \|f\|^{\clubsuit}_{B^{s,\tau}_{p,q}(\mathbb{R}^n)}$$

and $\|f\|_{B^{s,\tau}_{p,q}(\mathbb{R}^n)}$ are equivalent. Moreover, these assertions remain true if $\|f\|^{\clubsuit}_{F^{s,\tau}_{p,q}(\mathbb{R}^n)}$ is replaced by $\|f\|^{\spadesuit}_{B^{s,\tau}_{p,q}(\mathbb{R}^n)}$.

Remark 4.6.

(i) Further characterizations by differences for $B^{s,\tau}_{p,\infty}(\mathbb{R}^n)$ can be obtained from the identity $B^{s,1/u-1/p}_{u,\infty}(\mathbb{R}^n) = \mathcal{N}^s_{p\infty u}(\mathbb{R}^n)$ (see Corollary 3.3(ii)) in combination with Corollary 4.13 below.

(ii) For $\tau = 0$ we refer to Triebel [146, Sect. 3.5.3].

4.3.3 The Classes $A^{s,\tau}_{p,q}(\mathbb{R}^n)$ and Their Relations to Q Spaces

In the homogeneous situation the case of first order differences in a symmetrized version has a certain history, connected with Q spaces; see item (xx) in Sect. 1.3. Here we are studying the inhomogeneous counterpart.

Let $M=1$ and choose $p=q$. Then the following corollary is a simple conclusion of Theorem 4.7. Recall that

$$A^{s,\tau}_{p,p}(\mathbb{R}^n)=F^{s,\tau}_{p,p}(\mathbb{R}^n)=B^{s,\tau}_{p,p}(\mathbb{R}^n).$$

Corollary 4.4. *Let $p\in[1,\infty]$,*

$$0<s\le\max\{s,\,s+n\tau-n/p\}<1 \qquad \text{and} \qquad 0\le\tau<\frac{1}{p}+\frac{s}{n}.$$

Then $f\in A^{s,\tau}_{p,p}(\mathbb{R}^n)$ if, and only if $f\in L^p_\tau(\mathbb{R}^n)$ and

$$\|f\|^{\diamondsuit}_{A^{s,\tau}_{p,p}(\mathbb{R}^n)}\equiv\sup_{P\in\mathcal{Q}}\frac{1}{|P|^\tau}\left\{\int_{|h|<2(l(P)\wedge 1)}|h|^{-sp}\int_P|f(x+h)-f(x)|^p\,dx\frac{dh}{|h|^n}\right\}^{1/p}<\infty.$$

Furthermore, $\|f\|_{L^p_\tau(\mathbb{R}^n)}+\|f\|^{\diamondsuit}_{A^{s,\tau}_{p,p}(\mathbb{R}^n)}$ and $\|f\|_{A^{s,\tau}_{p,p}(\mathbb{R}^n)}$ are equivalent. If in addition $\tau\in[1/p,\infty)$, then it will be enough to extend the supremum in $\|f\|^{\diamondsuit}_{A^{s,\tau}_{p,p}(\mathbb{R}^n)}$ only over those dyadic cubes P such that $|P|\le 1$.

Proof. Our starting point is Theorem 4.7 with respect to $\|f\|^{\heartsuit}_{B^{s,\tau}_{p,p}(\mathbb{R}^n)}$. In $\|f\|^{\heartsuit}_{B^{s,\tau}_{p,p}(\mathbb{R}^n)}$ we may interchange the order of integration and obtain

$$\int_0^{2(l(P)\wedge 1)}t^{-sp}\int_P t^{-n}\int_{t/2\le|h|<t}|f(x+h)-f(x)|^p\,dh\,dx\frac{dt}{t}$$
$$\sim\int_{|h|<2(l(P)\wedge 1)}|h|^{-sp}\int_P|f(x+h)-f(x)|^p\,dx\frac{dh}{|h|^n}.$$

The proof is complete. □

Now it is only a small step to the next conclusion.

Corollary 4.5. *Let p,s and τ be as in Corollary 4.4.*

(i) *Then $f\in A^{s,\tau}_{p,p}(\mathbb{R}^n)$ if, and only if $f\in L^p_\tau(\mathbb{R}^n)$ and*

$$|||f|||^{\diamondsuit}_{A^{s,\tau}_{p,p}(\mathbb{R}^n)}\equiv\sup_{P\in\mathcal{Q}}\frac{1}{|P|^\tau}\left\{\int_P\int_P\frac{|f(x)-f(y)|^p}{|x-y|^{sp+n}}\,dx\,dy\right\}^{1/p}<\infty. \qquad (4.32)$$

Furthermore, $\|f\|_{L^p_\tau(\mathbb{R}^n)}+|||f|||^{\diamondsuit}_{A^{s,\tau}_{p,p}(\mathbb{R}^n)}$ and $\|f\|_{A^{s,\tau}_{p,p}(\mathbb{R}^n)}$ are equivalent.

(ii) *If in addition* $\tau \in [1/p,\infty)$ *then it will be enough to extend the supremum in (4.32) over those dyadic cubes* P *such that* $|P| \le 1$.

Proof. Step 1. Replacing $x+h$ by y in the definition of $\|f\|^{\lozenge}_{A^{s,\tau}_{p,p}(\mathbb{R}^n)}$ we obtain

$$|x-y| < 2(l(P)\wedge 1).$$

Now we distinguish two cases, small cubes and large cubes. Let P be a dyadic cube with $l(P) \le 1$. Then

$$\begin{aligned}&\frac{1}{|P|^\tau}\left\{\int_{|h|<2(l(P)\wedge 1)}|h|^{-sp}\int_P |f(x+h)-f(x)|^p\,dx\,\frac{dh}{|h|^n}\right\}^{1/p}\\ &\le \frac{1}{|P|^\tau}\left\{\int_{4P}\int_P \frac{|f(x)-f(y)|^p}{|x-y|^{sp+n}}\,dx\,dy\right\}^{1/p}\\ &\lesssim \sup_{Q\in\mathscr{Q}}\frac{1}{|Q|^\tau}\left\{\int_Q\int_Q \frac{|f(x)-f(y)|^p}{|x-y|^{sp+n}}\,dx\,dy\right\}^{1/p}.\end{aligned}$$

Now let P be a dyadic cube with $l(P) > 1$. Then it is obvious that

$$\int_{|h|<1}|h|^{-sp}\int_P |f(x+h)-f(x)|^p\,dx\,\frac{dh}{|h|^n} \le \int_{4P}\int_P \frac{|f(x)-f(y)|^p}{|x-y|^{sp+n}}\,dx\,dy.$$

Hence $\|f\|^{\lozenge}_{A^{s,\tau}_{p,p}(\mathbb{R}^n)} \lesssim |||f|||^{\lozenge}_{A^{s,\tau}_{p,p}(\mathbb{R}^n)}$ follows.

Step 2. We turn to the converse inequality. It is enough to consider large cubes. For small cubes we can argue as in Step 1. Let P be a dyadic cube with $l(P) > 1$. Then

$$\frac{1}{|P|^\tau}\left\{\int_P\int_P \frac{|f(x)-f(y)|^p}{|x-y|^{sp+n}}\,dx\,dy\right\}^{1/p} \lesssim I_1 + I_2,$$

where

$$\begin{aligned}I_1 &\equiv \frac{1}{|P|^\tau}\left\{\int_{|h|<1}\int_P \frac{|f(x+h)-f(x)|^p}{|h|^{sp+n}}\,dx\,dh\right\}^{1/p},\\ I_2 &\equiv \frac{1}{|P|^\tau}\left\{\int_{|h|>1}\int_P \frac{|f(x)-f(y)|^p}{|h|^{sp+n}}\,dx\,dh\right\}^{1/p}.\end{aligned}$$

The first term I_1 can be estimated by $\lesssim \|f\|^{\lozenge}_{A^{s,\tau}_{p,p}(\mathbb{R}^n)}$. The second term I_2 allows an estimate by $\|f\|_{L^p_\tau(\mathbb{R}^n)}$. This finishes the proof of (i). The modifications for proving (ii) are obvious. □

Remark 4.7. The above Corollary could be formulated also as follows: The spaces $A^{s,\tau}_{2,2}(\mathbb{R}^n)$ and $Q_\alpha(\mathbb{R}^n)\cap L^2_\tau(\mathbb{R}^n)$ coincide in the sense of equivalent norms as far as $0<s=\alpha<1$ and $\tau = \frac{1}{2}-\alpha/n \ge 0$.

4.3.4 The Characterization of $F^s_{\infty,q}(\mathbb{R}^n)$ by Differences

We concentrate on a combination of Proposition 2.4, we will use the identity

$$F^{s,1/q}_{q,q}(\mathbb{R}^n) = B^{s,1/q}_{q,q}(\mathbb{R}^n) = F^s_{\infty,q}(\mathbb{R}^n),$$

and Theorem 4.7 if $q \in [1,\infty)$ or Theorem 4.8 if $q \in (0,1)$. The case $q = \infty$ has been treated in Corollary 4.3.

First we study the case that $q \in [1,\infty)$.

Corollary 4.6. *Let $q \in [1,\infty)$ and $s \in (0,M)$ with $M \in \mathbb{N}$. Then $f \in F^s_{\infty,q}(\mathbb{R}^n)$ if, and only if $N_q(f) < \infty$ and*

$$|||f|||^{\spadesuit}_{F^s_{\infty,q}(\mathbb{R}^n)} \equiv \sup_{\{P\in\mathscr{Q}:\,|P|\le 1\}} \left\{ \frac{1}{|P|} \int_0^{2l(P)} t^{-sq} \sup_{t/2\le|h|<t} \int_P |\Delta_h^M f(x)|^q \, dx \frac{dt}{t} \right\}^{1/q} < \infty.$$

Further, $N_q(f) + |||f|||^{\spadesuit}_{F^s_{\infty,q}(\mathbb{R}^n)}$ and $\|f\|_{F^s_{\infty,q}(\mathbb{R}^n)}$ are equivalent.

We continue with $q \in (0,1)$.

Corollary 4.7. *Let $q \in (0,1)$ and*

$$\frac{n}{q} - n < s < M$$

with $M \in \mathbb{N}$. Then $f \in F^s_{\infty,q}(\mathbb{R}^n)$ if, and only if $N_1(f) < \infty$ and $|||f|||^{\spadesuit}_{F^s_{\infty,q}(\mathbb{R}^n)} < \infty$. Furthermore, $N_1(f) + |||f|||^{\spadesuit}_{F^s_{\infty,q}(\mathbb{R}^n)}$ and $\|f\|_{F^s_{\infty,q}(\mathbb{R}^n)}$ are equivalent.

Remark 4.8.

(i) Let $q \in (0,\infty)$. Since $F^s_{\infty,q}(\mathbb{R}^n) \subset L^\infty(\mathbb{R}^n)$ if $s > 0$, it is not difficult to prove that under the above restrictions, $f \in F^s_{\infty,q}(\mathbb{R}^n)$ if, and only if $f \in L^\infty(\mathbb{R}^n)$ and $|||f|||^{\spadesuit}_{F^s_{\infty,q}(\mathbb{R}^n)} < \infty$. Furthermore, $\|f\|_{L^\infty(\mathbb{R}^n)} + |||f|||^{\spadesuit}_{F^s_{\infty,q}(\mathbb{R}^n)}$ and $\|f\|_{F^s_{\infty,q}(\mathbb{R}^n)}$ are equivalent.

(ii) Of course, in both cases one could also work with

$$|||f|||^{\clubsuit}_{F^s_{\infty,q}(\mathbb{R}^n)} \equiv \sup_{\{P\in\mathscr{Q}:\,|P|\le 1\}} \left\{ \frac{1}{|P|} \int_0^{2l(P)} t^{-sq} \times \int_P \left[t^{-n} \int_{t/2\le|h|<t} |\Delta_h^M f(x)| dh \right]^q dx \frac{dt}{t} \right\}^{\frac{1}{q}}$$

or

$$|||f|||^{\heartsuit}_{F^s_{\infty,q}(\mathbb{R}^n)} \equiv \sup_{\{P\in\mathscr{Q}:\,|P|\le 1\}} \left\{ \frac{1}{|P|} \int_0^{2l(P)} t^{-sq} \int_P t^{-n} \int_{t/2\le|h|<t} |\Delta_h^M f(x)|^q dh\, dx \frac{dt}{t} \right\}^{\frac{1}{q}}.$$

We omit the details.

4.4 Characterizations via Oscillations

In this section we establish the oscillation characterization of $A^{s,\tau}_{p,q}(\mathbb{R}^n)$ which leads to the interpretation as local approximation spaces.

4.4.1 Preparations

Let $M \in \mathbb{Z}_+$. Denote by $\mathscr{P}_M(\mathbb{R}^n)$ the *set of all polynomials of total degree less than or equal to* M. For a fixed ball $B(x,t) = \{y \in \mathbb{R}^n : |x-y| < t\}$ and $u \in (0,\infty]$, define the local oscillation of f by setting, for all $x \in \mathbb{R}^n$ and $t \in (0,\infty)$,

$$\operatorname{osc}^M_u f(x,t) \equiv \inf \left(t^{-n} \int_{B(x,t)} |f(y) - P(y)|^u \, dy \right)^{1/u}, \tag{4.33}$$

where the infimum is taken over all polynomials $P(y) \in \mathscr{P}_M(\mathbb{R}^n)$ and suitable modification is made when $u = \infty$; see, for example, [146, Sect. 1.7.2].

Following Seeger [130], we deduce that differences can be easily dominated by oscillations. Recall, the abbreviations $\|f\|_{B^\clubsuit(Q)}$ and $\|f\|_{F^\clubsuit(Q)}$ have been introduced in Sect. 4.2.2.

Lemma 4.10. *Let* $M \in \mathbb{N}$, $Q \in \mathscr{Q}$, $p, q \in (0,\infty]$, $\tau \in [0,\infty)$, *and* $s > 0$.

(i) There exists a positive constant C such that for all $f \in L^1_{\mathrm{loc}}(\mathbb{R}^n)$,

$$\|f\|_{B^\clubsuit(Q)} \le C \frac{1}{|Q|^\tau} \left\{ \int_0^{2(l(Q)\wedge 1)} t^{-sq} \left[\int_Q \left(\operatorname{osc}^{M-1}_1 f(x,Mt)\right)^p dx \right]^{q/p} \frac{dt}{t} \right\}^{1/q}$$

(ii) There exists a positive constant C such that for all $f \in L^1_{\mathrm{loc}}(\mathbb{R}^n)$,

$$\|f\|_{F^\clubsuit(Q)} \le C \frac{1}{|Q|^\tau} \left\{ \int_Q \left[\int_0^{2(l(Q)\wedge 1)} t^{-sq} \left(\operatorname{osc}^{M-1}_1 f(x,Mt)\right)^q \frac{dt}{t} \right]^{p/q} dx \right\}^{1/p}.$$

Proof. For any $x \in \mathbb{R}^n$, let $P_t f \in \mathscr{P}_{M-1}(\mathbb{R}^n)$ be the best approximate of f in $L^1(B(x,Mt))$. It follows that

$$\begin{aligned}\Delta^M_h f(x) &= \Delta^M_h (f - P_t f)(x) \\ &= (-1)^M (f(x) - P_t f(x)) + \sum_{i=0}^{M-1} (-i)^i \binom{M}{i} (f - P_t f)(x + (M-i)h).\end{aligned}$$

Clearly, if $i=0,\dots,M-1$, then

$$\frac{1}{|Q|^\tau}\left\{\int_0^{2(l(Q)\wedge 1)} t^{-sq}\left[\int_Q\left(t^{-n}\int_{|h|<t}|(f-P_tf)(x+(M-i)h)|\,dh\right)^p dx\right]^{\frac{q}{p}}\frac{dt}{t}\right\}^{\frac{1}{q}}$$
$$\lesssim \frac{1}{|Q|^\tau}\left\{\int_0^{2(l(Q)\wedge 1)} t^{-sq}\left[\int_Q\left(t^{-n}\int_{B(x,Mt)}|(f-P_tf)(y)|\,dy\right)^p dx\right]^{\frac{q}{p}}\frac{dt}{t}\right\}^{\frac{1}{q}}$$
$$\lesssim \frac{1}{|Q|^\tau}\left\{\int_0^{2(l(Q)\wedge 1)} t^{-sq}\left[\int_Q\left(\mathrm{osc}_1^{M-1}f(x,Mt)\right)^p dx\right]^{\frac{q}{p}}\frac{dt}{t}\right\}^{\frac{1}{q}}.$$

Now we turn to the estimate of the term for $i=M$. Recall that for almost every $x\in\mathbb{R}^n$, $f(x)=\lim_{l\to\infty}P_{2^{-l}t}f(x)$ and

$$|P_tf(x)|\le\frac{1}{|B(x,t)|}\int_{B(x,t)}|f(y)|\,dy;$$

see [46, (2.3) and (2.7)]. Furthermore, if $\widetilde{P}\in\mathscr{P}_{M-1}(\mathbb{R}^n)$, then

$$\inf_{P\in\mathscr{P}_{M-1}(\mathbb{R}^n)}\left\{\int_B|f(y)+\widetilde{P}(y)-P(y)|\,dy\right\}=\inf_{P\in\mathscr{P}_{M-1}(\mathbb{R}^n)}\left\{\int_B|f(y)-P(y)|\,dy\right\}.$$

Consequently,

$$|f(x)-P_tf(x)|\le\sum_{l=0}^\infty|P_{2^{-l-1}t}f(x)-P_{2^{-l}t}f(x)|$$
$$\le 2\sum_{l=0}^\infty|P_{2^{-l}t}(f-P_{2^{-l}t}f)(x)|$$
$$\lesssim\sum_{l=0}^\infty\frac{1}{|B(x,2^{-l}t)|}\int_{B(x,2^{-l}t)}|f(y)-P_{2^{-l}t}f(y)|\,dy.$$

Next we employ (4.15) and obtain with $d=\min\{1,p,q\}$ that

$$\left\{\int_0^{2(l(Q)\wedge 1)} t^{-sq}\left[\int_Q\left(t^{-n}\int_{|h|<t}|(f-P_tf)(x)|\,dh\right)^p dx\right]^{\frac{q}{p}}\frac{dt}{t}\right\}^{\frac{d}{q}}$$
$$\lesssim\sum_{l=0}^\infty\left\{\int_0^{2(l(Q)\wedge 1)} t^{-sq}\left[\int_Q\left(\frac{1}{|B(x,2^{-l}t)|}\right.\right.\right.$$
$$\left.\left.\left.\times\int_{B(x,2^{-l}t)}|f(y)-P_{2^{-l}t}f(y)|\,dy\right)^p dx\right]^{\frac{q}{p}}\frac{dt}{t}\right\}^{\frac{d}{q}}$$

$$\lesssim \sum_{l=0}^{\infty} 2^{-lsd} \left\{ \int_0^{2(l(Q)\wedge 1)} \omega^{-sq} \left[\int_Q \left(\frac{1}{|B(x,\omega)|} \right. \right. \right.$$

$$\left. \left. \left. \times \int_{B(x,\omega)} |f(y) - P_\omega f(y)|\, dy \right)^p dx \right]^{\frac{q}{p}} \frac{d\omega}{\omega} \right\}^{\frac{d}{q}}$$

$$\lesssim \left\{ \int_0^{2(l(Q)\wedge 1)} t^{-sq} \left[\int_Q \left(\mathrm{osc}_1^{M-1} f(x,Mt) \right)^p dx \right]^{\frac{q}{p}} \frac{dt}{t} \right\}^{\frac{d}{q}},$$

since $s > 0$. Similarly, for the F-case, we employ (4.16) and proceed as above. This finishes the proof. □

Next we recall Whitney's approximation theorem in a version given in [70, Theorem A.1] but traced to Brudnyi [27] and Nevskii [107]. We refer also to the appendix in [70] for some further remarks concerning the rich history of this result.

Proposition 4.1. *Let $p \in (0,\infty]$, $M \in \mathbb{N}$, $\rho \in (0,1]$ and $f \in L^p_{\mathrm{loc}}(\mathbb{R}^n)$. Then there exists a positive constant $C = C(M,\rho,p,n)$ such that for any cube Q with side length a there is a polynomial $P \in \mathscr{P}_{M-1}(\mathbb{R}^n)$ satisfying*

$$\int_Q |f(y) - P(y)|^p\, dy \le C a^{-n} \int_{|h|<\rho a} \int_Q |\Delta_h^M f(y)|^p\, dy dz. \tag{4.34}$$

As an immediate consequence of (4.34) we obtain the inequality

$$\mathrm{osc}_p^{M-1} f(x,t) \lesssim \left(t^{-2n} \int_{|h|<\rho t} \int_Q |\Delta_h^M f(y)|^p\, dy dh \right)^{1/p}$$

$$\lesssim t^{-n/p} \sup_{|h|<\rho t} \left(\int_Q |\Delta_h^M f(y)|^p\, dy \right)^{1/p}, \tag{4.35}$$

where $B(x,t) \subset Q$ and Q has side length $2t$. Observe, $\mathrm{osc}_p^{M-1} f(x,t)$ is monotone with respect to p. In view of this property and with $p \ge 1$, Lemma 4.10 in combination with (4.35) yields characterizations by means of oscillations as long as the space is characterized by $\|f\|^{\clubsuit}$ and $\|f\|^{\spadesuit}$ simultaneously.

Since we know characterizations of $F^{s,\tau}_{p,q}(\mathbb{R}^n)$ by means of $\|f\|^{\spadesuit}_{F^{s,\tau}_{p,q}(\mathbb{R}^n)}$ only under very restrictive conditions (see Theorem 4.6), we proceed in a different way. We shall compare oscillations and wavelet coefficients. Most of the arguments will be the same as used in the comparison of differences and wavelet coefficients.

We need some more abbreviations. For $P \in \mathscr{Q}$ and $f \in L^1_{\mathrm{loc}}(\mathbb{R}^n)$ we define

$$\|f\|_{\overline{B}(P)} \equiv \frac{1}{|P|^\tau} \left\{ \int_0^{2(l(P)\wedge 1)} t^{-sq} \left[\int_P (\mathrm{osc}_1^{M-1} f(x,t))^p\, dx \right]^{q/p} \frac{dt}{t} \right\}^{1/q},$$

and

$$\|f\|_{\overline{F}(P)} \equiv \frac{1}{|P|^{\tau}} \left\{ \int_P \left[\int_0^{2(l(P)\wedge 1)} t^{-sq} \left(\operatorname{osc}_1^{M-1} f(x,t)\right)^q \frac{dt}{t} \right]^{p/q} dx \right\}^{1/p}.$$

Lemma 4.11. (i) *Let p, q, s, τ be as in Lemma 4.7(i). Then there exists a positive constant C such that for all $Q \in \mathcal{Q}$ and $f \in L^1_{\mathrm{loc}}(\mathbb{R}^n)$,*

$$\|f - W_{j_Q} f\|_{\overline{B}(Q)} \leq C \frac{1}{|Q|^{\tau}} \left\{ \sum_{\omega=(j_Q\vee 0)}^{\infty} 2^{\omega(s+n/2)q} \sum_{i=1}^{2^n-1} \left(2^{-\omega n} \sum_{r \in \mathcal{I}_{Q,\omega}} |a_{i,\omega,r}|^p \right)^{q/p} \right\}^{1/q}.$$

(ii) *Let p, q, s, τ be as in Lemma 4.7(ii). Then there exists a positive constant C such that for all $Q \in \mathcal{Q}$ and $f \in L^1_{\mathrm{loc}}(\mathbb{R}^n)$,*

$$\|f - W_{j_Q} f\|_{\overline{F}(Q)} \leq C \frac{1}{|Q|^{\tau}} \left\| \left[\sum_{\omega=(j_Q\vee 0)}^{\infty} \sum_{i=1}^{2^n-1} \sum_{r \in \mathcal{I}_{Q,\omega}} 2^{wsq} |a_{i,\omega,r} \widetilde{\chi}_{Q_{\omega r}}|^q \right]^{1/q} \right\|_{L^p(Q)}.$$

Proof. We follow the proof of Lemma 4.7 and in particular, employ the same notions as there. Instead the characterization by differences we employ the oscillation characterization of $A^s_{p,q}(\mathbb{R}^n)$ (see [146, Theorem 1.7.3]) in combination with the wavelet characterization of $A^s_{p,q}(\mathbb{R}^n)$ (see Sect. 4.2.1). As above this yields

$$\begin{aligned} \|f - W_{j_Q} f\|_{\overline{F}(Q)} &\leq \frac{1}{|Q|^{\tau}} \|g\|_{F^s_{p,q}(\mathbb{R}^n)} \\ &\sim \frac{1}{|Q|^{\tau}} \left\| \left[\sum_{\omega=(j_Q\vee 0)}^{\infty} \sum_{i=1}^{2^n-1} \sum_{r \in \mathcal{I}_{Q,\omega}} 2^{wsq} |a_{i,\omega,r} \widetilde{\chi}_{Q_{\omega r}}|^q \right]^{1/q} \right\|_{L^p(Q)}, \end{aligned}$$

and similar estimates are true for the B-case. This finishes the proof of Lemma 4.11. □

Now we turn to the counterpart of Lemma 4.8.

Lemma 4.12. *Let $Q \in \mathcal{Q}$ and p, q, M, s, τ be as in Lemma 4.8. Then there exists a positive constant C such that for all $f \in L^1_{\mathrm{loc}}(\mathbb{R}^n)$,*

$$\|W_{j_Q} f\|_{\overline{A}(Q)} \leq C \|f\|_{\overset{\blacktriangle}{A}{}^{s,\tau}_{p,q}(\mathbb{R}^n)}.$$

Proof. We follow the proof of Lemma 4.8. Instead of (4.28) we use the inequality

$$\left| g(y) - \sum_{|\alpha|\le M-1} \frac{1}{\alpha!}\frac{\partial^\alpha g}{\partial y^\alpha}(x)(y-x)^\alpha \right| \lesssim \|g\|_{C^M(\mathbb{R}^n)} |y-x|^M$$

with respect to $g = \phi_{0,k}$ and $g = \psi_{i,j,k}$. No further ideas are needed. Details are left to the reader. □

4.4.2 Oscillations and Besov-Type Spaces

Most of our arguments are based on the characterization by differences. So we keep the structure of displaying the results as there. We shall use the abbreviations

$$\|f\|^{\clubsuit,\mathrm{os}}_{A^{s,\tau}_{p,q}(\mathbb{R}^n)} \equiv \sup_{P\in\mathcal{Q}} \|f\|_{\overline{A}(P)},$$

$$\|f\|^{\clubsuit,\mathrm{osl}}_{A^{s,\tau}_{p,q}(\mathbb{R}^n)} \equiv \sup_{\{P\in\mathcal{Q}:\,|P|\le 1\}} \|f\|_{\overline{A}(P)}$$

with $A \in \{B,F\}$.

Theorem 4.10. *Let $p \in [1,\infty]$, $q \in (0,\infty]$,*

$$0 < s \le \max\{s,\, s+n\tau - n/p\} < M$$

with $M \in \mathbb{N}$ and

$$0 \le \tau < \frac{s}{n} + \frac{1}{p}.$$

(i) *Then $f \in B^{s,\tau}_{p,q}(\mathbb{R}^n)$ if, and only if $f \in L^p_\tau(\mathbb{R}^n)$ and $\|f\|^{\clubsuit,\mathrm{os}}_{B^{s,\tau}_{p,q}(\mathbb{R}^n)} < \infty$, where*

$$\|f\|^{\clubsuit,\mathrm{os}}_{B^{s,\tau}_{p,q}(\mathbb{R}^n)} \equiv \sup_{P\in\mathcal{Q}} \frac{1}{|P|^\tau}\left\{\int_0^{2(l(P)\wedge 1)} t^{-sq} \right.$$
$$\left. \times \left[\int_P \left(\mathrm{osc}_1^{M-1} f(x,t)\right)^p dx\right]^{\frac{q}{p}} \frac{dt}{t}\right\}^{\frac{1}{q}}. \quad (4.36)$$

Further, $\|f\|_{L^p_\tau(\mathbb{R}^n)} + \|f\|^{\clubsuit,\mathrm{os}}_{B^{s,\tau}_{p,q}(\mathbb{R}^n)}$ and $\|f\|_{B^{s,\tau}_{p,q}(\mathbb{R}^n)}$ are equivalent.

(ii) *Let additionally $\tau \in [1/p,\infty)$. Then $f \in B^{s,\tau}_{p,q}(\mathbb{R}^n)$ if, and only if $N_p(f) < \infty$ and $\|f\|^{\clubsuit,\mathrm{osl}}_{B^{s,\tau}_{p,q}(\mathbb{R}^n)} < \infty$. Further, $N_p(f) + \|f\|^{\clubsuit,\mathrm{osl}}_{B^{s,\tau}_{p,q}(\mathbb{R}^n)}$ and $\|f\|_{B^{s,\tau}_{p,q}(\mathbb{R}^n)}$ are equivalent.*

Proof. The inequality

$$\|f\|_{L^p_\tau(\mathbb{R}^n)} + \|f\|^{\clubsuit,\mathrm{os}}_{B^{s,\tau}_{p,q}(\mathbb{R}^n)} \lesssim \|f\|_{B^{s,\tau}_{p,q}(\mathbb{R}^n)}$$

follows from Theorem 4.7 and (4.35). The inequality

$$\|f\|_{B^{s,\tau}_{p,q}(\mathbb{R}^n)} \lesssim \|f\|_{L^p_\tau(\mathbb{R}^n)} + \|f\|^{\clubsuit,\mathrm{os}}_{B^{s,\tau}_{p,q}(\mathbb{R}^n)}$$

is a consequence of Theorem 4.7 and Lemma 4.10. □

Remark 4.9. The same proof yields the following: if we replace $\|f\|^{\clubsuit,\mathrm{os}}_{B^{s,\tau}_{p,q}(\mathbb{R}^n)}$ by

$$\|f\|^{\heartsuit,\mathrm{os}}_{B^{s,\tau}_{p,q}(\mathbb{R}^n)} \equiv \sup_{P\in\mathscr{Q}} \frac{1}{|P|^\tau} \left\{ \int_0^{2(l(P)\wedge 1)} t^{-sq} \left[\int_P \left(\mathrm{osc}^{M-1}_p f(x,t)\right)^p dx \right]^{q/p} \frac{dt}{t} \right\}^{1/q},$$

then all assertions in Theorem 4.10 remain true. Also in this case we may replace $\|f\|^{\heartsuit,\mathrm{os}}_{B^{s,\tau}_{p,q}(\mathbb{R}^n)}$ by $\|f\|^{\heartsuit,\mathrm{osl}}_{B^{s,\tau}_{p,q}(\mathbb{R}^n)}$, defined in analogy to $\|f\|^{\clubsuit,\mathrm{osl}}_{B^{s,\tau}_{p,q}(\mathbb{R}^n)}$, if $\tau \ge 1/p$.

Next we consider the case $p\in(0,1)$ and $\tau\in[1/p,\infty)$.

Theorem 4.11. *Let $p\in(0,1)$, $q\in(0,\infty]$,*

$$\sigma_p < s \le \max\{s, s+n\tau-n/p\} < M$$

with $M\in\mathbb{N}$ and

$$\frac{1}{p} \le \tau < 1+\frac{s}{n}.$$

Then $f\in B^{s,\tau}_{p,q}(\mathbb{R}^n)$ if, and only if $N_1(f)<\infty$ and $\|f\|^{\clubsuit,\mathrm{osl}}_{B^{s,\tau}_{p,q}(\mathbb{R}^n)}<\infty$. Further, $N_1(f)+\|f\|^{\clubsuit,\mathrm{osl}}_{B^{s,\tau}_{p,q}(\mathbb{R}^n)}$ and $\|f\|_{B^{s,\tau}_{p,q}(\mathbb{R}^n)}$ are equivalent.

Proof. We combine Theorem 4.8, Lemmas 4.10–4.12 and Theorem 4.1. □

Finally we study the case $p\in(0,1)$ and $\tau\in[0,1/p)$.

Theorem 4.12. *Let $p\in(0,1)$, $q\in(0,\infty]$,*

$$\sigma_p < s \le \max\{s, s+n\tau-n/p\} < M$$

with $M\in\mathbb{N}$ and

$$0\le\tau<\min\left\{1+\frac{s}{n},\frac{1}{p}\right\}.$$

Let $\sigma_p<s_0<s$. Then $f\in B^{s,\tau}_{p,q}(\mathbb{R}^n)$ if, and only if $\Phi_{\tau,p}(f)<\infty$,

$$\sup_{\{P\in\mathscr{Q},\,|P|\ge1\}} \frac{\|f\|_{B^{s_0}_{p,\infty}(2P)}}{|P|^\tau} < \infty,$$

and $\|f\|^{\clubsuit,\mathrm{os}}_{B^{s,\tau}_{p,q}(\mathbb{R}^n)} < \infty$. *Further,*

$$\sup_{\{P\in\mathcal{Q},\,|P|\geq 1\}} \frac{\|f\|_{B^{s_0}_{p,\infty}(2P)}}{|P|^\tau} + \Phi_{\tau,p}(f) + \|f\|_{L^p_\tau(\mathbb{R}^n)} + \|f\|^{\clubsuit,\mathrm{os}}_{B^{s,\tau}_{p,q}(\mathbb{R}^n)}$$

and $\|f\|_{B^{s,\tau}_{p,q}(\mathbb{R}^n)}$ *are equivalent.*

Proof. We combine Theorem 4.9, Lemmas 4.10–4.12 and Theorem 4.1. □

Remark 4.10.

(i) Further characterizations by oscillations for $B^{s,\tau}_{p,\infty}(\mathbb{R}^n)$ can be obtained from the identity $B^{s,1/u-1/p}_{u,\infty}(\mathbb{R}^n) = \mathcal{N}^s_{p\infty u}(\mathbb{R}^n)$ (see Corollary 3.3(ii)) in combination with Corollary 4.17 below.

(ii) For $\tau = 0$ we refer to Triebel [146, Sect. 3.5.2]. Some more references can be found in [146, Sect. 1.7].

4.4.3 Oscillations and Triebel-Lizorkin-Type Spaces

Now we are in the position to formulate the results with respect to $F^{s,\tau}_{p,q}(\mathbb{R}^n)$.

Theorem 4.13. *Let* $p \in [1,\infty)$, $q \in (0,\infty]$,

$$\sigma_{1,q} < s \leq \max\{s, s+n\tau-n/p\} < M$$

with $M \in \mathbb{N}$ *and* $\tau \in [0, \tau_{s,p,q})$.

(i) *Then* $f \in F^{s,\tau}_{p,q}(\mathbb{R}^n)$ *if, and only if* $f \in L^p_\tau(\mathbb{R}^n)$ *and* $\|f\|^{\clubsuit,\mathrm{os}}_{F^{s,\tau}_{p,q}(\mathbb{R}^n)} < \infty$. *Further,* $\|f\|_{L^p_\tau(\mathbb{R}^n)} + \|f\|^{\clubsuit,\mathrm{os}}_{F^{s,\tau}_{p,q}(\mathbb{R}^n)}$ *and* $\|f\|_{F^{s,\tau}_{p,q}(\mathbb{R}^n)}$ *are equivalent.*

(ii) *Let additionally* $\tau \in [1/p,\infty)$. *Then* $f \in F^{s,\tau}_{p,q}(\mathbb{R}^n)$ *if, and only if* $N_p(f) < \infty$ *and* $\|f\|^{\clubsuit,\mathrm{osl}}_{F^{s,\tau}_{p,q}(\mathbb{R}^n)} < \infty$. *Further,* $N_p(f) + \|f\|^{\clubsuit,\mathrm{osl}}_{F^{s,\tau}_{p,q}(\mathbb{R}^n)}$ *and* $\|f\|_{F^{s,\tau}_{p,q}(\mathbb{R}^n)}$ *are equivalent.*

Proof. We combine Theorem 4.13, Lemma 4.10, Theorem 4.1(ii) and Lemmas 4.11, 4.12. □

Next we consider the case $p \in (0,1)$ and $\tau \in [1/p,\infty)$.

Theorem 4.14. *Let* $p \in (0,1)$, $q \in (0,\infty]$,

$$\sigma_{p,q} < s \leq \max\{s, s+n\tau-n/p\} < M$$

with $M \in \mathbb{N}$ *and* $\tau \in [1/p, \tau_{s,p,q})$. *Then* $f \in F^{s,\tau}_{p,q}(\mathbb{R}^n)$ *if, and only if* $N_1(f) < \infty$ *and* $\|f\|^{\clubsuit,\mathrm{osl}}_{F^{s,\tau}_{p,q}(\mathbb{R}^n)} < \infty$. *Further,* $N_1(f) + \|f\|^{\clubsuit,\mathrm{osl}}_{F^{s,\tau}_{p,q}(\mathbb{R}^n)}$ *and* $\|f\|_{F^{s,\tau}_{p,q}(\mathbb{R}^n)}$ *are equivalent.*

Proof. We combine Theorem 4.14, Lemma 4.10, Theorem 4.1(ii) and Lemmas 4.11, 4.12. □

Finally we study the case $p \in (0,1)$ and $\tau \in [0,1/p)$.

Theorem 4.15. *Let $p \in (0,1)$, $q \in (0,\infty]$,*

$$\sigma_{p,q} < s \leq \max\{s, s+n\tau - n/p\} < M$$

with $M \in \mathbb{N}$ and $\tau \in [0, \min\{\tau_{s,p,q}, 1/p\})$. Let $\sigma_p < s_0 < s$. Then $f \in F^{s,\tau}_{p,q}(\mathbb{R}^n)$ if, and only if $\Phi_{\tau,p}(f) < \infty$,

$$\sup_{\{P \in \mathcal{Q}, |P| \geq 1\}} \frac{\|f\|_{B^{s_0}_{p,\infty}(2P)}}{|P|^\tau} < \infty,$$

and $\|f\|^{\clubsuit,\mathrm{os}}_{F^{s,\tau}_{p,q}(\mathbb{R}^n)} < \infty$. Further,

$$\sup_{\{P \in \mathcal{Q}, |P| \geq 1\}} \frac{\|f\|_{B^{s_0}_{p,\infty}(2P)}}{|P|^\tau} + \Phi_{\tau,p}(f) + \|f\|_{L^p_\tau(\mathbb{R}^n)} + \|f\|^{\clubsuit,\mathrm{os}}_{F^{s,\tau}_{p,q}(\mathbb{R}^n)}$$

and $\|f\|_{F^{s,\tau}_{p,q}(\mathbb{R}^n)}$ are equivalent.

Proof. We combine Theorem 4.14, Lemma 4.10, Theorem 4.1(ii), Lemmas 4.11 and 4.12. □

Remark 4.11.
(i) Further characterizations by oscillations for $F^{s,\tau}_{p,q}(\mathbb{R}^n)$ can be obtained from the identity $F^{s,1/u-1/p}_{u,q}(\mathbb{R}^n) = \mathcal{E}^s_{pqu}(\mathbb{R}^n)$ (see Corollary 3.3(i)) in combination with Corollary 4.16 below.
(ii) For $\tau = 0$ we refer to Seeger [130] and Triebel [146, Sect. 3.5.2].
(iii) A technical remark. Let $C \geq 2$ be a fixed positive constant. All the theorems in Sects. 4.3 and 4.4 remain true if we replace $2(l(P) \wedge 1)$ by $C(l(P) \wedge 1)$.

4.4.4 Oscillations and $F^s_{\infty,q}(\mathbb{R}^n)$

We concentrate on a combination of Proposition 2.4, we will use the identity $F^{s,1/q}_{q,q}(\mathbb{R}^n) = B^{s,1/q}_{q,q}(\mathbb{R}^n) = F^s_{\infty,q}(\mathbb{R}^n)$, and Theorem 4.10 if $q \in [1,\infty)$ or Theorem 4.11 if $q \in (0,1)$. The case $q = \infty$ is contained in Theorem 4.10.

First we study $q \in [1,\infty)$.

Corollary 4.8. *Let $q \in [1,\infty)$ and $s \in (0,M)$ with $M \in \mathbb{N}$. Then $f \in F^s_{\infty,q}(\mathbb{R}^n)$ if, and only if $N_q(f) < \infty$ and*

$$\|f\|_{F^s_{\infty,q}(\mathbb{R}^n)}^{\heartsuit,\mathrm{osl}} \equiv \sup_{\{P\in\mathscr{Q}:\,|P|\le 1\}} \left\{\frac{1}{|P|}\int_0^{2l(P)} t^{-sq}\left(\mathrm{osc}_q^{M-1} f(x,t)\right)^q \frac{dt}{t}\right\}^{1/q} < \infty.$$

Further, $N_q(f) + \|f\|_{F^s_{\infty,q}(\mathbb{R}^n)}^{\heartsuit,\mathrm{osl}}$ *and* $\|f\|_{F^s_{\infty,q}(\mathbb{R}^n)}$ *are equivalent.*

We continue with $q \in (0,1)$.

Corollary 4.9. *Let* $q \in (0,1)$ *and*

$$\frac{n}{q} - n < s < M$$

with $M \in \mathbb{N}$. *Then* $f \in F^s_{\infty,q}(\mathbb{R}^n)$ *if, and only if* $N_1(f) < \infty$ *and*

$$\|f\|_{F^s_{\infty,q}(\mathbb{R}^n)}^{\clubsuit,\mathrm{osl}} \equiv \sup_{\{P\in\mathscr{Q}:\,|P|\le 1\}} \left\{\frac{1}{|P|}\int_0^{2l(P)} t^{-sq}\left(\mathrm{osc}_1^{M-1} f(x,t)\right)^q \frac{dt}{t}\right\}^{1/q} < \infty.$$

Furthermore, $N_1(f) + \|f\|_{F^s_{\infty,q}(\mathbb{R}^n)}^{\clubsuit,\mathrm{osl}}$ *and* $\|f\|_{F^s_{\infty,q}(\mathbb{R}^n)}$ *are equivalent.*

Remark 4.12. Let $q \in (0,\infty)$. Since $F^s_{\infty,q}(\mathbb{R}^n) \subset L^\infty(\mathbb{R}^n)$ if $s > 0$, it is not difficult to prove that under the above restrictions, $f \in F^s_{\infty,q}(\mathbb{R}^n)$ if, and only if $f \in L^\infty(\mathbb{R}^n)$ and $\|f\|_{F^s_{\infty,q}(\mathbb{R}^n)}^{\clubsuit,\mathrm{osl}} < \infty$.

4.5 The Hedberg-Netrusov Approach to Spaces of Besov-Triebel-Lizorkin Type

In [70] Hedberg and Netrusov developed an axiomatic approach to function spaces of Besov-Triebel-Lizorkin type. This general theory covers our spaces $F^{s,\tau}_{p,q}(\mathbb{R}^n)$ and $B^{s,\tau}_{p,q}(\mathbb{R}^n)$, at least under some restrictions. We shall consider two different realizations of the Hedberg-Netrusov approach.

4.5.1 Some Preliminaries

Recall, Morrey spaces $\mathscr{M}^p_u(\mathbb{R}^n)$, $0 < u \le p \le \infty$ have been defined in item (xvi) in Sect. 1.3. Let $q \in (0,\infty]$ and $s \in \mathbb{R}$. Then we define two types of quasi-Banach spaces of locally Lebesgue-integrable functions. By $\ell^s_q(\mathscr{M}^p_u(\mathbb{R}^n))$ we denote the collection of all sequences $\{f_j\}_{j=0}^\infty$ of those functions such that

$$\|\{f_j\}_{j=0}^\infty\|_{\ell^s_q(\mathscr{M}^p_u(\mathbb{R}^n))} \equiv \left(\sum_{j=0}^\infty 2^{jsq}\|f_j\|^q_{\mathscr{M}^p_u(\mathbb{R}^n)}\right)^{1/q} < \infty.$$

Furthermore, by $\mathscr{M}^p_u(\mathbb{R}^n)(\ell^s_q)$ we denote the collection of all sequences $\{f_j\}_{j=0}^\infty$ of locally Lebesgue-integrable functions such that

$$\|\{f_j\}_{j=0}^\infty\|_{\mathscr{M}^p_u(\mathbb{R}^n)(\ell^s_q)} \equiv \left\| \left(\sum_{j=0}^\infty 2^{jsq} |f_j|^q \right)^{1/q} \right\|_{\mathscr{M}^p_u(\mathbb{R}^n)} < \infty.$$

Of course, both $\ell^s_q(\mathscr{M}^p_u(\mathbb{R}^n))$ and $\mathscr{M}^p_u(\mathbb{R}^n)(\ell^s_q)$, are quasi-Banach spaces. With $d = \min\{1, u, q\}$ it holds

$$\|\{f_j\}_{j=0}^\infty + \{g_j\}_{j=0}^\infty\|^d_{\ell^s_q(\mathscr{M}^p_u(\mathbb{R}^n))} \le \|\{f_j\}_{j=0}^\infty\|^d_{\ell^s_q(\mathscr{M}^p_u(\mathbb{R}^n))} + \|\{g_j\}_{j=0}^\infty\|^d_{\ell^s_q(\mathscr{M}^p_u(\mathbb{R}^n))},$$

and

$$\|\{f_j\}_{j=0}^\infty + \{g_j\}_{j=0}^\infty\|^d_{\mathscr{M}^p_u(\mathbb{R}^n)(\ell^s_q)} \le \|\{f_j\}_{j=0}^\infty\|^d_{\mathscr{M}^p_u(\mathbb{R}^n)(\ell^s_q)} + \|\{g_j\}_{j=0}^\infty\|^d_{\mathscr{M}^p_u(\mathbb{R}^n)(\ell^s_q)}$$

for all sequences $\{f_j\}_{j=0}^\infty$ and $\{g_j\}_{j=0}^\infty$ of locally Lebesgue-integrable functions. The maximal operator

$$\mathrm{M}_r f(x) \equiv \left(\sup_{a>0} a^{-n} \int_{B(0,a)} |f(x+y)|^r \, dy \right)^{1/r}$$

is bounded on these two types of spaces if r is chosen in an appropriate way. Indeed, with $w \equiv p/r$ we find

$$\begin{aligned}
\|\mathrm{M}_r f\|_{\mathscr{M}^p_u(\mathbb{R}^n)} &= \left[\sup_B |B|^{\frac{1}{w}-\frac{r}{u}} \left(\int_B [\mathrm{M}(|f|^r)(x)]^{u/r} \, dx \right)^{r/u} \right]^{1/r} \\
&= \left[\|\mathrm{M}(|f|^r)\|_{\mathscr{M}^w_{u/r}(\mathbb{R}^n)} \right]^{1/r} \\
&\lesssim \left[\||f|^r\|_{\mathscr{M}^w_{u/r}(\mathbb{R}^n)} \right]^{1/r} \\
&\lesssim \|f\|_{\mathscr{M}^p_u(\mathbb{R}^n)}
\end{aligned}$$

holds if $0 < r < u \le p \le \infty$; see Chiarenza and Frasca [39]. Immediately one obtains

$$\|\{\mathrm{M}_r f_j\}_{j=0}^\infty\|_{\ell^s_q(\mathscr{M}^p_u(\mathbb{R}^n))} \lesssim \|\{f_j\}_{j=0}^\infty\|_{\ell^s_q(\mathscr{M}^p_u(\mathbb{R}^n))} \tag{4.37}$$

under the same restrictions. The vector-valued situation has been treated by Tang and Xu [139]. We have

$$\|\{\mathrm{M}_r f_j\}_{j=0}^\infty\|_{\mathscr{M}^p_u(\mathbb{R}^n)(\ell^s_q)} \le \|\{f_j\}_{j=0}^\infty\|_{\mathscr{M}^p_u(\mathbb{R}^n)(\ell^s_q)}, \tag{4.38}$$

if $0 < r < \min\{u, q\} \le u \le p \le \infty$. Finally, we have to estimate the norms of the shift operators. Let

$$S_+(\{f_j\}_{j=0}^\infty) \equiv \{f_{j+1}\}_{j=0}^\infty \quad \text{and} \quad S_-(\{f_j\}_{j=0}^\infty) \equiv \{f_{j-1}\}_{j=0}^\infty,$$

$f_{-1} = 0$. Obviously,

$$\| (S_+)^N \,|\mathscr{L}(E)\| \lesssim 2^{-Ns}$$

and

$$\| (S_-)^N \,|\mathscr{L}(E)\| \lesssim 2^{Ns},$$

where E stands either for $\ell_q^s(\mathscr{M}_u^p(\mathbb{R}^n))$ or for $\mathscr{M}_u^p(\mathbb{R}^n)(\ell_q^s)$, and with constants behind $\lesssim$ independent of N. These observations can be summarized by saying that both types of spaces belong to the class $S(s,s,r)$ with r as above; see [70, p. 6].

4.5.2 Characterizations by Differences

Now we employ [70, Proposition 1.1.12, Theorem 1.1.14] with $E = \mathscr{M}_u^p(\mathbb{R}^n)(\ell_q^s)$. Recall that the Triebel-Lizorkin-Morrey spaces $\mathscr{E}_{pqu}^s(\mathbb{R}^n)$ have been defined in item (xxvi) in Sect. 1.3.

Corollary 4.10. *Let $v \in (0,\infty]$, $u \in (0,\infty)$, $p \in [u,\infty]$, and $s \in \mathbb{R}$ such that*

$$r \in (0, \min\{u,q\}) \quad \text{and} \quad s > n \max\left\{\frac{1}{r} - 1, \frac{1}{r} - \frac{1}{v}\right\}.$$

Then the following assertions are equivalent for functions in $L_{\text{loc}}^r(\mathbb{R}^n)$:

(i) $f \in \mathscr{E}_{pqu}^s(\mathbb{R}^n)$;
(ii) $f \in L_{\text{loc}}^v(\mathbb{R}^n)$ *and*

$$\|f\|_{\mathscr{E}_{pqu}^s(\mathbb{R}^n)}^\sharp \equiv \Big\| \|f\|_{L^v(B(x,1))} \Big\|_{\mathscr{M}_u^p(\mathbb{R}^n)} + \left\| \left(\sum_{j=1}^\infty 2^{j(s+\frac{n}{v})q} \left[\int_{B(x,2^{-j})} |\Delta_h^M f(x)|^v \, dh \right]^{q/v} \right)^{1/q} \right\|_{\mathscr{M}_u^p(\mathbb{R}^n)} < \infty.$$

The quasi-norms $\|f\|_{\mathscr{E}_{pqu}^s(\mathbb{R}^n)}$ and $\|f\|_{\mathscr{E}_{pqu}^s(\mathbb{R}^n)}^\sharp$ are equivalent.

For us it will be convenient to reformulate this a bit. We shall use the abbreviation

$$b_{v,t}(x) \equiv \left(\frac{1}{t^n} \int_{B(x,t)} |\Delta_h^M f(x)|^v \, dh \right)^{1/v}.$$

Then, using the monotonicity of $b_{v,t}$ with respect to t we find

$$\left\| \left(\sum_{j=1}^{\infty} 2^{j(s+\frac{n}{v})q} \left[\int_{B(x,2^{-j})} |\Delta_h^M f(x)|^v \, dh \right]^{q/v} \right)^{1/q} \right\|_{\mathscr{M}_u^p(\mathbb{R}^n)} \sim \left\| \left(\int_0^1 t^{-sq} \, b_{v,t}^q(x) \, \frac{dt}{t} \right)^{1/q} \right\|_{\mathscr{M}_u^p(\mathbb{R}^n)} .$$

Replacing $\mathscr{M}_u^p(\mathbb{R}^n)$ by its original definition we get the following supplement to Corollary 4.10.

Corollary 4.11. *Let u, p, q, v, r, s be as in Corollary 4.10. Then the following assertions are equivalent for functions in $L_{\mathrm{loc}}^r(\mathbb{R}^n)$:*

(i) $f \in \mathscr{E}_{pqu}^s(\mathbb{R}^n)$;
(ii) $f \in L_{\mathrm{loc}}^v(\mathbb{R}^n)$ *and*

$$\|f\|^{\natural}_{\mathscr{E}_{pqu}^s(\mathbb{R}^n)} \equiv \sup_{P \in \mathscr{Q}} \frac{1}{|P|^{\frac{1}{u}-\frac{1}{p}}} \left(\int_P \left[\int_{B(x,1)} |f(y)|^v \, dy \right]^{u/v} dx \right)^{1/u} + \sup_{P \in \mathscr{Q}} \frac{1}{|P|^{\frac{1}{u}-\frac{1}{p}}} \left(\int_P \left[\int_0^1 t^{-sq} \, b_{v,t}^q(x) \, \frac{dt}{t} \right]^{u/q} dx \right)^{1/u} < \infty .$$

The quasi-norms $\|f\|_{\mathscr{E}_{pqu}^s(\mathbb{R}^n)}$ and $\|f\|^{\natural}_{\mathscr{E}_{pqu}^s(\mathbb{R}^n)}$ are equivalent.

Remark 4.13. In view of the identity $F_{u,q}^{s,1/u-1/p}(\mathbb{R}^n) = \mathscr{E}_{pqu}^s(\mathbb{R}^n)$ (see Corollary 3.3(i)), the above corollary supplements the results of Sect. 4.3.1.

The next corollary deals with Besov-Morrey spaces $\mathscr{N}_{pqu}^s(\mathbb{R}^n)$; see item (xxv) in Sect. 1.3. Again we employ Proposition 1.1.12 and Theorem 1.1.14 in [70], this time with $E = \ell_q^s(\mathscr{M}_u^p(\mathbb{R}^n))$.

Corollary 4.12. *Let $s \in \mathbb{R}$, $v \in (0,\infty]$ and $0 < r < u \le p \le \infty$ such that*

$$s > n \max\left\{ \frac{1}{r} - 1, \frac{1}{r} - \frac{1}{v} \right\} .$$

Then the following assertions are equivalent for functions in $L_{\mathrm{loc}}^r(\mathbb{R}^n)$:

(i) $f \in \mathscr{N}_{pqu}^s(\mathbb{R}^n)$;
(ii) $f \in L_{\mathrm{loc}}^v(\mathbb{R}^n)$ *and*

$$\|f\|^{\sharp}_{\mathscr{N}_{pqu}^s(\mathbb{R}^n)} \equiv \left\| \, \|f\|_{L^v(B(x,1))} \right\|_{\mathscr{M}_u^p(\mathbb{R}^n)} + \left(\sum_{j=1}^{\infty} 2^{j(s+\frac{n}{v})q} \left\| \left(\int_{B(x,2^{-j})} |\Delta_h^M f(x)|^v \, dh \right)^{1/v} \right\|^q_{\mathscr{M}_u^p(\mathbb{R}^n)} \right)^{1/q} < \infty .$$

The quasi-norms $\|f\|_{\mathscr{N}^s_{pqu}(\mathbb{R}^n)}$ *and* $\|f\|^{\sharp}_{\mathscr{N}^s_{pqu}(\mathbb{R}^n)}$ *are equivalent.*

As above this can be reformulated a bit.

Corollary 4.13. *Let* u, p, q, v, r, s *be as in Corollary 4.12. Then the following assertions are equivalent for functions in* $L^r_{\mathrm{loc}}(\mathbb{R}^n)$*:*

(i) $f \in \mathscr{N}^s_{pqu}(\mathbb{R}^n)$;
(ii) $f \in L^v_{\mathrm{loc}}(\mathbb{R}^n)$ *and*

$$\|f\|^{\natural}_{\mathscr{N}^s_{pqu}(\mathbb{R}^n)} \equiv \sup_{P\in\mathscr{Q}} \frac{1}{|P|^{\frac{1}{u}-\frac{1}{p}}} \left(\int_P \left[\int_{B(x,1)} |f(y)|^v \, dy \right]^{u/v} dx \right)^{1/u}$$
$$+ \sup_{P\in\mathscr{Q}} \frac{1}{|P|^{\frac{1}{u}-\frac{1}{p}}} \left(\int_0^1 t^{-sq} \left[\int_P |b_{v,t}(x)|^u dx \right]^{q/u} \frac{dt}{t} \right)^{1/q} < \infty.$$

The quasi-norms $\|f\|_{\mathscr{N}^s_{pqu}(\mathbb{R}^n)}$ *and* $\|f\|^{\natural}_{\mathscr{N}^s_{pqu}(\mathbb{R}^n)}$ *are equivalent.*

Remark 4.14. In view of the identity $B^{s,1/u-1/p}_{u,\infty}(\mathbb{R}^n) = \mathscr{N}^s_{p\infty u}(\mathbb{R}^n)$ (see Corollary 3.3(i)), the above corollary supplements the results of Sect. 4.3.2.

4.5.3 Characterizations by Oscillations

Corollary 4.10 and 4.12 have direct counterparts in [70] which we now recall.

Corollary 4.14. *Let* v, u, r, p, s *be as in Corollary 4.10. Then the following assertions are equivalent for functions in* $L^r_{\mathrm{loc}}(\mathbb{R}^n)$*:*

(i) $f \in \mathscr{E}^s_{pqu}(\mathbb{R}^n)$;
(ii) $f \in L^v_{\mathrm{loc}}(\mathbb{R}^n)$ *and* $\|f\|^{\#}_{\mathscr{E}^s_{pqu}(\mathbb{R}^n)} < \infty$*, where*

$$\|f\|^{\#}_{\mathscr{E}^s_{pqu}(\mathbb{R}^n)} \equiv \left\| \, \|f\|_{L^v(B(\cdot,1))} \right\|_{\mathscr{M}^p_u(\mathbb{R}^n)} + \left\| \left(\sum_{j=1}^{\infty} 2^{jsq} \left[\mathrm{osc}_v f(\cdot, 2^{-j}) \right]^q \right)^{1/q} \right\|_{\mathscr{M}^p_u(\mathbb{R}^n)}.$$

The quasi-norms $\|f\|_{\mathscr{E}^s_{pqu}(\mathbb{R}^n)}$ *and* $\|f\|^{\#}_{\mathscr{E}^s_{pqu}(\mathbb{R}^n)}$ *are equivalent.*

Corollary 4.15. *Let* v, u, r, p, s *be as in Corollary 4.12. Then the following assertions are equivalent for functions in* $L^r_{\mathrm{loc}}(\mathbb{R}^n)$*:*

(i) $f \in \mathscr{N}^s_{pqu}(\mathbb{R}^n)$;

(ii) $f \in L^v_{\mathrm{loc}}(\mathbb{R}^n)$ *and* $\|f\|^{\#}_{\mathcal{N}^s_{pqu}(\mathbb{R}^n)} < \infty$, *where*

$$\|f\|^{\#}_{\mathcal{N}^s_{pqu}(\mathbb{R}^n)} \equiv \left\| \|f\|_{L^v(B(\cdot,1))} \right\|_{\mathcal{M}^p_u(\mathbb{R}^n)} + \left(\sum_{j=1}^{\infty} 2^{jsq} \| \operatorname{osc}_v f(\cdot, 2^{-j}) \|^q_{\mathcal{M}^p_u(\mathbb{R}^n)} \right)^{1/q}.$$

The quasi-norms $\|f\|_{\mathcal{N}^s_{pqu}(\mathbb{R}^n)}$ *and* $\|f\|^{\#}_{\mathcal{N}^s_{pqu}(\mathbb{R}^n)}$ *are equivalent.*

The quantities $\operatorname{osc}_u f(x,t)$ are not monotone in general. However, there exists a positive constant C such that for all $f \in L^v_{\mathrm{loc}}(\mathbb{R}^n)$ and all $x \in \mathbb{R}^n$,

$$\operatorname{osc}_v f(x,t) \le C \operatorname{osc}_v f(x,\bar{t}) \qquad \text{with} \qquad t < \bar{t} < 2t.$$

This is sufficient for establishing the counterparts of Corollaries 4.11 and 4.13.

Corollary 4.16. *Let v, u, r, p, s be as in Corollary 4.10. Then the following assertions are equivalent for functions in* $L^r_{\mathrm{loc}}(\mathbb{R}^n)$*:*

(i) $f \in \mathcal{E}^s_{pqu}(\mathbb{R}^n)$;
(ii) $f \in L^v_{\mathrm{loc}}(\mathbb{R}^n)$ *and*

$$\begin{aligned} |||f|||^{\#}_{\mathcal{E}^s_{pqu}(\mathbb{R}^n)} &\equiv \sup_{P \in \mathcal{Q}} \frac{1}{|P|^{\frac{1}{u}-\frac{1}{p}}} \left(\int_P \left[\int_{B(x,1)} |f(y)|^v \, dy \right]^{u/v} dx \right)^{1/u} \\ &+ \left\| \left(\int_0^1 t^{-sq} [\operatorname{osc}_v f(x,t)]^q \frac{dt}{t} \right)^{1/q} \right\|_{\mathcal{M}^p_u(\mathbb{R}^n)} < \infty. \end{aligned}$$

The quasi-norms $\|f\|_{\mathcal{E}^s_{pqu}(\mathbb{R}^n)}$ *and* $|||f|||^{\#}_{\mathcal{E}^s_{pqu}(\mathbb{R}^n)}$ *are equivalent.*

Remark 4.15. In view of the identity $F^{s,1/u-1/p}_{u,q}(\mathbb{R}^n) = \mathcal{E}^s_{pqu}(\mathbb{R}^n)$ (see Corollary 3.3(i)), the above corollary supplements the results of Sect. 4.4.3.

Corollary 4.17. *Let v, u, r, p, s be as in Corollary 4.12. Then the following assertions are equivalent for functions in* $L^r_{\mathrm{loc}}(\mathbb{R}^n)$*:*

(i) $f \in \mathcal{N}^s_{pqu}(\mathbb{R}^n)$;
(ii) $f \in L^v_{\mathrm{loc}}(\mathbb{R}^n)$ *and*

$$\begin{aligned} |||f|||^{\#}_{\mathcal{N}^s_{pqu}(\mathbb{R}^n)} &\equiv \sup_{P \in \mathcal{Q}} \frac{1}{|P|^{\frac{1}{u}-\frac{1}{p}}} \left(\int_P \left[\int_{B(x,1)} |f(y)|^v \, dy \right]^{u/v} dx \right)^{1/u} \\ &+ \left(\int_0^1 t^{-sq} \| \operatorname{osc}_v f(x,t) \|^q_{\mathcal{M}^p_u(\mathbb{R}^n)} \frac{dt}{t} \right)^{1/q} < \infty. \end{aligned}$$

The quasi-norms $\|f\|_{\mathcal{N}^s_{pqu}(\mathbb{R}^n)}$ *and* $|||f|||^{\#}_{\mathcal{N}^s_{pqu}(\mathbb{R}^n)}$ *are equivalent.*

Remark 4.16.
(i) In view of the identity

$$B_{u,\infty}^{s,1/u-1/p}(\mathbb{R}^n) = \mathscr{N}_{p\infty u}^{s}(\mathbb{R}^n)$$

(see Corollary 3.3(i)), the above corollary supplements the results of Sect. 4.4.2.
(ii) In [70] also the characterizations of $\mathscr{E}_{pqu}^{s}(\mathbb{R}^n)$ and $\mathscr{N}_{pqu}^{s}(\mathbb{R}^n)$ in terms of atoms are given.

4.6 A Characterization of $A_{p,q}^{s,\tau}(\mathbb{R}^n)$ via a Localization of $A_{p,q}^{s}(\mathbb{R}^n)$

In this section we always assume that

$$0 < s < 1 \qquad \text{and} \qquad 0 < s + n\tau - \frac{n}{p} < 1. \tag{4.39}$$

Our main tool will be the characterization of $A_{p,q}^{s,\tau}(\mathbb{R}^n)$ by means of first order differences; see Sect. 4.3. Under these restrictions we establish a characterization of $A_{p,q}^{s,\tau}(\mathbb{R}^n)$ via a localization of $A_{p,q}^{s}(\mathbb{R}^n)$, an idea which we picked up from [5], where such characterizations are investigated for $F_{\infty,2}^{s}(\mathbb{R}^n)$. Restricted to this section, we define

$$\|f\|_B \equiv \left\{\int_0^2 t^{-sq}\left\|\left(t^{-n}\int_{t/2\le|h|<t}|f(\cdot+h)-f(\cdot)|^p\,dh\right)^{1/p}\right\|_{L^p(\mathbb{R}^n)}^q\frac{dt}{t}\right\}^{1/q},$$

$$\|f\|_F^{\clubsuit} \equiv \left\|\left\{\int_0^2 t^{-sq}\left(t^{-n}\int_{t/2\le|h|<t}|f(\cdot+h)-f(\cdot)|\,dh\right)^q\frac{dt}{t}\right\}^{1/q}\right\|_{L^p(\mathbb{R}^n)}$$

and

$$\|f\|_F^{\heartsuit} \equiv \left\|\left\{\int_0^2 t^{-sq}t^{-n}\int_{t/2\le|h|<t}|f(\cdot+h)-f(\cdot)|^q\,dh\frac{dt}{t}\right\}^{1/q}\right\|_{L^p(\mathbb{R}^n)}.$$

Then [145, Theorem 2.5.9] implies that $\|f\|_{L^p(\mathbb{R}^n)}+\|f\|_A$ and $\|f\|_{A_{p,q}^{s}(\mathbb{R}^n)}$ are equivalent quasi-norms in $A_{p,q}^{s}(\mathbb{R}^n)$ ($s>\sigma_{p,q}$ if $A_{p,q}^{s}(\mathbb{R}^n)$=$F_{p,q}^{s}(\mathbb{R}^n)$). Further, recall that by Proposition 2.6,

$$A_{p,q}^{s,\tau}(\mathbb{R}^n) \subset \mathscr{Z}^{s+n\tau-n/p}(\mathbb{R}^n) = C^{s+n\tau-n/p}(\mathbb{R}^n)$$

since $0<s+n\tau-n/p<1$. Thus, we only deal with Hölder continuous functions.

4.6.1 A Characterization of $B^{s,\tau}_{p,q}(\mathbb{R}^n)$

We need some preparations.

Lemma 4.13. *Let $p, q \in (0,\infty]$, $\tau \in [0,\infty)$ and s be as in (4.39). Let*

$$\Psi \in B^s_{p,q}(\mathbb{R}^n) \cap L^\infty(\mathbb{R}^n)$$

be a function such that

$$\operatorname{supp}\Psi \subset [-1,1]^n \quad \text{and} \quad \int_{\mathbb{R}^n} \Psi(y)\,dy \equiv C_\Psi > 0. \tag{4.40}$$

For each dyadic cube $Q = 2^{-j}([0,1)^n + k)$, we set

$$\Psi_Q \equiv |Q|^{-1/2}\Psi(2^j(\cdot - x_Q))$$

and

$$f_Q \equiv C_\Psi^{-1}\,|Q|^{-1/2}\int_{\mathbb{R}^n} f(y)\Psi_Q(y)\,dy,$$

where x_Q denotes the center of Q. Then there exists a positive constant C such that for all $f \in C^{s+n\tau-n/p}(\mathbb{R}^n)$,

$$\|(f-f_Q)\Psi_Q\|_B \le C|Q|^{\tau-1/2}\left(\|f\|^{\heartsuit}_{B^{s,\tau}_{p,q}(\mathbb{R}^n)} + \|f\|_{C^{s+n\tau-n/p}(\mathbb{R}^n)}\right). \tag{4.41}$$

Here $\|f\|^{\heartsuit}_{B^{s,\tau}_{p,q}(\mathbb{R}^n)}$ is defined as in Sect. 4.1.3 with $M = 1$.

Proof. Observe that by (4.40),

$$\int_{\mathbb{R}^n}(f(y)-f_Q)\Psi_Q(y)\,dy = \int_{\mathbb{R}^n} f(y)\Psi_Q(y)\,dy - f_Q\int_{\mathbb{R}^n}\Psi_Q(y)\,dy = 0. \tag{4.42}$$

Let

$$P_Q \equiv \{x \in \mathbb{R}^n : |2^j x_i - k_i| \le 1, i = 1,\cdots,n\}.$$

Then $\operatorname{supp}\Psi_Q \subset P_Q$ and $|P_Q| \sim |Q|$. As a consequence of (4.40) and (4.42), we obtain

$$\begin{aligned} |f(x)-f_Q| &= \left|f(x) - f_Q - (C_\Psi)^{-1}|Q|^{-1/2}\int_{\mathbb{R}^n}(f(y)-f_Q)\Psi_Q(y)\,dy\right| \\ &\le (C_\Psi)^{-1}|Q|^{-1/2}\int_{\mathbb{R}^n}|f(x)-f(y)||\Psi_Q(y)|\,dy \end{aligned} \tag{4.43}$$

for arbitrary $x \in \mathbb{R}^n$. Since $0 < s + n\tau - n/p < 1$, the definition of $C^{s+n\tau-n/p}(\mathbb{R}^n)$ (see item (iv) in Sect. 1.3) implies that

$$|f(x)-f(y)| \le \|f\|_{C^{s+n\tau-n/p}(\mathbb{R}^n)}|x-y|^{s+n\tau-n/p}.$$

Thus, if $x \in 3P_Q$, by the support condition of Φ, we have

$$|f(x)-f_Q| \lesssim |Q|^{(s+n\tau-n/p)/n}\|f\|_{C^{s+n\tau-n/p}(\mathbb{R}^n)}\|\Psi\|_{L^\infty(\mathbb{R}^n)}. \tag{4.44}$$

Notice that for all $x \in \mathbb{R}^n$,

$$\begin{aligned}
&\Delta_h^1[(f-f_Q)\Psi_Q](x)\\
&= (f(x+h)-f_Q)\,\Psi_Q(x+h)-(f(x)-f_Q)\,\Psi_Q(x) &(4.45)\\
&= \Psi_Q(x)\,(f(x+h)-f(x))+(f(x+h)-f_Q)\,(\Psi_Q(x+h)-\Psi_Q(x)). &(4.46)
\end{aligned}$$

The formula (4.45) will be used in case $t \le 2(l(Q)\wedge 1)$, whereas the decomposition (4.46) will be applied if $2l(Q) < t < 2\ (l(Q) < 1)$. Oriented on these decompositions we introduced the following notation

$$\mathrm{I}_1 \equiv \left\{\int_0^{2(l(Q)\wedge 1)} t^{-sq}\left[\int_{\mathbb{R}^n} t^{-n}\int_{t/2\le|h|<t}|\Psi_Q(x)|^p|f(x+h)-f(x)|^p\,dh\,dx\right]^{\frac{q}{p}}\frac{dt}{t}\right\}^{\frac{1}{q}},$$

$$\mathrm{I}_2 \equiv \left\{\int_0^{2(l(Q)\wedge 1)} \cdots t^{-n}\int_{t/2\le|h|<t}|f(x+h)-f_Q|^p|\Psi_Q(x+h)-\Psi_Q(x)|^p dh\cdots\right\}^{\frac{1}{q}},$$

$$\mathrm{I}_3 \equiv \left\{\int_{2l(Q)}^{2} \cdots t^{-n}\int_{t/2\le|h|<t}|f(x+h)-f_Q|^p|\Psi_Q(x+h)|^p\,dh\cdots\right\}^{\frac{1}{q}}$$

and

$$\mathrm{I}_4 \equiv \left\{\int_{2l(Q)}^{2} \cdots t^{-n}\int_{t/2\le|h|<t}|f(x)-f_Q|^p|\Psi_Q(x)|^p\,dh\cdots\right\}^{\frac{1}{q}},$$

where when $l(Q) \ge 1$, I_3 and I_4 are void. Then it follows that

$$\|(f-f_Q)\Psi_Q\|_B \lesssim \mathrm{I}_1+\mathrm{I}_2+\mathrm{I}_3+\mathrm{I}_4. \tag{4.47}$$

Obviously,

$$\begin{aligned}
\mathrm{I}_1 &\le |Q|^{-1/2}\|\Psi\|_{L^\infty(\mathbb{R}^n)}\\
&\quad\times\left\{\int_0^{2(l(Q)\wedge 1)} t^{-sq}\left[\int_{P_Q} t^{-n}\int_{t/2\le|h|<t}|\Delta_h^1 f(x)|^p\,dh\,dx\right]^{q/p}\frac{dt}{t}\right\}^{1/q}\\
&\lesssim |Q|^{\tau-1/2}\|\Psi\|_{L^\infty(\mathbb{R}^n)}\|f\|^{\heartsuit}_{B_{p,q}^{s,\tau}(\mathbb{R}^n)}. &(4.48)
\end{aligned}$$

Next we turn to I_2. Since $|h| < t < 2(l(Q)\wedge 1)$, for either $x \in P_Q$ or $x+h \in P_Q$, we conclude that in any case x, $x+h \in 3P_Q$. Hence we may employ (4.44) and find

$$\begin{aligned}
\mathrm{I}_2 &\lesssim |Q|^{(s+n\tau-n/p)/n}\|f\|_{C^{s+n\tau-n/p}(\mathbb{R}^n)}\,\|\Psi\|_{L^\infty(\mathbb{R}^n)}\\
&\quad\times\left\{\int_0^{2(l(Q)\wedge 1)} t^{-sq}\left[\int_{\mathbb{R}^n} t^{-n}\int_{t/2\le|h|<t}|\Delta_h^1\Psi_Q(x)|^p\,dh\,dx\right]^{q/p}\frac{dt}{t}\right\}^{1/q}\\
&\sim |Q|^{(s+n\tau-n/p)/n}\|f\|_{C^{s+n\tau-n/p}(\mathbb{R}^n)}\,\|\Psi\|_{L^\infty(\mathbb{R}^n)}\,J,
\end{aligned}$$

where

$$J\equiv\left\{\int_0^{2(l(Q)\wedge 1)} t^{-sq}\left[\int_{\mathbb{R}^n} t^{-n}\int_{t/2\le|h|<t}|\Delta_h^1\Psi_Q(x)|^p\,dh\,dx\right]^{q/p}\frac{dt}{t}\right\}^{1/q}.$$

Let $Q\equiv Q_{j,k}$. Using a transformation of coordinates we can estimate J

$$\begin{aligned}
J &= |Q|^{1/p-1/2}\left\{\int_0^{2(l(Q)\wedge 1)} t^{-sq}\left[\int_{\mathbb{R}^n} t^{-n}\int_{t/2\le|h|<t}|\Delta_{2^jh}^1\Psi(x)|^p\,dh\,dx\right]^{q/p}\frac{dt}{t}\right\}^{1/q}\\
&= |Q|^{1/p-1/2}\left\{\int_0^{2(l(Q)\wedge 1)}\cdots(2^jt)^{-n}\int_{2^jt/2\le|h|<2^jt}|\Delta_h^1\Psi(x)|^p\,dh\cdots\right\}^{1/q}\\
&= |Q|^{1/p-1/2}2^{js}\left\{\int_0^{2^{j+1}(l(Q)\wedge 1)}\cdots\left(t^{-n}\int_{t/2\le|h|<t}|\Delta_h^1\Psi(x)|\,dh\right)^p\cdots\right\}^{1/q}\\
&\le |Q|^{1/p-1/2}2^{js}\|\Psi\|_{B^s_{p,q}(\mathbb{R}^n)}.
\end{aligned}$$

Inserting this estimate in our previous one we get

$$\mathrm{I}_2\lesssim|Q|^{\tau-1/2}\|f\|_{C^{s+n\tau-n/p}(\mathbb{R}^n)}\,\|\Psi\|_{B^s_{p,q}(\mathbb{R}^n)}\|\Psi\|_{L^\infty(\mathbb{R}^n)}.\tag{4.49}$$

To estimate I_3 we will have an advantage from working with $\|f\|^{\heartsuit}_{B^{s,\tau}_{p,q}(\mathbb{R}^n)}$. Observe that we can apply (4.44). Consequently,

$$\begin{aligned}
\mathrm{I}_3 &\lesssim |Q|^{(s+n\tau-n/p)/n}\|f\|_{C^{s+n\tau-n/p}(\mathbb{R}^n)}\\
&\quad\times\left\{\int_{2l(Q)}^{2} t^{-sq}\left[\int_{\mathbb{R}^n} t^{-n}\int_{t/2\le|h|<t}|\Psi_Q(x+h)|^p\,dh\,dx\right]^{q/p}\frac{dt}{t}\right\}^{1/q}\\
&\lesssim |Q|^{(s+n\tau-n/p)/n}\|f\|_{C^{s+n\tau-n/p}(\mathbb{R}^n)}\,\|\Psi_Q\|_{L^p(\mathbb{R}^n)}\left\{\int_{2l(Q)}^{2}t^{-sq}\frac{dt}{t}\right\}^{1/q}\\
&\lesssim |Q|^{(s+n\tau-n/p)/n}\,|Q|^{\frac{1}{p}-\frac{1}{2}}\,|Q|^{-s/n}\,\|f\|_{C^{s+n\tau-n/p}(\mathbb{R}^n)}\\
&\lesssim |Q|^{\tau-\frac{1}{2}}\,\|f\|_{C^{s+n\tau-n/p}(\mathbb{R}^n)}.
\end{aligned}\tag{4.50}$$

In the same way one derives

$$\mathrm{I}_4 \lesssim |Q|^{\tau-\frac{1}{2}} \|f\|_{C^{s+n\tau-n/p}(\mathbb{R}^n)}. \tag{4.51}$$

Combining (4.48)–(4.51) with (4.47) we get the estimate (4.41), which completes the proof of Lemma 4.13. □

Remark 4.17. If $\operatorname{supp}\Psi \subset [-N,N]^n$ for some N, then the assertion (4.41) remains true, probably with a different constant, but this is not of relevance for us.

Lemma 4.14. *Let $p,q\in(0,\infty]$, $\tau\in[0,\infty)$ and $s\in(0,1)$. Let $\|\cdot\|^{\heartsuit}_{A^{s,\tau}_{p,q}(\mathbb{R}^n)}$ be defined as in Sect. 4.1.3 with $M=1$. Let Ψ be a locally integrable function satisfying*

$$\Psi(x)=1 \quad \textit{if} \quad x\in[-3\sqrt{n},3\sqrt{n}]^n. \tag{4.52}$$

Then there exists a positive constant C such that for all $f\in L^1_{\mathrm{loc}}(\mathbb{R}^n)$,

$$\|f\|^{\heartsuit}_{B^{s,\tau}_{p,q}(\mathbb{R}^n)} \le C \sup_{Q\in\mathcal{Q}} \frac{|Q|^{1/2}}{|Q|^{\tau}} \left\{\int_0^{2(l(Q)\wedge 1)} t^{-sq}\left[\int_Q \times t^{-n}\int_{t/2\le|h|<t} |\Delta^1_h((f-f_Q)\Psi_Q)(x)|^p\,dh\,dx\right]^{q/p}\frac{dt}{t}\right\}^{1/q}$$

and

$$\|f\|^{\heartsuit}_{F^{s,\tau}_{p,q}(\mathbb{R}^n)} \le C \sup_{Q\in\mathcal{Q}} \frac{|Q|^{1/2}}{|Q|^{\tau}} \left\{\int_Q\left[\int_0^{2(l(Q)\wedge 1)} t^{-sq} \times t^{-n}\int_{t/2\le|h|<t} |\Delta^1_h((f-f_Q)\Psi_Q)(x)|^q\,dh\,\frac{dt}{t}\right]^{p/q} dx\right\}^{1/p}.$$

A similar estimate holds for $\|f\|^{\clubsuit}_{F^{s,\tau}_{p,q}(\mathbb{R}^n)}$.

Proof. Since

$$\Psi_Q(x)=\Psi_Q(x+h)=|Q|^{-1/2},$$

we then have that for all $x\in Q$ and $t<2(l(Q)\wedge 1)$,

$$\begin{aligned}\int_{t/2\le|h|<t}|\Delta^1_h f(x)|\,dh &= \int_{t/2\le|h|<t}|\Delta^1_h(f-f_Q)(x)|\,dh\\ &= |Q|^{1/2}\int_{t/2\le|h|<t}|\Delta^1_h((f-f_Q)\Psi_Q)(x)|\,dh,\end{aligned}$$

which completes the proof of Lemma 4.14. □

From Theorem 4.7, Lemmas 4.13, 4.14 and Remark 4.17, we deduce the following conclusion.

Theorem 4.16. *Let $s \in \mathbb{R}$, $p \in [1,\infty)$, $q \in (0,\infty]$ and $\tau \in [0,\infty)$ such that*

$$n \max\left\{0, \tau - \frac{1}{p}\right\} < s < 1 \qquad \text{and} \qquad 0 < s + n\tau - \frac{n}{p} < 1. \tag{4.53}$$

Let $\Psi \in B^s_{p,q}(\mathbb{R}^n) \cap L^\infty(\mathbb{R}^n)$ be a function satisfying $C_\Psi > 0$ (see (4.40)) and (4.52).

(i) *Then $f \in B^{s,\tau}_{p,q}(\mathbb{R}^n)$ if, and only if $f \in L^p_\tau(\mathbb{R}^n)$ and*

$$\|f\|^\bigstar_{B^{s,\tau}_{p,q}(\mathbb{R}^n)} \equiv \sup_{Q\in\mathcal{Q}} |Q|^{1/2-\tau} \|(f - f_Q)\Psi_Q\|_{B^s_{p,q}(\mathbb{R}^n)} < \infty. \tag{4.54}$$

Furthermore, $\|f\|_{L^p_\tau(\mathbb{R}^n)} + \|f\|^\bigstar_{B^{s,\tau}_{p,q}(\mathbb{R}^n)}$ and $\|f\|_{B^{s,\tau}_{p,q}(\mathbb{R}^n)}$ are equivalent.

(ii) *Let in addition $\tau \in [1/p,\infty)$. Then $f \in B^{s,\tau}_{p,q}(\mathbb{R}^n)$ if, and only if $N_p(f) < \infty$ and*

$$|||f|||^\bigstar_{B^{s,\tau}_{p,q}(\mathbb{R}^n)} \equiv \sup_{\{Q\in\mathcal{Q}:\,|Q|\le 1\}} |Q|^{1/2-\tau} \|(f - f_Q)\Psi_Q\|_{B^s_{p,q}(\mathbb{R}^n)} < \infty. \tag{4.55}$$

Furthermore, $N_p(f) + |||f|||^\bigstar_{B^{s,\tau}_{p,q}(\mathbb{R}^n)}$ and $\|f\|_{B^{s,\tau}_{p,q}(\mathbb{R}^n)}$ are equivalent.

Proof. Step 1. Let $f \in B^{s,\tau}_{p,q}(\mathbb{R}^n)$. By Propositions 2.7 and 2.1, we know that $B^{s,\tau}_{p,q}(\mathbb{R}^n) \subset L^p_\tau(\mathbb{R}^n)$. From Proposition 2.6 we derive that

$$B^{s,\tau}_{p,q}(\mathbb{R}^n) \subset C^{s+n\tau-n/p}(\mathbb{R}^n).$$

Next we employ Lemma 4.13, Remark 4.17 in combination with Theorem 4.7(i) and obtain

$$\|(f - f_Q)\Psi_Q\|_B \lesssim |Q|^{\tau-1/2} \|f\|_{B^{s,\tau}_{p,q}(\mathbb{R}^n)}.$$

Recall that the quasi-norms $\|f\|_{L^p(\mathbb{R}^n)} + \|f\|_B$ and $\|f\|_{B^s_{p,q}(\mathbb{R}^n)}$ are equivalent in $B^s_{p,q}(\mathbb{R}^n)$. Thus, we only need to estimate $\|(f - f_Q)\Psi_Q\|_{L^p(\mathbb{R}^n)}$. We shall use the notation from the proof of Lemma 4.13. In fact, since $\operatorname{supp}\Psi_Q \subset P_Q$, we see that

$$\|(f - f_Q)\Psi_Q\|_{L^p(\mathbb{R}^n)} \le |Q|^{-1/2} \|\Psi\|_{L^\infty(\mathbb{R}^n)} \left(\int_{P_Q} |f(x) - f_Q|^p\,dx\right)^{1/p}.$$

When $|Q| \le 1$, by (4.44) and $s > 0$, we have

$$\begin{aligned}\|(f - f_Q)\Psi_Q\|_{L^p(\mathbb{R}^n)} &\lesssim |Q|^{-1/2}|Q|^{1/p}|Q|^{(s+n\tau-n/p)/n}\|f\|_{C^{s+n\tau-n/p}(\mathbb{R}^n)}\\ &\lesssim |Q|^{\tau-1/2}\|f\|_{C^{s+n\tau-n/p}(\mathbb{R}^n)}.\end{aligned}$$

When $|Q| > 1$, using (4.43), we obtain

$$|f(x) - f_Q| \lesssim |Q|^{-1} \|\Psi\|_{L^\infty(\mathbb{R}^n)} \int_{P_Q} |f(x) - f(y)|\,dy.$$

Because of $p \geq 1$ we can apply Hölder's inequality and find

$$\begin{aligned}
\|(f - f_Q)\Psi_Q\|_{L^p(\mathbb{R}^n)} &\lesssim |Q|^{-3/2} \left[\int_{P_Q} \left(\int_{P_Q} |f(x) - f(y)|\,dy\right)^p dx\right]^{1/p} \\
&\lesssim |Q|^{\frac{1}{p}-\frac{3}{2}} \left[\int_{P_Q} |f(x)|^p\,dx\right]^{1/p} + |Q|^{\frac{1}{p}-\frac{3}{2}} \left(\int_{P_Q} |f(y)|^p\,dy\right)^{1/p} \\
&\lesssim |Q|^{\tau-1/2} \|f\|_{L^p_\tau(\mathbb{R}^n)}.
\end{aligned}$$

This proves one direction of the claim in (i).

Step 2. We suppose $f \in L^p_\tau(\mathbb{R}^n)$ and $\|f\|^{\bigstar}_{B^{s,\tau}_{p,q}(\mathbb{R}^n)} < \infty$. Applying Lemma 4.14 in combination with Theorem 4.7(i) we conclude that $f \in B^{s,\tau}_{p,q}(\mathbb{R}^n)$ and

$$\|f\|_{B^{s,\tau}_{p,q}(\mathbb{R}^n)} \lesssim \|f\|_{L^p_\tau(\mathbb{R}^n)} + \|f\|^{\bigstar}_{B^{s,\tau}_{p,q}(\mathbb{R}^n)} < \infty,$$

which completes the proof of (i).

Step 3. Proof of (ii). This time we work with Theorem 4.7(ii) and argue as above. Only one further comment is needed. Because $f \in C(\mathbb{R}^n)$ we have $N_p(f) \leq \|f\|_{C(\mathbb{R}^n)}$. □

The most interesting example is given by the choice $\Psi = \mathscr{X}$, where $\mathscr{X}$ denotes the characteristic function of the cube $[-1,1]^n$. This function does not satisfy the condition (4.52). However, there are some simple modifications to justify such a choice. We shall work with a modified system of dyadic cubes in this context. Let

$$P_{j,k} \equiv \{x \in \mathbb{R}^n : 2^{-j}(k_i - 1) \leq x_i < 2^{-j}(k_i + 1),\, i = 1, \ldots, n\}, \quad j \in \mathbb{Z}, k \in \mathbb{Z}^n,$$

and denote by $\mathscr{Q}^*$ the collection of all such cubes.

It is well known that $\mathscr{X} \in B^s_{p,q}(\mathbb{R}^n)$ if and only if either $s < 1/p$ and q is arbitrary or $s = 1/p$ and $q = \infty$; see [119, Lemma 2.3.1/3].

Corollary 4.18. *Let $p \in [1,\infty)$, $q \in (0,\infty]$, and s and τ be as in (4.53)*

(i) Let $s \in (0, 1/p)$. Then $f \in B^{s,\tau}_{p,q}(\mathbb{R}^n)$ if, and only if $f \in L^p_\tau(\mathbb{R}^n)$ and

$$\|f\|^{\blacksquare}_{B^{s,\tau}_{p,q}(\mathbb{R}^n)} \equiv \sup_{P \in \mathscr{Q}^*} |P|^{1/2-\tau} \left\| \left(f - \frac{1}{|P|}\int_P f(y)\,dy\right) \mathscr{X}_P \right\|_{B^s_{p,q}(\mathbb{R}^n)} < \infty, \quad (4.56)$$

where $\mathscr{X}_P$ is defined as in Lemma 4.13. Furthermore, $\|f\|_{L^p_\tau(\mathbb{R}^n)} + \|f\|^{\blacksquare}_{B^{s,\tau}_{p,q}(\mathbb{R}^n)}$ *and* $\|f\|_{B^{s,\tau}_{p,q}(\mathbb{R}^n)}$ *are equivalent.*

(ii) *Let $p \in (1,\infty)$. Then the assertions in (i) remain true if $s = 1/p$ and $q = \infty$.*

Proof. Let K be a natural number such that $2^K \geq 3\sqrt{n}$. Then, let Ψ be the characteristic function of the cube centered at the origin and having side-length $2K$. Under the given restrictions on s we can apply Theorem 4.16 with respect to this function Ψ. Next, observe

$$\sup_{Q\in\mathscr{Q}} |Q|^{1/2-\tau}\|(f-f_Q)\Psi_Q\|_{B^s_{p,q}(\mathbb{R}^n)}$$
$$\sim \sup_{P\in\mathscr{Q}^*} |P|^{1/2-\tau}\left\|\left(f-\frac{1}{|P|}\int_P f(y)\,dy\right)\mathscr{X}_P\right\|_{B^s_{p,q}(\mathbb{R}^n)},$$

which proves the claim. □

4.6.2 A Characterization of $F^{s,\tau}_{p,q}(\mathbb{R}^n)$

We proceed as in the previous subsection.

Lemma 4.15. *Let $p \in (1,\infty)$, $q \in (0,\infty)$, $\tau \in [0,\infty)$ and s be as in (4.39). Let*

$$\Psi \in F^s_{p,q}(\mathbb{R}^n) \cap L^\infty(\mathbb{R}^n)$$

be a function such that (4.40) is satisfied. Then there exists a positive constant C such that for all $f \in C^{s+n\tau-n/p}(\mathbb{R}^n)$,

$$\|(f-f_Q)\Psi_Q\|^{\clubsuit}_F \leq C|Q|^{\tau-1/2}\left(\|f\|^{\clubsuit}_{F^{s,\tau}_{p,q}(\mathbb{R}^n)} + \|f\|_{C^{s+n\tau-n/p}(\mathbb{R}^n)}\right). \tag{4.57}$$

If $q \in [1,\infty)$, (4.57) *also holds with $\|(f-f_Q)\Psi_Q\|^{\clubsuit}_F$ and $\|f\|^{\clubsuit}_{F^{s,\tau}_{p,q}(\mathbb{R}^n)}$ replaced, respectively, by $\|(f-f_Q)\Psi_Q\|^{\heartsuit}_F$ and $\|f\|^{\heartsuit}_{F^{s,\tau}_{p,q}(\mathbb{R}^n)}$. Here $\|f\|^{\clubsuit}_{F^{s,\tau}_{p,q}(\mathbb{R}^n)}$ and $\|f\|^{\heartsuit}_{F^{s,\tau}_{p,q}(\mathbb{R}^n)}$ are defined as in Sect. 4.1.3 with $M = 1$.*

Proof. By similarity, we only give the proof for $\|(f-f_Q)\Psi_Q\|^{\clubsuit}_F$. We comment on the modifications needed in comparison with the proof of Lemma 4.13. Again we employ (4.45) and (4.46). This leads to the following definitions of I_1 through I_4:

$$\mathrm{I}_1 \equiv \left\|\left\{\int_0^{2(l(Q)\wedge 1)} t^{-sq}\left(t^{-n}\int_{t/2\leq|h|<t}|\Psi_Q(x)||f(x+h)-f(x)|\,dh\right)^q\frac{dt}{t}\right\}^{\frac{1}{q}}\right\|_{L^p(\mathbb{R}^n)},$$

$$\mathrm{I}_2 \equiv \left\| \left\{ \int_0^{2(l(Q)\wedge 1)} \cdots \left(t^{-n} \int_{t/2\le|h|<t} |f(x+h)-f_Q| \right.\right.\right.$$
$$\left.\left.\left. \times |\Psi_Q(x+h)-\Psi_Q(x)|\,dh \right)^q \cdots \right\}^{\frac{1}{q}} \right\|_{L^p(\mathbb{R}^n)},$$

$$\mathrm{I}_3 \equiv \left\| \left\{ \int_{2l(Q)}^{2} \cdots \left(t^{-n} \int_{t/2\le|h|<t} |f(x+h)-f_Q|\,|\Psi_Q(x+h)|\,dh \right)^q \cdots \right\}^{\frac{1}{q}} \right\|_{L^p(\mathbb{R}^n)}$$

and

$$\mathrm{I}_4 \equiv \left\| \left\{ \int_{2l(Q)}^{2} \cdots \left(t^{-n} \int_{t/2\le|h|<t} |f(x)-f_Q|\,|\Psi_Q(x)|\,dh \right)^q \cdots \right\}^{\frac{1}{q}} \right\|_{L^p(\mathbb{R}^n)},$$

where, as above, when $l(Q)\ge 1$, then I_3 and I_4 are void. Again we have

$$\|(f-f_Q)\Psi_Q\|_F^{\clubsuit} \lesssim \mathrm{I}_1+\mathrm{I}_2+\mathrm{I}_3+\mathrm{I}_4\,.$$

Concerning the estimates of I_1 and I_2 we can argue as in the proof of Lemma 4.13 and obtain

$$\mathrm{I}_1 \lesssim |Q|^{\tau-1/2}\,\|\Psi\|_{L^\infty(\mathbb{R}^n)}\,\|f\|^{\clubsuit}_{F^{s,\tau}_{p,q}(\mathbb{R}^n)}$$

as well as

$$\mathrm{I}_2 \lesssim |Q|^{\tau-1/2}\,\|f\|_{C^{s+n\tau-n/p}(\mathbb{R}^n)}\,\|\Psi\|_{F^s_{p,q}(\mathbb{R}^n)}\|\Psi\|_{L^\infty(\mathbb{R}^n)}\,.$$

To estimate I_3 we will use the Hardy-Littlewood maximal function and obtain by using (4.44) that

$$\mathrm{I}_3 \lesssim |Q|^{(s+n\tau-n/p)/n}\|f\|_{C^{s+n\tau-n/p}(\mathbb{R}^n)}$$
$$\times \left\| \left\{ \int_{2l(Q)}^{2} t^{-sq} \left(t^{-n} \int_{t/2\le|h|<t} |\Psi_Q(x+h)|\,dh \right)^q \frac{dt}{t} \right\}^{1/q} \right\|_{L^p(\mathbb{R}^n)}$$
$$\lesssim |Q|^{(s+n\tau-n/p)/n}\|f\|_{C^{s+n\tau-n/p}(\mathbb{R}^n)} \left\| (\mathrm{M}|\Psi_Q|)(x) \left\{ \int_{2l(Q)}^{2} t^{-sq}\frac{dt}{t} \right\}^{1/q} \right\|_{L^p(\mathbb{R}^n)}$$
$$\lesssim \|\Psi\|_{L^p(\mathbb{R}^n)}\,|Q|^{\tau-\frac12}\,\|f\|_{C^{s+n\tau-n/p}(\mathbb{R}^n)},$$

because of $1<p<\infty$. In the same way one derives

$$\mathrm{I}_4 \lesssim |Q|^{\tau-\frac12}\,\|f\|_{C^{s+n\tau-n/p}(\mathbb{R}^n)}.$$

Collecting the estimates of $\mathrm{I}_1-\mathrm{I}_4$ we have proved (4.57). □

Remark 4.18. As in the B-case, if $\operatorname{supp}\Psi\subset[-N,N]^n$ for some N, then the assertion (4.57) remains true, probably with a different constant.

From Lemmas 4.15, 4.14 and Remark 4.18, we now deduce the counterpart of Theorem 4.16.

Theorem 4.17. *Let $s \in \mathbb{R}$, $p \in (1,\infty)$, $q \in (0,\infty]$ and $\tau \in [0,\infty)$ such that*

$$n \max\left\{0, \frac{1}{q}-1, \tau-\frac{1}{p}\right\} < s < 1 \qquad \text{and} \qquad 0 < s+n\tau-\frac{n}{p} < 1. \tag{4.58}$$

Let $\Psi \in F^s_{p,q}(\mathbb{R}^n) \cap L^\infty(\mathbb{R}^n)$ be a function satisfying $C_\Psi > 0$ (see (4.40)) and (4.52).

(i) *Then $f \in F^{s,\tau}_{p,q}(\mathbb{R}^n)$ if, and only if $f \in L^p_\tau(\mathbb{R}^n)$ and*

$$\|f\|^\bigstar_{F^{s,\tau}_{p,q}(\mathbb{R}^n)} \equiv \sup_{Q\in\mathcal{Q}} |Q|^{1/2-\tau} \|(f-f_Q)\Psi_Q\|_{F^s_{p,q}(\mathbb{R}^n)} < \infty.$$

Furthermore, $\|f\|_{L^p_\tau(\mathbb{R}^n)} + \|f\|^\bigstar_{F^{s,\tau}_{p,q}(\mathbb{R}^n)}$ and $\|f\|_{F^{s,\tau}_{p,q}(\mathbb{R}^n)}$ are equivalent.

(ii) *Let in addition $\tau \in [1/p,\infty)$. Then $f \in F^{s,\tau}_{p,q}(\mathbb{R}^n)$ if, and only if $N_p(f) < \infty$ and*

$$|||f|||^\bigstar_{F^{s,\tau}_{p,q}(\mathbb{R}^n)} \equiv \sup_{\{Q\in\mathcal{Q}:\,|Q|\le 1\}} |Q|^{1/2-\tau} \|(f-f_Q)\Psi_Q\|_{F^s_{p,q}(\mathbb{R}^n)} < \infty.$$

Furthermore, $N_p(f) + |||f|||^\bigstar_{F^{s,\tau}_{p,q}(\mathbb{R}^n)}$ and $\|f\|_{F^{s,\tau}_{p,q}(\mathbb{R}^n)}$ are equivalent.

Proof. The only difference in comparison with the proof of Theorem 4.16 consists in the fact that we use Theorem 4.3 instead of Theorem 4.7. □

It is well known that $\mathcal{X} \in F^s_{p,q}(\mathbb{R}^n)$ if, and only if either $s < 1/p$ and q is arbitrary; see [119, Lemma 2.3.1/3]. Therefore we obtain the counterpart of Corollary 4.18.

Corollary 4.19. *Let $p \in (1,\infty)$, $q \in (0,\infty]$, and s and τ be as in (4.58). Let in addition $s < 1/p$. Then $f \in F^{s,\tau}_{p,q}(\mathbb{R}^n)$ if, and only if $f \in L^p_\tau(\mathbb{R}^n)$ and*

$$\|f\|^\blacksquare_{F^{s,\tau}_{p,q}(\mathbb{R}^n)} \equiv \sup_{P\in\mathcal{Q}^*} |P|^{1/2-\tau} \left\| \left(f - \frac{1}{|P|}\int_P f(y)\,dy\right)\mathcal{X}_P \right\|_{F^s_{p,q}(\mathbb{R}^n)} < \infty.$$

Furthermore, $\|f\|_{L^p_\tau(\mathbb{R}^n)} + \|f\|^\blacksquare_{F^{s,\tau}_{p,q}(\mathbb{R}^n)}$ and $\|f\|_{F^{s,\tau}_{p,q}(\mathbb{R}^n)}$ are equivalent.

4.6.3 A Characterization of $F^s_{\infty,q}(\mathbb{R}^n)$

We employ Proposition 2.4 and Theorem 4.17(ii).

Theorem 4.18. *Let $p \in (1,\infty)$, $q \in (0,\infty)$ and*

$$n \max\left\{0, \frac{1}{q}-1\right\} < s < 1. \tag{4.59}$$

Let $\Psi \in F_{\infty,q}^{s}(\mathbb{R}^n)$ be a function satisfying $C_\Psi > 0$ (see (4.40)) and (4.52). Then $f \in F_{\infty,q}^{s}(\mathbb{R}^n)$ if, and only if $N_p(f) < \infty$ and

$$\|f\|^{\bigstar}_{F_{\infty,q}^{s}(\mathbb{R}^n)} \equiv \sup_{\{Q\in\mathcal{Q}:\,|Q|\le 1\}} |Q|^{1/2-\tau} \|(f - f_Q)\Psi_Q\|_{F_{p,q}^{s}(\mathbb{R}^n)} < \infty.$$

Furthermore, $N_p(f) + \|f\|^{\bigstar}_{F_{\infty,q}^{s}(\mathbb{R}^n)}$ and $\|f\|_{F_{\infty,q}^{s}(\mathbb{R}^n)}$ are equivalent.

Remark 4.19.

(i) It is a bit surprising that $F_{\infty,q}^{s}(\mathbb{R}^n)$ can be characterized by quantities, using $F_{p,q}^{s}(\mathbb{R}^n)$, and p can be chosen as we want (within $(1,\infty)$).

(ii) As mentioned at the beginning of this section we have taken over the idea for those characterizations from the paper [5]. There the homogeneous situation with $p = q = 2$ is investigated.

Chapter 5
Pseudo-Differential Operators

The main purpose of this chapter is to obtain the boundedness on $A^{s,\tau}_{p,q}(\mathbb{R}^n)$ of all pseudo-differential operators of type (1,1) with inhomogeneous symbols. The smooth molecular decomposition characterizations of $A^{s,\tau}_{p,q}(\mathbb{R}^n)$ play an important role in this chapter.

5.1 Pseudo-Differential Operators of Class $\mathscr{S}^{\mu}_{1,1}(\mathbb{R}^n)$

We begin with recalling the following class of inhomogeneous symbols, which is a special case of Hörmander class of symbols; see, for example, [73, 74] and [146, Chap. 6].

Definition 5.1. Let $\mu \in \mathbb{R}$. A smooth function a defined on $\mathbb{R}^n \times \mathbb{R}^n$ is called to belong to the class $\mathscr{S}^{\mu}_{1,1}(\mathbb{R}^n)$, if a satisfies the following differential inequalities that for all $\alpha, \beta \in \mathbb{Z}^n_+$,

$$\sup_{x,\,\xi\in\mathbb{R}^n} (1+|\xi|)^{-\mu-|\alpha|+|\beta|}|\partial^{\alpha}_x \partial^{\beta}_{\xi} a(x,\xi)| < \infty.$$

The main result of this section is the following.

Theorem 5.1. *Let* $s \in \mathbb{R}$ *and* $p,q \in (0,\infty]$. *Further we assume* $\tau \in [0,\tau_{s,p,q})$ *if* $A^{s,\tau}_{p,q}(\mathbb{R}^n) = F^{s,\tau}_{p,q}(\mathbb{R}^n)$ *and* $\tau \in [0,\tau_{s,p})$ *if* $A^{s,\tau}_{p,q}(\mathbb{R}^n) = B^{s,\tau}_{p,q}(\mathbb{R}^n)$. *Let* $\mu \in \mathbb{R}$, $a \in \mathscr{S}^{\mu}_{1,1}(\mathbb{R}^n)$ *and* $a(x,D)$ *be the corresponding pseudo-differential operator such that*

$$a(x,D)(f)(x) \equiv \int_{\mathbb{R}^n} e^{ix\xi} a(x,\xi)\widehat{f}(\xi)\,d\xi$$

for all smooth molecules f *for* $A^{s+\mu,\tau}_{p,q}(\mathbb{R}^n)$ *and* $x \in \mathbb{R}^n$.

(i) *If* $s > \sigma_{p,q}$ ($s > \sigma_p$ *if* $A^{s,\tau}_{p,q}(\mathbb{R}^n) = B^{s,\tau}_{p,q}(\mathbb{R}^n)$), *then* $a(x,D)$ *is a continuous linear mapping from* $A^{s+\mu,\tau}_{p,q}(\mathbb{R}^n)$ *to* $A^{s,\tau}_{p,q}(\mathbb{R}^n)$.

D. Yang et al., *Morrey and Campanato Meet Besov, Lizorkin and Triebel*, Lecture Notes in Mathematics 2005, DOI 10.1007/978-3-642-14606-0_5,

(ii) *If $s \le \sigma_{p,q}$ ($s \le \sigma_p$ if $A^{s,\tau}_{p,q}(\mathbb{R}^n) = B^{s,\tau}_{p,q}(\mathbb{R}^n)$), assume further that its formal adjoint $a(x,D)^*$ satisfies*

$$a(x,D)^*(x^\beta) \in \mathscr{P}(\mathbb{R}^n) \tag{5.1}$$

for all $\beta \in \mathbb{Z}^n_+$ with $|\beta| \le N$, where N is as in Definition 3.2. Then $a(x,D)$ is a continuous linear mapping from $A^{s+\mu,\tau}_{p,q}(\mathbb{R}^n)$ to $A^{s,\tau}_{p,q}(\mathbb{R}^n)$.

Proof. For simplicity, in what follows, we write $T \equiv a(x,D)$.

Let $f \in A^{s+\mu,\tau}_{p,q}(\mathbb{R}^n)$, Φ, Ψ, φ and ψ be as in Lemma 2.3. Then by the Calderón reproducing formula, we have

$$f = \sum_{l(Q)=1} \langle f, \Phi_Q\rangle \Psi_Q + \sum_{j=1}^{\infty} \sum_{l(Q)=2^{-j}} \langle f, \varphi_Q\rangle \psi_Q$$

in $\mathscr{S}'(\mathbb{R}^n)$. Set $t_Q \equiv \langle f, \Phi_Q\rangle$ if $l(Q) = 1$ and $t_Q \equiv \langle f, \varphi_Q\rangle$ if $l(Q) < 1$. Then from the φ-transform characterization of $A^{s+\mu,\tau}_{p,q}(\mathbb{R}^n)$ obtained in Theorem 2.1, we deduce that $\{t_Q\}_{l(Q)\le 1} \in a^{s+\mu,\tau}_{p,q}(\mathbb{R}^n)$ with

$$\|\{t_Q\}_{l(Q)\le 1}\|_{a^{s+\mu,\tau}_{p,q}(\mathbb{R}^n)} \lesssim \|f\|_{A^{s+\mu,\tau}_{p,q}(\mathbb{R}^n)},$$

which further implies that

$$\|\{|Q|^{-\mu/n} t_Q\}_{l(Q)\le 1}\|_{a^{s,\tau}_{p,q}(\mathbb{R}^n)} \lesssim \|f\|_{A^{s+\mu,\tau}_{p,q}(\mathbb{R}^n)}. \tag{5.2}$$

We claim that

$$T(f) \equiv \sum_{l(Q)\le 1} t_Q T(\psi_Q)$$

in $\mathscr{S}'(\mathbb{R}^n)$ and satisfies

$$\|T(f)\|_{A^{s,\tau}_{p,q}(\mathbb{R}^n)} \lesssim \|f\|_{A^{s+\mu,\tau}_{p,q}(\mathbb{R}^n)},$$

where and in what follows, when $l(Q) = 1$, ψ is replaced by Ψ. To see this, by (5.2) and the smooth molecular decomposition characterizations of $A^{s,\tau}_{p,q}(\mathbb{R}^n)$ (see Theorem 3.2), it suffices to prove that for all $Q \in \mathscr{Q}$ with $l(Q) \le 1$, $|Q|^{\mu/n} T(\psi_Q)$ is a multiple of a smooth synthesis molecule for $A^{s,\tau}_{p,q}(\mathbb{R}^n)$ supported near Q.

The proof is similar to that in [69]. For the reader's convenience, we give some details. For all dyadic cubes $Q = Q_{jk}$ and $x \in \mathbb{R}^n$, we set

$$T_Q(\psi)(x) \equiv \int_{\mathbb{R}^n} e^{ix\xi} a(2^{-j}(x+k), 2^j\xi)\widehat{\psi}(\xi)\,d\xi.$$

Then an argument, using a change of variables, yields that for all $x \in \mathbb{R}^n$,

$$T(\psi_Q)(x) = \int_{\mathbb{R}^n} e^{ix\xi} a(x,\xi)\widehat{\psi_Q}(\xi)\,d\xi = 2^{jn/2} T_Q(\psi)(2^j x - k).$$

Recall that $J = n/\min\{1, p, q\}$ if $A^{s,\tau}_{p,q}(\mathbb{R}^n)$ means $F^{s,\tau}_{p,q}(\mathbb{R}^n)$, and $J = n/\min\{1, p\}$ if $A^{s,\tau}_{p,q}(\mathbb{R}^n)$ means $B^{s,\tau}_{p,q}(\mathbb{R}^n)$. Fix a multi-index γ and $M \in \mathbb{N}$ such that

$$M > \max\{J, J-s\}/2.$$

Then for all $x \in \mathbb{R}^n$,

$$\partial^\gamma T_Q(\psi)(x) = \int_{\mathbb{R}^n} e^{ix\xi} \sum_{\delta \le \gamma} C_\delta (i\xi)^\delta \partial_x^{\gamma-\delta}[a(2^{-j}(x+k), 2^j\xi)]\widehat{\psi}(\xi)\,d\xi$$

for certain constants C_δ, where $\delta \le \gamma$ means that $\delta_i \le \gamma_i$ for all $i \in \{1, \cdots, n\}$. By $(I-\Delta_\xi)^M(e^{ix\xi}) = (1+|x|^2)^M e^{ix\xi}$ and an integration by parts, we obtain that for all $x \in \mathbb{R}^n$,

$$\begin{aligned} &\partial^\gamma T_Q(\psi)(x) \\ &= \int_{\mathbb{R}^n} e^{ix\xi} \frac{(I-\Delta_\xi)^M}{(1+|x|^2)^M} \sum_{\delta \le \gamma} C_\delta (i\xi)^\delta \partial_x^{\gamma-\delta} a(2^{-j}(x+k), 2^j\xi)\widehat{\psi}(\xi)\,d\xi. \end{aligned} \tag{5.3}$$

Leibniz's rule and Definition 5.1 yield that for all $\xi \in \mathbb{R}^n$,

$$\begin{aligned} &\left|(I-\Delta_\xi)^M \partial_x^{\gamma-\delta}[(i\xi)^\delta a(2^{-j}(x+k), 2^j\xi)\widehat{\psi}(\xi)]\right| \\ &\lesssim \sum_{|\alpha+\beta| \le 2M} \left|\partial_\xi^\beta \partial_x^{\gamma-\delta}[a(2^{-j}(x+k), 2^j\xi)]\right| \left|\partial_\xi^\alpha[(i\xi)^\delta \widehat{\psi}(\xi)]\right| \\ &\lesssim \sum_{|\alpha+\beta| \le 2M} 2^{j|\beta|} 2^{-j(|\gamma|-|\delta|)} (1+|2^j\xi|)^{\mu+|\gamma|-|\delta|-|\beta|} \left|\partial_\xi^\alpha[(i\xi)^\delta \widehat{\psi}(\xi)]\right|. \end{aligned}$$

If $l(Q) = 1$, then $j = 0$. The above estimate and (2.1) imply that the right-hand side of (5.3) is pointwise bounded by $C(\gamma)(1+|x|^2)^{-M}$ for some positive constant $C(\gamma)$. If $l(Q) < 1$, then $j \ge 1$ and for all $1/2 \le |\xi| \le 2$, $1+|2^j\xi| \sim |2^j\xi|$. This fact together with the above estimate and the support condition of $\widehat{\psi}$ also yields that there exists a positive constant $C(\gamma)$ such that the right-hand side of (5.3) is pointwise bounded by $C(\gamma)2^{j\mu}(1+|x|^2)^{-M}$. Thus, for all $x \in \mathbb{R}^n$, we have

$$|\partial^\gamma T_Q(\psi)(x)| \le C(\gamma)2^{j\mu}(1+|x|^2)^{-M}$$

for all $Q \in \mathscr{Q}$ with $l(Q) \le 1$. Using dilation and translation then deduces that for all $x \in \mathbb{R}^n$,

$$\begin{aligned} |\partial^\gamma T(\psi_Q)(x)| &\le C(\gamma)2^{j\mu}2^{j|\gamma|}2^{jn}(1+|2^jx-k|^2)^{-M} \\ &\le C(\gamma)|Q|^{-\mu/n-1/2-|\gamma|/n}(1+l(Q)^{-1}|x-x_Q|)^{-2M}. \end{aligned}$$

Therefore, a constant multiple of $|Q|^{\mu/n}T(\psi_Q)$ satisfies (3.4) through (3.7).

Notice that if $s > J-n$, we allow us not to postulate the vanishing moment condition (3.3). Then $|Q|^{\mu/n}T(\psi_Q)$ is a multiple of a smooth synthesis molecule for $A^{s,\tau}_{p,q}(\mathbb{R}^n)$ supported near Q. If $s \le J-n$, we must check the vanishing moment condition for $T(\psi_Q)$ when $l(Q) < 1$. In fact, by (2.2) and the hypothesis (5.1), we see that for all $\beta \in \mathbb{Z}_+^n$ with $|\beta| \le N$,

$$\int_{\mathbb{R}^n} x^\beta T(\psi_Q)(x)\,dx = \langle x^\beta, T(\psi_Q)\rangle = \langle T^*(x^\beta), \psi_Q\rangle = 0.$$

Thus, in the case when $s \le J - n$, $|Q|^{\mu/n} T(\psi_Q)$ is also a multiple of a smooth synthesis molecule for $A^{s,\tau}_{p,q}(\mathbb{R}^n)$ supported near Q, which completes the proof of Theorem 5.1. □

Remark 5.1.

(i) Since we only need to control a finite number of derivatives of molecules we obtain the estimate

$$\begin{aligned}&\|a(x,D)|A^{s,\tau}_{p,q}(\mathbb{R}^n) \to A^{s,\tau}_{p,q}(\mathbb{R}^n)\| \\ &\lesssim \max_{|\alpha|,|\beta|\le M} \sup_{x,\xi} (1+|\xi|)^{-\mu-|\alpha|+|\beta|} |\partial_x^\alpha \partial_\xi^\beta a(x,\xi)|\end{aligned} \tag{5.4}$$

for some $M \equiv M(s,p,q,\tau)$, where $\|a(x,D)|A^{s,\tau}_{p,q}(\mathbb{R}^n) \to A^{s,\tau}_{p,q}(\mathbb{R}^n)\|$ denotes the operator norm of $a(x,D)$ from $A^{s,\tau}_{p,q}(\mathbb{R}^n)$ to $A^{s,\tau}_{p,q}(\mathbb{R}^n)$.

(ii) The counterpart of Theorem 5.1 for the homogeneous Besov-type space $\dot{B}^{s,\tau}_{p,q}(\mathbb{R}^n)$ and Triebel-Lizorkin-type space $\dot{F}^{s,\tau}_{p,q}(\mathbb{R}^n)$ were already obtained in [127, Theorem 1.5].

(iii) The proof given above uses ideas of Grafakos and Torres [69], which itself has been based on [61, 66, 140, 141].

(iv) The boundedness of pseudo-differential operators of the "exotic" class $\mathscr{S}^{\mu}_{1,1}(\mathbb{R}^n)$ has its own history. Here we only mention the contributions of Meyer [98] (boundedness on $H^s_p(\mathbb{R}^n)$, $s > 0$, $1 < p < \infty$), Bourdaud [16] (boundedness on $B^s_{p,q}(\mathbb{R}^n)$, $s > 0$, $1 \le p,q \le \infty$), Runst [117] and Torres [140]. The last two authors have dealt with the general case of Besov-Triebel-Lizorkin spaces including values of p and q less than 1.

As an immediate consequence of Theorem 5.1, we have the following conclusion.

Corollary 5.1. *Let $\gamma \in \mathbb{Z}_+^n$ and s, p, q and τ be as in Theorem 5.1. Then the operator $\partial^\gamma : A^{s+|\gamma|,\tau}_{p,q}(\mathbb{R}^n) \to A^{s,\tau}_{p,q}(\mathbb{R}^n)$ is continuous.*

Form Theorem 5.1 and the smooth atomic decomposition characterization of $A^{s,\tau}_{p,q}(\mathbb{R}^n)$, we also deduce the following result.

Corollary 5.2. *Let s, p, q and τ be as in Theorem 5.1. Assume that $l \in \mathbb{Z}_+$ such that $s+2l > \sigma_{p,g}$ if $A^{s,\tau}_{p,q}(\mathbb{R}^n) = F^{s,\tau}_{p,q}(\mathbb{R}^n)$, and $s+2l > \sigma_p$ if $A^{s,\tau}_{p,q}(\mathbb{R}^n) = B^{s,\tau}_{p,q}(\mathbb{R}^n)$. Then any $f \in A^{s,\tau}_{p,q}(\mathbb{R}^n)$ can be represented as $f = (I + (-\Delta)^l)h$ with $h \in A^{s+2l,\tau}_{p,q}(\mathbb{R}^n)$ and*

$$C^{-1}\|f\|_{A^{s,\tau}_{p,q}(\mathbb{R}^n)} \le \|h\|_{A^{s+2l,\tau}_{p,q}(\mathbb{R}^n)} \le C\|f\|_{A^{s,\tau}_{p,q}(\mathbb{R}^n)},$$

where C is a positive constant independent of f and h.

Proof. We first show that the operator $I+(-\Delta)^l$ is continuous from $A^{s+2l,\tau}_{p,q}(\mathbb{R}^n)$ to $A^{s,\tau}_{p,q}(\mathbb{R}^n)$. Let $h\in A^{s+2l,\tau}_{p,q}(\mathbb{R}^n)$. By Theorem 3.3 and Remark 3.1, there exist a sequence $t\equiv\{t_Q\}_{l(Q)\le 1}\subset\mathbb{C}$ satisfying

$$\|t\|_{a^{s+2l,\tau}_{p,q}(\mathbb{R}^n)}\lesssim\|h\|_{A^{s+2l,\tau}_{p,q}(\mathbb{R}^n)}$$

and a family $\{a_Q\}_{l(Q)\le 1}$ of smooth atoms for $A^{s+2l,\tau}_{p,q}(\mathbb{R}^n)$ such that $h=\sum_{l(Q)\le 1}t_Qa_Q$ in $\mathscr{S}'(\mathbb{R}^n)$, where the smooth atom a_Q has the regularity condition that

$$\|\partial^\beta a_Q\|_{L^\infty(\mathbb{R}^n)}\le|Q|^{-1/2-|\beta|/n}$$

if $|\beta|\le\widetilde{K}$, and the moment condition that $\int_{\mathbb{R}^n}x^\beta a_Q(x)\,dx=0$ if $|\beta|\le\widetilde{N}$, where $\widetilde{K}\ge\max\{\lfloor s+n\tau+1\rfloor,0\}+2l$ and $\widetilde{N}\ge\max\{\lfloor J-n-s\rfloor,-1\}+2l$. In view of the actual construction in [64, p. 132], we see that t_Qa_Q is obtained canonically for all $h\in A^{s+2l,\tau}_{p,q}(\mathbb{R}^n)$.

We now claim that

$$f\equiv(I+(-\Delta)^l)h\equiv\sum_{l(Q)\le 1}t_Q(I+(-\Delta)^l)a_Q$$

converges in $\mathscr{S}'(\mathbb{R}^n)$ and satisfies

$$\|f\|_{A^{s,\tau}_{p,q}(\mathbb{R}^n)}\lesssim\|h\|_{A^{s+2l,\tau}_{p,q}(\mathbb{R}^n)}.$$

To this end, by the inequality that

$$\|\{|Q|^{-\frac{2l}{n}}t_Q\}_{l(Q)\le 1}\|_{a^{s,\tau}_{p,q}(\mathbb{R}^n)}=\|t\|_{a^{s+2l,\tau}_{p,q}(\mathbb{R}^n)}\lesssim\|h\|_{A^{s+2l,\tau}_{p,q}(\mathbb{R}^n)}$$

and Theorem 3.3 again, it suffices to prove that for each $Q\in\mathscr{Q}$ with $l(Q)\le 1$, $b_Q\equiv|Q|^{\frac{2l}{n}}(I+(-\Delta)^l)a_Q$ is a constant multiple of a smooth atom for $A^{s,\tau}_{p,q}(\mathbb{R}^n)$ supported near Q.

Obviously, b_Q satisfies the support condition (3.13). On the other hand, since for all $\beta\in\mathbb{Z}^n_+$ with $|\beta|\le\widetilde{K}$,

$$\|\partial^\beta a_Q\|_{L^\infty(\mathbb{R}^n)}\le|Q|^{-1/2-|\beta|/n},$$

then for all $\gamma\in\mathbb{Z}^n_+$ with $|\beta|\le\widetilde{K}-2l$ and $Q\in\mathscr{Q}$ with $l(Q)\le 1$,

$$\begin{aligned}\|\partial^\gamma b_Q\|_{L^\infty(\mathbb{R}^n)}&=|Q|^{\frac{2l}{n}}\|\partial^\gamma(I+(-\Delta)^l)a_Q\|_{L^\infty(\mathbb{R}^n)}\\&\le|Q|^{\frac{2l}{n}}\left\{|Q|^{-1/2-|\gamma|/n}+|Q|^{-1/2-|\gamma|/n-2l/n}\right\}\\&\lesssim|Q|^{-1/2-|\gamma|/n}.\end{aligned}$$

Similarly, by the moment condition of a_Q, we obtain that

$$\int_{\mathbb{R}^n}x^\gamma b_Q(x)\,dx=0\quad\text{if}\quad|\gamma|\le\widetilde{N}-2l.$$

Thus, a constant multiple of b_Q satisfies the regularity condition (3.15) and the moment condition (3.14), which proves the previous claim and further implies that $I+(-\Delta)^l$ is continuous from $A^{s+2l,\tau}_{p,q}(\mathbb{R}^n)$ to $A^{s,\tau}_{p,q}(\mathbb{R}^n)$.

To finish the proof of Corollary 5.2, we need to show that $I+(-\Delta)^l$ is a surjective operator. Let $f\in A^{s,\tau}_{p,q}(\mathbb{R}^n)$ and set

$$a(x,\xi)\equiv(1+|\xi|^{2l})^{-1}$$

for all x, $\xi\in\mathbb{R}^n$. It is easy to see that $a\in\mathscr{S}^{-2l}_{1,1}(\mathbb{R}^n)$. By $s+2l>J-n$ and Theorem 5.1, the corresponding operator $a(x,D)$ is a continuous linear mapping from $A^{s,\tau}_{p,q}(\mathbb{R}^n)$ to $A^{s+2l,\tau}_{p,q}(\mathbb{R}^n)$. Set $h\equiv a(x,D)f$. Then $h\in A^{s+2l,\tau}_{p,q}(\mathbb{R}^n)$ and

$$f\equiv(I+(-\Delta)^l)h,$$

which completes the proof of Corollary 5.2. □

Remark 5.2. Corollaries 5.1 and 5.2 will be of certain use in the next chapter.

In addition, by Theorem 5.1, we also obtain the so-called lifting properties for the spaces $A^{s,\tau}_{p,q}(\mathbb{R}^n)$. Let $\sigma\in\mathbb{R}$. Recall that the lifting operator I_σ is defined by

$$\widehat{I_\sigma f}\equiv(1+|\cdot|^2)^{\sigma/2}\widehat{f},\quad f\in\mathscr{S}'(\mathbb{R}^n);$$

see, for example, [145, p. 58]. It is well known that I_σ is a one-to-one mapping from $\mathscr{S}'(\mathbb{R}^n)$ onto itself.

Notice that

$$a(x,\xi)\equiv(1+|\xi|^2)^{\sigma/2}\in\mathscr{S}^{\sigma}_{1,1}(\mathbb{R}^n).$$

Applying Theorem 5.1, we have the following result.

Proposition 5.1. *Let $\sigma\in\mathbb{R}$ and s, p, q and τ be as in Theorem 5.1. Then the operator I_σ maps $A^{s,\tau}_{p,q}(\mathbb{R}^n)$ isomorphically onto $A^{s-\sigma,\tau}_{p,q}(\mathbb{R}^n)$.*

We remark that Proposition 5.1 when $\tau=0$ generalizes the classic conclusion in [145, Theorem 2.3.8].

5.2 Composition of Functions in $A^{s,\tau}_{p,q}(\mathbb{R}^n)$

Let $f:\ \mathbb{R}\to\mathbb{R}$ be a smooth function such that $f(0)=0$. Then there is a well-known connection between mapping properties of the nonlinear composition operator

$$T_f: g\to f\circ g,\qquad g\in A^{s,\tau}_{p,q}(\mathbb{R}^n),$$

and the boundedness of pseudo-differential operators from the class $\mathscr{S}^0_{1,1}(\mathbb{R}^n)$. We follow [98] and [100, Sect. 16.2].

Let ψ, $\{\psi^j\}_{j\in\mathbb{Z}_+}$ and φ_j be defined as in (1.1) and (1.2). Observe

$$\sum_{j=0}^{M} \psi^j(x) = \psi(2^{-M}x) \to 1 \qquad \text{if} \quad M \to \infty .$$

For $g \in C(\mathbb{R}^n)$ we define

$$\Delta_j g \equiv \varphi_j * g \qquad \text{and} \qquad S_j g \equiv \mathscr{F}^{-1}[\psi(2^{-j}\xi)\mathscr{F}g(\xi)] .$$

Then the composition $f \circ g$ can be written as

$$f \circ g = f \circ S_0 g + (f \circ S_1 g - f \circ S_0 g) + \ldots + (f \circ S_{j+1} g - f \circ S_j g) + \ldots .$$

The convergence of the latter telescopic series follows from the inequality

$$|f \circ g(x) - f \circ S_j g(x)| \le \left(\sup_{\substack{|u|,|v|\le \|g\|_\infty \\ u\neq v}} \frac{|f(u)-f(v)|}{|u-v|} \right) |g(x) - g_j(x)|$$

combined with the uniform convergence of $S_j g$ to g. With

$$m_j(x) \equiv \int_0^1 f'(S_j g(x) + t\Delta_j g(x))\, dt ,$$

we can rewrite $f \circ g$ as

$$f \circ g(x) = f \circ S_0 g + \sum_{j=0}^{\infty} m_j(x)\Delta_j g(x) . \tag{5.5}$$

Lemma 5.1. *Let $g \in C(\mathbb{R}^n)$. Then the linear operator*

$$\mathscr{L} g(x) \equiv \sum_{j=0}^{\infty} m_j(x)\Delta_j g(x)$$

with symbol

$$a(x,\xi) \equiv \sum_{j=0}^{\infty} m_j(x)\, \psi^j(\xi)$$

belongs to $\mathscr{S}^0_{1,1}(\mathbb{R}^n)$.

Proof. Let $j \in \mathbb{N}$ and let $2^{j-1} \le |\xi| \le 2^j$. Then only $\psi^{j-1}(\xi)$ and $\psi^j(\xi)$ can be different from 0. Thus, $a(x,\xi)$ is finite and

$$\begin{aligned} |\partial_x^\alpha \partial_\xi^\beta a(x,\xi)| &= \left| \sum_{\ell=j-1}^{j} \partial_x^\alpha m_\ell(x)\, \partial_\xi^\beta \psi^\ell(\xi) \right| \\ &\lesssim 2^{-j|\beta|} \max_{j-1\le \ell\le j} \left| \partial_x^\alpha m_\ell(x) \left[\partial^\beta \psi(2^{-\ell}\xi) - 2^{|\beta|}\partial^\beta \psi(2^{-\ell+1}\xi) \right] \right| . \end{aligned}$$

Let

$$b_j(x,t) \equiv S_j g(x) + t\,\Delta_j g(x)\,.$$

Then

$$|\partial^\gamma b_j(x,t)| \lesssim 2^{j|\gamma|}\, \|g\|_{L^\infty(\mathbb{R}^n)}$$

with constants behind $\lesssim$ independent of $x \in \mathbb{R}^n$, $t \in (0,1)$, $j \in \mathbb{N}$ and $g \in C(\mathbb{R}^n)$. Faa die Bruno's formula yields

$$\begin{aligned} |\partial_x^\alpha m_j(x)| &\lesssim \sum_{k=1}^{|\alpha|} |f^{(k+1)}(b_j(x,t))| \sum_{\gamma^1+\ldots+\gamma^k=\alpha} \partial^{\gamma^1} b_j(x,t) \ldots \partial^{\gamma^k} b_j(x,t) \\ &\lesssim 2^{j|\alpha|}\, \|f\|_{C^{|\alpha|+1}(B(0,\|g\|_{L^\infty(\mathbb{R}^n)}))} \left(\|g\|_{L^\infty(\mathbb{R}^n)} + \|g\|^{|\alpha|}_{L^\infty(\mathbb{R}^n)}\right). \end{aligned}$$

With an obvious modification for $|\xi| \le 1$ we have found the estimate

$$\begin{aligned} |\partial_x^\alpha \partial_\xi^\beta a(x,\xi)| \; &\lesssim \|f\|_{C^{|\alpha|+1}(B(0,\|g\|_{L^\infty(\mathbb{R}^n)}))} \\ &\quad \times \left(\|g\|_{L^\infty(\mathbb{R}^n)} + \|g\|^{|\alpha|}_{L^\infty(\mathbb{R}^n)}\right) (1+|\xi|)^{|\alpha|-|\beta|}. \end{aligned} \tag{5.6}$$

This proves the lemma. □

Remark 5.3. Lemma 5.1 is in principle known, we refer to [100, Lemma 16.2/1,2]. However, we repeated the proof since we missed a reference for the estimate (5.6), which we need for the proof of the next theorem.

We shall work with functions f which are infinitely differentiable, i. e. $f \in C^\infty(\mathbb{R})$.

Theorem 5.2. *Assume that $p, q \in (0,\infty]$. Let either $s \in (\sigma_{p,q},\infty)$ and $\tau \in [0,\tau_{s,p,q})$ if $A^{s,\tau}_{p,q}(\mathbb{R}^n) = F^{s,\tau}_{p,q}(\mathbb{R}^n)$ or let $s \in (\sigma_p,\infty)$ and $\tau \in [0,\tau_{s,p})$ if $A^{s,\tau}_{p,q}(\mathbb{R}^n) = B^{s,\tau}_{p,q}(\mathbb{R}^n)$. Let $f \in C^\infty(\mathbb{R})$ and $f(0) = 0$. Then, for all real-valued functions $g \in A^{s,\tau}_{p,q}(\mathbb{R}^n) \cap C(\mathbb{R}^n)$, the function $f \circ g$ also belongs to $A^{s,\tau}_{p,q}(\mathbb{R}^n) \cap C(\mathbb{R}^n)$. The associated operator*

$$T_f : A^{s,\tau}_{p,q}(\mathbb{R}^n) \cap C(\mathbb{R}^n) \to A^{s,\tau}_{p,q}(\mathbb{R}^n) \cap C(\mathbb{R}^n)$$

is bounded.

Proof. From Lemma 5.1, Theorem 5.1, (5.4) and (5.6), we deduce that

$$\|\mathscr{L} g\|_{A^{s,\tau}_{p,q}(\mathbb{R}^n)} \lesssim \|f\|_{C^{M+1}(B(0,\|g\|_{L^\infty(\mathbb{R}^n)}))} \left(\|g\|_{L^\infty(\mathbb{R}^n)} + \|g\|^{M}_{L^\infty(\mathbb{R}^n)}\right) \|g\|_{A^{s,\tau}_{p,q}(\mathbb{R}^n)}$$

for some $M \equiv M(s,p,q,\tau) \in \mathbb{N}$. By (5.5), it remains to show that $f \circ S_0 g \in A^{s,\tau}_{p,q}(\mathbb{R}^n)$ and to estimate $\|f \circ S_0 g\|_{A^{s,\tau}_{p,q}(\mathbb{R}^n)}$. Of course, $f \circ S_0 g \in C^\infty(\mathbb{R}^n)$. Since g is bounded, also $S_0 g$ is bounded and we have the obvious estimate

$$\|f \circ S_0 g\|_{C(\mathbb{R}^n)} \le \|f\|_{C(B(0,\|S_0 g\|_{L^\infty(\mathbb{R}^n)}))} \le \|f\|_{C(B(0,c\,\|g\|_{L^\infty(\mathbb{R}^n)}))},$$

where c is a positive constant independent of f and g. To estimate $\|f\circ S_0 g\|_{A^{s,\tau}_{p,q}(\mathbb{R}^n)}$ we shall apply the characterizations by differences; see Sect. 4.3. By the elementary embeddings in Proposition 2.1 it will enough to derive an estimate of $\|f\circ S_0 g\|_{B^{s_1,\tau}_{p,\infty}(\mathbb{R}^n)}$ for some $s_1 > s$. Now we make use of an argument which we have applied also in the proof of Lemma 4.4. Let f and g be fixed. From the regularity of f and $S_0 g$ it is clear that $f\circ S_0 g \in B^{s_1}_{p,\infty}(P)$ for any dyadic cube P. For the same reasons we also have $S_0 g \in B^{s_1}_{p,\infty}(P)$. We associate to P an extension $\mathscr{E}_P(S_0 g)$ of the restriction of $S_0 g$ to P such that

$$\|\mathscr{E}_P(S_0 g)\|_{B^{s_1}_{p,\infty}(\mathbb{R}^n)} \le 2\|S_0 g\|_{B^{s_1}_{p,\infty}(P)}\,.$$

Of course, $f\circ\mathscr{E}_P(S_0 g)$ is an extension of the restriction of $(f\circ(S_0 g))$ to P. To have a more precise notation we shall write $a_t(f)$ instead of a_t; see (4.12). Then it follows from known estimates of composition operators on Besov spaces (see [119, Theorem 5.3.4/2]) that

$$\begin{aligned}
&\frac{1}{|P|^\tau}\sup_{0<t<2(l(P)\wedge 1)} t^{-s_1}\left(\int_P [a_t(f\circ\mathscr{E}_P(S_0 g))(x)]^p\,dx\right)^{1/p}\\
&\lesssim \frac{1}{|P|^\tau}\,\|f\circ\mathscr{E}_P(S_0 g)\|_{B^{s_1}_{p,\infty}(\mathbb{R}^n)}\\
&\lesssim \|f\|_{C^{s_2}(B(0,c\|g\|_{L^\infty(\mathbb{R}^n)}))}\frac{1}{|P|^\tau}\,\|\mathscr{E}_P(S_0 g)\|_{B^{s_1}_{p,\infty}(\mathbb{R}^n)}\,\|\mathscr{E}_P(S_0 g)\|^{s_2-1}_{L^\infty(\mathbb{R}^n)}\\
&\lesssim \|f\|_{C^{s_2}(B(0,c\|g\|_{L^\infty(\mathbb{R}^n)}))}\|g\|^{s_2-1}_{L^\infty(\mathbb{R}^n)}\,\frac{1}{|P|^\tau}\,\|S_0 g\|_{B^{s_1}_{p,\infty}(P)}\\
&\lesssim \|f\|_{C^{s_2}(B(0,c\|g\|_{L^\infty(\mathbb{R}^n)}))}\|g\|^{s_2-1}_{L^\infty(\mathbb{R}^n)}\,\|S_0 g\|_{B^{s_1,\tau}_{p,\infty}(\mathbb{R}^n)}
\end{aligned}$$

for some $s_2 > \max\{s_1, 1\}$. Since

$$\|S_0 g\|_{B^{s_1,\tau}_{p,\infty}(\mathbb{R}^n)} \lesssim \|S_0 g\|_{B^{s,\tau}_{p,\infty}(\mathbb{R}^n)} \lesssim \|g\|_{B^{s,\tau}_{p,\infty}(\mathbb{R}^n)},$$

we finally have obtained

$$\|f\circ S_0 g\|^{\clubsuit}_{B^{s_1,\tau}_{p,\infty}(\mathbb{R}^n)} \lesssim \|f\|_{C^{s_2}(B(0,c\|g\|_{L^\infty(\mathbb{R}^n)}))}\|g\|^{s_2-1}_{L^\infty(\mathbb{R}^n)}\,\|g\|_{B^{s,\tau}_{p,\infty}(\mathbb{R}^n)}\,.$$

Having a look at Theorems 4.7, 4.8 and 4.9, we need to estimate also $\|f\circ S_0 g\|_{L^p_\tau(\mathbb{R}^n)}$, $N_1(f\circ S_0 g)$, and

$$\sup_{\{P\in\mathscr{Q}:\,|P|\ge 1\}} \|f\circ S_0 g\|_{B^{s_0}_{p,\infty}(2P)}/|P|^\tau,$$

where $\sigma_p < s_0 < s$. The estimate of the last term follows from an argument as above. Finally, the estimates of $\|f \circ S_0 g\|_{L^p_\tau(\mathbb{R}^n)}$ and $N_1(f \circ S_0 g)$ are obtained by using the local Lipschitz continuity of f and $f(0) = 0$. We have

$$\begin{aligned}\|f \circ S_0 g - f(0)\|_{L^p_\tau(\mathbb{R}^n)} &\leq \|f\|_{C^1(B(0,c\|g\|_{L^\infty(\mathbb{R}^n)}))} \|S_0 g\|_{L^p_\tau(\mathbb{R}^n)} \\ &\lesssim \|f\|_{C^1(B(0,c\|g\|_{L^\infty(\mathbb{R}^n)}))} \|S_0 g\|_{A^{s,\tau}_{p,q}(\mathbb{R}^n)} \\ &\lesssim \|f\|_{C^1(B(0,c\|g\|_{L^\infty(\mathbb{R}^n)}))} \|g\|_{A^{s,\tau}_{p,q}(\mathbb{R}^n)},\end{aligned}$$

where we used Propositions 2.7, 2.1 inbetween. In the same way one can estimate $N_1(f \circ S_0 g)$. This proves $f \circ g \in A^{s,\tau}_{p,q}(\mathbb{R}^n)$ for all $g \in A^{s,\tau}_{p,q}(\mathbb{R}^n)$ and at the same time, that bounded subsets of $A^{s,\tau}_{p,q}(\mathbb{R}^n)$ are mapped into bounded subsets of $A^{s,\tau}_{p,q}(\mathbb{R}^n)$. □

Remark 5.4.

(i) As we have mentioned above, we followed Meyer [98] and Meyer and Coifman [100, Sect. 16.2], where the case of Bessel potential spaces

$$H^s_p(\mathbb{R}^n) = F^{s,0}_{p,2}(\mathbb{R}^n)$$

with $p \in (1,\infty)$, was treated.

(ii) In case $\tau = 0$ much more is known about the properties of those composition operators. Surveys can be found in the monographs, Appell and Zabrejko [6] (first order Sobolev spaces, L^p-spaces), [118], [119, Chap. 5] and in the paper of Bourdaud [17]. For the latest progress we refer to [20, 21] and [22].

Chapter 6
Key Theorems

In this chapter we focus on some crucial problems including pointwise multipliers, diffeomorphisms and traces, which govern the theory of the spaces $A^{s,\tau}_{p,q}(\mathbb{R}^n)$ to a large extent. These problems are of vital importance both for function spaces treated for their own sake and for applications to partial differential equations. Following Triebel's monograph [146], we call these assertions key theorems, since these theorems are the basis for the definitions of Besov-type spaces and Triebel-Lizorkin-type spaces on domains. An important tool used in this chapter is the smooth atomic decomposition characterization of $A^{s,\tau}_{p,q}(\mathbb{R}^n)$ in Theorem 3.3.

6.1 Pointwise Multipliers

Pointwise multiplication in Besov and Triebel-Lizorkin spaces has been studied extensively in the last 30 years; see, for example, [95, 114, 119, 145, 146] and [96]. The two monographs [95, 96] by Maz'ya and Shaposnikova are completely devoted to this subject. However, the authors restrict their interest essentially to the Bessel-potential spaces $F^s_{p,2}(\mathbb{R}^n)$, $1 < p < \infty$, and the Slobodeckij spaces $B^s_{p,p}(\mathbb{R}^n)$, $1 \le p \le \infty$.

Let X and Y be two quasi-Banach spaces of functions (distributions). Then the basic question consists in descriptions of the associated multiplier space $M(X,Y)$ given by

$$M(X,Y) \equiv \{f : f \cdot g \in Y \text{ for all } g \in X\}.$$

This space is equipped with the induced quasi-norm

$$\|f\|_{M(X,Y)} \equiv \sup_{\|g\|_X \le 1} \|f \cdot g\|_Y .$$

Here, in this lecture note, we will be concerned with the easier problem of proving embeddings into $M(X) \equiv M(X,X)$ with $X \equiv A^{s,\tau}_{p,q}(\mathbb{R}^n)$.

We shall present three different approaches. The first one uses atoms and the lifting operator I_σ and is the most general, i. e., it can be applied for all p, q, s and some τ (depending on s and p). The second one uses the Hardy-Littlewood maximal

D. Yang et al., *Morrey and Campanato Meet Besov, Lizorkin and Triebel*,
Lecture Notes in Mathematics 2005, DOI 10.1007/978-3-642-14606-0_6,

function. As a consequence we have to restrict us to spaces $X = F^{s,\tau}_{p,q}(\mathbb{R}^n)$ with $\tau < 1/p$. However, it leads to a very natural extension of some classical assertions, known for many years in case $\tau = 0$. Our third approach uses characterizations by differences. Here we concentrate on $\tau = 1/p$ leaving possible further applications aside.

6.1.1 Smooth Functions are Pointwise Multipliers for $A^{s,\tau}_{p,q}(\mathbb{R}^n)$

We shall prove embeddings of the type

$$C^m(\mathbb{R}^n) \subset M(A^{s,\tau}_{p,q}(\mathbb{R}^n)) \qquad \text{if} \qquad m = m(s,p,q,\tau,n)$$

is large enough. Here $C^m(\mathbb{R}^n)$ is defined in item (iii) in Sect. 1.3.

Theorem 6.1. *Let s, p, q and τ be as in Theorem 5.1. If $m \in \mathbb{N}$ is sufficiently large, then there exists a positive constant $C(m)$ such that for all $g \in C^m(\mathbb{R}^n)$ and all $f \in A^{s,\tau}_{p,q}(\mathbb{R}^n)$,*

$$\|gf\|_{A^{s,\tau}_{p,q}(\mathbb{R}^n)} \le C(m) \left(\sum_{|\alpha| \le m} \|\partial^\alpha g\|_{L^\infty(\mathbb{R}^n)} \right) \|f\|_{A^{s,\tau}_{p,q}(\mathbb{R}^n)}. \tag{6.1}$$

Proof. First we prove (6.1) under the assumption $s > J - n$. Let $f \in A^{s,\tau}_{p,q}(\mathbb{R}^n)$ and $g \in C^m(\mathbb{R}^n)$ with $m \ge \max\{\lfloor s + n\tau + 1 \rfloor, 0\}$. By Theorem 3.3, we can write f as

$$f = \sum_{l(Q) \le 1} t_Q a_Q$$

in $\mathscr{S}'(\mathbb{R}^n)$, where each a_Q is a smooth atom for $A^{s,\tau}_{p,q}(\mathbb{R}^n)$ supported near Q and the sequence $t \equiv \{t_Q\}_{l(Q)\le 1} \subset \mathbb{C}$ satisfies

$$\|t\|_{a^{s,\tau}_{p,q}(\mathbb{R}^n)} \lesssim \|f\|_{A^{s,\tau}_{p,q}(\mathbb{R}^n)}.$$

Set $b_Q \equiv g a_Q$, then $\operatorname{supp} b_Q \subset 3Q$. To show $gf \in A^{s,\tau}_{p,q}(\mathbb{R}^n)$, by Theorem 3.3 again, it suffices to prove that each b_Q is a constant multiple of a smooth atom for $A^{s,\tau}_{p,q}(\mathbb{R}^n)$ supported near Q. By the assumption $s > J - n$, there is no need to postulate any moment condition on b_Q. Thus, we focus on the regularity condition of b_Q. Indeed, since $l(Q) \le 1$ and $m \ge \max\{\lfloor s + n\tau + 1 \rfloor, 0\}$, for all $\gamma \in \mathbb{Z}^n_+$ with $|\gamma| \le \max\{\lfloor s + n\tau + 1 \rfloor, 0\}$, we have

$$\begin{aligned} \|\partial^\gamma b_Q\|_{L^\infty(\mathbb{R}^n)} &= \|\partial^\gamma (g a_Q)\|_{L^\infty(\mathbb{R}^n)} \\ &\le \sum_{\alpha \le \gamma} \|\partial^\alpha g\|_{L^\infty(\mathbb{R}^n)} \|\partial^{\gamma-\alpha} a_Q\|_{L^\infty(\mathbb{R}^n)} \end{aligned}$$

$$\leq \sum_{\alpha\leq\gamma} \|\partial^\alpha g\|_{L^\infty(\mathbb{R}^n)} |Q|^{-\frac{1}{2}-\frac{|\gamma|-|\alpha|}{n}}$$
$$\leq \left(\sum_{|\alpha|\leq m} \|\partial^\alpha g\|_{L^\infty(\mathbb{R}^n)}\right) |Q|^{-\frac{1}{2}-\frac{|\gamma|}{n}},$$

which implies that (6.1) holds provided $s > J - n$.

Now we consider the case when $s \leq J - n$. Fix $l \in \mathbb{N}$ such that $s + 2l > J - n$. Then by Corollary 5.2, any $f \in A^{s,\tau}_{p,q}(\mathbb{R}^n)$ can be represented as $f \equiv (I + (-\Delta)^l)h$ with $h \in A^{s+2l,\tau}_{p,q}(\mathbb{R}^n)$ and $\|h\|_{A^{s+2l,\tau}_{p,q}(\mathbb{R}^n)} \sim \|f\|_{A^{s,\tau}_{p,q}(\mathbb{R}^n)}$. Similarly to the argument in [146, p. 204], we have

$$gf = (I + (-\Delta)^l)(gh) + \sum_{|\alpha|<2l} \partial^\alpha (g_\alpha h),$$

where each g_α is the summation of terms of type $\partial^\beta g$ with $|\beta| \leq 2l$. Then by Corollaries 5.1, 5.2 and Proposition 2.1(ii), we obtain

$$\|gf\|_{A^{s,\tau}_{p,q}(\mathbb{R}^n)} \lesssim \sum_{|\alpha|\leq 2l} \|g_\alpha h\|_{A^{s+2l,\tau}_{p,q}(\mathbb{R}^n)}. \tag{6.2}$$

Let $m \in \mathbb{N}$ such that

$$m - 2l \geq \lfloor s + 2l + n\tau + 1 \rfloor.$$

Then applying the previous proved result to the right-hand side of (6.2) yields that for all $|\alpha| \leq 2l$,

$$\|g_\alpha h\|_{A^{s+2l,\tau}_{p,q}(\mathbb{R}^n)} \lesssim \left(\sum_{|\beta|\leq m-2l} \|\partial^\beta g_\alpha\|_{L^\infty(\mathbb{R}^n)}\right) \|h\|_{A^{s+2l,\tau}_{p,q}(\mathbb{R}^n)}$$
$$\lesssim \left(\sum_{|\beta|\leq m} \|\partial^\beta g\|_{L^\infty(\mathbb{R}^n)}\right) \|f\|_{A^{s,\tau}_{p,q}(\mathbb{R}^n)},$$

which together with (6.2) implies (6.1), and then completes the proof of Theorem 6.1. □

Remark 6.1. Theorem 6.1 generalizes some classical results on Besov spaces and Triebel-Lizorkin spaces by taking $\tau = 0$; see, for example, [146, Theorem 4.2.2].

6.1.2 Pointwise Multipliers and Paramultiplication

In this part we shall improve Theorem 6.1 for certain parameter constellations. We follow a well-known strategy. The product of f and g will be decomposed into three parts (involving paraproducts). These three parts can be estimated by applying some tools from Fourier analysis, in particular Marschall's inequality in combination with the dyadic ball criterion. The tools rely on the boundedness of the Hardy-Littlewood maximal function. As a consequence we investigate principally the multipliers of the spaces $\mathscr{E}^s_{pqu}(\mathbb{R}^n)$. However, based on their identification with $F^{s,\tau}_{u,q}(\mathbb{R}^n)$, $\tau = 1/u - 1/p$, this leads to corresponding results for $M(F^{s,\tau}_{p,q}(\mathbb{R}^n))$. This procedure causes the restriction to $\tau < 1/p$.

6.1.2.1 Marschall's Inequality and the Dyadic Ball Criterion

This subsection contains some preliminaries. Inspired by Marschall's paper [94], we shall give a version of his pointwise estimate of pseudo-differential operators $b(x,D)$, that is suitable for us.

In Marschall's inequality the symbol is estimated via the norm of a homogeneous Besov space $\dot{B}^s_{p,q}(\mathbb{R}^n) \equiv \dot{B}^{s,0}_{p,q}(\mathbb{R}^n)$. We refer to Chap. 8 for the definitions of these spaces. It will be convenient for us to work with a homogeneous dyadic partition of unity. Let ψ and ψ^1 be as in (1.1) and define

$$\phi_1 \equiv \psi^1 \qquad \text{and} \qquad \phi_k \equiv \phi_1(2^{-k+1}\cdot), \qquad k \in \mathbb{Z}.$$

Then it follows that

$$1 = \sum_{k=-\infty}^{\infty} \phi_k \qquad \text{on} \quad \mathbb{R}^n \setminus \{0\}.$$

Proposition 6.1. *Let a symbol $b \in C_0^\infty(\mathbb{R}^n)$ and a function $f \in C^\infty(\mathbb{R}^n)$ be given such that, for $A > 0$ and $R \geq 1$,*

$$\operatorname{supp} \mathscr{F} f \subset B(0,AR) \quad \textit{and} \quad \operatorname{supp} b \subset B(0,A). \tag{6.3}$$

Let $t \in (0,1]$. Then there exists a positive constant C such that

$$|b(D)u(x)| \leq C(RA)^{\frac{n}{t}-n} \, \| b \|_{\dot{B}^{n/t}_{1,t}(\mathbb{R}^n)} \left[\mathrm{M}(|f|^t)\right]^{1/t}(x). \tag{6.4}$$

Here C can be taken as a function of t only.

Proof. Since convolutions in $\mathscr{S}(\mathbb{R}^n) * \mathscr{S}'(\mathbb{R}^n)$ are mapped to products by the Fourier transformation,

$$b(D)u(x) = \mathscr{F}^{-1}(b\mathscr{F}u)(x) = \frac{1}{(2\pi)^{n/2}} \int_{\mathbb{R}^n} \mathscr{F}^{-1}b(x-y)u(y)\,dy.$$

With x fixed, $y \mapsto \mathscr{F}^{-1}b(x-y)u(y)$ has its spectrum in

$$B(0,A) + B(0,RA) \subset B(0,(R+1)A).$$

Let us recall the Nikol'skij inequality in a form stated in [145, Sect. 1.3.2]. Let $0 < p \le q \le \infty$, $b \in (0,\infty)$ and $\varphi \in \mathscr{S}(\mathbb{R}^n)$ satisfying $\operatorname{supp} \widehat{\varphi} \subset \overline{B(0,b)}$, then

$$\| \varphi \|_{L^q(\mathbb{R}^n)} \lesssim b^{n(1/p-1/q)} \| \varphi \|_{L^p(\mathbb{R}^n)}$$

with a constant independent on φ and b. Applying this inequality with $q=1$ and $0 < t \le 1$ we obtain that

$$\begin{aligned} |b(D)u(x)| &\le \int_{\mathbb{R}^n} |\mathscr{F}^{-1}b(x-y)u(y)|\,dy \\ &\lesssim (RA)^{\frac{n}{t}-n} \| \mathscr{F}^{-1}b(x-\cdot)\,u \|_{L^t(\mathbb{R}^n)} \\ &\lesssim (RA)^{\frac{n}{t}-n} \left(\sum_{k\in\mathbb{Z}} \| \phi_k(x-\cdot)\,\mathscr{F}^{-1}b(x-\cdot)\,u \|_{L^t(\mathbb{R}^n)}^t \right)^{1/t}. \end{aligned} \tag{6.5}$$

By the obvious estimate

$$\sup_{y\in\mathbb{R}^n} |\phi_k(y)\,\mathscr{F}^{-1}b(y)| \le \int_{\mathbb{R}^n} \left| \mathscr{F}^{-1}(\phi_k \mathscr{F}^{-1}b)(\eta) \right| d\eta \equiv b_k\,,$$

one finds

$$\int_{B(x,2^{k+1})} |\phi_k(x-y)\,\mathscr{F}^{-1}b(x-y)\,f(y)|^t\,dy \lesssim b_k^t\, 2^{kn}\,\mathrm{M}(|f|^t)(x).$$

Inserting this into (6.5) we obtain the desired inequality, since

$$\sum_{k\in\mathbb{Z}} \left(2^{kn/t} \| \mathscr{F}^{-1}[\phi_k \mathscr{F} b] \|_{L^1(\mathbb{R}^n)} \right)^t = \| b \|_{B^{n/t}_{1,t}(\mathbb{R}^n)}^t\,,$$

which completes the proof. □

Remark 6.2. Proposition 6.1 is a simplified version of an inequality proved in [81].

Now we turn to the dyadic ball criterion for the spaces $F^{s,\tau}_{p,q}(\mathbb{R}^n)$.

Proposition 6.2. *Let $p \in (0,\infty), q \in (0,\infty]$, $\tau \in [0,1/p)$ and $s \in (\sigma_{p,q},\infty)$. Suppose $\{u_j\}_{j=0}^\infty \subset \mathscr{S}'(\mathbb{R}^n)$ such that*

$$\operatorname{supp} \mathscr{F} u_j \subset B(0,2^{j+2})\,, \qquad j \in \mathbb{Z}_+\,, \tag{6.6}$$

and

$$A \equiv \sup_{P\in\mathscr{Q}} \frac{1}{|P|^\tau} \left[\int_P \left(\sum_{j=0}^\infty 2^{jsq} |u_j(x)|^q \right)^{p/q} dx \right]^{1/p} < \infty. \tag{6.7}$$

Then $u \equiv \sum_{j=0}^{\infty} u_j$ *converges in* $\mathscr{S}'(\mathbb{R}^n)$ *and its limit* u *belongs to* $F_{p,q}^{s,\tau}(\mathbb{R}^n)$. *Furthermore,*

$$\|u\|_{F_{p,q}^{s,\tau}(\mathbb{R}^n)} \le CA, \tag{6.8}$$

where C *is a positive constant independent of* $\{u_j\}_{j=0}^{\infty}$ *and* A.

Proof. The proof will be based on the identity $F_{p,q}^{s,\tau}(\mathbb{R}^n) = \mathscr{E}_{r,q,p}^{s}(\mathbb{R}^n)$, $1/r = 1/p - \tau$; see Corollary 3.3.

Step 1. Temporarily we assume that the sequence $\{u_j\}_{j=0}^{\infty}$ is finite, i. e., $u_j = 0$ if $j > N$ for some $N \in \mathbb{N}$. Since u_j is a smooth function it belongs to $L^r(P)$ for all $P \in \mathscr{Q}$ and all $r \in (0,\infty]$. Let $\{\varphi_k\}_{k=0}^{\infty}$ be the system defined in (1.2). Since $\operatorname{supp} \mathscr{F}\varphi_k \subset B(0, 2^{k+1})$, Marschall's inequality yields

$$\begin{aligned} |\varphi_k * u_{j+k-3}(x)| &\lesssim \|\mathscr{F}^{-1}\varphi_k\|_{\dot{B}_{1,t}^{n/t}(\mathbb{R}^n)} (2^{j+k})^{\frac{n}{t}-n} \left[\mathrm{M}(|u_{j+k-3}|^t)\right]^{1/t}(x) \\ &\lesssim 2^{j(\frac{n}{t}-n)} \left[\mathrm{M}(|u_{j+k-3}|^t)\right]^{1/t}(x), \end{aligned}$$

where the constants behind $\lesssim$ do not depend on x, j and k. Let $u_j \equiv 0$, if $j < 0$. Next we observe that, using (6.6),

$$\varphi_k * u = \sum_{j=(k-3)\vee 0}^{\infty} \varphi_k * u_j = \sum_{j=0}^{\infty} \varphi_k * u_{j+k-3}\,.$$

Let $d = \min\{1, p, q\}$. Thus, the vector-valued maximal inequality of Tang and Xu (see (4.38)) implies that

$$\begin{aligned} &\|u\|_{\mathscr{E}_{r,q,p}^{s}(\mathbb{R}^n)}^d \\ &\lesssim \sum_{j=0}^{\infty} \sup_{P\in\mathscr{Q}} \frac{1}{|P|^{\tau d}} \left[\int_P \left(\sum_{k=0}^{\infty} 2^{ksq} |\varphi_k * u_{j+k-3}(x)|^q\right)^{p/q} dx\right]^{d/q} \\ &\lesssim \sum_{j=0}^{\infty} 2^{j(\frac{n}{t}-n)d} \sup_{P\in\mathscr{Q}} \frac{1}{|P|^{\tau d}} \left[\int_P \left(\sum_{k=0}^{\infty} 2^{ksq} \left[\mathrm{M}(|u_{j+k-3}|^t)\right]^{q/t}(x)\right)^{p/q} dx\right]^{d/q} \\ &\lesssim \sum_{j=0}^{\infty} 2^{j(\frac{n}{t}-n)d} \sup_{P\in\mathscr{Q}} \frac{1}{|P|^{\tau d}} \left[\int_P \left(\sum_{k=0}^{\infty} 2^{ksq} |u_{j+k-3}|^q\right)^{p/q} dx\right]^{d/q} \\ &\lesssim \sum_{j=0}^{\infty} 2^{j(\frac{n}{t}-n)d} 2^{-jsd} A^d. \end{aligned} \tag{6.9}$$

For t approaching $\min\{1, p, q\}$ the condition $s > \sigma_{p,q}$ becomes sufficient for estimating $\|u\|_{\mathscr{E}_{r,q,p}^{s}(\mathbb{R}^n)}$ by CA. Because of the mentioned coincidence we obtain the same conclusion with $\|u\|_{\mathscr{E}_{r,q,p}^{s}(\mathbb{R}^n)}$ replaced by $\|u\|_{F_{p,q}^{s,\tau}(\mathbb{R}^n)}$.

Step 2. We remove the restriction to finite sequences. Let $q<\infty$. Then, by applying the methods of Step 1, we get

$$\left\|\sum_{j=N}^{\infty} u_j\right\|_{F^{s,\tau}_{p,q}(\mathbb{R}^n)} \to 0 \qquad \text{if} \quad N\to\infty .$$

Thus, $\{\sum_{j=0}^{L} u_j\}_{L\in\mathbb{N}}$ is convergent in $F^{s,\tau}_{p,q}(\mathbb{R}^n)$. Now, let $q=\infty$. Then, by repeating these arguments with s replaced by $s-\varepsilon$, $\varepsilon>0$, and q replaced by 1, we obtain the convergence of $\{\sum_{j=0}^{L} u_j\}_{L\in\mathbb{N}}$ in $F^{s-\varepsilon,\tau}_{p,1}(\mathbb{R}^n)$ and therefore in $\mathscr{S}'(\mathbb{R}^n)$; see Proposition 2.3. Furthermore, Step 1 combined with the Fatou property of $F^{s,\tau}_{p,q}(\mathbb{R}^n)$ (see Proposition 2.8) yields $u\in F^{s,\tau}_{p,q}(\mathbb{R}^n)$ and at the same time (6.8). □

Later on we also need a supplement dealing with dyadic annuli instead of dyadic balls.

Proposition 6.3. *Let $p\in(0,\infty)$, $q\in(0,\infty]$, $\tau\in[0,1/p)$ and $s\in\mathbb{R}$. Suppose that the sequence $\{u_j\}_{j=0}^{\infty}\subset\mathscr{S}'(\mathbb{R}^n)$ such that* $\operatorname{supp}\mathscr{F}u_0\subset B(0,4)$,

$$\operatorname{supp}\mathscr{F}u_j\subset B(0,2^{j+2})\setminus B(0,2^{j-3}), \qquad j\in\mathbb{N}, \tag{6.10}$$

and

$$A\equiv\sup_{P\in\mathscr{Q}}\frac{1}{|P|^{\tau}}\left[\int_P\left(\sum_{j=0}^{\infty}2^{jsq}|u_j(x)|^q\right)^{p/q}dx\right]^{1/p}<\infty. \tag{6.11}$$

Then $u\equiv\sum_{j=0}^{\infty}u_j$ converges in $\mathscr{S}'(\mathbb{R}^n)$ and its limit u belongs to $F^{s,\tau}_{p,q}(\mathbb{R}^n)$. Furthermore,

$$\|u\|_{F^{s,\tau}_{p,q}(\mathbb{R}^n)}\le CA, \tag{6.12}$$

where C is a positive constant independent of $\{u_j\}_j$ and A.

Proof. Observe that

$$\varphi_k*u=\sum_{j=\max\{k-2,0\}}^{k+2}\varphi_k*u_j, \qquad k\in\mathbb{Z}_+ .$$

Based on this identity we can proceed as in the proof of Proposition 6.2. Since the sum with respect to j in the last line of the estimate (6.9) has always less than 6 summands, there is no need for a restriction with respect to s. □

6.1.2.2 The Decomposition of the Product

Let ψ, $\{\psi^j\}_{j\in\mathbb{Z}_+}$ and $\{\varphi_j\}_{j\in\mathbb{Z}_+}$ be as in (1.1) and (1.2), respectively. Let $\varphi_{-1}\equiv 0$. For $f\in\mathscr{S}'(\mathbb{R}^n)$ we put

$$S^jf(x)\equiv\sum_{\ell=0}^{j}(\varphi_\ell*f)(x)=\mathscr{F}^{-1}[\psi(2^{-j}\xi)\mathscr{F}f(\xi)](x).$$

Using these smooth approximations with respect to f and g, we define the product of these distributions as

$$f \cdot g \equiv \lim_{j\to\infty} S^j f \cdot S^j g,$$

whenever this limit exists in $\mathscr{S}'(\mathbb{R}^n)$. For a further discussion of this definition we refer to [80] and [119, Chap. 4].

Related to this definition we introduce the following operators:

$$\Pi_1(f,g) = \sum_{k=2}^{\infty} S^{k-2} f \cdot (\varphi_k * g), \tag{6.13}$$

$$\Pi_2(f,g) = \sum_{k=0}^{\infty} [(\varphi_{k-1} * f) + (\varphi_k * f) + (\varphi_{k+1} * f)] \cdot (\varphi_k * g), \tag{6.14}$$

and

$$\Pi_3(f,g) = \sum_{k=2}^{\infty} (\varphi_k * f) \cdot S^{k-2} g = \Pi_1(g,f). \tag{6.15}$$

It follows that

$$f \cdot g = \Pi_1(f,g) + \Pi_2(f,g) + \Pi_3(f,g), \tag{6.16}$$

whenever these three limits exist in $\mathscr{S}'(\mathbb{R}^n)$. The advantage of this decomposition consists in

$$\operatorname{supp} \mathscr{F} (S^{k-2} f \cdot (\varphi_k * g)) \subset \{\xi : 2^{k-3} \le |\xi| \le 2^{k+1}\}, \quad k = 2,3,\ldots \tag{6.17}$$

and

$$\operatorname{supp} \mathscr{F} \left(\sum_{\ell=k-1}^{k+1} (\varphi_\ell * f) \cdot (\varphi_k * g) \right) \subset \{\xi : |\xi| \le 5 \cdot 2^k\}, \quad k = 0,1,\ldots. \tag{6.18}$$

This means, we can apply either the dyadic ball criterion or Proposition 6.3 in connection with these operators.

Remark 6.3. The splitting technique from formula (6.16) has been invented independently by Peetre [114] and Triebel [143].

6.1.2.3 Multiplication by Hölder Continuous Functions

Theorem 6.2. *Let $s \in \mathbb{R}$, $p \in (0,\infty)$, $q \in (0,\infty]$ and $\tau \in [0,1/p)$. Suppose that*

$$\rho > \max\left\{0, |s|, \frac{n}{p} - n - s\right\}. \tag{6.19}$$

Then the embedding $C^\rho(\mathbb{R}^n) \subset M(F^{s,\tau}_{p,q}(\mathbb{R}^n))$ holds.

Proof. It will be enough to estimate $\Pi_1(f,g)$, $\Pi_2(f,g)$ and $\Pi_3(f,g)$ for $f \in C^\rho(\mathbb{R}^n)$ and $g \in F^{s,\tau}_{p,q}(\mathbb{R}^n)$. Again we shall employ the identity $F^{s,\tau}_{p,q}(\mathbb{R}^n) = \mathscr{E}^s_{r,q,p}(\mathbb{R}^n)$ and $1/r = 1/p - \tau$; see Corollary 3.3.

Step 1. Estimate of Π_1. Recall that the convolution inequality

$$\sup_{k\in\mathbb{Z}_+} \sup_{x\in\mathbb{R}^n} |S^k f(x)| \lesssim \|f\|_{L^\infty(\mathbb{R}^n)}.$$

Applying this convolution inequality and Proposition 6.3 with $u_0 = u_1 = 0$ and $u_{k+2} \equiv S^{k-2} f \cdot (\varphi_k * g)$, $k \geq 0$, we find

$$\begin{aligned}
\|\Pi_1(f,g)\|_{F^{s,\tau}_{p,q}(\mathbb{R}^n)} &\lesssim \sup_{P\in\mathscr{Q}} \frac{1}{|P|^\tau} \left[\int_P \left(\sum_{k=2}^\infty 2^{ksq}|u_k(x)|^q\right)^{p/q} dx\right]^{1/p} \\
&\lesssim \|f\|_{L^\infty(\mathbb{R}^n)} \sup_{P\in\mathscr{Q}} \frac{1}{|P|^\tau} \left[\int_P \left(\sum_{k=2}^\infty 2^{ksq}|(\varphi_k * g)(x)|^q\right)^{p/q} dx\right]^{1/p} \\
&\lesssim \|f\|_{L^\infty(\mathbb{R}^n)} \|g\|_{\mathscr{E}^s_{r,q,p}(\mathbb{R}^n)} \\
&\lesssim \|f\|_{L^\infty(\mathbb{R}^n)} \|g\|_{F^{s,\tau}_{p,q}(\mathbb{R}^n)}. \qquad (6.20)
\end{aligned}$$

Step 2. Estimate of Π_2. Recall that the embedding $F^{s+\rho,\tau}_{p,\infty}(\mathbb{R}^n) \subset F^{s,\tau}_{p,q}(\mathbb{R}^n)$; see Proposition 2.1. This time we have to use the dyadic ball criterion. For simplicity we put

$$u_k \equiv (\varphi_k * f) \cdot (\varphi_k * g), \quad k \in \mathbb{Z}_+.$$

Then we obtain

$$\begin{aligned}
&\|\Pi_2(f,g)\|_{F^{s,\tau}_{p,q}(\mathbb{R}^n)} \\
&\quad\lesssim \|\Pi_2(f,g)\|_{F^{s+\rho,\tau}_{p,\infty}(\mathbb{R}^n)} \\
&\quad\lesssim \sup_{P\in\mathscr{Q}} \frac{1}{|P|^\tau} \left(\int_P \left[\sup_{k\in\mathbb{Z}_+} 2^{k(s+\rho)}|(\varphi_k * f)(x) \cdot (\varphi_k * g)(x)|\right]^p dx\right)^{1/p} \\
&\quad\lesssim \left(\sup_{k\in\mathbb{Z}_+} 2^{k\rho}\|\varphi_k * f\|_{L^\infty(\mathbb{R}^n)}\right) \sup_{P\in\mathscr{Q}} \frac{1}{|P|^\tau} \left[\int_P \left(\sup_{k\in\mathbb{Z}_+} 2^{ks}|(\varphi_k * g)(x)|\right)^p dx\right]^{1/p} \\
&\quad\lesssim \|f\|_{B^\rho_{\infty,\infty}(\mathbb{R}^n)} \|g\|_{F^{s,\tau}_{p,\infty}(\mathbb{R}^n)},
\end{aligned}$$

since

$$s + \rho > n \max\left\{0, \frac{1}{p} - 1\right\}.$$

Step 3. Estimate of Π_3. Let $\rho' < \rho$ be a number which also satisfies (6.19). Again we can apply Proposition 6.3. This yields

$$\begin{aligned}
&\|\Pi_3(f,g)\|_{F^{s,\tau}_{p,q}(\mathbb{R}^n)} \\
&\lesssim \sup_{P\in\mathscr{Q}} \frac{1}{|P|^\tau} \left[\int_P \left(\sum_{k=2}^{\infty} 2^{ksq} |(\varphi_k * f)(x)\cdot S^{k-2}g(x)|^q\right)^{p/q} dx\right]^{1/p} \\
&\lesssim \sup_{P\in\mathscr{Q}} \frac{\|f\|_{B^\rho_{\infty,\infty}(\mathbb{R}^n)}}{|P|^\tau} \left[\int_P \left(\sum_{k=2}^{\infty} 2^{k(s-\rho)q} |S^{k-2}g(x)|^q\right)^{p/q} dx\right]^{1/p} \\
&\lesssim \sup_{P\in\mathscr{Q}} \frac{\|f\|_{B^\rho_{\infty,\infty}(\mathbb{R}^n)}}{|P|^\tau} \left[\int_P \left(\sup_{k=2,3,\ldots} 2^{k(s-\rho')} \sum_{j=0}^{k-2} |(\varphi_j * g)(x)|\right)^{p} dx\right]^{1/p} \\
&\lesssim \|f\|_{B^\rho_{\infty,\infty}(\mathbb{R}^n)} \|g\|_{F^{s-\rho',\tau}_{p,1}(\mathbb{R}^n)} \\
&\lesssim \|f\|_{B^\rho_{\infty,\infty}(\mathbb{R}^n)} \|g\|_{F^{s,\tau}_{p,q}(\mathbb{R}^n)}
\end{aligned}$$

because of $\rho' > 0$. Summarizing Step 1 through Step 3 we have proved the embedding

$$B^\rho_{\infty,\infty}(\mathbb{R}^n) \subset M(F^{s,\tau}_{p,q}(\mathbb{R}^n)).$$

However, because of $C^\rho(\mathbb{R}^n) \subset B^\rho_{\infty,\infty}(\mathbb{R}^n)$ this is sufficient. □

Remark 6.4. For $\tau = 0$ this is a well-known result; we refer to [145, Corollary 2.8.2], [64] and [119, Sect. 4.7.1].

6.1.2.4 Multiplication Algebras

This time we study the question under which conditions do we have the embedding $X \subset M(X)$. Essentially the same methods as used in the proof of Theorem 6.2 apply.

Theorem 6.3. *Let $p\in(0,\infty)$, $q\in(0,\infty]$, $\tau\in[0,1/p)$ and $s\in(\sigma_{p,q},\infty)$. Then there exists a positive constant C such that for all $f,g \in F^{s,\tau}_{p,q}(\mathbb{R}^n)\cap L^\infty(\mathbb{R}^n)$,*

$$\|f\cdot g\|_{F^{s,\tau}_{p,q}(\mathbb{R}^n)} \le C\left(\|f\|_{L^\infty(\mathbb{R}^n)}\|g\|_{F^{s,\tau}_{p,q}(\mathbb{R}^n)} + \|g\|_{L^\infty(\mathbb{R}^n)}\|f\|_{F^{s,\tau}_{p,q}(\mathbb{R}^n)}\right). \tag{6.21}$$

Proof. The estimate of $\Pi_1(f,g)$, given in (6.20), is totally sufficient for our purpose. Since $\Pi_3(f,g) = \Pi_1(g,f)$ we also get the estimate of Π_3 on this way. Finally, we have to deal with Π_2. Similarly as in Step 2 of the proof of Theorem 6.2 we find

$$\begin{aligned}
&\|\Pi_2(f,g)\|_{F^{s,\tau}_{p,q}(\mathbb{R}^n)} \\
&\lesssim \sup_{P\in\mathscr{Q}} \frac{1}{|P|^\tau} \left(\int_P \left[\sum_{k=0}^{\infty} 2^{ksq} |(\varphi_k * f)(x)\cdot(\varphi_k * g)(x)|^q\right]^{p/q} dx\right)^{1/p}
\end{aligned}$$

$$\lesssim \left(\sup_{k=0,1,\ldots} \|\varphi_k * f\|_{L^\infty(\mathbb{R}^n)} \right)$$

$$\times \sup_{P\in\mathcal{Q}} \frac{1}{|P|^\tau} \left(\int_P \left[\sum_{k=0}^{\infty} 2^{ksq} |(\varphi_k * g)(x)|^q \right]^{p/q} dx \right)^{1/p}$$

$$\lesssim \|f\|_{B^0_{\infty,\infty}(\mathbb{R}^n)} \|g\|_{F^{s,\tau}_{p,\infty}(\mathbb{R}^n)},$$

where we could apply Proposition 6.2 since $s > \sigma_{p,q}$. The proof is completed by taking into account the embedding $L^\infty(\mathbb{R}^n) \subset B^0_{\infty,\infty}(\mathbb{R}^n)$. □

Remark 6.5.

(i) The estimate (6.21) implies that the spaces $F^{s,\tau}_{p,q}(\mathbb{R}^n) \cap L^\infty(\mathbb{R}^n)$ are algebras with respect to pointwise multiplication.

(ii) For $\tau = 0$ we refer to [119, Theorem 4.6.4/2].

Combining Theorem 6.3 with Proposition 2.6 we get the following conclusion concerning the algebra properties of $F^{s,\tau}_{p,q}(\mathbb{R}^n)$.

Corollary 6.1. *Let $s \in \mathbb{R}$, $p \in (0,\infty)$, $q \in (0,\infty]$ and $\tau \in [0,1/p)$ such that*

$$s > n \max\left\{ \frac{1}{p} - \tau, \frac{1}{q} - 1 \right\}.$$

Then $F^{s,\tau}_{p,q}(\mathbb{R}^n)$ is an algebra with respect to pointwise multiplication.

Remark 6.6.

(i) For $\tau = 0$ this question had some history. For the Bessel potential spaces $H^s_p(\mathbb{R}^n) = F^{s,0}_{p,2}(\mathbb{R}^n)$, $p \in (1,\infty)$, it was settled by Strichartz [135]. This was extended by Triebel in [144, Sect. 2.6.2], Kalyabin [84,86] and Franke [59]; see also [119, Theorem 4.6.4/1].

(ii) Characterizations of $M(W^m_p(\mathbb{R}^n))$, $H^s_p(\mathbb{R}^n)$ and $M(B^s_{p,p}(\mathbb{R}^n))$ can be found in [95, 96]. For a characterization of $M(F^s_{p,q}(\mathbb{R}^n))$, $s > n/p$, we refer to Franke [59] and [119, Theorem 4.9.1/1].

6.1.3 A Characterization of $M(F^s_{\infty,q}(\mathbb{R}^n))$

The methods used in the previous subsection do partly not apply to the spaces $F^s_{\infty,q}(\mathbb{R}^n)$, since we always require $\tau < 1/p$; see Proposition 2.4. However, it is quite easy to prove the following.

Theorem 6.4. *Let $q \in (0,\infty]$ and $s \in (\sigma_{1,q},\infty)$. Then*

$$M(F^s_{\infty,q}(\mathbb{R}^n)) = F^s_{\infty,q}(\mathbb{R}^n)$$

in the sense of equivalent quasi-norms.

Proof. Step 1. We shall prove

$$F^s_{\infty,q}(\mathbb{R}^n) \subset M(F^s_{\infty,q}(\mathbb{R}^n)).$$

There is an elementary approach based on Corollaries 4.6, 4.7 if $q < \infty$, and Theorem 4.7 ($q = \infty$). We employ the formula

$$\Delta_h^M (f \cdot g)(x) = \sum_{k=0}^{M} c_k \, (\Delta_h^k f)(x) \, (\Delta_h^{M-k} g)(x + kh),$$

where $c_k = c_k(M)$ are certain constants depending only on M. Choosing M such that

$$\frac{n}{q} - n < s < \frac{M}{2}$$

then either $k > s > n/q - n$ or $M - k > s > n/q - n$. Let $k_0 \in \mathbb{Z}_+$ be chosen such that $k_0 \le s < k_0 + 1$. Thus, if $0 < q < \infty$,

$$\begin{aligned} &\sup_{t/2 \le |h| < t} \int_P |\Delta_h^M (f \cdot g)(x)|^q dx \\ &\lesssim \sum_{k=0}^{k_0} \|f\|_{L^\infty(\mathbb{R}^n)} \sup_{t/2 \le |h| < t} \int_P |\Delta_h^{M-k} g(x + kh)|^q dx \\ &\quad + \sum_{k=k_0+1}^{M} \|g\|_{L^\infty(\mathbb{R}^n)} \sup_{t/2 \le |h| < t} \int_P |\Delta_h^k f(x)|^q dx . \end{aligned}$$

For $t < l(P) \le 1$ we have

$$\sup_{t/2 \le |h| < t} \int_P |\Delta_h^{M-k} g(x + kh)|^q dx \lesssim \sup_{t/2 \le |h| < t} \int_{(M+1)P} |\Delta_h^{M-k} g(x)|^q dx .$$

Altogether, in view of Corollaries 4.6, 4.7, this proves

$$|||f \cdot g|||^{\spadesuit}_{F^s_{\infty,q}(\mathbb{R}^n)} \lesssim \|f\|_{L^\infty(\mathbb{R}^n)} \, |||g|||^{\spadesuit}_{F^s_{\infty,q}(\mathbb{R}^n)} + \|g\|_{L^\infty(\mathbb{R}^n)} \, |||f|||^{\spadesuit}_{F^s_{\infty,q}(\mathbb{R}^n)} .$$

The estimate of $N_q(f)$ (or $N_1(f)$) and the modifications, needed for $q = \infty$, are obvious.

Step 2. The function $g \equiv 1$ belongs to $F^s_{\infty,q}(\mathbb{R}^n)$. Thus, if $f \cdot g \in F^s_{\infty,q}(\mathbb{R}^n)$ for all $g \in F^s_{\infty,q}(\mathbb{R}^n)$, necessarily we get $f \in F^s_{\infty,q}(\mathbb{R}^n)$. In addition we have

$$\|f\|_{M(F^s_{\infty,q}(\mathbb{R}^n))} \ge \frac{\|f\|_{F^s_{\infty,q}(\mathbb{R}^n)}}{\|1\|_{F^s_{\infty,q}(\mathbb{R}^n)}} .$$

This proves the theorem. □

Remark 6.7.

(i) The assertions of the Theorem do not extend to $s = 0$. E.g, if $q = 2$, the correct description of $M(\text{bmo})$ was found by Janson [75]. For a description of $M(B^0_{\infty,\infty}(\mathbb{R}^n))$ we refer to [87].

(ii) Theorem 6.4 implies that the spaces $F^s_{\infty,q}(\mathbb{R}^n)$ are algebras with respect to pointwise multiplication, at least, if $s > \sigma_{1,q}$. For $q \geq 1$ a different proof of this fact can be found in [93].

6.1.4 A Characterization of $M(F^s_{p,q}(\mathbb{R}^n))$, $s < n/p$

As said above, in case $\tau = 0$ much more is known; see, for example, [131]. Of certain relevance for this lecture note is the description of $M(F^s_{p,q}(\mathbb{R}^n))$, $p \in (0,1)$, $s \in (\sigma_{p,q}, n/p)$, given by Netrusov [106].

Theorem 6.5. *Let $p \in (0,1]$, $q \in (0,\infty]$ and $s \in (\sigma_{p,q}, n/p)$. Then $f \in M(F^s_{p,q}(\mathbb{R}^n))$ if, and only if $f \in L^\infty(\mathbb{R}^n)$, f can be represented in $\mathcal{S}'(\mathbb{R}^n)$ in the form*

$$f = \sum_{j=0}^{\infty} f_j, \qquad \operatorname{supp} \mathcal{F} f_j \subset B(0, 2^{j+1}) \setminus B(0, 2^{j-1}),$$

such that

$$\sup_{j \in \mathbb{Z}_+} \sup_{x \in \mathbb{R}^n} 2^{j(\frac{n}{p} - s)} \left(\int_{B(x, 2^{-j})} \left[\sum_{k=j}^{\infty} 2^{ksq} |f_k(x)|^q \right]^{p/q} dx \right)^{1/p} < \infty .$$

Remark 6.8. Netrusov [106] did not publish a proof of this remarkable result. A proof under more restrictive conditions can be found in [132].

For $p = q = 1$ some more simple characterizations have been found by Maz'ya and Shaposnikova; see [95, Sect. 3.4.2]

Theorem 6.6. *Let $s = m + \sigma$, where m is a nonnegative integer and σ is a real number with $\sigma \in (0,1)$. Then $f \in M(F^s_{1,1}(\mathbb{R}^n))$ if, and only if $f \in L^\infty(\mathbb{R}^n)$ and*

$$\sup_{0<r<1} \sup_{x \in \mathbb{R}^n} r^{s-n} \sum_{|\alpha| \leq m} \left(\int_{B(x,r)} |\partial^\alpha f(y)|\, dy \right.$$
$$\left. + \int_{B(x,r)} \int_{B(x,r)} \frac{|\partial^\alpha f(y) - \partial^\alpha f(x)|}{|y - x|^{n+\sigma}} dy\, dx \right) < \infty . \quad (6.22)$$

We would like to reformulate Theorem 6.6. For this reason we recall the definition of the space $F^{s,\tau}_{p,q,\text{unif}}(\mathbb{R}^n)$. Let ψ be as in (1.1). A distribution f belongs to $F^{s,\tau}_{p,q,\text{unif}}(\mathbb{R}^n)$ if

$$\|f\|_{F^{s,\tau}_{p,q,\mathrm{unif}}(\mathbb{R}^n)} \equiv \sup_{\lambda\in\mathbb{Z}^n} \|f\,\psi(\cdot-\lambda)\|_{F^{s,\tau}_{p,q}(\mathbb{R}^n)} < \infty.$$

Let $m=0$ in Theorem 6.6. Then, as an immediate conclusion of Corollary 4.5, we obtain the following.

Corollary 6.2. *Let $s\in(0,1)$. Then $f\in M(F^s_{1,1}(\mathbb{R}^n))$ if, and only if*

$$f\in L^\infty(\mathbb{R}^n)\cap F^{s,\tau}_{1,1,\mathrm{unif}}(\mathbb{R}^n),$$

where $\tau=1-s/n$.

Remark 6.9. We conjecture that

$$M(F^s_{p,q}(\mathbb{R}^n)) = L^\infty(\mathbb{R}^n)\cap F^{s,\tau}_{p,q,\mathrm{unif}}(\mathbb{R}^n), \qquad \tau=\frac{1}{p}-\frac{s}{n},$$

holds under the restrictions of Theorem 6.5.

6.2 Diffeomorphisms

Let $m\in\mathbb{N}$ and $BC(\mathbb{R}^n)$ denote the collection of all complex-valued bounded and continuous functions in $\mathbb{R}^n$. We begin with recalling the notion of diffeomorphisms; see, for example, [146, p. 206].

A one-to-one mapping $y=\psi(x)$ of $\mathbb{R}^n$ onto $\mathbb{R}^n$ is called an *m-diffeomorphism* if the components ψ_j of $\psi=(\psi_1,\cdots,\psi_n)$ have classical derivatives up to order m with $\partial^\alpha\psi_j\in BC(\mathbb{R}^n)$ if $0<|\alpha|\le m$, and $|\det\psi_*(x)|\ge c>0$ for some positive constant c and all $x\in\mathbb{R}^n$, where ψ_* stands for the Jacobian matrix of ψ. We remark that if ψ is an m-diffeomorphism, then its inverse ψ^{-1} is also an m-diffeomorphism.

The mapping ψ is called a *diffeomorphism* if it is an m-diffeomorphism for any $m\in\mathbb{N}$.

Let ψ be a diffeomorphism. It is well known that for any $f\in\mathscr{S}(\mathbb{R}^n)$,

$$f\circ\psi\equiv f(\psi(\cdot))$$

makes sense. If ψ is only an m-diffeomorphism and $f\in A^{s,\tau}_{p,q}(\mathbb{R}^n)$, we define $f\circ\psi$ via smooth atoms for $A^{s,\tau}_{p,q}(\mathbb{R}^n)$. We wish to know that whether the mapping

$$D_\psi(f)\equiv f\circ\psi$$

is a linear and bounded operator from $A^{s,\tau}_{p,q}(\mathbb{R}^n)$ onto itself. Based on the smooth atomic decomposition characterization of $A^{s,\tau}_{p,q}(\mathbb{R}^n)$ in Theorem 3.3, we have the following conclusion.

Theorem 6.7. *Let $m \in \mathbb{N}$, ψ be an m-diffeomorphism and s, p, q and τ be as in Theorem 5.1. If $m \in \mathbb{N}$ is sufficiently large, then D_ψ is an isomorphic mapping of $A^{s,\tau}_{p,q}(\mathbb{R}^n)$ onto itself.*

Proof. Notice that ψ is an m-diffeomorphism if and only if its inverse ψ^{-1} is an m-diffeomorphism. To show D_ψ is an isomorphic mapping of $A^{s,\tau}_{p,q}(\mathbb{R}^n)$ onto itself, we only need to prove that D_ψ is a linear and continuous operator.

We first consider the case when $s > J - n$. Let $f \in A^{s,\tau}_{p,q}(\mathbb{R}^n)$. By Theorem 3.3, we can write f as

$$f = \sum_{l(Q)\le 1} t_Q a_Q$$

in $\mathscr{S}'(\mathbb{R}^n)$, where each a_Q is a smooth atom for $A^{s,\tau}_{p,q}(\mathbb{R}^n)$ supported near Q and the sequence $t \equiv \{t_Q\}_{l(Q)\le 1} \subset \mathbb{C}$ satisfies

$$\|t\|_{a^{s,\tau}_{p,q}(\mathbb{R}^n)} \lesssim \|f\|_{A^{s,\tau}_{p,q}(\mathbb{R}^n)}.$$

For all $l(Q) \le 1$, set $m_Q \equiv a_Q(\psi(\cdot))$. We claim that

$$D_\psi(f) \equiv \sum_{l(Q)\le 1} t_Q a_Q(\psi(\cdot)) = \sum_{l(Q)\le 1} t_Q m_Q$$

converges in $\mathscr{S}'(\mathbb{R}^n)$ and satisfies

$$\|D_\psi(f)\|_{A^{s,\tau}_{p,q}(\mathbb{R}^n)} \lesssim \|f\|_{A^{s,\tau}_{p,q}(\mathbb{R}^n)}.$$

By Theorem 3.3 again, it suffices to prove that each m_Q is also a multiple of a smooth atom for $A^{s,\tau}_{p,q}(\mathbb{R}^n)$ supported near a new dyadic cube $\widetilde{Q}$ with $|\widetilde{Q}| \sim |Q|$.

By the assumption $s > J - n$, no moment conditions (3.3) are necessary. Since $|\det \psi_*(x)| \ge c > 0$, it is easy to check that there exists a $\widetilde{Q} \in \mathscr{Q}$ such that $|\widetilde{Q}| \sim |Q|$ and $\psi^{-1}(3Q) \subset 3\widetilde{Q}$. Thus,

$$\operatorname{supp} m_Q = \operatorname{supp}(a_Q(\psi(\cdot))) \subset 3\widetilde{Q}.$$

Let $m \ge \max\{\lfloor s + n\tau + 1 \rfloor, 0\}$. By the regularity condition of a_Q and $l(Q) \le 1$, we see that for all $|\gamma| \le \max\{\lfloor s + n\tau + 1 \rfloor, 0\}$

$$\begin{aligned} \|\partial^\gamma(a_Q(\psi(\cdot)))\|_{L^\infty(\mathbb{R}^n)} &\le \sum_{\alpha\le\gamma} C(\alpha,\psi)\|\partial^\alpha a_Q\|_{L^\infty(\mathbb{R}^n)} \\ &\le \sum_{\alpha\le\gamma} C(\alpha,\psi)|Q|^{-1/2-|\alpha|/n} \\ &\le C(\gamma,\psi)|Q|^{-1/2-|\gamma|/n}. \end{aligned}$$

Therefore, each m_Q is a multiple of a smooth atom for $A^{s,\tau}_{p,q}(\mathbb{R}^n)$, which further implies that D_ψ is a linear and continuous operator from $A^{s,\tau}_{p,q}(\mathbb{R}^n)$ to itself when $s > J - n$.

Now we consider the case when $s \le J - n$. In this case, we use an argument similar to that used in Step 3 of the proof of [146, Theorem 4.3.2]. For convenience of the reader, we give the details. Indeed, if we fix $l \in \mathbb{N}$ such that $s + 2l > J - n$, as in the proof of Theorem 6.1, we represent $f \in A^{s,\tau}_{p,q}(\mathbb{R}^n)$ as

$$f \equiv (I + (-\Delta)^l)h$$

with $h \in A^{s+2l,\tau}_{p,q}(\mathbb{R}^n)$ and

$$\|h\|_{A^{s+2l,\tau}_{p,q}(\mathbb{R}^n)} \sim \|f\|_{A^{s,\tau}_{p,q}(\mathbb{R}^n)}.$$

Then, if $m \ge 2l$,

$$f(x) = (I + (-\Delta)^l)h(\psi \circ \psi^{-1}(x)) = \sum_{|\alpha| \le 2l} C_\alpha(x)(\partial^\alpha h \circ \psi)(\psi^{-1}(x))$$

for some bounded and continuous functions C_α. Then, if we choose m as in the proof of Theorem 6.1 in the case $s \le J - n$, by Theorem 6.1, Corollary 5.1 and Proposition 2.1(ii), we have

$$\begin{aligned}
\|f \circ \psi\|_{A^{s,\tau}_{p,q}(\mathbb{R}^n)} &\lesssim \sum_{|\alpha| \le 2l} \|C_\alpha \partial^\alpha h \circ \psi\|_{A^{s,\tau}_{p,q}(\mathbb{R}^n)} \\
&\lesssim \sum_{|\alpha| \le 2l} \|\partial^\alpha h \circ \psi\|_{A^{s,\tau}_{p,q}(\mathbb{R}^n)} \\
&\lesssim \|h \circ \psi\|_{A^{s+2l,\tau}_{p,q}(\mathbb{R}^n)},
\end{aligned}$$

which together with the previous proved result when $s > J - n$ and the fact

$$\|h\|_{A^{s+2l,\tau}_{p,q}(\mathbb{R}^n)} \sim \|f\|_{A^{s,\tau}_{p,q}(\mathbb{R}^n)}$$

yields the desired result, and then completes the proof of Theorem 6.7. □

Theorem 6.7 generalizes the classical results on Besov spaces and Triebel-Lizorkin spaces by taking $\tau = 0$; see, for example, [146, Proposition 4.3.1, Remark 4.3.1 and Theorem 4.3.2].

6.3 Traces

The trace theorem is of crucial interest for boundary value problems of elliptic differential operators. Let $x = (x_1, \cdots, x_n) \in \mathbb{R}^n$ and $x' \equiv (x_1, \cdots, x_{n-1}) \in \mathbb{R}^{n-1}$. We are interested in properties of the trace operator

$$\mathrm{Tr}: \quad f(x) \to f(x', 0). \tag{6.23}$$

For $\tau = 0$ such problems have been treated extensively; see, for example, [145, Sect. 2.7.2] and [62,64]. In this section, we deal with the corresponding problem for the spaces $A^{s,\tau}_{p,q}(\mathbb{R}^n)$. It is easy to see that (6.23) makes sense for all smooth atoms f for $A^{s,\tau}_{p,q}(\mathbb{R}^n)$. We follow the approach of Frazier and Jawerth [62,64]. They showed the usefulness of atomic characterizations in connection with the trace problem.

6.3.1 Traces of Functions in $A^{s,\tau}_{p,q}(\mathbb{R}^n)$

In this section, to emphasize the dimension n, we denote by $\mathscr{Q}(\mathbb{R}^n)$ the collection of all dyadic cubes in $\mathbb{R}^n$ and by $\mathscr{Q}_j(\mathbb{R}^n)$ the collection of all $Q \in \mathscr{Q}(\mathbb{R}^n)$ with $l(Q) = 2^{-j}$ for all $j \in \mathbb{Z}$.

The main result of this section is the following theorem. The proof is similar to those given for [127, Theorems 1.3 and 1.4].

Theorem 6.8. *Let $n \geq 2$, $p, q \in (0,\infty]$,*

$$s \in (1/p + (n-1)[1/\min\{1,p\} - 1], \infty),$$

J be as in Definition 3.2 and $\tau \in [0, 1/p + (s+n-J)/n)$. Then Tr *is a linear, continuous and surjective operator from $B^{s,\tau}_{p,q}(\mathbb{R}^n)$ to $B^{s-\frac{1}{p},\frac{n\tau}{n-1}}_{p,q}(\mathbb{R}^{n-1})$ and from $F^{s,\tau}_{p,q}(\mathbb{R}^n)$ to $F^{s-\frac{1}{p},\frac{n\tau}{n-1}}_{p,p}(\mathbb{R}^{n-1})$.*

Our proof of Theorem 6.8 will take full advantage of the smooth atomic decomposition characterizations of $A^{s,\tau}_{p,q}(\mathbb{R}^n)$ and $A^{s-\frac{1}{p},\frac{n\tau}{n-1}}_{p,q}(\mathbb{R}^{n-1})$. Neither of which requires any moment condition because of the assumption on s.

Proof of Theorem 6.8. By similarity, we only consider the Besov-type spaces. Let $f \in B^{s,\tau}_{p,q}(\mathbb{R}^n)$. By Theorem 3.3, we write

$$f = \sum_{\{Q \in \mathscr{Q}(\mathbb{R}^n):\, l(Q) \leq 1\}} t_Q a_Q$$

in $\mathscr{S}'(\mathbb{R}^n)$, where each a_Q is a smooth atom for $B^{s,\tau}_{p,q}(\mathbb{R}^n)$ supported near Q and the coefficient sequence $t \equiv \{t_Q\}_{\{Q \in \mathscr{Q}(\mathbb{R}^n):\, l(Q) \leq 1\}} \subset \mathbb{C}$ satisfies

$$\|t\|_{b^{s,\tau}_{p,q}(\mathbb{R}^n)} \lesssim \|f\|_{B^{s,\tau}_{p,q}(\mathbb{R}^n)}.$$

Precisely, the smooth function a_Q satisfies the support condition (3.13) and the regularity conditions (3.15) for all $|\gamma| \leq \max\{\lfloor s + n\tau + 1 \rfloor, 0\}$. Since

$$s \in (1/p + (n-1)(1/\min\{1,p\} - 1), \infty),$$

the moment condition (3.14) is an empty condition.

Recall that $t_Q a_Q$ is obtained canonically for $f \in B^{s,\tau}_{p,q}(\mathbb{R}^n)$. Then the definition of $\mathrm{Tr}(f)$ can be rephrased as

$$\mathrm{Tr}(f)(x',0) \equiv \sum_{\{Q\in\mathscr{Q}(\mathbb{R}^n):\,l(Q)\le 1\}} t_Q \mathrm{Tr}(a_Q)(x',0).$$

We now verify that the summation in the right-hand side of the above equality converges in $\mathscr{S}'(\mathbb{R}^{n-1})$ and satisfies

$$\|\mathrm{Tr}(f)\|_{B^{s-\frac{1}{p},\frac{n\tau}{n-1}}_{p,q}(\mathbb{R}^{n-1})} \lesssim \|f\|_{B^{s,\tau}_{p,q}(\mathbb{R}^n)}.$$

Since $\operatorname{supp} a_Q \subset 3Q$, then if $i \notin \{0,1,2\}$,

$$a_{Q'\times[(i-1)l(Q'),il(Q'))}(\cdot',0) \equiv 0.$$

Thus, the summation

$$\sum_{\{Q\in\mathscr{Q}(\mathbb{R}^n):\,l(Q)\le 1\}} t_Q \mathrm{Tr}(a_Q)(\cdot',0)$$

can be re-written as

$$\sum_{i=0}^{2} \sum_{\substack{Q'\in\mathscr{Q}(\mathbb{R}^{n-1})\\ l(Q')\le 1}} t_{Q'\times[(i-1)l(Q'),il(Q'))} a_{Q'\times[(i-1)l(Q'),il(Q'))}(\cdot',0). \tag{6.24}$$

To show that (6.24) converges in $\mathscr{S}'(\mathbb{R}^{n-1})$, by Theorem 3.3 again, it is sufficient to prove that each

$$b_{Q'} \equiv [l(Q')]^{1/2} a_{Q'\times[(i-1)l(Q'),il(Q'))}(\cdot',0)$$

is a smooth atom for $B^{s-\frac{1}{p},\frac{n\tau}{n-1}}_{p,q}(\mathbb{R}^{n-1})$ supported near Q' and for all $i \notin \{0,1,2\}$,

$$\left\| \left\{ [l(Q')]^{-1/2} t_{Q'\times[(i-1)l(Q'),il(Q'))} \right\}_{\{Q'\in\mathscr{Q}(\mathbb{R}^{n-1}):\,l(Q')\le 1\}} \right\|_{b^{s-\frac{1}{p},\frac{n\tau}{n-1}}_{p,q}(\mathbb{R}^{n-1})} < \infty.$$

By similarity, we only consider the case when $i=1$. It immediately deduces from the corresponding properties of a_Q that $b_{Q'}$ satisfies (3.13) and (3.15), namely, $b_{Q'}$ is a smooth atom for $B^{s-\frac{1}{p},\frac{n\tau}{n-1}}_{p,q}(\mathbb{R}^{n-1})$ supported near Q'. On the other hand,

$$\begin{aligned}
&\left\| \left\{ [l(Q')]^{-1/2} t_{Q'\times[0,l(Q'))} \right\}_{\{Q'\in\mathscr{Q}(\mathbb{R}^{n-1}):\,l(Q')\le 1\}} \right\|_{b^{s-\frac{1}{p},\frac{n\tau}{n-1}}_{p,q}(\mathbb{R}^{n-1})} \\
&= \sup_{P'\in\mathscr{Q}(\mathbb{R}^{n-1})} \frac{1}{|P'|^{\frac{n\tau}{n-1}}} \left\{ \sum_{j=j_{P'}}^{\infty} \left(\sum_{\substack{Q'\subset P',\,l(Q')\le 1\\ Q'\in\mathscr{Q}(\mathbb{R}^{n-1})}} \frac{[l(Q')]^{-p/2}|t_{Q'\times[0,l(Q'))}|^p}{[l(Q')]^{ps-1+(n-1)(p/2-1)}} \right)^{q/p} \right\}^{1/q}
\end{aligned}$$

$$= \sup_{P' \in \mathscr{Q}(\mathbb{R}^{n-1})} \frac{1}{|P'|^{\frac{n\tau}{n-1}}} \left\{ \sum_{j=j_{P'}}^{\infty} \left(\sum_{\substack{Q' \subset P', l(Q') \le 1 \\ Q' \in \mathscr{Q}(\mathbb{R}^{n-1})}} \frac{|t_{Q' \times [0, l(Q'))}|^p}{[l(Q')]^{ps+pn/2-n}} \right)^{q/p} \right\}^{1/q}$$
$$\le \|t\|_{b^{s,\tau}_{p,q}(\mathbb{R}^n)}.$$

Therefore, by Theorem 3.3, we obtain that (6.24) converges in $\mathscr{S}'(\mathbb{R}^{n-1})$ and

$$\|\mathrm{Tr}(f)\|_{B^{s-\frac{1}{p},\frac{n\tau}{n-1}}_{p,q}(\mathbb{R}^{n-1})} \lesssim \|t\|_{b^{s,\tau}_{p,q}(\mathbb{R}^n)} \lesssim \|f\|_{B^{s,\tau}_{p,q}(\mathbb{R}^n)}.$$

We now show that Tr is surjective. To this end, by Theorem 3.3 again, any $f \in B^{s-\frac{1}{p},\frac{n\tau}{n-1}}_{p,q}(\mathbb{R}^{n-1})$ can be represented as

$$f = \sum_{\{Q' \in \mathscr{Q}(\mathbb{R}^{n-1}): l(Q') \le 1\}} \lambda_{Q'} a_{Q'}$$

in $\mathscr{S}'(\mathbb{R}^{n-1})$, where each $a_{Q'}$ is a smooth atom for $B^{s-\frac{1}{p},\frac{n\tau}{n-1}}_{p,q}(\mathbb{R}^{n-1})$ supported near Q' and the coefficient sequence

$$\lambda \equiv \{\lambda_{Q'}\}_{\{Q' \in \mathscr{Q}(\mathbb{R}^{n-1}): l(Q') \le 1\}} \subset \mathbb{C}$$

satisfies

$$\|\lambda\|_{b^{s-\frac{1}{p},\frac{n\tau}{n-1}}_{p,q}(\mathbb{R}^{n-1})} \lesssim \|f\|_{B^{s-\frac{1}{p},\frac{n\tau}{n-1}}_{p,q}(\mathbb{R}^{n-1})}.$$

Let $\varphi \in C_c^\infty(\mathbb{R})$ with $\operatorname{supp} \varphi \subset (-\frac{1}{2}, \frac{1}{2})$ and $\varphi(0) = 1$. For all $Q' \in \mathscr{Q}(\mathbb{R}^{n-1})$ and all $x \in \mathbb{R}$, set $\varphi_{Q'}(x) \equiv \varphi(2^{-\log_2 l(Q')} x)$ and

$$F \equiv \sum_{\{Q' \in \mathscr{Q}(\mathbb{R}^{n-1}):\, l(Q') \le 1\}} \lambda_{Q'} a_{Q'} \otimes \varphi_{Q'}.$$

It is easy to check that each $[l(Q')]^{-1/2} a_{Q'} \otimes \varphi_{Q'}$ is a smooth atom for $B^{s,\tau}_{p,q}(\mathbb{R}^n)$ supported near $Q' \times [0, l(Q'))$. Moreover,

$$\|\{[l(Q')]^{\frac{1}{2}} \lambda_{Q'}\}_{\{Q' \in \mathscr{Q}(\mathbb{R}^{n-1}): l(Q') \le 1\}}\|_{b^{s,\tau}_{p,q}(\mathbb{R}^n)} \lesssim \|f\|_{B^{s-\frac{1}{p},\frac{n\tau}{n-1}}_{p,q}(\mathbb{R}^{n-1})}.$$

Then Theorem 3.3 implies that $F \in B^{s,\tau}_{p,q}(\mathbb{R}^n)$ and

$$\|F\|_{B^{s,\tau}_{p,q}(\mathbb{R}^n)} \lesssim \|f\|_{B^{s-\frac{1}{p},\frac{n\tau}{n-1}}_{p,q}(\mathbb{R}^{n-1})};$$

furthermore, $\mathrm{Tr}(F) = f$, which shows that Tr is surjective, and then, completes the proof of Theorem 6.8. □

Remark 6.10.

(i) We would like to mention that Theorem 6.8 generalizes the classical trace theorems for Besov spaces and Triebel-Lizorkin spaces by taking $\tau = 0$; see, for example, [13, 77, 108, 109], [145, Sect. 2.7.2] or [146, Sect. 4.4].

(ii) The counterpart of Theorem 6.8 for the homogeneous Besov-type space $\dot{B}^{s,\tau}_{p,q}(\mathbb{R}^n)$ and Triebel-Lizorkin-type space $\dot{F}^{s,\tau}_{p,q}(\mathbb{R}^n)$ were already obtained in [127].

(iii) Limiting situations for $\tau = 0$, i. e. $s = \frac{1}{p} + (n-1)\max\{0, 1/p - 1\}$, are investigated in [28, 64, 112], [146, Sect. 4.4.3] and [56].

6.3.2 *Traces of Functions in $F^s_{\infty,q}(\mathbb{R}^n)$ and Some Consequences*

In view of Proposition 2.4, Theorem 6.8 yields that Tr is a linear, continuous and surjective operator from

$$F^s_{\infty,q}(\mathbb{R}^n) = F^{s,1/p}_{p,q}(\mathbb{R}^n) \to F^{s-1/p, \frac{1}{p} + \frac{1}{p(n-1)}}_{p,p}(\mathbb{R}^{n-1}),$$

as long as $s > \sigma_{p,q}$ (then τ can be chosen to be $1/p$) and

$$s > \frac{1}{p} + (n-1)\max\left\{0, \frac{1}{p} - 1\right\}.$$

However, the range spaces of $F^s_{\infty,q}(\mathbb{R}^n)$ under Tr are well known. We refer to Marschall [93] and Frazier and Jawerth [64, Theorem 11.2] (in combination with the comments at the end of Sect. 12 in [64]). There it is proved that

$$\mathrm{Tr}(F^s_{\infty,q}(\mathbb{R}^n)) = F^s_{\infty,\infty}(\mathbb{R}^{n-1}) = B^s_{\infty,\infty}(\mathbb{R}^{n-1}) = \mathscr{Z}^s(\mathbb{R}^{n-1}).$$

Thus, we got two different characterizations. This is stated as a supplement to Proposition 2.4.

Lemma 6.1. *Let $s \in \mathbb{R}$ and $p \in (0, \infty)$ such that*

$$s > \frac{1}{p} + n\max\left\{0, \frac{1}{p} - 1\right\}.$$

Then

$$F^{s-1/p, \frac{1}{p} + \frac{1}{pn}}_{p,p}(\mathbb{R}^n) = \mathscr{Z}^s(\mathbb{R}^n)$$

in the sense of equivalent quasi-norms.

This procedure can be iterated by taking into account $\mathrm{Tr}(\mathscr{Z}^s(\mathbb{R}^n)) = \mathscr{Z}^s(\mathbb{R}^{n-1})$. Furthermore, it can be combined with Proposition 2.6.

Theorem 6.9. *Let* $0 < p < p_0 < \infty$, $k \in \mathbb{N}$ *and*

$$s > \frac{k}{p} + n \max\left\{0, \frac{1}{p} - 1\right\}.$$

(i) *Then*

$$F_{p,q}^{s-k/p, \frac{1}{p}\frac{n+k}{n}}(\mathbb{R}^n) = \mathscr{L}^s(\mathbb{R}^n) \qquad \text{if} \qquad p \le q \le \infty, \tag{6.25}$$

and

$$F_{p_0,q}^{s-\frac{k+n}{p}+\frac{n}{p_0}, \frac{1}{p}\frac{n+k}{n}}(\mathbb{R}^n) = \mathscr{L}^s(\mathbb{R}^n) \qquad \text{if} \qquad 0 < q \le \infty, \tag{6.26}$$

in the sense of equivalent quasi-norms.

(ii) *We have*

$$B_{p,q}^{s-k/p, \frac{1}{p}\frac{n+k}{n}}(\mathbb{R}^n) = \mathscr{L}^s(\mathbb{R}^n) \qquad \text{if} \qquad p \le q \le \infty, \tag{6.27}$$

and

$$B_{p_0,q}^{s-\frac{k+n}{p}+\frac{n}{p_0}, \frac{1}{p}\frac{n+k}{n}}(\mathbb{R}^n) = \mathscr{L}^s(\mathbb{R}^n) \qquad \text{if} \qquad p \le q \le \infty, \tag{6.28}$$

in the sense of equivalent quasi-norms.

Proof. The iteration of the trace argument yields, in case $p \in (0,\infty)$, $k \in \mathbb{N}$ and

$$s > \frac{k}{p} + n \max\left\{0, \frac{1}{p} - 1\right\},$$

the coincidence of the spaces

$$F_{p,p}^{s-k/p, \frac{1}{p}\frac{n+k}{n}}(\mathbb{R}^n) = \mathscr{L}^s(\mathbb{R}^n)$$

in the sense of equivalent quasi-norms. Let $0 < p \le p_1 < \infty$. Now we apply Propositions 2.1, 2.6 and Corollary 2.2 and obtain

$$\begin{aligned}
\mathscr{L}^s(\mathbb{R}^n) = F_{p,p}^{s-k/p, \frac{1}{p}\frac{n+k}{n}}(\mathbb{R}^n) &\subset B_{p,\infty}^{s-k/p, \frac{1}{p}\frac{n+k}{n}}(\mathbb{R}^n) \\
&\subset B_{p_1,\infty}^{s-\frac{k+n}{p}+\frac{n}{p_1}, \frac{1}{p}\frac{n+k}{n}}(\mathbb{R}^n) \subset \mathscr{L}^s(\mathbb{R}^n),
\end{aligned}$$

since

$$\left(s - \frac{k+n}{p} + \frac{n}{p_1}\right) + n\tau - \frac{n}{p_1} = s.$$

Taking $p_1 = p$ we have proved (6.25) and (6.27). Now let $p_1 = p_0 > p$. Because of

$$F_{p,p}^{s-k/p, \frac{1}{p}\frac{n+k}{n}}(\mathbb{R}^n) \subset F_{p_0,q}^{s-\frac{k+n}{p}+\frac{n}{p_0}, \frac{1}{p}\frac{n+k}{n}}(\mathbb{R}^n) \subset \mathscr{L}^s(\mathbb{R}^n)$$

(see Corollary 2.2), also (6.26) is proved. Finally, (6.28) is a consequence of

$$F_{p,p}^{s-k/p,\frac{1}{p}\frac{n+k}{n}}(\mathbb{R}^n) = B_{p,p}^{s-k/p,\frac{1}{p}\frac{n+k}{n}}(\mathbb{R}^n) \subset B_{p_0,p}^{s-\frac{k+n}{p}+\frac{n}{p_0},\frac{1}{p}\frac{n+k}{n}}(\mathbb{R}^n) \subset \mathscr{Z}^s(\mathbb{R}^n);$$

see again Corollary 2.2. □

Remark 6.11.

(i) We believe that Theorem 6.9 is not the final answer to the questions: (α) For which set of parameters p, q, τ, we have $F_{p,q}^{s,\tau}(\mathbb{R}^n) = \mathscr{Z}^s(\mathbb{R}^n)$; and (β) For which set of parameters p, q, τ, we have $B_{p,q}^{s,\tau}(\mathbb{R}^n) = \mathscr{Z}^s(\mathbb{R}^n)$.

(ii) Differently from [146], wherein Triebel established the mapping properties of pointwise multipliers, trace properties and the theorem on diffeomorphisms for Besov spaces and Triebel-Lizorkin spaces via the local mean characterizations of these spaces, in this chapter, we establish these key properties for $B_{p,q}^{s,\tau}(\mathbb{R}^n)$ and $F_{p,q}^{s,\tau}(\mathbb{R}^n)$ via the smooth atomic decomposition characterizations of these spaces.

Recently, Drihem [52] independently introduced the spaces $\mathscr{L}_{p,q}^{\lambda,s}(\mathbb{R}^n)$, for all $s \in \mathbb{R}$, $\lambda \in [0,\infty)$ and $p, q \in (0,\infty)$, and obtained their maximal function and local mean characterizations. As in Triebel [146], these characterizations provide another possible way to obtain the key properties for these spaces. Recall that the spaces $\mathscr{L}_{p,q}^{\lambda,s}(\mathbb{R}^n)$ when $p, q \in [1,\infty)$ were originally introduced by El Baraka [49, 50]. It is easy to see that the spaces $\mathscr{L}_{p,q}^{\lambda,s}(\mathbb{R}^n)$ for all $s \in \mathbb{R}$, $\lambda \in [0,\infty)$ and $p, q \in (0,\infty)$ are just the Besov-type spaces $B_{p,q}^{s,\lambda/(nq)}(\mathbb{R}^n)$. We also point out that the maximal function and local mean characterizations for homogeneous spaces $\dot{B}_{p,q}^{s,\tau}(\mathbb{R}^n)$ and $\dot{F}_{p,q}^{s,\tau}(\mathbb{R}^n)$ were established in [167].

6.4 Spaces on $\mathbb{R}_+^n$ and Smooth Domains

In this section, we introduce the Besov-type spaces and Triebel-Lizorkin-type spaces on $\mathbb{R}_+^n$ and Ω, where

$$\mathbb{R}_+^n \equiv \{x = (x', x_n) : x' \in \mathbb{R}^{n-1}, x_n > 0\}$$

and Ω stands for a bounded C^∞ domain in $\mathbb{R}^n$; see, for example, [145, Sect. 3.2.1]. We remark that domain always stands for an open set.

6.4.1 Spaces on $\mathbb{R}_+^n$

Let $D(\mathbb{R}_+^n)$ be the *set of all* $C^\infty(\mathbb{R}_+^n)$ *functions with compact supports in* $\mathbb{R}_+^n$ and denoted by $D'(\mathbb{R}_+^n)$ its *topological dual*. The spaces on $\mathbb{R}_+^n$ are defined as follows.

Definition 6.1. Let $s \in \mathbb{R}$, $\tau \in [0,\infty)$ and $p, q \in (0,\infty]$. The space $A^{s,\tau}_{p,q}(\mathbb{R}^n_+)$ is defined to be the restriction of $A^{s,\tau}_{p,q}(\mathbb{R}^n)$ on $\mathbb{R}^n_+$, quasi-normed by

$$\|f\|_{A^{s,\tau}_{p,q}(\mathbb{R}^n_+)} \equiv \inf \|g\|_{A^{s,\tau}_{p,q}(\mathbb{R}^n)},$$

where the infimum is taken over all $g \in A^{s,\tau}_{p,q}(\mathbb{R}^n)$ with $g|_{\mathbb{R}^n_+} = f$ in the sense of $D'(\mathbb{R}^n_+)$.

From Definition 6.1, we deduce that if

$$\sum_{m=1}^{\infty} \|f_m\|^{\min\{1,p,q\}}_{A^{s,\tau}_{p,q}(\mathbb{R}^n_+)} < \infty,$$

then

$$\sum_{m=1}^{\infty} f_m \in A^{s,\tau}_{p,q}(\mathbb{R}^n_+),$$

which further yields that the space $A^{s,\tau}_{p,q}(\mathbb{R}^n_+)$ is a quasi-Banach space. In particular, $A^{s,\tau}_{p,q}(\mathbb{R}^n_+)$ is a Banach space if $p, q \in [1,\infty]$.

Let $m \in \mathbb{N}$. Denoted by $C^m(\mathbb{R}^n_+)$ the set of all functions f on $\mathbb{R}^n_+$ such that $f = g|_{\mathbb{R}^n_+}$ for some functions $g \in C^m(\mathbb{R}^n)$ such that

$$\|\partial^\alpha g\|_{L^\infty(\mathbb{R}^n)} \sim \|\partial^\alpha f\|_{L^\infty(\mathbb{R}^n_+)}$$

for all $|\alpha| \le m$. We also have the following pointwise multiplication assertion for $A^{s,\tau}_{p,q}(\mathbb{R}^n_+)$, which is an immediate corollary of Definition 6.1 and Theorem 6.1. We omit the details.

Theorem 6.10. *Let $m \in \mathbb{N}$ and s, τ, p, q be as in Theorem 6.1. If m is sufficiently large, then there exists a positive constant $C(m)$ such that for all $g \in C^m(\mathbb{R}^n_+)$ and all $f \in A^{s,\tau}_{p,q}(\mathbb{R}^n_+)$,*

$$\|gf\|_{A^{s,\tau}_{p,q}(\mathbb{R}^n_+)} \le C(m) \sum_{|\alpha| \le m} \|\partial^\alpha g\|_{L^\infty(\mathbb{R}^n_+)} \|f\|_{A^{s,\tau}_{p,q}(\mathbb{R}^n_+)}.$$

Next we establish the lifting property for $A^{s,\tau}_{p,q}(\mathbb{R}^n_+)$. Recall that Franke and Runst [60] (see also [119, Proposition 2.4.3]) constructed a family $\{J_\sigma\}_{\sigma\in\mathbb{R}}$ of isomorphisms mapping $A^s_{p,q}(\mathbb{R}^n)$ to $A^{s-\sigma}_{p,q}(\mathbb{R}^n)$ such that

(i) J_σ and $J_{-\sigma}$ are inverse to each other;
(ii) If $f \in \mathscr{S}'(\mathbb{R}^n)$ is supported in $\mathbb{R}^{n-1} \times (-\infty,0]$, so is $J_\sigma f$.

Notice that the classical lifting operator I_σ does not satisfy the above condition (ii). Following Sawano [124], for $\varepsilon \in [0,\infty)$, we define a holomorphic function ψ_ε on $\mathbb{C}$ by setting, for all $z \in \mathbb{C}$,

$$\psi_\varepsilon(z) \equiv \int_{-\infty}^{0} \eta(t) e^{-i\varepsilon t z}\, dt - iz,$$

where $\eta \in \mathscr{S}(\mathbb{R})$ is a positive real-valued function supported in $(-\infty,0)$ with integral 2. Let $\mathbb{H} \equiv \{z \in \mathbb{C} : \mathrm{Im}(z) > 0\}$ and $\overline{\mathbb{H}} \equiv \{z \in \mathbb{C} : \mathrm{Im}(z) \geq 0\}$. Define a function $\phi^\sigma : \mathbb{R}^{n-1} \times \overline{\mathbb{H}} \to \mathbb{C}$ by

$$\phi^\sigma(x', z_n) \equiv (1+|x'|^2)^{\sigma/2}\left[\psi_\varepsilon\left(\frac{z_n}{(1+|x'|^2)^{\sigma/2}}\right)\right]^\sigma.$$

It was proved in [124, Lemma 4.3] that for all $\alpha \in \mathbb{Z}_+^n$, there exists a positive constant $C(\alpha)$ such that for all $(x', z_n) \in \mathbb{R}^{n-1} \times \overline{\mathbb{H}}$,

$$|\partial^\alpha \phi^1(x', z_n)| \leq C(\alpha)[(1+|x'|^2)^{1/2} + |z_n|]^{1-|\alpha|}.$$

Especially, when $|\alpha| = 0$,

$$|\phi^1(x', z_n)| \sim (1+|x'|^2)^{1/2} + |z_n|.$$

Denote again by ϕ^σ the restriction of ϕ^σ to $\mathbb{R}^n$. The above observations imply that $\phi^\sigma \in \mathscr{S}^\sigma_{1,1}(\mathbb{R}^n)$. Define J_σ by setting, for all $\xi \in \mathbb{R}^n$,

$$\widehat{J_\sigma f}(\xi) \equiv \phi^\sigma(\xi)\widehat{f}(\xi).$$

Then from Theorem 5.1, we deduce that Proposition 5.1 is still true with I_σ replaced by J_σ.

Proposition 6.4. *Let $\sigma \in \mathbb{R}$, s, p, q and τ be as in Theorem 5.1. Then the operator J_σ maps $A^{s,\tau}_{p,q}(\mathbb{R}^n)$ isomorphically onto $A^{s-\sigma,\tau}_{p,q}(\mathbb{R}^n)$.*

Recall that if $f \in \mathscr{S}'(\mathbb{R}^n)$ is supported in $\mathbb{R}^{n-1} \times (-\infty, 0]$, so is $J_\sigma f$; see [60] or [124, Proposition 4.6]. We then have the following result, which is an immediate corollary of Proposition 6.4.

Proposition 6.5. *Let $\sigma \in \mathbb{R}$, s, p, q and τ be as in Theorem 5.1. Let $f \in A^{s,\tau}_{p,q}(\mathbb{R}^n_+)$. Then $J_\sigma f \equiv J_\sigma g|_{\mathbb{R}^n_+}$ does not depend on the choice of the representative $g \in A^{s,\tau}_{p,q}(\mathbb{R}^n)$ of f and J_σ maps $A^{s,\tau}_{p,q}(\mathbb{R}^n_+)$ isomorphically onto $A^{s-\sigma,\tau}_{p,q}(\mathbb{R}^n_+)$.*

The restriction operator re is a linear and bounded operator from $A^{s,\tau}_{p,q}(\mathbb{R}^n)$ onto $A^{s,\tau}_{p,q}(\mathbb{R}^n_+)$. It is natural to ask whether there exists a linear and bounded operator ext from $A^{s,\tau}_{p,q}(\mathbb{R}^n_+)$ into $A^{s,\tau}_{p,q}(\mathbb{R}^n)$ such that re $\circ$ ext is the identity in $A^{s,\tau}_{p,q}(\mathbb{R}^n_+)$. Extension problems for Besov spaces and Triebel-Lizorkin spaces have been studied in depth by Triebel; see, for example, [145, Sect. 2.9] and [146, Sect. 4.5].

Let $M \in \mathbb{Z}_+$ be large enough, $0 < \lambda_0 < \lambda_1 < \cdots < \lambda_M$ and $a_0, \cdots, a_M$ be real numbers such that for all $l \in \{0, \cdots, M\}$,

$$\sum_{k=0}^{M} a_k(-\lambda_k)^l = 0.$$

As in [146, Sect. 4.5.2], we define ext_M by setting, for all functions f on $\mathbb{R}^n_+$ and $x=(x',x_n)\in\mathbb{R}^n$,

$$\mathrm{ext}_M f(x) \equiv \begin{cases} f(x), & \text{if } x\in\mathbb{R}^n_+; \\ \sum_{k=0}^{M} a_k f(x',-\lambda_k x_n), & \text{if } x_n\le 0. \end{cases} \tag{6.29}$$

Then we have the following extension theorem. Similarly to [146, Sect. 4.5.2], its proof relies on the oscillation characterization in Theorems 4.10 and 4.13.

Theorem 6.11. *Let $p\in[1,\infty)$, $q\in(0,\infty]$, $s\in\mathbb{R}$ and $\tau\in[0,\infty)$. There exists a linear and bounded operator* ext *from $A^{s,\tau}_{p,q}(\mathbb{R}^n_+)$ into $A^{s,\tau}_{p,q}(\mathbb{R}^n)$ such that* re $\circ$ ext *is the identity in $A^{s,\tau}_{p,q}(\mathbb{R}^n_+)$.*

Proof. We first consider the case when

$$s>\max\{J-n,J-n+n(\tau-1/p)\},$$

where J is as in Definition 3.2.

Let $M>\max\{s,s+n(\tau-1/p)\}$ and ext_M be as in (6.29). We prove that ext_M is a linear and bounded extension operator from $A^{s,\tau}_{p,q}(\mathbb{R}^n_+)$ into $A^{s,\tau}_{p,q}(\mathbb{R}^n)$. Notice that the assumption

$$s>\max\{J-n,J-n+n(\tau-1/p)\}$$

together with Proposition 2.7 implies that $A^{s,\tau}_{p,q}(\mathbb{R}^n_+)\subset L^p_\tau(\mathbb{R}^n_+)$. Thus, $\mathrm{ext}_M f$ makes sense for $f\in A^{s,\tau}_{p,q}(\mathbb{R}^n_+)$.

As in the proof of [146, Theorem 4.5.2], we denote by $\mathrm{osc}^M_u f$ the oscillations based on $\mathbb{R}^n$ and $\mathrm{Osc}^M_u f$ the oscillations based on $\mathbb{R}^n_+$ in the sense of (4.33). By [146, p. 224, (7)–(9)], we have the following estimates: Let $x=(x',x_n)\in\mathbb{R}^n$ and $t\in(0,2]$. If $x_n>t$, then

$$\mathrm{osc}^M_1(\mathrm{ext}_M f)(x,t)=\mathrm{Osc}^M_1 f(x,t);$$

if $x_n<-t$, then

$$\mathrm{osc}^M_1(\mathrm{ext}_M f)(x,t)\lesssim\sum_{k=0}^{M}\mathrm{Osc}^M_1 f((x',-\lambda_k x_n),Ct)$$

for some positive constant C; if $|x_n|\le t$, then

$$\mathrm{osc}^M_1(\mathrm{ext}_M f)(x,t)\lesssim\mathrm{Osc}^M_1 f((x',|x_n|),Ct)$$

for some positive constant C. By Theorems 4.10 and 4.13, the definition of $\mathrm{ext}_M f$ and Remark 4.11, we obtain that

$$\begin{aligned}\|f\|_{B^{s,\tau}_{p,q}(\mathbb{R}^n_+)} &\le \|\mathrm{ext}_M f\|_{B^{s,\tau}_{p,q}(\mathbb{R}^n)}\\ &\lesssim \|f\|_{L^p_\tau(\mathbb{R}^n_+)} + \sup_{\substack{P\in\mathcal{Q}\\ P\subset\mathbb{R}^n_+}} \frac{1}{|P|^\tau}\left\{\int_0^{C(l(P)\wedge 1)} t^{-sq}\right.\\ &\qquad\left.\times\left(\int_P \left[\mathrm{Osc}_1^{M-1} f(x,Mt)\right]^p dx\right)^{q/p}\frac{dt}{t}\right\}^{1/q}\\ &\lesssim \|f\|_{B^{s,\tau}_{p,q}(\mathbb{R}^n_+)}\end{aligned}$$

and

$$\begin{aligned}\|f\|_{F^{s,\tau}_{p,q}(\mathbb{R}^n_+)} &\le \|\mathrm{ext}_M f\|_{F^{s,\tau}_{p,q}(\mathbb{R}^n)}\\ &\lesssim \|f\|_{L^p_\tau(\mathbb{R}^n_+)} + \sup_{\substack{P\in\mathcal{Q}\\ P\subset\mathbb{R}^n_+}} \frac{1}{|P|^\tau}\left\{\int_P\left(\int_0^{C(l(P)\wedge 1)} t^{-sq}\right.\right.\\ &\qquad\left.\left.\times\left[\mathrm{Osc}_1^{M-1} f(x,Mt)\right]^q \frac{dt}{t}\right)^{p/q} dx\right\}^{1/p}\\ &\lesssim \|f\|_{F^{s,\tau}_{p,q}(\mathbb{R}^n_+)}.\end{aligned}$$

Then $\mathrm{ext}\equiv\mathrm{ext}_M$ is the desired extension operator in the case

$$s>\max\{J-n, J-n+n(\tau-1/p)\}.$$

For the case when

$$s\le\max\{J-n, J-n+n(\tau-1/p)\},$$

choose $\sigma\in\mathbb{R}$ such that

$$s+\sigma>\max\{J-n, J-n+n(\tau-1/p)\}.$$

Let

$$M>\max\{s+\sigma, s+\sigma+n(\tau-1/p)\}.$$

From the proved conclusion and Proposition 6.5, we deduce that

$$\mathrm{ext}\equiv J_\sigma\circ\mathrm{ext}_M\circ J_{-\sigma}$$

is the desired extension operator, which completes the proof of Theorem 6.11. □

6.4.2 Spaces on Smooth Domains

We now deal with the spaces on a bounded C^∞ domain Ω in $\mathbb{R}^n$. Let $D(\Omega)$ be the set of all $C^\infty(\Omega)$ functions supported in Ω and denoted by $D'(\Omega)$ its topological dual. Observe that $\phi \in D(\Omega)$ can be extended to $\mathscr{S}(\mathbb{R}^n)$ by setting $\phi \equiv 0$ outside Ω. Then the restriction operator

$$\mathrm{Re} : \mathscr{S}'(\mathbb{R}^n) \to D'(\Omega)$$

can be defined naturally as an adjoint operator.

The Besov-type spaces and Triebel-Lizorkin-type spaces on Ω are defined as follows.

Definition 6.2. Let $s \in \mathbb{R}$, $\tau \in [0,\infty)$ and $p, q \in (0,\infty]$. The space $A^{s,\tau}_{p,q}(\Omega)$ is defined to be the restriction of $A^{s,\tau}_{p,q}(\mathbb{R}^n)$ on Ω, quasi-normed by

$$\|f\|_{A^{s,\tau}_{p,q}(\Omega)} \equiv \inf \|g\|_{A^{s,\tau}_{p,q}(\mathbb{R}^n)},$$

where the infimum is taken over all $g \in A^{s,\tau}_{p,q}(\mathbb{R}^n)$ with $g|_\Omega = f$ in the sense of $D'(\Omega)$.

The space $A^{s,\tau}_{p,q}(\Omega)$ is also a quasi-Banach space. Let $C^m(\Omega)$ be the set of all functions f on Ω such that $f = g|_\Omega$ for some functions $g \in C^m(\mathbb{R}^n)$ with

$$\|\partial^\alpha g\|_{L^\infty(\mathbb{R}^n)} \sim \|\partial^\alpha f\|_{L^\infty(\Omega)}$$

for all $|\alpha| \le m$. Similarly to Theorem 6.10, we also obtain the pointwise multiplication theorem for $A^{s,\tau}_{p,q}(\Omega)$.

Theorem 6.12. *Let $m \in \mathbb{N}$, s, τ, p, q be as in Theorem 6.1. If m is sufficiently large, then there exists a positive constant $C(m)$ such that for all $g \in BC^m(\Omega)$ and all $f \in A^{s,\tau}_{p,q}(\Omega)$,*

$$\|gf\|_{A^{s,\tau}_{p,q}(\Omega)} \le C(m) \sum_{|\alpha|\le m} \|\partial^\alpha g\|_{L^\infty(\Omega)} \|f\|_{A^{s,\tau}_{p,q}(\Omega)}.$$

Theorem 6.12 is an immediate corollary of Definition 6.2 and Theorem 6.1

To obtain the extension property for $A^{s,\tau}_{p,q}(\Omega)$, we need some preparations. Since Ω is bounded, there exists a finite collection $\{B_m\}_{m=1}^k$ of open balls and a C^∞ domain Ω_0 such that $\overline{\Omega_0} \subset \Omega$ and

$$\Omega \subset \left\{\Omega_0 \cup (\cup_{m=1}^k B_m)\right\}.$$

Furthermore, there exist k-diffeomorphisms $\psi_1, \cdots, \psi_k$ on $\mathbb{R}^n$ satisfying that, for all $m \in \{1,\cdots,k\}$,

$$\psi_m(B_m \cap \Omega) = \psi_m(B_m) \cap \mathbb{R}^n_+ \tag{6.30}$$

and

$$\psi_m(B_m \cap \partial\Omega) = \psi_m(B_m) \cap \partial\mathbb{R}^n_+;$$

see [146, Sect. 5.1.3] or [124, Sect. 5].

Let $\phi_0, \cdots, \phi_k \in C_c^\infty(\mathbb{R}^n)$ be the $C^\infty(\mathbb{R}^n)$ resolution of unity satisfying that $\operatorname{supp}\phi_0 \subset \Omega_0$, $\operatorname{supp}\phi_m \subset B_m$ for $m \in \{1,\cdots,k\}$ and $\sum_{m=0}^k \phi_m \equiv 1$ in a neighborhood of $\overline{\Omega}$.

Now we establish the extension theorem for $A^{s,\tau}_{p,q}(\Omega)$. The proof of Theorem 6.13 is similar to that for [124, Theorem 5.4]. For the sake of convenience of the reader, we give the details.

Theorem 6.13. *Let $p \in [1,\infty)$, $q \in (0,\infty]$, $s \in \mathbb{R}$ and $\tau \in [0,\infty)$. There exists a linear and bounded operator* Ext *from $A^{s,\tau}_{p,q}(\Omega)$ into $A^{s,\tau}_{p,q}(\mathbb{R}^n)$ such that* $\mathrm{Re} \circ \mathrm{Ext}$ *is the identity in $A^{s,\tau}_{p,q}(\Omega)$.*

Proof. Let ext be the extension operator obtained in Theorem 6.11. Let $k \in \mathbb{N}$ be sufficiently large. For each $m \in \{1,\cdots,k\}$, let $\overline{\phi}_m$ be a bump function such that $\overline{\phi}_m \equiv 1$ in a neighborhood of $\operatorname{supp}\phi_m$ and has support in B_m.

For $f \in A^{s,\tau}_{p,q}(\Omega)$, we choose a representation $g \in A^{s,\tau}_{p,q}(\mathbb{R}^n)$ of f such that $\|g\|_{A^{s,\tau}_{p,q}(\mathbb{R}^n)} \lesssim \|f\|_{A^{s,\tau}_{p,q}(\Omega)}$. Define

$$\mathrm{Ext}f \equiv \phi_0 \cdot g + \sum_{m=1}^k \overline{\phi_m} \cdot \left[\mathrm{ext}\left((\phi_m \cdot g) \circ \psi_m^{-1}|_{\mathbb{R}^n_+}\right)\right] \circ \psi_m.$$

From this and the support conditions of ϕ_m, we deduce that for all $\varphi \in \mathcal{S}(\mathbb{R}^n)$,

$$\begin{aligned}\langle \mathrm{Ext}f, \varphi\rangle &= \langle \phi_0 \cdot g, \varphi\rangle + \sum_{m=1}^k \left\langle \left[\mathrm{ext}\left((\phi_m \cdot g) \circ \psi_m^{-1}|_{\mathbb{R}^n_+}\right)\right] \circ \psi_m, \overline{\phi_m} \cdot \varphi \right\rangle \\ &= \langle \phi_0 \cdot g, \varphi\rangle + \sum_{m=1}^k \left\langle \mathrm{ext}\left((\phi_m \cdot g) \circ \psi_m^{-1}|_{\mathbb{R}^n_+}\right), |J(\psi_m^{-1})| \cdot [\overline{\phi_m} \cdot \varphi] \circ \psi_m^{-1} \right\rangle,\end{aligned}$$

which together with (6.30) implies that $\mathrm{Ext}f$ is independent of the choice of g.

It was also proved in [124] that for all test functions $h \in D(\Omega)$,

$$\langle \mathrm{Ext}f|_\Omega, h\rangle = \langle f, h\rangle.$$

In fact, let Eh be the extended function of h by setting $Eh \equiv h$ on Ω and $Eh \equiv 0$ outside Ω. Then $Eh \in \mathcal{S}(\mathbb{R}^n)$ and we have

$$\begin{aligned}\langle \mathrm{Ext}f|_\Omega, h\rangle &= \langle \mathrm{Ext}f, Eh\rangle \\ &= \langle \phi_0 \cdot g, Eh\rangle \\ &\quad + \sum_{m=1}^k \left\langle \mathrm{ext}\left((\phi_m \cdot g) \circ \psi_m^{-1}|_{\mathbb{R}^n_+}\right), |J(\psi_m^{-1})| \cdot [\overline{\phi_m} \cdot Eh] \circ \psi_m^{-1} \right\rangle,\end{aligned}$$

which together with the fact that $[\overline{\phi_m} \cdot Eh] \circ \psi_m^{-1}$ is supported in $\mathbb{R}^n_+$ yields that

$$\begin{aligned}\langle \mathrm{Ext} f|_{\Omega}, h\rangle &= \langle \phi_0 \cdot g, Eh\rangle + \sum_{m=1}^{k} \left\langle (\phi_m \cdot g)\circ \psi_m^{-1}, |J(\psi_m^{-1})| \cdot [\overline{\phi_m} \cdot Eh]\circ \psi_m^{-1}\right\rangle \\ &= \langle \phi_0 \cdot g, Eh\rangle + \sum_{m=1}^{k} \left\langle \phi_m \cdot g, \overline{\phi_m} \cdot Eh\right\rangle \\ &= \langle g, Eh\rangle \\ &= \langle f, h\rangle.\end{aligned}$$

Thus, $\mathrm{Ext} f|_{\Omega} = f$ in $D'(\Omega)$. By the definition of $\mathrm{Ext} f$, the pointwise multiplication property in Theorem 6.1, the diffeomorphism property in Theorem 6.7 and the extension conclusion in Theorem 6.11, we obtain that

$$\|\mathrm{Ext} f\|_{A^{s,\tau}_{p,q}(\mathbb{R}^n)} \lesssim \|g\|_{A^{s,\tau}_{p,q}(\mathbb{R}^n)} \lesssim \|f\|_{A^{s,\tau}_{p,q}(\Omega)},$$

which further yields that the operator Ext is the desired one, and then completes the proof of Theorem 6.13. □

Remark 6.12. Let Ω be a bounded Lipschitz domain. Then Rychkov [121] has proved the existence of a universal extension operator for all spaces $F^s_{p,q}(\Omega)$ and $B^s_{p,q}(\Omega)$, simultaneously.

Chapter 7
Inhomogeneous Besov-Hausdorff and Triebel-Lizorkin-Hausdorff Spaces

Similarly to [164, Sects. 4, 5] and [165, Sects. 5, 6], in this section, we introduce the inhomogeneous Besov-Hausdorff space $BH^{s,\tau}_{p,q}(\mathbb{R}^n)$ and the Triebel-Lizorkin-Hausdorff space $FH^{s,\tau}_{p,q}(\mathbb{R}^n)$, whose dual spaces are, respectively, certain Besov-type space and Triebel-Lizorkin-type space when $p \in (1,\infty)$, $q \in [1,\infty)$, $s \in \mathbb{R}$ and $\tau \in [0, \frac{1}{(p\vee q)'}]$. Recall that $(p \vee q)'$ denotes the conjugate index of $p \vee q$, namely, $\frac{1}{p\vee q} + \frac{1}{(p\vee q)'} = 1$. The spaces $BH^{s,\tau}_{p,q}(\mathbb{R}^n)$ and $FH^{s,\tau}_{p,q}(\mathbb{R}^n)$ have some properties similar to those of $A^{s,\tau}_{p,q}(\mathbb{R}^n)$, which include the φ-transform characterization, embedding properties, smooth atomic and molecular decompositions.

7.1 Tent Spaces

We begin with recalling the notion of Hausdorff capacities; see [1, 2, 163].

Definition 7.1. Let $d \in (0,\infty)$ and $E \subset \mathbb{R}^n$. The *d-dimensional Hausdorff capacity of E* is defined by

$$\Lambda_d^{(\infty)}(E) \equiv \inf\left\{\sum_j r_j^d : E \subset \bigcup_j B(x_j, r_j)\right\}, \tag{7.1}$$

where the infimum is taken over all covers of E by countable families of open balls with radius r_j.

The notion of $\Lambda_d^{(\infty)}$ in Definition 7.1 when $d = 0$ also makes sense, and $\Lambda_0^{(\infty)}$ is monotone, countably subadditive; however, $\Lambda_0^{(\infty)}$ does not vanish on the empty set, it has the property that for all sets $E \subset \mathbb{R}^n$, $\Lambda_0^{(\infty)}(E) \geq 1$ and $\Lambda_0^{(\infty)}(E) = 1$ if E is bounded.

D. Yang et al., *Morrey and Campanato Meet Besov, Lizorkin and Triebel*, Lecture Notes in Mathematics 2005, DOI 10.1007/978-3-642-14606-0_7,

A *dyadic version* of Hausdorff capacity, $\widetilde{\Lambda}_d^{(\infty)}$, was introduced in [163], which is defined by

$$\widetilde{\Lambda}_d^{(\infty)}(E) \equiv \inf\left\{\sum_j l(I_j)^d : E \subset \left(\bigcup_j I_j\right)^\circ\right\},$$

where now the infimum ranges only over covers of E by dyadic cubes $\{I_j\}_j$ and A° denotes the interior of the set A.

Recall that $\Lambda_d^{(\infty)}$ and $\widetilde{\Lambda}_d^{(\infty)}$ are equivalent, i. e., there exist positive, finite constants C_1 and C_2, only depending on the dimension n, such that

$$C_1\Lambda_d^{(\infty)}(E) \le \widetilde{\Lambda}_d^{(\infty)}(E) \le C_2\Lambda_d^{(\infty)}(E) \quad \text{for all } E \subset \mathbb{R}^n. \tag{7.2}$$

We also recall the notions of the *Choquet integral* with respect to the Hausdorff capacities $\Lambda_d^{(\infty)}$ and $\widetilde{\Lambda}_d^{(\infty)}$; see [1,2]. For any function $f : \mathbb{R}^n \mapsto [0,\infty]$, define

$$\int_{\mathbb{R}^n} f(x)\,d\Lambda_d^{(\infty)}(x) \equiv \int_0^\infty \Lambda_d^{(\infty)}(\{x \in \mathbb{R}^n : f(x) > \lambda\})\,d\lambda.$$

This functional is not sublinear, so sometimes we need to use an equivalent integral with respect to $\widetilde{\Lambda}_d^{(\infty)}$, which is sublinear, and satisfies *Fatou's lemma* that for all nonnegative $\widetilde{\Lambda}_d^{(\infty)}$-measurable functions $\{f_m\}_{m=1}^\infty$,

$$\int_{\mathbb{R}^n} \liminf_{m\to\infty} f_m\,d\widetilde{\Lambda}_d^{(\infty)} \le \liminf_{m\to\infty} \int_{\mathbb{R}^n} f_m\,d\widetilde{\Lambda}_d^{(\infty)}.$$

For any measurable function f on $\mathbb{R}^{n+1}_+$ and all $x \in \mathbb{R}^n$, we define the *nontangential maximal function* $Nf(x)$ by

$$Nf(x) \equiv \sup_{|y-x|<t} |f(y,t)|.$$

Set

$$\mathbb{R}^{n+1}_{\mathbb{Z}_+} \equiv \left\{(x,a) \in \mathbb{R}^{n+1}_+ : -\log_2 a \in \mathbb{Z}_+\right\}.$$

In what follows, for all functions F on $\mathbb{R}^{n+1}_{\mathbb{Z}_+}$, $x \in \mathbb{R}^n$ and $j \in \mathbb{Z}_+$, write

$$F^j(x) \equiv F(x, 2^{-j}).$$

We then introduce the following tent spaces.

Definition 7.2. Let $s \in \mathbb{R}$, $p \in (1,\infty)$, $q \in [1,\infty)$ and $\tau \in (0, \frac{1}{(p\vee q)'}]$.

(i) The *tent space* $BT^{s,\tau}_{p,q}(\mathbb{R}^{n+1}_{\mathbb{Z}_+})$ is defined to be the set of all functions f on $\mathbb{R}^{n+1}_{\mathbb{Z}_+}$ such that $\{f^j\}_{j\in\mathbb{Z}_+}$ are Lebesgue measurable and

$$\|f\|_{BT^{s,\tau}_{p,q}(\mathbb{R}^{n+1}_{\mathbb{Z}_+})} \equiv \inf_\omega \left\{\sum_{j=0}^\infty 2^{jsq}\|f^j[\omega^j]^{-1}\|^q_{L^p(\mathbb{R}^n)}\right\}^{1/q} < \infty,$$

where the infimum is taken over all nonnegative Borel measurable functions ω on $\mathbb{R}^{n+1}_+$ with

$$\int_{\mathbb{R}^n} [N\omega(x)]^{(p\vee q)'}\, d\Lambda^{(\infty)}_{n\tau(p\vee q)'}(x) \le 1, \tag{7.3}$$

and with the restriction that ω is allowed to vanish only where f vanishes.

(ii) The *tent space* $FT^{s,\tau}_{p,q}(\mathbb{R}^{n+1}_{\mathbb{Z}_+})$ $(q \neq 1)$ is defined to be the set of all functions f on $\mathbb{R}^{n+1}_{\mathbb{Z}_+}$ such that $\{f^j\}_{j\in\mathbb{Z}_+}$ are Lebesgue measurable and

$$\|f\|_{FT^{s,\tau}_{p,q}(\mathbb{R}^{n+1}_{\mathbb{Z}_+})} \equiv \inf_{\omega} \left\| \left\{ \sum_{j=0}^{\infty} 2^{jsq} |f^j|^q [\omega^j]^{-q} \right\}^{1/q} \right\|_{L^p(\mathbb{R}^n)} < \infty,$$

where the infimum is taken over all nonnegative Borel measurable functions ω on $\mathbb{R}^{n+1}_+$ with the same restrictions as in (i).

These tent spaces are applied later to determine the predual spaces of $B^{s,\tau}_{p,q}(\mathbb{R}^n)$ and $F^{s,\tau}_{p,q}(\mathbb{R}^n)$. We use $AT^{s,\tau}_{p,q}(\mathbb{R}^{n+1}_{\mathbb{Z}_+})$ to denote either $BT^{s,\tau}_{p,q}(\mathbb{R}^{n+1}_{\mathbb{Z}_+})$ or $FT^{s,\tau}_{p,q}(\mathbb{R}^{n+1}_{\mathbb{Z}_+})$.

Remark 7.1.

(i) The notion of $AT^{s,\tau}_{p,q}(\mathbb{R}^{n+1}_{\mathbb{Z}_+})$ can be extended to $\tau = 0$. In this case, (7.3) implies that ω has upper bound. In fact, for all nonnegative integers k, set

$$E_k \equiv \{x \in \mathbb{R}^n : [N\omega(x)]^{(p\vee q)'/q} > k\}.$$

Then by (7.3) and the monotone property of $\Lambda^{(\infty)}_0$, we have

$$\begin{aligned}
1 &\ge \int_{\mathbb{R}^n} [N\omega(x)]^{(p\vee q)'}\, d\Lambda^{(\infty)}_0(x)\\
&= \int_0^\infty \Lambda^{(\infty)}_0\left(\{x\in\mathbb{R}^n : [N\omega(x)]^{(p\vee q)'} > \lambda\}\right) d\lambda\\
&= \sum_{k=0}^{\infty} \int_k^{k+1} \Lambda^{(\infty)}_0\left(\{x\in\mathbb{R}^n : [N\omega(x)]^{(p\vee q)'} > \lambda\}\right) d\lambda\\
&\ge \sum_{k=0}^{\infty} \int_k^{k+1} \Lambda^{(\infty)}_0\left(\{x\in\mathbb{R}^n : [N\omega(x)]^{(p\vee q)'} > k+1\}\right) d\lambda\\
&= \sum_{k=0}^{\infty} \Lambda^{(\infty)}_0(E_{k+1}).
\end{aligned}$$

Notice that $\Lambda^{(\infty)}_0(E) \ge 1$ for any set E. The argument above yields that $\Lambda^{(\infty)}_0(E_k) = 0$ for all $k \in \mathbb{N}$, which implies that for all $x \in \mathbb{R}^n$ and $k \in \mathbb{Z}$, $\omega^k(x) \le 1$. Thus,

$$\|f\|_{BT^{s,\tau}_{p,q}(\mathbb{R}^{n+1}_{\mathbb{Z}_+})} \equiv \left\{ \sum_{k=0}^{\infty} 2^{-ksq} \left\| f^k \right\|^q_{L^p(\mathbb{R}^n)} \right\}^{1/q}$$

and

$$\|f\|_{FT^{s,\tau}_{p,q}(\mathbb{R}^{n+1}_{\mathbb{Z}_+})} \equiv \left\| \left\{ \sum_{k=0}^{\infty} 2^{-ksq} |f^k|^q \right\}^{1/q} \right\|_{L^p(\mathbb{R}^n)}.$$

(ii) It is easy to check that $\|\cdot\|_{AT^{s,\tau}_{p,q}(\mathbb{R}^{n+1}_{\mathbb{Z}_+})}$ is a quasi-norm, namely, there exists a nonnegative constant $\rho \in [0,1]$ such that for all $f_1, f_2 \in AT^{s,\tau}_{p,q}(\mathbb{R}^{n+1}_{\mathbb{Z}_+})$,

$$\|f_1+f_2\|_{AT^{s,\tau}_{p,q}(\mathbb{R}^{n+1}_{\mathbb{Z}_+})} \le 2^{\rho}(\|f_1\|_{AT^{s,\tau}_{p,q}(\mathbb{R}^{n+1}_{\mathbb{Z}_+})} + \|f_2\|_{AT^{s,\tau}_{p,q}(\mathbb{R}^{n+1}_{\mathbb{Z}_+})}). \tag{7.4}$$

In fact, let ω_1, ω_2 be nonnegative Borel measurable functions on $\mathbb{R}^{n+1}_+$ satisfying (7.3) such that

$$\left\{ \sum_{j=0}^{\infty} 2^{jsq} \|f^j [\omega^j]^{-1}\|^q_{L^p(\mathbb{R}^n)} \right\}^{1/q} \le 2\|f_i\|_{BT^{s,\tau}_{p,q}(\mathbb{R}^{n+1}_{\mathbb{Z}_+})}$$

for $i \in \{1,2\}$. Notice that $\omega \equiv 2^{-\frac{1}{(p\vee q)'}} \max\{\omega_1, \omega_2\}$ still satisfies (7.3). Then

$$\begin{aligned}\|f_1+f_2\|_{BT^{s,\tau}_{p,q}(\mathbb{R}^{n+1}_{\mathbb{Z}_+})} &\lesssim \left\{ \sum_{j=0}^{\infty} 2^{jsq} \|(f_1^j+f_2^j)[\omega^j]^{-1}\|^q_{L^p(\mathbb{R}^n)} \right\}^{1/q} \\ &\lesssim \|f_1\|_{BT^{s,\tau}_{p,q}(\mathbb{R}^{n+1}_{\mathbb{Z}_+})} + \|f_2\|_{BT^{s,\tau}_{p,q}(\mathbb{R}^{n+1}_{\mathbb{Z}_+})}.\end{aligned}$$

The proof of $FT^{s,\tau}_{p,q}(\mathbb{R}^{n+1}_{\mathbb{Z}_+})$ is similar.

(iii) If ω satisfies (7.3), then

$$\begin{aligned}[\omega(x,t)]^{(p\vee q)'} \Lambda^{(\infty)}_{n\tau(p\vee q)'}(B(x,t)) &= \int_{\mathbb{R}^n} [\omega(x,t)]^{(p\vee q)'} \chi_{B(x,t)}(y)\, d\Lambda^{(\infty)}_{n\tau(p\vee q)'}(y) \\ &\lesssim \int_{\mathbb{R}^n} (N\omega(y))^{(p\vee q)'}\, d\Lambda^{(\infty)}_{n\tau(p\vee q)'}(y) \\ &\lesssim 1,\end{aligned}$$

which together with $\Lambda^{(\infty)}_{n\tau(p\vee q)'}(B(x,t)) = t^{n\tau(p\vee q)'}$ further implies that $\omega(x,t) \lesssim t^{-n\tau}$.

(iv) Let $0 < a \le b \le 1/\tau$. We claim that $\int_{\mathbb{R}^n} [N\omega(x)]^a\, d\Lambda^{(\infty)}_{n\tau a}(x) < \infty$ induces

$$\int_{\mathbb{R}^n} [N\omega(x)]^b\, d\Lambda^{(\infty)}_{n\tau b}(x) < \infty.$$

To this end, without loss of generality, we may assume

$$\int_{\mathbb{R}^n} [N\omega(x)]^a\, d\Lambda^{(\infty)}_{n\tau a}(x) \le 1.$$

For all $l \in \mathbb{Z}$, set $E_l \equiv \{x \in \mathbb{R}^n : N\omega(x) > 2^l\}$. Then

$$1 \ge \int_{\mathbb{R}^n} [N\omega(x)]^a\, d\Lambda^{(\infty)}_{n\tau a}(x) \sim \sum_{l\in\mathbb{Z}} 2^{la} \Lambda^{(\infty)}_{n\tau a}(E_l).$$

For each $l \in \mathbb{Z}$, there exists a countable ball cover $\{B(x_{jl}, r_{jl})\}_j$ of E_l such that

$$\Lambda^{(\infty)}_{n\tau a}(E_l) \sim \sum_j r_{jl}^{n\tau a}.$$

Thus, $\sum_{l\in\mathbb{Z}} 2^{la} \sum_j r_{jl}^{n\tau a} \lesssim 1$. For all j and l, $2^l r_{jl}^{n\tau} \lesssim 1$. Then $2^{lb} r_{il}^{n\tau b} \lesssim 2^{la} r_{il}^{n\tau a}$ since $a \le b$. Therefore,

$$\begin{aligned}\int_{\mathbb{R}^n} [N\omega(x)]^b \, d\Lambda^{(\infty)}_{n\tau b}(x) &\sim \sum_{l\in\mathbb{Z}} 2^{lb} \Lambda^{(\infty)}_{n\tau b}(E_l)\\ &\lesssim \sum_{l\in\mathbb{Z}} 2^{lb} \sum_j r_{jl}^{n\tau b}\\ &\lesssim \sum_{l\in\mathbb{Z}} 2^{la} \sum_j r_{jl}^{n\tau a}\\ &\lesssim 1.\end{aligned}$$

This proves our claim.

Similarly to [164, 165], we have the following atoms for the spaces $AT^{s,\tau}_{p,q}(\mathbb{R}^{n+1}_{\mathbb{Z}_+})$.

Definition 7.3. Let $s \in \mathbb{R}$, $p \in (1, \infty)$, $q \in [1, \infty)$ and $\tau \in (0, \frac{1}{(p\vee q)'}]$. A function a on $\mathbb{R}^{n+1}_{\mathbb{Z}_+}$ is called an $AT^{s,\tau}_{p,q}(\mathbb{R}^{n+1}_{\mathbb{Z}_+})$-*atom* associated a ball B, if a is supported in

$$T(B) \equiv \{(x,t) \in \mathbb{R}^{n+1}_{\mathbb{Z}_+} : B(x,t) \subset B\}$$

and satisfies that

$$\int_{\mathbb{R}^n} \left[\sum_{j=0}^{\infty} 2^{jsq} |a^j(x)|^q \chi_{T(B)}(x, 2^{-j})\right]^{p/q} dx \le |B|^{-\tau p}$$

if $AT^{s,\tau}_{p,q}(\mathbb{R}^{n+1}_{\mathbb{Z}_+}) = FT^{s,\tau}_{p,q}(\mathbb{R}^{n+1}_{\mathbb{Z}_+})$, or

$$\sum_{j=0}^{\infty} 2^{jsq} \left[\int_{\mathbb{R}^n} |a^j(x)|^p \chi_{T(B)}(x, 2^{-j})\, dx\right]^{q/p} \le |B|^{-\tau q}$$

if $AT^{s,\tau}_{p,q}(\mathbb{R}^{n+1}_{\mathbb{Z}_+}) = BT^{s,\tau}_{p,q}(\mathbb{R}^{n+1}_{\mathbb{Z}_+})$.

Lemma 7.1. *Let $s \in \mathbb{R}$, $p \in (1, \infty)$, $q \in [1, \infty)$ and $\tau \in (0, \frac{1}{(p\vee q)'}]$. Then there exists a positive constant C such that all $AT^{s,\tau}_{p,q}(\mathbb{R}^{n+1}_{\mathbb{Z}_+})$-atoms a belong to $AT^{s,\tau}_{p,q}(\mathbb{R}^{n+1}_{\mathbb{Z}_+})$ with*

$$\|a\|_{AT^{s,\tau}_{p,q}(\mathbb{R}^{n+1}_{\mathbb{Z}_+})} \le C.$$

Proof. By similarity, we only consider the space $FT^{s,\tau}_{p,q}(\mathbb{R}^{n+1}_{\mathbb{Z}_+})$. Suppose a is an $FT^{s,\tau}_{p,q}(\mathbb{R}^{n+1}_{\mathbb{Z}_+})$-atom associated with a ball $B \equiv B(x_B, r_B)$. Let ε be a positive real number such that $n\tau+\varepsilon > n\tau(p\vee q)'$. We set

$$\omega(x,t) \equiv \left[\kappa r_B^{-n\tau(p\vee q)'} \min\left\{1, \left(\frac{r_B}{\sqrt{|x-x_B|^2+t^2}}\right)^{n\tau+\varepsilon}\right\}\right]^{1/(p\vee q)'},$$

where the positive constant κ will be determined later. Notice that for all $x\in\mathbb{R}^n$, the distance between the cone $\Gamma(x)$ and $(x_B,0)$ is $|x-x_B|/\sqrt{2}$. Thus the nontangential maximal function of ω is bounded by

$$N\omega(x) \le \left[\kappa r_B^{-n\tau(p\vee q)'} \min\left\{1, \left(\frac{\sqrt{2}r_B}{|x-x_B|}\right)^{n\tau+\varepsilon}\right\}\right]^{1/(p\vee q)'}.$$

Therefore, by $n\tau+\varepsilon > n\tau(p\vee q)'$,

$$\begin{aligned}
\kappa^{-1} &\int_{\mathbb{R}^n} [N\omega(x)]^{(p\vee q)'}\, d\Lambda^{(\infty)}_{n\tau(p\vee q)'}(x)\\
&\le \int_{\mathbb{R}^n} r_B^{-n\tau(p\vee q)'} \min\left\{1, \left(\frac{\sqrt{2}r_B}{|x-x_B|}\right)^{n\tau+\varepsilon}\right\} d\Lambda^{(\infty)}_{n\tau(p\vee q)'}(x)\\
&= \int_0^\infty \Lambda^{(\infty)}_{n\tau(p\vee q)'}\left(\left\{x\in\mathbb{R}^n : r_B^{-n\tau(p\vee q)'} \min\left\{1, \left(\frac{\sqrt{2}r_B}{|x-x_B|}\right)^{n\tau+\varepsilon}\right\} > \lambda\right\}\right) d\lambda\\
&\le \int_0^{r_B^{-n\tau(p\vee q)'}} \Lambda^{(\infty)}_{n\tau(p\vee q)'}\left(B\left(x_B, \sqrt{2}r_B(\lambda r_B^{n\tau(p\vee q)'})^{-1/(n\tau+\varepsilon)}\right)\right) d\lambda\\
&\le \int_0^{r_B^{-n\tau(p\vee q)'}} \left[\sqrt{2}r_B(\lambda r_B^{n\tau(p\vee q)'})^{-1/(n\tau+\varepsilon)}\right]^{n\tau(p\vee q)'} d\lambda\\
&= C,
\end{aligned}$$

where the constant C is independent of r_B. Choose $\kappa = C^{-1}$ to make ω satisfy (7.3). Notice that if $(x, 2^{-k}) \in T(B)$, then $[\omega^k(x)]^{-1} \sim r_B^{n\tau}$. Then we have

$$\begin{aligned}
&\int_{\mathbb{R}^n} \left\{\sum_{k=0}^\infty 2^{ksq}|a^k(x)|^q[\omega^k(x)]^{-q}\right\}^{p/q} dx\\
&\quad \sim r_B^{n\tau p} \int_{\mathbb{R}^n} \left\{\sum_{k\in\mathbb{Z}} 2^{ksq}|a^k(x)|^q \chi_{T(B)}(x, 2^{-k})\right\}^{p/q} dx\\
&\quad \lesssim 1,
\end{aligned}$$

which yields $a \in FT^{s,\tau}_{p,q}(\mathbb{R}^{n+1}_{\mathbb{Z}_+})$ and completes the proof of Lemma 7.1. □

To obtain the atomic decomposition characterization of $AT^{s,\tau}_{p,q}(\mathbb{R}^{n+1}_{\mathbb{Z}_+})$, we need the following lemma.

Lemma 7.2. *Let $s \in \mathbb{R}$, $p \in (1,\infty)$ and $\tau \in (0, \frac{1}{p'}]$. If $\{g_j\}_j \subset FT^{s,\tau}_{p,p}(\mathbb{R}^{n+1}_{\mathbb{Z}_+})$ and $\sum_j \|g_j\|_{FT^{s,\tau}_{p,p}(\mathbb{R}^{n+1}_{\mathbb{Z}_+})} < \infty$, then*

$$g \equiv \sum_j g_j \in FT^{s,\tau}_{p,p}(\mathbb{R}^{n+1}_{\mathbb{Z}_+})$$

and there exists a positive constant C, independent of $\{g_j\}_j$, such that

$$\|g\|_{FT^{s,\tau}_{p,p}(\mathbb{R}^{n+1}_{\mathbb{Z}_+})} \le C \sum_j \|g_j\|_{FT^{s,\tau}_{p,p}(\mathbb{R}^{n+1}_{\mathbb{Z}_+})}.$$

Proof. Without lost of generality, we may assume that $\lambda_j = \|g_j\|_{FT^{s,\tau}_{p,p}(\mathbb{R}^{n+1}_{\mathbb{Z}_+})} > 0$ for all j. Let $f_j \equiv \lambda_j^{-1} g_j$. Then $\|f_j\|_{FT^{s,\tau}_{p,p}(\mathbb{R}^{n+1}_{\mathbb{Z}_+})} = 1$ and $g = \sum_j \lambda_j f_j$. For any $\varepsilon > 0$, take $\omega_j \ge 0$ such that

$$\int_{\mathbb{R}^n} [N\omega_j(x)]^{p'} \, d\Lambda^{(\infty)}_{n\tau p'}(x) \le 1$$

and

$$\left\| \left\{ \sum_{k=0}^{\infty} 2^{ksp} |f_j^k|^p [\omega_j^k]^{-p} \right\}^{1/p} \right\|_{L^p(\mathbb{R}^n)} \le 1 + \varepsilon.$$

Since $p > 1$, then

$$|g|^p = \left| \sum_j \lambda_j f_j \right|^p \le \left(\sum_j \lambda_j |f_j|^p [\omega_j]^{-p} \right) \left(\sum_j \lambda_j [\omega_j]^{p'} \right)^{p/p'}.$$

Notice that $\sum_j \lambda_j < \infty$. Define

$$\omega = C_1^{1/p'} C_2^{-1/p'} \left(\sum_j \lambda_j \right)^{-1/p'} \left(\sum_j \lambda_j [\omega_j]^{p'} \right)^{1/p'},$$

where C_1 and C_2 are as in (7.2). Notice that the vanishing of ω implies the vanishing of all ω_j, which only happen whenever all the g_j vanish, namely, when g is zero. Then by (7.2), the subadditivity of the nontangential maximal function, and the sublinear property of the integral with respect to $\widetilde{\Lambda}^{(\infty)}_d$, we obtain

$$\begin{aligned}
\int_{\mathbb{R}^n} [N\omega(x)]^{p'} \, d\Lambda^{(\infty)}_{n\tau p'}(x) &\le C_1^{-1} \int_{\mathbb{R}^n} [N\omega(x)]^{p'} \, d\widetilde{\Lambda}^{(\infty)}_{n\tau p'}(x) \\
&\le C_2^{-1} \left(\sum_j \lambda_j \right)^{-1} \sum_j \lambda_j \int_{\mathbb{R}^n} [N\omega_j(x)]^{p'} \, d\widetilde{\Lambda}^{(\infty)}_{n\tau p'}(x) \\
&\le 1.
\end{aligned}$$

Furthermore, we have

$$\begin{aligned}
&\int_{\mathbb{R}^n} \sum_{k=0}^{\infty} 2^{ksp} |g^k(x)|^p [\omega^k(x)]^{-p} \, dx \\
&\lesssim \left(\sum_j \lambda_j\right)^{p/p'} \int_{\mathbb{R}^n} \sum_{k=0}^{\infty} 2^{ksp} \left[\sum_j \lambda_j |f_j^k(x)|^p [\omega_j^k(x)]^{-p}\right] dx \\
&\lesssim \left(\sum_j \lambda_j\right)^{p/p'} \sum_j \lambda_j \int_{\mathbb{R}^n} \sum_{k=0}^{\infty} 2^{ksp} |f_j^k(x)|^p [\omega_j^k(x)]^{-p} \, dx \\
&\lesssim \left(\sum_j \lambda_j\right)^{p} (1+\varepsilon)^p.
\end{aligned}$$

Therefore, $\|g\|_{FT^{s,\tau}_{p,p}(\mathbb{R}^{n+1}_{\mathbb{Z}_+})} \lesssim \sum_j \lambda_j$, which completes the proof of Lemma 7.2. □

As an important tool of this section, we need the following Lemma 4.1 in [44].

Lemma 7.3. *Let $d \in (0,n]$ and $\{I_j\}$ be a sequence of dyadic cubes in $\mathbb{R}^n$ such that $\sum_j |I_j|^{d/n} < \infty$. Then there exists a sequence $\{J_k\}$ of dyadic cubes with mutually disjoint interiors, $\cup_k J_k = \cup_j I_j$ and*

$$\sum_k |J_k|^{d/n} \le \sum_j |I_j|^{d/n}.$$

Moreover, if a set $O \subset (\cup_j I_j)$, then the tent

$$T(O) \subset \left[\bigcup_k T((J_k)^*)\right],$$

where $(J_k)^*$ is the cube with the same center as J_k but $5\sqrt{n}$ times the side length.

We then have the following atomic decomposition of $AT^{s,\tau}_{p,q}(\mathbb{R}^{n+1}_{\mathbb{Z}_+})$. The proof is similar to that of [164, Theorem 4.1(i)]; see also [44, Theorem 5.4].

Proposition 7.1. *Let $s \in \mathbb{R}$, $p \in (1,\infty)$, $q \in [1,\infty)$ and $\tau \in (0, \frac{1}{(p\vee q)'}]$. If $f \in AT^{s,\tau}_{p,q}(\mathbb{R}^{n+1}_{\mathbb{Z}_+})$, then there exists a sequence $\{a_m\}_m$ of $AT^{s,\tau}_{p,q}(\mathbb{R}^{n+1}_{\mathbb{Z}_+})$-atoms and an ℓ^1-sequence $\{\lambda_m\}_m \subset \mathbb{C}$ such that $f = \sum_m \lambda_m a_m$ pointwise and*

$$\sum_m |\lambda_m| \le C \|f\|_{AT^{s,\tau}_{p,q}(\mathbb{R}^{n+1}_{\mathbb{Z}_+})}.$$

In particular, if $p = q \in (1,\infty)$, then $f = \sum_m \lambda_m a_m$ also in $AT^{s,\tau}_{p,p}(\mathbb{R}^{n+1}_{\mathbb{Z}_+})$.

Conversely, if $p = q \in (1, \infty)$ *and there exist a sequence* $\{a_m\}_m$ *of* $AT^{s,\tau}_{p,p}(\mathbb{R}^{n+1}_{\mathbb{Z}_+})$*-atoms and an* ℓ^1*-sequence* $\{\lambda_m\}_m \subset \mathbb{C}$ *such that* $f = \sum_m \lambda_m a_m$ *pointwise, then* $f = \sum_m \lambda_m a_m$ *also in* $AT^{s,\tau}_{p,p}(\mathbb{R}^{n+1}_{\mathbb{Z}_+})$ *and*

$$\|f\|_{AT^{s,\tau}_{p,p}(\mathbb{R}^{n+1}_{\mathbb{Z}_+})} \le C \sum_m |\lambda_m|,$$

where C *is a positive constant independent of* f.

Proof. By similarity, we only consider the space $FT^{s,\tau}_{p,q}(\mathbb{R}^{n+1}_{\mathbb{Z}_+})$.

Let $f \in FT^{s,\tau}_{p,q}(\mathbb{R}^{n+1}_{\mathbb{Z}_+})$. Let ω be a nonnegative Borel measurable function satisfying (7.3) and

$$\int_{\mathbb{R}^n} \left(\sum_{k=0}^{\infty} 2^{ksq} |f^k(x)|^q [\omega^k(x)]^{-q} \right)^{p/q} dx \le 2\|f\|^p_{FT^{s,\tau}_{p,q}(\mathbb{R}^{n+1}_{\mathbb{Z}_+})}.$$

For each $l \in \mathbb{Z}$, let

$$E_l \equiv \{x \in \mathbb{R}^n : N\omega(x) > 2^l\}.$$

From (7.2) and (7.3), it follows that $\widetilde{\Lambda}^{(\infty)}_{n\tau(p\vee q)'}(E_l) < \infty$, which together with Lemma 7.3 and its proof in [44, p. 386–387] yields that there exists a sequence $\{I_{j,l}\}_j$ of dyadic cubes with disjoint interiors such that

$$\sum_j [l(I_{j,l})]^{n\tau(p\vee q)'} \le 2\widetilde{\Lambda}^{(\infty)}_{n\tau(p\vee q)'}(E_l)$$

and

$$T(E_l) \subset \bigcup_j S^*(I_{j,l}),$$

where

$$S^*(I_{j,l}) \equiv \left\{(y,t) \in \mathbb{R}^{n+1}_+ : y \in I_{j,l},\ 0 < t < 2\mathrm{diam}(I_{j,l})\right\}.$$

The advantage is that $\{S^*(I_{j,l})\}_j$ have disjoint interiors for different values of j. Define

$$T_{j,l} \equiv S^*(I_{j,l}) \bigcap \left\{ \bigcup_{m>l} \bigcup_i S^*(I_{i,m}) \right\}^c,$$

where for any set $E \subset \mathbb{R}^n$, $E^c \equiv \mathbb{R}^n \setminus E$. Then $T_{j,l}$ have disjoint interiors for different values of j or l. Notice that

$$\bigcup_l T(E_l) = \{(x,t) \in \mathbb{R}^{n+1}_+ :\ \omega(x,t) > 0\}$$

and for each j and l, $S^*(I_{j,l})$ is contained in an $(n+1)$-dimensional cube of side length $2\mathrm{diam}(I_{j,l})$. By an argument similar to that in [44, p. 396], we know that $\cup_l \cup_j T_{j,l}$ contains

$$\{(x,t) \in \mathbb{R}^{n+1}_+ :\ \omega(x,t) > 0\} \setminus T_\infty,$$

where T_∞ is a set of zero $n\tau(p\vee q)'$-Hausdorff capacity and hence also zero $(n+1)$-dimensional Lebesgue measure. This observation further implies that $f=\sum f\chi_{T_{j,l}}$ a. e. on $\mathbb{R}^{n+1}_{\mathbb{Z}_+}$ (or more precisely, quasi-everywhere with respect to $n\tau(p\vee q)'$-Hausdorff capacity).

Recall that $I^*_{j,l}=5\sqrt{n}I_{j,l}$. Let

$$a_{j,l}\equiv f\chi_{T_{j,l}}\left\{[l(I^*_{j,l})]^{n\tau p}\int_{\mathbb{R}^n}\left[\sum_{k=0}^{\infty}2^{ksq}|f^k(x)|^q\chi_{T_{j,l}}(x,2^{-k})\right]^{p/q}dx\right\}^{-1/p},$$

and

$$\lambda_{j,l}\equiv\left\{[l(I^*_{j,l})]^{n\tau p}\int_{\mathbb{R}^n}\left[\sum_{k=0}^{\infty}2^{ksq}|f^k(x)|^q\chi_{T_{j,l}}(x,2^{-k})\right]^{p/q}dx\right\}^{1/p}.$$

We see that $f=\sum_{j,l}\lambda_{j,l}a_{j,l}$ pointwise. Since $S^*(I_{j,l})\subset T(B_{j,l})$, where $B_{j,l}$ is the ball with the same center as $I_{j,l}$ and radius $5\sqrt{n}l(I_{j,l})/2$, then $\operatorname{supp}a_{j,l}\subset T(B_{j,l})$. It is easy to see that each $a_{j,l}$ is an $FT^{s,\tau}_{p,q}(\mathbb{R}^{n+1}_{\mathbb{Z}_+})$-atom.

Next we verify that $\{\lambda_{j,l}\}_{j,l}$ is ℓ^1-summable. Notice that $\omega\le 2^{l+1}$ on $T_{j,l}\subset(T(E_{l+1}))^c$. When $p\ge q$, by Hölder's inequality and (2.11),

$$\begin{aligned}\sum_{j,l}|\lambda_{j,l}|&\le\sum_{j,l}2^{(l+1)}[l(I^*_{j,l})]^{n\tau}\\&\quad\times\left\{\int_{\mathbb{R}^n}\left[\sum_{k=0}^{\infty}2^{ksq}|f^k(x)|^q[w^k(x)]^{-q}\chi_{T_{j,l}}(x,2^{-k})\right]^{\frac{p}{q}}dx\right\}^{\frac1p}\\&\le\left\{\sum_{j,l}2^{(l+1)p'}l(I^*_{j,l})^{n\tau p'}\right\}^{\frac{1}{p'}}\\&\quad\times\left\{\int_{\mathbb{R}^n}\sum_{j,l}\left[\sum_{k=0}^{\infty}2^{ksq}|f^k(x)|^q[\omega^k(x)]^{-q}\chi_{T_{j,l}}(x,2^{-k})\right]^{\frac{p}{q}}dx\right\}^{\frac1p}\\&\lesssim\|f\|_{FT^{s,\tau}_{p,q}(\mathbb{R}^{n+1}_{\mathbb{Z}_+})}\left\{\sum_l 2^{lp'}\Lambda^{(\infty)}_{n\tau p'}(E_l)\right\}^{\frac{1}{p'}}\\&\lesssim\|f\|_{FT^{s,\tau}_{p,q}(\mathbb{R}^{n+1}_{\mathbb{Z}_+})}\left\{\int_{\mathbb{R}^n}[N\omega(x)]^{p'}\,d\Lambda^{(\infty)}_{n\tau p'}(x)\right\}^{\frac{1}{p'}}\\&\lesssim\|f\|_{FT^{s,\tau}_{p,q}(\mathbb{R}^{n+1}_{\mathbb{Z}_+})}.\end{aligned}$$

When $p<q$, by Hölder's inequality and Minkowski's inequality,

$$\begin{aligned}\sum_{j,l}|\lambda_{j,l}| &\le \left\{\sum_{j,l}2^{(l+1)q'}[l(I^*_{j,l})]^{n\tau q'}\right\}^{\frac{1}{q'}}\\ &\quad\times\left\{\sum_{j,l}\left[\int_{\mathbb{R}^n}\left(\sum_{k=0}^{\infty}2^{ksq}|f^k(x)|^q[\omega^k(x)]^{-q}\chi_{T_{j,l}}(x,2^{-k})\right)^{\frac{p}{q}}dx\right]^{\frac{q}{p}}\right\}^{\frac{1}{q}}\\ &\le \left\{\sum_{j,l}2^{(l+1)q'}[l(I^*_{j,l})]^{n\tau q'}\right\}^{\frac{1}{q'}}\\ &\quad\times\left\{\int_{\mathbb{R}^n}\left(\sum_{j,l}\sum_{k=0}^{\infty}2^{ksq}|f^k(x)|^q[\omega^k(x)]^{-q}\chi_{T_{j,l}}(x,2^{-k})\right)^{\frac{p}{q}}dx\right\}^{\frac{1}{p}}\\ &\lesssim \|f\|_{FT^{s,\tau}_{p,q}(\mathbb{R}^{n+1}_{\mathbb{Z}_+})}\left\{\sum_l 2^{lq'}\Lambda^{(\infty)}_{n\tau q'}(E_l)\right\}^{\frac{1}{q'}}\\ &\lesssim \|f\|_{FT^{s,\tau}_{p,q}(\mathbb{R}^{n+1}_{\mathbb{Z}_+})}\left\{\int_{\mathbb{R}^n}[N\omega(x)]^{q'}\,d\Lambda^{(\infty)}_{n\tau q'}(x)\right\}^{\frac{1}{q'}}\\ &\lesssim \|f\|_{FT^{s,\tau}_{p,q}(\mathbb{R}^{n+1}_{\mathbb{Z}_+})}.\end{aligned}$$

In particular, if $p=q\in(1,\infty)$, by Lemma 7.2, $f=\sum_j\lambda_j a_j$ also in $FT^{s,\tau}_{p,q}(\mathbb{R}^{n+1}_{\mathbb{Z}_+})$.

On the other hand, assume that $p=q\in(1,\infty)$ and there exist a sequence $\{a_j\}_j$ of $FT^{s,\tau}_{p,q}(\mathbb{R}^{n+1}_{\mathbb{Z}_+})$-atoms and an ℓ^1-sequence $\{\lambda_j\}_j\subset\mathbb{C}$ such that $f=\sum_j\lambda_j a_j$ pointwise. By Lemma 7.2 again, we obtain that the summation $f=\sum_j\lambda_j a_j$ converges in $FT^{s,\tau}_{p,q}(\mathbb{R}^{n+1}_{\mathbb{Z}_+})$, which completes the proof of Proposition 7.1. □

For all $f\in AT^{s,\tau}_{p,p}(\mathbb{R}^{n+1}_{\mathbb{Z}_+})$, set

$$|||f|||_{AT^{s,\tau}_{p,p}(\mathbb{R}^{n+1}_{\mathbb{Z}_+})}\equiv\inf\left\{\sum_m|\lambda_m|:\ f=\sum_m\lambda_m a_m\right\},\tag{7.5}$$

where the infimum is taken over all possible atomic decomposition of f. Proposition 7.1 implies that the norm $|||\cdot|||_{AT^{s,\tau}_{p,p}(\mathbb{R}^{n+1}_{\mathbb{Z}_+})}$ is equivalent to the quasi-norm $\|\cdot\|_{AT^{s,\tau}_{p,p}(\mathbb{R}^{n+1}_{\mathbb{Z}_+})}$, which together with Lemma 7.2 further yields that $AT^{s,\tau}_{p,p}(\mathbb{R}^{n+1}_{\mathbb{Z}_+})$ becomes a Banach space under the norm $|||\cdot|||_{AT^{s,\tau}_{p,p}(\mathbb{R}^{n+1}_{\mathbb{Z}_+})}$.

As the dual spaces of $AT^{s,\tau}_{p,q}(\mathbb{R}^{n+1}_{\mathbb{Z}_+})$, we now introduce the following two classes of tent spaces.

Definition 7.4. Let $s\in\mathbb{R}$, $p\in(1,\infty)$, $q\in(1,\infty]$ and $\tau\in(0,\infty)$. The *tent space* $AW^{s,\tau}_{p,q}(\mathbb{R}^{n+1}_{\mathbb{Z}_+})$ is defined to be the set of all functions f on $\mathbb{R}^{n+1}_{\mathbb{Z}_+}$ such that $\{f^k\}_{k\in\mathbb{Z}_+}$

are Lebesgue measurable and $\|f\|_{AW^{s,\tau}_{p,q}(\mathbb{R}^{n+1}_{\mathbb{Z}_+})} < \infty$, where when $AW^{s,\tau}_{p,q}(\mathbb{R}^{n+1}_{\mathbb{Z}_+}) = BW^{s,\tau}_{p,q}(\mathbb{R}^{n+1}_{\mathbb{Z}_+})$,

$$\|f\|_{BW^{s,\tau}_{p,q}(\mathbb{R}^{n+1}_{\mathbb{Z}_+})} \equiv \sup_B \frac{1}{|B|^\tau} \left\{ \sum_{k=0}^{\infty} 2^{ksq} \left[\int_{\mathbb{R}^n} |f^k(x)|^p \chi_{T(B)}(x, 2^{-k})\, dx \right]^{q/p} \right\}^{1/q},$$

and when $AW^{s,\tau}_{p,q}(\mathbb{R}^{n+1}_{\mathbb{Z}_+}) = FW^{s,\tau}_{p,q}(\mathbb{R}^{n+1}_{\mathbb{Z}_+})$,

$$\|f\|_{FW^{s,\tau}_{p,q}(\mathbb{R}^{n+1}_{\mathbb{Z}_+})} \equiv \sup_B \frac{1}{|B|^\tau} \left\{ \int_{\mathbb{R}^n} \left[\sum_{k=0}^{\infty} 2^{ksq} |f^k(x)|^q \chi_{T(B)}(x, 2^{-k}) \right]^{p/q} dx \right\}^{1/p},$$

and the supremum runs over all balls B in $\mathbb{R}^n$.

We need the following technical lemma.

Lemma 7.4. *Let $s \in \mathbb{R}$, $p \in (1,\infty)$, $q \in (1,\infty]$, $\tau \in (0,\infty)$ and $a \in (0,\infty)$. Then there exists a positive constant C such that for all $f \in AW^{s,\tau}_{p,q}(\mathbb{R}^{n+1}_{\mathbb{Z}_+})$ and nonnegative Borel measurable functions ω on $\mathbb{R}^{n+1}_+$, when $p \le q$,*

$$\sum_{k=0}^{\infty} 2^{ksq} \left\{ \int_{\mathbb{R}^n} |f^k(x)|^p [\omega^k(x)]^{ap}\, dx \right\}^{q/p} \le C \|f\|^q_{BW^{s,\tau}_{p,q}(\mathbb{R}^{n+1}_{\mathbb{Z}_+})} \int_{\mathbb{R}^n} [N\omega(x)]^{aq}\, d\Lambda^{(\infty)}_{n\tau q}(x)$$

and

$$\int_{\mathbb{R}^n} \left\{ \sum_{k=0}^{\infty} 2^{ksq} |f^k(x)|^q [\omega^k(x)]^{aq} \right\}^{p/q} dx \le C \|f\|^p_{FW^{s,\tau}_{p,q}(\mathbb{R}^{n+1}_{\mathbb{Z}_+})} \int_{\mathbb{R}^n} [N\omega(x)]^{ap}\, d\Lambda^{(\infty)}_{n\tau p}(x);$$

when $p > q$,

$$\sum_{k=0}^{\infty} 2^{ksq} \left\{ \int_{\mathbb{R}^n} |f^k(x)|^p [\omega^k(x)]^{ap}\, dx \right\}^{q/p} \le C \|f\|^q_{BW^{s,\tau}_{p,q}(\mathbb{R}^{n+1}_{\mathbb{Z}_+})} \left\{ \int_{\mathbb{R}^n} [N\omega(x)]^{ap}\, d\Lambda^{(\infty)}_{n\tau p}(x) \right\}^{q/p}$$

and

$$\int_{\mathbb{R}^n} \left\{ \sum_{k=0}^{\infty} 2^{ksq} |f^k(x)|^q [\omega^k(x)]^{aq} \right\}^{p/q} dx \le C \|f\|^p_{FW^{s,\tau}_{p,q}(\mathbb{R}^{n+1}_{\mathbb{Z}_+})} \left\{ \int_{\mathbb{R}^n} [N\omega(x)]^{aq}\, d\Lambda^{(\infty)}_{n\tau q}(x) \right\}^{p/q}.$$

Proof. By similarity, we only consider $FW^{s,\tau}_{p,q}(\mathbb{R}^{n+1}_{\mathbb{Z}_+})$.

For all $l \in \mathbb{Z}$, set

$$O_l \equiv \{x \in \mathbb{R}^n : N\omega(x) > 2^l\}.$$

Without loss of generality, we may assume that the integrals on the right-hand side of the desired inequalities are finite. Hence $\Lambda^{(\infty)}_{n\tau(p\wedge q)}(O_l) < \infty$.

Let $\{I^l_j\}_j$ be some dyadic cube covering of O_l with

$$\sum_j |I^l_j|^{\tau(p\wedge q)} \lesssim \Lambda^{(\infty)}_{n\tau(p\wedge q)}(O_l).$$

Then Lemma 7.3 tells us that there exists a sequence $\{J^l_i\}_i$ of dyadic cubes with mutually disjoint interiors such that

$$\sum_i |J^l_i|^{\tau(p\wedge q)} \le \sum_j |I^l_j|^{\tau(p\wedge q)}$$

and

$$T(O_l) \subset \left[\bigcup_i T((J^l_i)^*)\right].$$

Notice that if $\omega^k(y) > 2^l$, then $N\omega(x) > 2^l$ for all $x \in B(y, 2^{-k})$, and hence $(y, 2^{-k}) \in T(O_l)$. We have

$$A_l \equiv \{(y, 2^{-k}) \in \mathbb{R}^{n+1}_+ : 2^l < \omega^k(y) \le 2^{l+1}\} \subset T(O_l). \tag{7.6}$$

When $p \le q$, by (2.11), (7.6) and Definition 7.4, we have

$$\begin{aligned}
&\int_{\mathbb{R}^n} \left\{\sum_{k=0}^{\infty} 2^{ksq}|f^k(x)|^q[\omega^k(x)]^{aq}\right\}^{p/q} dx \\
&= \int_{\mathbb{R}^n} \left\{\sum_{l\in\mathbb{Z}}\sum_{k=0}^{\infty} 2^{ksq}|f^k(x)|^q[\omega^k(x)]^{aq}\chi_{A_l}(x, 2^{-k})\right\}^{p/q} dx \\
&\lesssim \sum_{l\in\mathbb{Z}} 2^{lap} \int_{\mathbb{R}^n} \left[\sum_{k=0}^{\infty} 2^{ksq}|f^k(x)|^q\chi_{A_l}(x, 2^{-k})\right]^{p/q} dx \\
&\lesssim \sum_{l\in\mathbb{Z}} 2^{lap} \sum_i \int_{\mathbb{R}^n} \left[\sum_{k=0}^{\infty} 2^{ksq}|f^k(x)|^q\chi_{T((J^l_i)^*)}(x, 2^{-k})\right]^{p/q} dx \\
&\lesssim \sum_{l\in\mathbb{Z}} 2^{lap/q} \sum_i [l((J^l_i)^*)]^{n\tau p} \|f\|^p_{FW^{s,\tau}_{p,q}(\mathbb{R}^{n+1}_{\mathbb{Z}_+})} \\
&\lesssim \sum_{l\in\mathbb{Z}} 2^{lap} \sum_j [l(I^l_j)]^{n\tau p} \|f\|^p_{FW^{s,\tau}_{p,q}(\mathbb{R}^{n+1}_{\mathbb{Z}_+})}
\end{aligned}$$

$$\lesssim \sum_{l\in\mathbb{Z}} 2^{lap}\Lambda_{n\tau p}^{(\infty)}(O_l)\|f\|^p_{FW^{s,\tau}_{p,q}(\mathbb{R}^{n+1}_{\mathbb{Z}_+})}$$
$$\lesssim \|f\|^p_{FW^{s,\tau}_{p,q}(\mathbb{R}^{n+1}_{\mathbb{Z}_+})}\int_{\mathbb{R}^n}[N\omega(x)]^{ap}\,d\Lambda_{n\tau p}^{(\infty)}(x).$$

When $p>q$, by Minkowski's inequality, (7.6) and Definition 7.4, we obtain

$$\left\{\int_{\mathbb{R}^n}\left[\sum_{k=0}^{\infty}2^{ksq}|f^k(x)|^q[\omega^k(x)]^{aq}\right]^{p/q}dx\right\}^{1/p}$$
$$\lesssim \left\{\sum_{l\in\mathbb{Z}}2^{laq}\left[\int_{\mathbb{R}^n}\left(\sum_{k=0}^{\infty}2^{ksq}|f^k(x)|^q\chi_{A_l}(x,2^{-k})\right)^{p/q}dx\right]^{q/p}\right\}^{1/q}$$
$$\lesssim \left\{\sum_{l\in\mathbb{Z}}2^{laq}\sum_i\left[\int_{\mathbb{R}^n}\left(\sum_{k=0}^{\infty}2^{ksq}|f^k(x)|^q\chi_{T((J_i^l)^*)}(x,2^{-k})\right)^{p/q}dx\right]^{q/p}\right\}^{1/q}$$
$$\lesssim \left\{\sum_{l\in\mathbb{Z}}2^{laq}\sum_i[l((J_i^l)^*)]^{n\tau q}\|f\|^q_{FW^{s,\tau}_{p,q}(\mathbb{R}^{n+1}_{\mathbb{Z}_+})}\right\}^{1/q}$$
$$\lesssim \|f\|_{FW^{s,\tau}_{p,q}(\mathbb{R}^{n+1}_{\mathbb{Z}_+})}\left\{\sum_{l\in\mathbb{Z}}2^{laq}\Lambda_{n\tau q}^{(\infty)}(O_l)\right\}^{1/q}$$
$$\lesssim \|f\|_{FW^{s,\tau}_{p,q}(\mathbb{R}^{n+1}_{\mathbb{Z}_+})}\left\{\int_{\mathbb{R}^n}[N\omega(x)]^{aq}\,d\Lambda_{n\tau q}^{(\infty)}(x)\right\}^{1/q},$$

which completes the proof of Lemma 7.4. □

In the following theorem, we establish the dual relation between the tent spaces $AT^{s,\tau}_{p,q}(\mathbb{R}^{n+1}_{\mathbb{Z}_+})$ and $AW^{s,\tau}_{p,q}(\mathbb{R}^{n+1}_{\mathbb{Z}_+})$.

Theorem 7.1. *Let $s\in\mathbb{R}$, $p\in(1,\infty)$, $q\in[1,\infty)$ and $\tau\in(0,\frac{1}{(p\vee q)'}]$. Then the dual space of $AT^{s,\tau}_{p,q}(\mathbb{R}^{n+1}_{\mathbb{Z}_+})$ is $AW^{-s,\tau}_{p',q'}(\mathbb{R}^{n+1}_{\mathbb{Z}_+})$ under the following pairing*

$$\langle f,g\rangle=\int_{\mathbb{R}^n}\sum_{k=0}^{\infty}f^k(x)g^k(x)\,dx. \tag{7.7}$$

Proof. By similarity, we only consider $FT^{s,\tau}_{p,q}(\mathbb{R}^{n+1}_{\mathbb{Z}_+})$.

We first show that each function $g\in FW^{-s,\tau}_{p',q'}(\mathbb{R}^{n+1}_{\mathbb{Z}_+})$ induces a bounded linear functional on $FT^{s,\tau}_{p,q}(\mathbb{R}^{n+1}_{\mathbb{Z}_+})$ via the pairing in (7.7). Indeed, let ω be a nonnegative Borel measurable function on $\mathbb{R}^{n+1}_+$ satisfying (7.3). Then by Lemma 7.4, we have

$$\left\{\int_{\mathbb{R}^n}\left[\sum_{k=0}^{\infty}2^{-ksq'}|g^k(x)|^{q'}[\omega^k(x)]^{q'}\right]^{p'/q'}dx\right\}^{1/p'}\lesssim\|g\|_{FW^{-s,\tau}_{p',q'}(\mathbb{R}^{n+1}_{\mathbb{Z}_+})}.$$

Therefore, for all $f \in FT^{s,\tau}_{p,q}(\mathbb{R}^{n+1}_{\mathbb{Z}_+})$, by Hölder's inequality, we have

$$\begin{aligned}
&\left|\int_{\mathbb{R}^n} \sum_{k=0}^{\infty} f^k(x) g^k(x)\,dx\right| \\
&\le \int_{\mathbb{R}^n} \left[\sum_{k=0}^{\infty} 2^{ksq} |f^k(x)|^q [\omega^k(x)]^{-q}\right]^{1/q} \left[\sum_{k=0}^{\infty} 2^{-ksq'} |g^k(x)|^{q'} [\omega^k(x)]^{q'}\right]^{1/q'} dx \\
&\le \left\{\int_{\mathbb{R}^n} \left[\sum_{k=0}^{\infty} 2^{ksq} |f^k(x)|^q [\omega^k(x)]^{-q}\right]^{p/q} dx\right\}^{1/p} \\
&\quad \times \left\{\int_{\mathbb{R}^n} \left[\sum_{k=0}^{\infty} 2^{-ksq'} |g^k(x)|^{q'} [\omega^k(x)]^{q'}\right]^{p'/q'} dx\right\}^{1/p'} \\
&\lesssim \left\{\int_{\mathbb{R}^n} \left[\sum_{k=0}^{\infty} 2^{ksq} |f^k(x)|^q [\omega^k(x)]^{-q}\right]^{p/q} dx\right\}^{1/p} \|g\|_{FW^{-s,\tau}_{p',q'}(\mathbb{R}^{n+1}_{\mathbb{Z}_+})}.
\end{aligned}$$

Taking the infimum over all admissible ω gives the desired conclusion.

Next we prove the converse. Let L be a bounded linear functional on $FT^{s,\tau}_{p,q}(\mathbb{R}^{n+1}_{\mathbb{Z}_+})$. Fix a ball $B \equiv B(x_B, r_B)$ in $\mathbb{R}^n$. For $\varepsilon \in (0, r_B)$, define

$$T^{\varepsilon}(B) \equiv T(B) \cap \{(x,t) : \varepsilon \le t \le 1\}.$$

If f is supported in $T^{\varepsilon}(B)$ with $f \in L^p(\ell^q(T^{\varepsilon}(B)))$, namely,

$$\int_{\mathbb{R}^n} \left\{\sum_{k=0}^{\infty} |f^k(x)|^q \chi_{T^{\varepsilon}(B)}(x, 2^{-k})\right\}^{p/q} dx < \infty,$$

then fixing ω as in the proof of Lemma 7.1, we have $(\omega(x,t))^{-1} \sim r_B^{n\tau}$ for all $(x,t) \in T(B)$, and

$$\begin{aligned}
\|f\|^p_{FT^{s,\tau}_{p,q}(\mathbb{R}^{n+1}_{\mathbb{Z}_+})} &\lesssim r_B^{n\tau p} \int_{\mathbb{R}^n} \left[\sum_{k=0}^{\infty} 2^{ksq} |f^k(x)|^q \chi_{T^{\varepsilon}(B)}(x, 2^{-k})\right]^{p/q} dx \\
&\lesssim r_B^{n\tau p} \left(\frac{1}{\varepsilon^{(s\vee 0)p}} + 1\right) \int_{\mathbb{R}^n} \left\{\sum_{k=0}^{\infty} |f^k(x)|^q \chi_{T^{\varepsilon}(B)}(x, 2^{-k})\right\}^{p/q} dx.
\end{aligned}$$

Hence L induces a bounded linear functional on $L^p(\ell^q(T^{\varepsilon}(B)))$, and acts via the inner-product with a unique function $g_B \in L^{p'}(\ell^{q'}(T^{\varepsilon}(B)))$ (see [145, p. 177]). For all $j \in \mathbb{N}$, taking $B_j = B(0, j)$ and $\varepsilon_j = 2^{-j}$, we get a unique $g_{B_j} \in L^{p'}(\ell^{q'}(T^{2^{-j}}(B_j)))$

for each j. Moreover, by the uniqueness, $g_{B_{j+1}} = g_{B_j}$ on $T^{2^{-j}}(B_j)$; letting $j \to \infty$, we get a unique function g on $\mathbb{R}^n \times \mathbb{Z}_+$ that is locally in $L^{p'}(\ell^{q'}(\mathbb{R}^n \times \mathbb{Z}_+))$, and such that

$$L(f) = \int_{\mathbb{R}^n} \sum_{k=0}^{\infty} f^k(x) g^k(x)\, dx, \tag{7.8}$$

whenever $f \in FT^{s,\tau}_{p,q}(\mathbb{R}^{n+1}_{\mathbb{Z}_+})$ with support in some finite tent $T^{\varepsilon}(B)$. We claim that the subspace of such f is dense in $FT^{s,\tau}_{p,q}(\mathbb{R}^{n+1}_{\mathbb{Z}_+})$. In fact, for any $f \in FT^{s,\tau}_{p,q}(\mathbb{R}^{n+1}_{\mathbb{Z}_+})$, set $f_j \equiv f\chi_{T^{2^{-j}}(B_j)}$, then $f_j \to f$ pointwise as $j \to \infty$. Notice that $|f - f_j| \le 2|f|$. By Lebesgue's dominated convergence theorem we obtain that $f_j \to f$ in $FT^{s,\tau}_{p,q}(\mathbb{R}^{n+1}_{\mathbb{Z}_+})$ as $j \to \infty$. Define $g(x,t) \equiv g^k(x)$ when $t \in [2^{-k}, 2^{-k+1})$ for all $k \in \mathbb{Z}_+$. Thus, if we can show that $g \in FW^{-s,\tau}_{p',q'}(\mathbb{R}^{n+1}_{\mathbb{Z}_+})$, then by taking limits we will get the representation of L via the pairing (7.7).

To verify $g \in FW^{-s,\tau}_{p',q'}(\mathbb{R}^{n+1}_{\mathbb{Z}_+})$, fix a ball $B \subset \mathbb{R}^n$. For every $\varepsilon > 0$, set

$$\begin{aligned} f_\varepsilon(x,t) &\equiv t^{sq'} |g(x,t)|^{q'-1} \chi_{T^\varepsilon(B)}(x,t)\, \mathrm{sgn}\, g(x,t) \\ &\quad \times \left(\sum_{k=0}^{\infty} 2^{-ksq'} |g^k(x)|^{q'} \chi_{T^\varepsilon(B)}(x, 2^{-k}) \right)^{\frac{p'}{q'}-1}, \end{aligned}$$

where $\mathrm{sgn}\, g(x,t) \equiv 1$ when $g(x,t) > 0$, $\mathrm{sgn}\, g(x,t) \equiv -1$ when $g(x,t) < 0$ and $\mathrm{sgn}\, g(x,t) \equiv 0$ when $g(x,t) = 0$. Then f_ε is supported in $T^\varepsilon(B)$.

Recall that if we choose ω as in the proof of Lemma 7.1, then for all $(x,t) \in T(B)$, $[\omega(x,t)]^{-1} \sim |B|^\tau$. Therefore,

$$\begin{aligned} |L(f_\varepsilon)| &\le \|L\| \|f_\varepsilon\|_{FT^{s,\tau}_{p,q}(\mathbb{R}^{n+1}_{\mathbb{Z}_+})} \\ &\lesssim \|L\| \left\{ \int_{\mathbb{R}^n} \left(\sum_{k=0}^{\infty} 2^{ksq} |f^k_\varepsilon(x)|^q \chi_{T^\varepsilon(B)(x,2^{-k})} [\omega^k(x)]^{-q} \right)^{p/q} dx \right\}^{1/p} \\ &\sim \|L\| |B|^\tau \left\{ \int_{\mathbb{R}^n} \left(\sum_{k=0}^{\infty} 2^{-ksq'} |g^k(x)|^{q'} \chi_{T^\varepsilon(B)(x,2^{-k})} \right)^{p'/q'} dx \right\}^{1/p}, \end{aligned}$$

which together with the fact that

$$L(f_\varepsilon) = \int_{\mathbb{R}^n} \sum_{k=0}^{\infty} f^k_\varepsilon(x) g^k(x)\, dx = \int_{\mathbb{R}^n} \left(\sum_{k=0}^{\infty} 2^{-ksq'} |g^k(x)|^{q'} \chi_{T^\varepsilon(B)(x,2^{-k})} \right)^{p'/q'} dx$$

yields

$$|B|^{-\tau} \left\{ \int_{\mathbb{R}^n} \left(\sum_{k=0}^{\infty} 2^{-ksq'} |g^k(x)|^{q'} \chi_{T^\varepsilon(B)(x,2^{-k})} \right)^{p'/q'} dx \right\}^{1/p'} \lesssim \|L\|.$$

Notice that the above inequality is true for all $\varepsilon > 0$ with a constant independent of ε. We get the same inequality for the integral over $T(B)$, which is independent of the choice of B. Then taking infimum over all balls B in $\mathbb{R}^n$, we see that $g \in FW^{-s,\tau}_{p',q'}(\mathbb{R}^{n+1}_{\mathbb{Z}_+})$, which completes the proof of Theorem 7.1. □

By Remark 7.1(i), Theorem 7.1 is also correct for $\tau = 0$.

Recall that Proposition 7.1 implies that $AT^{s,\tau}_{p,p}(\mathbb{R}^{n+1}_{\mathbb{Z}_+})$ is a Banach space. When $p = q$, we also determine the predual space of $AT^{s,\tau}_{p,p}(\mathbb{R}^{n+1}_{\mathbb{Z}_+})$. Denote by ${}_0AW^{s,\tau}_{p,p}(\mathbb{R}^{n+1}_{\mathbb{Z}_+})$ the closure of all functions in $AW^{s,\tau}_{p,p}(\mathbb{R}^{n+1}_{\mathbb{Z}_+})$ with compact support. We then have the following conclusion.

Theorem 7.2. *Let $s \in \mathbb{R}$, $p \in (1,\infty)$ and $\tau \in (0, \frac{1}{p}]$. Then the dual space of the tent space ${}_0AW^{s,\tau}_{p,p}(\mathbb{R}^{n+1}_{\mathbb{Z}_+})$ is $AT^{-s,\tau}_{p',p'}(\mathbb{R}^{n+1}_{\mathbb{Z}_+})$ under the pairing (7.7).*

To prove this theorem, we need some technical lemmas.

Lemma 7.5. *Let $s \in \mathbb{R}$, $p \in (1,\infty)$ and $\tau \in (0, \frac{1}{p'}]$. Then there exists a positive constant C such that for all $f \in AT^{s,\tau}_{p,p}(\mathbb{R}^{n+1}_{\mathbb{Z}_+})$,*

$$C^{-1}\|f\|_{AT^{s,\tau}_{p,p}(\mathbb{R}^{n+1}_{\mathbb{Z}_+})} \le \sup_{\|g\|_{AW^{-s,\tau}_{p',p'}(\mathbb{R}^{n+1}_{\mathbb{Z}_+})}\le 1,\ g \text{ has compact support}} \left\{\left|\int_{\mathbb{R}^n}\sum_{k=0}^{\infty} f^k(x)g^k(x)\,dx\right|\right\}$$
$$\le C\|f\|_{AT^{s,\tau}_{p,p}(\mathbb{R}^{n+1}_{\mathbb{Z}_+})}.$$

Proof. Recall that the norm $|||\cdot|||_{AT^{s,\tau}_{p,p}(\mathbb{R}^{n+1}_{\mathbb{Z}_+})}$ is equivalent to $\|\cdot\|_{AT^{s,\tau}_{p,p}(\mathbb{R}^{n+1}_{\mathbb{Z}_+})}$. By Theorem 7.1 and the Hahn-Banach theorem, there exists an $h \in AW^{-s,\tau}_{p',p'}(\mathbb{R}^{n+1}_{\mathbb{Z}_+})$ with $\|h\|_{AW^{-s,\tau}_{p',p'}(\mathbb{R}^{n+1}_{\mathbb{Z}_+})} \le 1$ such that

$$\|f\|_{AT^{s,\tau}_{p,p}(\mathbb{R}^{n+1}_{\mathbb{Z}_+})} \sim |||f|||_{AT^{s,\tau}_{p,p}(\mathbb{R}^{n+1}_{\mathbb{Z}_+})} \sim \left|\int_{\mathbb{R}^n}\sum_{k=0}^{\infty} f^k(x)h^k(x)\,dx\right|.$$

For $j \in \mathbb{N}$, let

$$g(x,2^{-k}) \equiv h(x,2^{-k})\chi_{\{|x|\le j,\,1/j\le 2^{-k}\le 1\}}(x,2^{-k}).$$

Then

$$\|g\|_{AW^{-s,\tau}_{p',p'}(\mathbb{R}^{n+1}_{\mathbb{Z}_+})} \le \|h\|_{AW^{-s,\tau}_{p',p'}(\mathbb{R}^{n+1}_{\mathbb{Z}_+})} \le 1$$

and g has compact support. Furthermore, Lebesgue's dominated convergence theorem implies that if j is large enough, then

$$\|f\|_{AT^{s,\tau}_{p,p}(\mathbb{R}^{n+1}_{\mathbb{Z}_+})} \sim |||f|||_{AT^{s,\tau}_{p,p}(\mathbb{R}^{n+1}_{\mathbb{Z}_+})} \sim \left|\int_{\mathbb{R}^n}\sum_{k=0}^{\infty} f^k(x)g^k(x)\,dx\right|,$$

which completes the proof of Lemma 7.5. □

The proof of the following lemma is a modification of [42, Lemma 4.2].

Lemma 7.6. *Let $p \in (1,\infty)$, $\tau \in (0, \frac{1}{p'}]$ and $\{f_m\}_{m\in\mathbb{N}}$ be a uniformly bounded sequence in $AT^{0,\tau}_{p,p}(\mathbb{R}^{n+1}_{\mathbb{Z}_+})$. Then there exist a function $f \in AT^{0,\tau}_{p,p}(\mathbb{R}^{n+1}_{\mathbb{Z}_+})$ and a subsequence $\{f_{m_i}\}_{i\in\mathbb{N}}$ of $\{f_m\}_{m\in\mathbb{N}}$ such that for all $g \in AW^{0,\tau}_{p',p'}(\mathbb{R}^{n+1}_{\mathbb{Z}_+})$ with compact support,*

$$\langle f_{m_i}, g\rangle \to \langle f, g\rangle$$

as $i \to \infty$, where $\langle f, g\rangle$ is defined as in (7.7), and

$$\|f\|_{AT^{0,\tau}_{p,p}(\mathbb{R}^{n+1}_{\mathbb{Z}_+})} \le C \sup_{m\in\mathbb{N}} \|f_m\|_{AT^{0,\tau}_{p,p}(\mathbb{R}^{n+1}_{\mathbb{Z}_+})}$$

with C being a positive constant independent of f.

Proof. Without loss of generality, we may assume that $\|f_m\|_{AT^{0,\tau}_{p,p}(\mathbb{R}^{n+1}_{\mathbb{Z}_+})} \le 1$ for all $m \in \mathbb{N}$. By Proposition 7.1 and its proof, each f_m has an atomic decomposition representation

$$f_m = \sum_{j\in\mathbb{Z}} \sum_{Q\in I_j^{(m)}} \lambda_{m,j,Q} a_{m,j,Q}$$

in $AT^{0,\tau}_{p,p}(\mathbb{R}^{n+1}_{\mathbb{Z}_+})$, where $I_j^{(m)} \subset \mathcal{Q}(\mathbb{R}^n)$, $\lambda_m \equiv \{\lambda_{m,j,Q}\}_{j\in\mathbb{Z},\, Q\in I_j^{(m)}} \subset \mathbb{C}$ satisfies that

$$\sum_{j\in\mathbb{Z}} \sum_{Q\in I_j^{(m)}} |\lambda_{m,j,Q}| \lesssim 1$$

and each $a_{m,j,Q}$ is an $A\dot{T}^{0,\tau}_{p}(\mathbb{R}^{n+1}_{\mathbb{Z}_+})$-atom supported in $T(B_Q)$, where and in what follows, for all $Q \in \mathcal{Q}(\mathbb{R}^n)$, $B_Q \equiv B(c_Q, 5\sqrt{n}l(Q)/2)$.

For all $m \in \mathbb{N}$, define a sequence $\widetilde{\lambda}_m \equiv \{\widetilde{\lambda}_{m,j,Q}\}_{j\in\mathbb{Z},Q\in\mathcal{Q}(\mathbb{R}^n)} \subset \mathbb{C}$ by setting, for all $j \in \mathbb{Z}$, $\widetilde{\lambda}_{m,j,Q} \equiv \lambda_{m,j,Q}$ when $Q \in I_j^{(m)}$ and $\widetilde{\lambda}_{m,j,Q} \equiv 0$ otherwise, and a set $\{\widetilde{a}_{m,j,Q}\}_{j\in\mathbb{Z},Q\in\mathcal{Q}(\mathbb{R}^n)}$ of functions on $\mathbb{R}^{n+1}_{\mathbb{Z}}$ by setting, for all $j \in \mathbb{Z}$, $\widetilde{a}_{m,j,Q} \equiv a_{m,j,Q}$ when $Q \in I_j^{(m)}$ and $\widetilde{a}_{m,j,Q} \equiv 0$ otherwise. We see that for each $m \in \mathbb{N}$,

$$\|\widetilde{\lambda}_m\|_{\ell^1} = \sum_{j\in\mathbb{Z}} \sum_{Q\in\mathcal{Q}(\mathbb{R}^n)} |\widetilde{\lambda}_{m,j,Q}| = \sum_{j\in\mathbb{Z}} \sum_{Q\in I_j^{(m)}} |\lambda_{m,j,Q}| \lesssim 1$$

and each $\widetilde{a}_{m,j,Q}$ is still an $AT^{0,\tau}_{p,p}(\mathbb{R}^{n+1}_{\mathbb{Z}_+})$-atom supported in $T(B_Q)$. Moreover,

$$f_m = \sum_{j\in\mathbb{Z}} \sum_{Q\in\mathcal{Q}(\mathbb{R}^n)} \widetilde{\lambda}_{m,j,Q} \widetilde{a}_{m,j,Q}$$

in $AT^{0,\tau}_{p,p}(\mathbb{R}^{n+1}_{\mathbb{Z}_+})$.

Since

$$\sum_{j\in\mathbb{Z}}\sum_{Q\in\mathcal{Q}(\mathbb{R}^n)}|\widetilde{\lambda}_{m,j,Q}|\lesssim 1$$

holds for all $m\in\mathbb{N}$, a diagonalization argument yields that there exist a sequence

$$\lambda\equiv\{\lambda_{j,Q}\}_{j\in\mathbb{Z},\,Q\in\mathcal{Q}(\mathbb{R}^n)}\in\ell^1$$

and a subsequence $\{\widetilde{\lambda}_{m_i}\}_{i\in\mathbb{N}}$ of $\{\widetilde{\lambda}_m\}_{m\in\mathbb{N}}$ such that $\widetilde{\lambda}_{m_i,j,Q}\to\lambda_{j,Q}$ as $i\to\infty$ for all $j\in\mathbb{Z}$ and $Q\in\mathcal{Q}(\mathbb{R}^n)$, and $\|\lambda\|_{\ell^1}\lesssim 1$.

On the other hand, recall that $\operatorname{supp}\widetilde{a}_{m,j,Q}\subset T(B_Q)$ for all $m\in\mathbb{N}$ and $j\in\mathbb{Z}$. From Definition 7.3, it follows that $\{\|\widetilde{a}_{m,j,Q}\|_{L^p(\ell^p(T(B_Q)))}\}_{m\in\mathbb{N}}$ is a uniformly bounded sequence in $L^p(\ell^p(T(B_Q)))$, where $L^p(\ell^p(T(B_Q)))$ consists of all functions on $T(B_Q)$ equipped with the norm that

$$\|F\|_{L^p(\ell^p(T(B_Q)))}\equiv\left\{\int_{\mathbb{R}^n}\sum_{i=0}^{\infty}|F(x,2^{-j})|^p\chi_{T(B_Q)}(x,2^{-j})\,dx\right\}^{1/p}.$$

Then by the Alaoglu theorem, there exist a unique function $a_{j,Q}\in L^p(\ell^p(T(B_Q)))$ and a subsequence of $\{\widetilde{a}_{m_i,j,Q}\}_{i\in\mathbb{N}}$, denoted by $\{\widetilde{a}_{m_i,j,Q}\}_{i\in\mathbb{N}}$ again, such that for all functions $g\in L^{p'}(\ell^{p'}(T(B_Q)))$,

$$\langle\widetilde{a}_{m_i,j,Q},g\rangle\to\langle a_{j,Q},g\rangle$$

as $i\to\infty$ and each $a_{j,Q}$ is also a constant multiple of an $AT^{0,\tau}_{p,p}(\mathbb{R}^{n+1}_{\mathbb{Z}_+})$-atom supported in $T(2B_Q)$ with the constant independent of j and Q. Applying a diagonalization argument again, we conclude that there exists a subsequence, denoted by $\{\widetilde{a}_{m_i,j,Q}\}_{i\in\mathbb{N}}$ again, such that for all $g\in L^{p'}(\ell^{p'}(T(B_Q)))$,

$$\langle\widetilde{a}_{m_i,j,Q},g\rangle\to\langle a_{j,Q},g\rangle$$

as $i\to\infty$ for all $j\in\mathbb{Z}$ and $Q\in\mathcal{Q}(\mathbb{R}^n)$. Let

$$f\equiv\sum_{j\in\mathbb{Z}}\sum_{Q\in\mathcal{Q}(\mathbb{R}^n)}\lambda_{j,Q}a_{j,Q}.$$

By Proposition 7.1, we see that $f\in AT^{0,\tau}_{p,p}(\mathbb{R}^{n+1}_{\mathbb{Z}_+})$ and

$$\|f\|_{AT^{0,\tau}_{p,p}(\mathbb{R}^{n+1}_{\mathbb{Z}_+})}\lesssim 1.$$

Let $g\in AW^{0,\tau}_{p',p'}(\mathbb{R}^{n+1}_{\mathbb{Z}_+})$ such that

$$\operatorname{supp}g\subset B(0,2^M)\times\{2^{-M},2^{-M+1},\cdots,2^M\}$$

for some $M \in \mathbb{N}$. Without loss of generality, we may assume that $\|g\|_{AW^{0,\tau}_{p',p'}(\mathbb{R}^{n+1}_{\mathbb{Z}_+})} = 1$. We need to show that $\langle f_{m_i}, g\rangle \to \langle f, g\rangle$ as $i \to \infty$. It is easy to see that

$$\|g\|_{L^{p'}(\ell^{p'}(T(B(0,2^M))))} \lesssim \|g\|_{AW^{0,\tau}_{p',p'}(\mathbb{R}^{n+1}_{\mathbb{Z}_+})} \sim 1.$$

Thus,

$$\langle \widetilde{a}_{m_i,j,Q}, g\rangle \to \langle a_{j,Q}, g\rangle$$

as $i \to \infty$ for all $j \in \mathbb{Z}$ and $Q \in \mathcal{Q}(\mathbb{R}^n)$.

Recall that $\|a\|_{AT^{0,\tau}_{p,p}(\mathbb{R}^{n+1}_{\mathbb{Z}_+})} \le \widetilde{C}$ for all $AT^{0,\tau}_{p,p}(\mathbb{R}^{n+1}_{\mathbb{Z}_+})$-atoms a, where $\widetilde{C}$ is a positive constant independent of a. By

$$\sum_{j\in\mathbb{Z}} \sum_{Q\in\mathcal{Q}(\mathbb{R}^n)} |\widetilde{\lambda}_{m_i,j,Q}| \lesssim 1,$$

we see that for any $\varepsilon > 0$, there exists an $L \in \mathbb{N}$ such that

$$\sum_{\{j\in\mathbb{Z}:\ |j|>L\}} \sum_{\{Q\in\mathcal{Q}(\mathbb{R}^n):\ |j_Q|>L \text{ or } Q\subsetneq[-2^L,2^L)^n\}} |\widetilde{\lambda}_{m_i,j,Q}| < \varepsilon/\widetilde{C}$$

and hence

$$\begin{aligned}
&\sum_{\substack{j\in\mathbb{Z}\\ |j|>L}} \sum_{\substack{Q\in\mathcal{Q}(\mathbb{R}^n)\\ |j_Q|>L \text{or} Q\subsetneq[-2^L,2^L)^n}} |\widetilde{\lambda}_{m_i,j,Q}||\langle \widetilde{a}_{m_i,j,Q}, g\rangle| \\
&\quad\le \sum_{\substack{j\in\mathbb{Z}\\ |j|>L}} \sum_{\substack{Q\in\mathcal{Q}(\mathbb{R}^n)\\ |j_Q|>L \text{or} Q\subsetneq[-2^L,2^L)^n}} |\widetilde{\lambda}_{m_i,j,Q}|\|\widetilde{a}_{m_i,j,Q}\|_{AT^{0,\tau}_{p,p}(\mathbb{R}^{n+1}_{\mathbb{Z}_+})} \|g\|_{AW^{0,\tau}_{p',p'}(\mathbb{R}^{n+1}_{\mathbb{Z}_+})} \\
&\quad\le \widetilde{C} \sum_{\substack{j\in\mathbb{Z}\\ |j|>L}} \sum_{\substack{Q\in\mathcal{Q}(\mathbb{R}^n)\\ |j_Q|>L \text{or} Q\subsetneq[-2^L,2^L)^n}} |\widetilde{\lambda}_{m_i,j,Q}| \\
&\quad< \varepsilon.
\end{aligned}$$

Similarly, by $\sum_{j\in\mathbb{Z}} \sum_{Q\in\mathcal{Q}(\mathbb{R}^n)} |\lambda_{j,Q}| \lesssim 1$, there exists an $L \in \mathbb{N}$ such that

$$\sum_{\substack{j\in\mathbb{Z}\\ |j|>L}} \sum_{\substack{Q\in\mathcal{Q}(\mathbb{R}^n)\\ |j_Q|>L \text{or} Q\subsetneq[-2^L,2^L)^n}} |\lambda_{j,Q}||\langle a_{j,Q}, g\rangle| < \varepsilon,$$

which yields

$$\lim_{i\to\infty} \langle f_{m_i}, g\rangle = \langle f, g\rangle$$

and completes the proof of Lemma 7.6. $\square$

We are now ready to prove Theorem 7.2.

Proof of Theorem 7.2. By Theorem 7.1 and the definition of ${}_0AW^{s,\tau}_{p,p}(\mathbb{R}^{n+1}_{\mathbb{Z}_+})$, we have that

$$ {}_0AW^{s,\tau}_{p,p}(\mathbb{R}^{n+1}_{\mathbb{Z}_+}) \subset AW^{s,\tau}_{p,p}(\mathbb{R}^{n+1}_{\mathbb{Z}_+}) = (AT^{-s,\tau}_{p',p'}(\mathbb{R}^{n+1}_{\mathbb{Z}_+}))^*, $$

which implies that

$$ AT^{-s,\tau}_{p',p'}(\mathbb{R}^{n+1}_{\mathbb{Z}_+}) \subset (AT^{-s,\tau}_{p',p'}(\mathbb{R}^{n+1}_{\mathbb{Z}_+})^{**} \subset ({}_0AW^{s,\tau}_{p,p}(\mathbb{R}^{n+1}_{\mathbb{Z}_+}))^*. $$

To show

$$ AT^{-s,\tau}_{p',p'}(\mathbb{R}^{n+1}_{\mathbb{Z}_+}) \subset ({}_0AW^{s,\tau}_{p,p}(\mathbb{R}^{n+1}_{\mathbb{Z}_+}))^*, $$

we first claim that if this is true when $s = 0$, then it is also true for all $s \in \mathbb{R}$. To see this, for all $u \in \mathbb{R}$, define an operator A_u by setting, for all functions f on $\mathbb{R}^{n+1}_{\mathbb{Z}_+}$, $x \in \mathbb{R}^n$ and $j \in \mathbb{Z}_+$,

$$ (A_u f)(x, 2^{-j}) \equiv 2^{ju} f(x, 2^{-j}). $$

Obviously, A_u is an isometric isomorphism from $AW^{s,\tau}_{p,p}(\mathbb{R}^{n+1}_{\mathbb{Z}_+})$ to $AW^{s+u,\tau}_{p,p}(\mathbb{R}^{n+1}_{\mathbb{Z}_+})$ and from $AT^{s,\tau}_{p,p}(\mathbb{R}^{n+1}_{\mathbb{Z}_+})$ to $AT^{s+u,\tau}_{p,p}(\mathbb{R}^{n+1}_{\mathbb{Z}_+})$. If $L \in ({}_0AW^{s,\tau}_{p,p}(\mathbb{R}^{n+1}_{\mathbb{Z}_+}))^*$, then

$$ L \circ A_s \in ({}_0AW^{0,\tau}_{p,p}(\mathbb{R}^{n+1}_{\mathbb{Z}_+}))^* $$

and hence, by the above assumption, there exists a function $g \in AT^{0,\tau}_{p',p'}(\mathbb{R}^{n+1}_{\mathbb{Z}_+})$ such that

$$ L \circ A_s(F) = \int_{\mathbb{R}^n} \sum_{j=0}^{\infty} f^j(x) g^j(x)\, dx $$

for all $F \in {}_0AW^{0,\tau}_{p,p}(\mathbb{R}^{n+1}_{\mathbb{Z}_+})$. Notice that $A_s \circ A_{-s}$ is the identity on ${}_0AW^{s,\tau}_{p,p}(\mathbb{R}^{n+1}_{\mathbb{Z}_+})$ and A_{-s} is an isometric isomorphism from ${}_0AW^{s,\tau}_{p,p}(\mathbb{R}^{n+1}_{\mathbb{Z}_+})$ onto ${}_0AW^{0,\tau}_{p,p}(\mathbb{R}^{n+1}_{\mathbb{Z}_+})$. Therefore,

$$ L(f) = L \circ A_s \circ A_{-s}(f) = \int_{\mathbb{R}^n} \sum_{j=0}^{\infty} (A_{-s} f)^j(x) g^j(x)\, dx = \int_{\mathbb{R}^n} \sum_{j=0}^{\infty} f^j(x) (A_{-s} g)^j(x)\, dx $$

for all $f \in {}_0AW^{s,\tau}_{p,p}(\mathbb{R}^{n+1}_{\mathbb{Z}_+})$. Since $g \in AT^{0,\tau}_{p',p'}(\mathbb{R}^{n+1}_{\mathbb{Z}_+})$, we have $A_{-s} g \in AT^{-s,\tau}_{p',p'}(\mathbb{R}^{n+1}_{\mathbb{Z}_+})$ and

$$ \|A_{-s} g\|_{AT^{-s,\tau}_{p',p'}(\mathbb{R}^{n+1}_{\mathbb{Z}_+})} = \|g\|_{AT^{0,\tau}_{p',p'}(\mathbb{R}^{n+1}_{\mathbb{Z}_+})}. $$

Thus, the above claim is true.

Next we prove that

$$ ({}_0AW^{0,\tau}_{p,p}(\mathbb{R}^{n+1}_{\mathbb{Z}_+}))^* \subset AT^{0,\tau}_{p',p'}(\mathbb{R}^{n+1}_{\mathbb{Z}_+}). $$

To this end, we choose $L \in ({}_0AW^{0,\tau}_{p,p}(\mathbb{R}^{n+1}_{\mathbb{Z}_+}))^*$. It suffices to show that there exists a $g \in AT^{0,\tau}_{p',p'}(\mathbb{R}^{n+1}_{\mathbb{Z}_+})$ such that for all $f \in AW^{0,\tau}_{p,p}(\mathbb{R}^{n+1}_{\mathbb{Z}_+})$ with compact support, L has a form as in (7.7). In fact, for $f \in AW^{0,\tau}_{p,p}(\mathbb{R}^{n+1}_{\mathbb{Z}_+})$ with compact support, if $\langle h, f\rangle = 0$ holds for all $h \in AT^{0,\tau}_{p',p'}(\mathbb{R}^{n+1}_{\mathbb{Z}_+})$, then Theorem 7.1 implies that f must be the zero element of $AW^{0,\tau}_{p,p}(\mathbb{R}^{n+1}_{\mathbb{Z}_+})$. Thus, $AT^{0,\tau}_{p',p'}(\mathbb{R}^{n+1}_{\mathbb{Z}_+})$ is a total set of linear functionals on ${}_0AW^{0,\tau}_{p,p}(\mathbb{R}^{n+1}_{\mathbb{Z}_+})$.

To complete the proof of Theorem 7.2, we need the following functional analysis result (see [48, p. 439, Exercise 41]): *Let $\mathscr{X}$ be a locally convex linear topological space and $\mathscr{Y}$ be a linear subspace of $\mathscr{X}^*$. Then $\mathscr{Y}$ is $\mathscr{X}$-dense in $\mathscr{X}^*$ if and only if $\mathscr{Y}$ is a total set of functionals on $\mathscr{X}$*. From this functional result and the fact that $AT^{0,\tau}_{p',p'}(\mathbb{R}^{n+1}_{\mathbb{Z}_+})$ is a total set of linear functionals on ${}_0AW^{0,\tau}_{p,p}(\mathbb{R}^{n+1}_{\mathbb{Z}_+})$, we deduce that $AT^{0,\tau}_{p',p'}(\mathbb{R}^{n+1}_{\mathbb{Z}_+})$ is weak $*$-dense in $({}_0AW^{0,\tau}_{p,p}(\mathbb{R}^{n+1}_{\mathbb{Z}_+}))^*$. Then there exists a sequence $\{g^{(m)}\}_{m\in\mathbb{N}}$ in $AT^{0,\tau}_{p',p'}(\mathbb{R}^{n+1}_{\mathbb{Z}_+})$ such that

$$\langle g^{(m)}, f\rangle \to L(f)$$

as $m \to \infty$ for all f in ${}_0AW^{0,\tau}_{p,p}(\mathbb{R}^{n+1}_{\mathbb{Z}_+})$. Applying the Banach-Steinhaus theorem, we conclude that the sequence $\{\|g^{(m)}\|_{AT^{0,\tau}_{p',p'}(\mathbb{R}^{n+1}_{\mathbb{Z}_+})}\}_{m\in\mathbb{N}}$ is uniformly bounded. Then by Lemmas 7.6 and 7.5, we obtain a subsequence $\{g^{(m_i)}\}_{i\in\mathbb{N}}$ and $g \in AT^{0,\tau}_{p',p'}(\mathbb{R}^{n+1}_{\mathbb{Z}_+})$ such that

$$L(f) = \lim_{i\to\infty} \langle g^{(m_i)}, f\rangle = \langle g, f\rangle$$

for all $f \in AW^{0,\tau}_{p,p}(\mathbb{R}^{n+1}_{\mathbb{Z}_+})$ with compact support and

$$\begin{aligned}\|g\|_{AT^{0,\tau}_{p',p'}(\mathbb{R}^{n+1}_{\mathbb{Z}_+})} &\lesssim \sup_{\substack{\|f\|_{AW^{0,\tau}_{p',p'}(\mathbb{R}^{n+1}_{\mathbb{Z}_+})}\le 1\\ f \text{ has compact support}}} |\langle g, f\rangle| \\ &\lesssim \sup_{\substack{\|f\|_{AW^{0,\tau}_{p',p'}(\mathbb{R}^{n+1}_{\mathbb{Z}_+})}\le 1\\ f \text{ has compact support}}} |L(f)| \\ &\lesssim \|L\|_{({}_0AW^{0,\tau}_{p',p'}(\mathbb{R}^{n+1}_{\mathbb{Z}_+}))^*},\end{aligned}$$

which completes the proof of Theorem 7.2. □

Remark 7.2. It is still unclear whether or not Theorem 7.2 is true for the spaces ${}_0AT^{s,\tau}_{p,q}(\mathbb{R}^{n+1}_{\mathbb{Z}_+})$ and $AW^{-s,\tau}_{p',q'}(\mathbb{R}^{n+1}_{\mathbb{Z}_+})$ when $p \neq q$. The difficulty lies in the fact that the space $AT^{s,\tau}_{p,q}(\mathbb{R}^{n+1}_{\mathbb{Z}_+})$ when $p \neq q$ is only known to be a quasi-Banach space so far. Thus, Lemma 7.5 in the case that $p \neq q$ seems not available, due to the Hahn-Banach theorem is not valid for these spaces.

7.2 Besov-Hausdorff Spaces and Triebel-Lizorkin-Hausdorff Spaces

In this section, we determine the predual spaces of $A^{s,\tau}_{p,q}(\mathbb{R}^n)$. Let Φ and φ satisfy, respectively, (2.1) and (2.2). We define an operator ρ_φ by setting

$$\rho_\varphi(f)(x,2^{-j}) \equiv \varphi_j * f(x)$$

for all $f \in \mathscr{S}'(\mathbb{R}^n)$, $x \in \mathbb{R}^n$ and $j \in \mathbb{Z}_+$, where when $j=0$, φ_0 is replaced by Φ. Conversely. for all functions F on $\mathbb{R}^{n+1}_{\mathbb{Z}_+}$ and $x \in \mathbb{R}^n$, we define a map π_φ by

$$\pi_\varphi(F)(x) \equiv \sum_{k=0}^{\infty} \int_{\mathbb{R}^n} F(y,2^{-k})\varphi_k(x-y)\,dy = \sum_{k=0}^{\infty} \int_{\mathbb{R}^n} F^k(y)\varphi_k(x-y)\,dy, \tag{7.9}$$

which makes sense due to the following technical lemma.

Lemma 7.7. *Let $s \in \mathbb{R}$, $p \in (1,\infty)$, $q \in (1,\infty]$ and $\tau \in [0,\infty)$, Φ and φ satisfy, respectively, (2.1) and (2.2), and for all $\xi \in \mathbb{R}^n$,*

$$|\widehat{\Phi}(\xi)|^2 + \sum_{j=1}^{\infty} |\widehat{\varphi}(2^{-j}\xi)|^2 = 1.$$

Then π_φ is a bounded and surjective linear operator from $AW^{s,\tau}_{p,q}(\mathbb{R}^{n+1}_{\mathbb{Z}_+})$ to $A^{s,\tau}_{p,q}(\mathbb{R}^n)$.

Proof. By similarity, we only give the proof for the space $FW^{s,\tau}_{p,q}(\mathbb{R}^{n+1}_{\mathbb{Z}_+})$. Let $F \in FW^{s,\tau}_{p,q}(\mathbb{R}^{n+1}_{\mathbb{Z}_+})$. Notice that there exists a constant $\gamma > 1$ such that for all cubes P, $P \times (0, l(P)] \subset T(\gamma P)$. Therefore, we have

$$\|F\|_{FW^{s,\tau}_{p,q}(\mathbb{R}^{n+1}_{\mathbb{Z}_+})} \sim \sup_{P \in \mathcal{Q}} \frac{1}{|P|^\tau} \left\{ \int_P \left[\sum_{j=j_P \vee 0}^{\infty} 2^{jsq} |F^j(x)|^q \right]^{p/q} dx \right\}^{1/p}. \tag{7.10}$$

We claim that (7.9) holds in $\mathscr{S}'(\mathbb{R}^n)$. For all $m \in \mathbb{Z}_+$ and $k \in \mathbb{Z}^n$, setting $R_{-1} \equiv \emptyset$, $R_m \equiv [-2^{m+1}, 2^{m+1})^n$ and

$$\chi_{R_m \setminus R_{m-1}}(k) \equiv \chi_{\{k \in \mathbb{Z}^n:\ Q_{0k} \subset R_m \setminus R_{m-1}\}}(k),$$

we then have $\sum_{k \in \mathbb{Z}^n} \chi_{R_m \setminus R_{m-1}}(k) \lesssim 2^{mn}$. Then for all $\phi \in \mathscr{S}(\mathbb{R}^n)$ and $i \in \mathbb{Z}_+$, by Lemma 2.4 and (7.10), we obtain

$$\begin{aligned}
&\int_{\mathbb{R}^n} |\phi * \varphi_i(x-y)||F^i(y)|\,dy \\
&\quad \lesssim \sum_{m=0}^{\infty} \sum_{k \in \mathbb{Z}^n} \chi_{R_m \setminus R_{m-1}}(k) \int_{x+Q_{0k}} \frac{2^{-iM}}{(1+|x-y|)^{n+M}} |F^i(y)|\,dy
\end{aligned}$$

$$
\begin{aligned}
&\lesssim \sum_{m=0}^{\infty} \sum_{k\in\mathbb{Z}^n} \chi_{R_m\setminus R_{m-1}}(k) 2^{-iM} 2^{-m(n+M)} 2^{-is} \\
&\quad \times \left(\int_{x+Q_{0k}} 2^{isp} |F^i(y)|^p \, dy\right)^{1/p} \\
&\lesssim 2^{-iM} 2^{-is} \|F\|_{FW^{s,\tau}_{p,q}(\mathbb{R}^{n+1}_{\mathbb{Z}_+})},
\end{aligned}
$$

where M can be any positive number. If we choose $M > \max\{0, -s\}$, then, as $l \to \infty$,

$$
\sum_{|i|>l} \int_{\mathbb{R}^n} |\phi * \varphi_i(x-y)||F^i(y)| \, dy \to 0,
$$

which implies that (7.9) holds in $\mathscr{S}'(\mathbb{R}^n)$.

Now we verify that

$$
\|\pi_\varphi(F)\|_{F^{s,\tau}_{p,q}(\mathbb{R}^n)} \lesssim \|F\|_{FW^{s,\tau}_{p,q}(\mathbb{R}^{n+1}_{\mathbb{Z}_+})}.
$$

By the above claim and Lemma 2.4, we see that

$$
\begin{aligned}
&\|\pi_\varphi(F)\|_{F^{s,\tau}_{p,q}(\mathbb{R}^n)} \\
&\quad = \sup_{P\in\mathscr{Q}} \frac{1}{|P|^\tau} \left\{ \int_P \left[\sum_{j=j_P\vee 0}^{\infty} 2^{jsq} |\varphi_j * \pi_\varphi(F)(x)| \right]^{p/q} dx \right\}^{1/p} \\
&\quad \le \sup_{P \text{dyadic}} \frac{1}{|P|^\tau} \left\{ \int_P \left[\sum_{j=j_P\vee 0}^{\infty} \left(\sum_{i=0}^{\infty} \int_{\mathbb{R}^n} 2^{js} |\varphi_j * \phi_i(x-y)||F^i(y)| \, dy \right)^q \right]^{p/q} dx \right\}^{1/p} \\
&\quad \lesssim \sup_{P\in\mathscr{Q}} \frac{1}{|P|^\tau} \left\{ \int_P \left[\sum_{j=j_P\vee 0}^{\infty} \right.\right. \\
&\qquad \left.\left. \times \left(\sum_{i=0}^{\infty} \int_{\mathbb{R}^n} \frac{2^{js} 2^{-|i-j|M} 2^{-(i\wedge j)M} |F^i(y)|}{(2^{-(i\wedge j)} + |x-y|)^{n+M}} \, dy \right)^q \right]^{p/q} dx \right\}^{1/p}.
\end{aligned}
$$

Similarly to the proof of Lemma 4.1, applying (7.10), we have

$$
\|\pi_\varphi(F)\|_{F^{s,\tau}_{p,q}(\mathbb{R}^n)} \lesssim \|F\|_{FW^{s,\tau}_{p,q}(\mathbb{R}^{n+1}_{\mathbb{Z}_+})};
$$

see also [164, p. 2797]. Thus, π_φ is a bounded linear operator from $FW^{s,\tau}_{p,q}(\mathbb{R}^{n+1}_{\mathbb{Z}_+})$ to $F^{s,\tau}_{p,q}(\mathbb{R}^n)$.

By the Calderón reproducing formula in Lemma 2.3, the composite operator $\pi_\varphi \rho_\varphi$ is the identity on $F^{s,\tau}_{p,q}(\mathbb{R}^n)$, which implies the surjectivity of π_φ, and hence completes the proof of Lemma 7.7. □

From Lemma 7.7, we also deduce that

$$\|f\|_{F^{s,\tau}_{p,q}(\mathbb{R}^n)} \sim \|\rho_\varphi(f)\|_{FW^{s,\tau}_{p,q}(\mathbb{R}^{n+1}_{\mathbb{Z}_+})}.$$

We now introduce the following spaces.

Definition 7.5. Let $s \in \mathbb{R}$, $p \in (1,\infty)$, $q \in [1,\infty)$ and $\tau \in [0, \frac{1}{(p\vee q)'}]$, Φ and φ satisfy, respectively, (2.1) and (2.2).

(i) The *Besov-Hausdorff space* $BH^{s,\tau}_{p,q}(\mathbb{R}^n)$ is defined to be the set of all $f \in \mathcal{S}'(\mathbb{R}^n)$ such that

$$\|f\|_{BH^{s,\tau}_{p,q}(\mathbb{R}^n)} \equiv \|\rho_\varphi(f)\|_{BT^{s,\tau}_{p,q}(\mathbb{R}^{n+1}_{\mathbb{Z}_+})} < \infty.$$

(ii) The *Triebel-Lizorkin-Hausdorff space* $FH^{s,\tau}_{p,q}(\mathbb{R}^n)$ $(q \neq 1)$ is defined to be the set of all $f \in \mathcal{S}'(\mathbb{R}^n)$ such that

$$\|f\|_{FH^{s,\tau}_{p,q}(\mathbb{R}^n)} \equiv \|\rho_\varphi(f)\|_{FT^{s,\tau}_{p,q}(\mathbb{R}^{n+1}_{\mathbb{Z}_+})} < \infty.$$

For simplicity, we use $AH^{s,\tau}_{p,q}(\mathbb{R}^n)$ to denote either $BH^{s,\tau}_{p,q}(\mathbb{R}^n)$ or $FH^{s,\tau}_{p,q}(\mathbb{R}^n)$.

Remark 7.3.
(i) From (7.4), we deduce that $\|\cdot\|_{AH^{s,\tau}_{p,q}(\mathbb{R}^n)}$ is a quasi-norm.
(ii) By Remark 7.1(i), when $\tau = 0$, $AH^{s,\tau}_{p,q}(\mathbb{R}^n) = A^s_{p,q}(\mathbb{R}^n)$.

To show that the space $AH^{s,\tau}_{p,q}(\mathbb{R}^n)$ is independent of the choices of Φ and φ, we need a technical lemma. For all $\beta \in [1,\infty)$ and $x \in \mathbb{R}^n$, define the β-nontangential maximal function $N_\beta f$ of a measurable function f on $\mathbb{R}^{n+1}_+$ by

$$N_\beta f(x) \equiv \sup_{|y-x|<\beta t} |f(y,t)|.$$

In Definition 7.2, with (7.3) replaced by

$$\int_{\mathbb{R}^n} [N_\beta \omega(x)]^{(p\vee q)'} \, d\Lambda^{(\infty)}_{n\tau(p\vee q)'}(x) \le 1, \tag{7.11}$$

we obtain another tent space, denoted by $AT^{s,\tau}_{p,q}(\beta, \mathbb{R}^{n+1}_{\mathbb{Z}_+})$. Then, obviously,

$$AT^{s,\tau}_{p,q}(1, \mathbb{R}^{n+1}_{\mathbb{Z}_+}) = AT^{s,\tau}_{p,q}(\mathbb{R}^{n+1}_{\mathbb{Z}_+}).$$

Generally, we have the following result.

Lemma 7.8. *Let $\beta \in [1,\infty)$ and s, τ, p, q be as in Definition 7.2. Then*

$$AT^{s,\tau}_{p,q}(\beta, \mathbb{R}^{n+1}_{\mathbb{Z}_+}) = AT^{s,\tau}_{p,q}(\mathbb{R}^{n+1}_{\mathbb{Z}_+})$$

and there exists a positive constant C such that for all functions $F \in AT^{s,\tau}_{p,q}(\mathbb{R}^{n+1}_{\mathbb{Z}_+})$ and $\beta \in [1,\infty)$,

$$\|F\|_{AT^{s,\tau}_{p,q}(\mathbb{R}^{n+1}_{\mathbb{Z}_+})} \le \|F\|_{AT^{s,\tau}_{p,q}(\beta,\mathbb{R}^{n+1}_{\mathbb{Z}_+})} \le C\beta^{n(\lfloor (p\vee q)'\rfloor+2)/q}\|F\|_{AT^{s,\tau}_{p,q}(\mathbb{R}^{n+1}_{\mathbb{Z}_+})}.$$

Proof. The first inequality is trivial, so that we only need to verify the second one. Again we only consider the space $FT^{s,\tau}_{p,q}(\mathbb{R}^{n+1}_{\mathbb{Z}_+})$.

Let $F \in FT^{s,\tau}_{p,q}(\mathbb{R}^{n+1}_{\mathbb{Z}_+})$ and ω be a nonnegative Borel measurable function on $\mathbb{R}^{n+1}_+$ satisfying (7.3) such that

$$\left\|\left\{\sum_{k=0}^{\infty} 2^{ksq}|F^k|^q[\omega^k]^{-q}\right\}^{1/q}\right\|_{L^p(\mathbb{R}^n)} \le 2\|F\|_{F^{s,\tau}_{p,q}(\mathbb{R}^n)}.$$

For any $\beta \in [1,\infty)$, obviously,

$$B(x,\beta t) \subset \left(\bigcup_{y\in B(x,\beta t)} B(y,t/5)\right).$$

By [71, p. 2, Theorem 1.2], there exists a set $\{a_j\}_{j=1}^{\lfloor 5^n(a+1)^n\rfloor} \subset B(0,\beta t)$ such that

$$B(x,\beta t) \subset \left(\bigcup_{j=1}^{\lfloor 5^n(a+1)^n\rfloor} B(x+a_j,t)\right).$$

From this, it follows that for all $x \in \mathbb{R}^n$,

$$N_\beta\omega(x) \le \sum_{j=1}^{\lfloor 5^n(\beta+1)^n\rfloor} \sup_{|y-x-a_j|<t} \omega(y,t) = \sum_{j=1}^{\lfloor 5^n(\beta+1)^n\rfloor} N\omega(x+a_j),$$

which together with the geometry property of the Hausdorff capacity and (7.3) yields that

$$\begin{aligned}
&\int_{\mathbb{R}^n} [N_\beta\omega(x)]^{(p\vee q)'}\,d\Lambda^{(\infty)}_{n\tau(p\vee q)'}(x)\\
&\lesssim (\lfloor 5^n(\beta+1)^n\rfloor)^{\lfloor (p\vee q)'\rfloor+1}\sum_{j=1}^{\lfloor 5^n(\beta+1)^n\rfloor}\int_{\mathbb{R}^n}[N\omega(x+a_j)]^{(p\vee q)'}\,d\Lambda^{(\infty)}_{n\tau(p\vee q)'}(x)\\
&\lesssim (\lfloor 5^n(\beta+1)^n\rfloor)^{\lfloor (p\vee q)'\rfloor+2}.
\end{aligned}$$

Set

$$\widetilde{\omega} \equiv \kappa\omega/(\lfloor 5^n(\beta+1)^n\rfloor)^{\lfloor (p\vee q)'\rfloor+2},$$

where the positive constant κ is chosen such that $\widetilde{\omega}$ satisfies (7.3). Then we obtain

$$\begin{aligned}\|F\|_{FT^{s,\tau}_{p,q}(\beta,\mathbb{R}^{n+1}_{\mathbb{Z}_+})} &\le \left\| \left\{ \sum_{k=0}^{\infty} 2^{ksq} |F^k|^q [\widetilde{\omega}^k]^{-q} \right\}^{1/q} \right\|_{L^p(\mathbb{R}^n)} \\ &\lesssim (\lfloor 5^n(\beta+1)^n \rfloor)^{(\lfloor (p\vee q)' \rfloor + 2)/q} \|F\|_{FT^{s,\tau}_{p,q}(\mathbb{R}^{n+1}_{\mathbb{Z}_+})},\end{aligned}$$

which completes the proof of Lemma 7.8. □

Proposition 7.2. *Let $s \in \mathbb{R}$, $p \in (1,\infty)$, $q \in [1,\infty)$ and $\tau \in [0, \frac{1}{(p\vee q)'}]$. Then the space $AH^{s,\tau}_{p,q}(\mathbb{R}^n)$ is independent of the choices of Φ and φ.*

Proof. We only consider $FH^{s,\tau}_{p,q}(\mathbb{R}^n)$ and $\tau > 0$. Let Φ and φ satisfy, respectively, (2.1) and (2.2). Then there exist Ψ and ψ satisfying, respectively, (2.1) and (2.2) such that (2.6) holds. To emphasize φ and ψ, in this proof, we use $\|\cdot\|_{FH^{s,\tau}_{p,q}(\varphi,\mathbb{R}^n)}$ and $\|\cdot\|_{FH^{s,\tau}_{p,q}(\psi,\mathbb{R}^n)}$ to replace $\|\cdot\|_{FH^{s,\tau}_{p,q}(\mathbb{R}^n)}$. By symmetry, it suffices to show that

$$\|f\|_{FH^{s,\tau}_{p,q}(\psi,\mathbb{R}^n)} \lesssim \|f\|_{FH^{s,\tau}_{p,q}(\varphi,\mathbb{R}^n)}$$

for all $f \in FH^{s,\tau}_{p,q}(\varphi,\mathbb{R}^n)$.

Let $\widetilde{\omega}$ be a nonnegative Borel measurable function satisfying (7.3) and

$$\left\| \left\{ \sum_{j=0}^{\infty} 2^{jsq} |\varphi_j * f|^q [\widetilde{\omega}^j]^{-q} \right\}^{1/q} \right\|_{L^p(\mathbb{R}^n)} \le 2\|f\|_{FH^{s,\tau}_{p,q}(\varphi,\mathbb{R}^n)}. \tag{7.12}$$

Set $\psi_k \equiv 0$ if $k < 0$. Notice that $\psi_k * \widetilde{\psi}_j \equiv 0$ when $|k-j| > 1$. Then by Lemma 2.4, for all $x \in \mathbb{R}^n$, we have

$$\begin{aligned}\left| \sum_{j=0}^{\infty} \psi_k * \widetilde{\psi}_j * \varphi_j * f \right| &\le \sum_{j=(k-1)\vee 0}^{k+1} \int_{\mathbb{R}^n} |\psi_k * \widetilde{\psi}_j(x-y)| |\varphi_j * f(y)|\,dy \\ &\lesssim \sum_{j=(k-1)\vee 0}^{k+1} \int_{\mathbb{R}^n} \frac{2^{-|k-j|M} 2^{-(k\wedge j)M}}{(2^{-(k\wedge j)} + |x-y|)^{n+M}} |\varphi_j * f(y)|\,dy \\ &\lesssim \sum_{j=(k-1)\vee 0}^{k+1} 2^{-|k-j|M} \left(2^{(k\wedge j)n} \int_{|x-y|<2^{-(k\wedge j)}} |\varphi_j * f(y)|\,dy \right. \\ &\quad \left. + \sum_{l=1}^{\infty} 2^{-lM - ln + (k\wedge j)n} \int_{2^{l-(k\wedge j)-1} \le |x-y| < 2^{l-(k\wedge j)}} |\varphi_j * f(y)|\,dy \right),\end{aligned}$$

where $M \in \mathbb{N}$ will be determined later.

Thus by Lemma 2.3 and Remark 7.1(ii), we see that

$$
\begin{aligned}
&\|f\|_{FH^{s,\tau}_{p,q}(\psi,\mathbb{R}^n)}\\
&\lesssim \inf_{\omega}\left\|\left\{\sum_{k=0}^{\infty}2^{ksq}\left[\sum_{j=(k-1)\vee 0}^{k+1}2^{-|k-j|M}2^{(k\wedge j)n}\right.\right.\right.\\
&\quad\left.\left.\left.\times\int_{|\cdot-y|<2^{-(k\wedge j)}}|\varphi_j*f(y)|[\omega^k(\cdot)]^{-1}\,dy\right]^q\right\}^{1/q}\right\|_{L^p(\mathbb{R}^n)}\\
&\quad+\sum_{l=1}^{\infty}2^{-lM+l\rho}\inf_{\omega}\left\|\left\{\sum_{k=0}^{\infty}2^{ksq}\left[\sum_{j=(k-1)\vee 0}^{k+1}2^{-|k-j|M}2^{-ln}2^{(k\wedge j)n}\right.\right.\right.\\
&\quad\left.\left.\left.\times\int_{|\cdot-y|<2^{l-(k\wedge j)}}|\varphi_j*f(y)|[\omega^k(\cdot)]^{-1}\,dy\right]^q\right\}^{1/q}\right\|_{L^p(\mathbb{R}^n)}\\
&\equiv \mathrm{J}_0+\sum_{l=1}^{\infty}2^{-lM+l\rho}\mathrm{J}_l,
\end{aligned}
$$

where ρ is same as in (7.4).

For all $l\in\mathbb{Z}_+$ and $(x,t)\in\mathbb{R}^{n+1}_+$, set

$$\omega_l(x,t)\equiv\frac{\kappa}{2^{(l+3)n(\lfloor(p\vee q)'\rfloor+2)}}\sup\left\{\widetilde{\omega}(y,s):\ |y-x|<2^{l+3}s\text{ and }t/4\le s\le 4t\right\},$$

where $\kappa\in(0,\infty)$ will be determined later. Then, for all $x\in\mathbb{R}^n$,

$$
\begin{aligned}
N\omega_l(x)&=\frac{\kappa}{2^{(l+3)n(\lfloor(p\vee q)'\rfloor+2)}}\sup_{|y-x|<t}\sup\left\{\widetilde{\omega}(z,s):\ |y-z|<2^{l+3}s\text{ and }\frac{t}{4}\le s\le 4t\right\}\\
&\le\frac{\kappa}{2^{(l+3)n(\lfloor(p\vee q)'\rfloor+2)}}\sup_{|z-x|<2^{l+3}s}\widetilde{\omega}(z,s)\\
&=\frac{\kappa}{2^{(l+3)n(\lfloor(p\vee q)'\rfloor+2)}}N_{2^{l+3}}\widetilde{\omega}(x).
\end{aligned}
$$

If we choose $\kappa\in(0,\infty)$, independent of l and x, small enough, by an argument same as in the proof of Lemma 7.8, we see that ω_l satisfies (7.3).

Notice that

$$[\omega_l^k(x)]^{-1}\widetilde{\omega}^j(y)\lesssim 2^{(l+3)n(\lfloor(p\vee q)'\rfloor+2)}$$

when $|x-y|<2^{l-(k\wedge j)}$ and $k-1\le j\le k+1$. Thus by the Fefferman-Stein vector-valued inequality (see [58]), we have

$$
\begin{aligned}
\mathrm{J}_l\le&\left\|\left\{\sum_{k=0}^{\infty}2^{ksq}\left[\sum_{j=(k-1)\vee 0}^{k+1}2^{-|k-j|M}2^{-ln}2^{(k\wedge j)n}\right.\right.\right.\\
&\left.\left.\left.\times\int_{|\cdot-y|<2^l2^{-(k\wedge j)}}|\varphi_j*f(y)|[\omega_l^k(\cdot)]^{-1}\,dy\right]^q\right\}^{1/q}\right\|_{L^p(\mathbb{R}^n)}
\end{aligned}
$$

$$\lesssim 2^{ln(\lfloor (p\vee q)'\rfloor+2)} \left\| \left\{ \sum_{k=0}^{\infty} 2^{ksq} \left[\sum_{j=(k-1)\vee 0}^{k+1} 2^{-|k-j|M} 2^{-ln} 2^{(k\wedge j)n} \right.\right.\right.$$

$$\left.\left.\left. \times \int_{|\cdot - y|<2^l 2^{-(k\wedge j)}} |\varphi_j * f(y)| [\widetilde{w}^j(y)]^{-1/q}\, dy \right]^q \right\}^{1/q} \right\|_{L^p(\mathbb{R}^n)}$$

$$\lesssim 2^{ln(\lfloor (p\vee q)'\rfloor+2)}$$

$$\times \left\| \left\{ \sum_{k=0}^{\infty} 2^{ksq} \left[\sum_{j=(k-1)\vee 0}^{k+1} 2^{-|k-j|M} 2^{-js} \mathrm{M}\left(2^{js} |\varphi_j * f| [\widetilde{w}^j]^{-1}\right) \right]^q \right\}^{1/q} \right\|_{L^p(\mathbb{R}^n)}$$

$$\lesssim 2^{l(\lfloor (p\vee q)'\rfloor+2)} \left\| \left\{ \sum_{j=0}^{\infty} \left[\mathrm{M}\left(2^{js} |\varphi_j * f| [\widetilde{w}^j]^{-1}\right)\right]^q \right\}^{1/q} \right\|_{L^p(\mathbb{R}^n)}$$

$$\lesssim 2^{ln(\lfloor (p\vee q)'\rfloor+2)} \left\| \left\{ \sum_{j=0}^{\infty} 2^{jsq} |\varphi_j * f|^q [\widetilde{w}^j]^{-q} \right\}^{1/q} \right\|_{L^p(\mathbb{R}^n)}$$

$$\lesssim 2^{ln(\lfloor (p\vee q)'\rfloor+2)} \|f\|_{FH^{s,\tau}_{p,q}(\varphi,\mathbb{R}^n)}.$$

Therefore, choosing $M > n(\lfloor (p\vee q)'\rfloor + 2) + \rho$, we obtain

$$\begin{aligned} \|f\|_{FH^{s,\tau}_{p,q}(\psi,\mathbb{R}^n)} &\lesssim \mathrm{J}_0 + \sum_{l=1}^{\infty} 2^{-lM+l\rho} \mathrm{J}_l \\ &\lesssim \sum_{l=0}^{\infty} 2^{-lM+l\rho} 2^{ln(\lfloor (p\vee q)'\rfloor+2)} \|f\|_{FH^{s,\tau}_{p,q}(\varphi,\mathbb{R}^n)} \\ &\lesssim \|f\|_{FH^{s,\tau}_{p,q}(\varphi,\mathbb{R}^n)}, \end{aligned}$$

which completes the proof of Proposition 7.2. □

Lemma 7.9. *Let $s\in\mathbb{R}$, $p\in(1,\infty)$, $q\in[1,\infty)$ and $\tau\in[0,\frac{1}{(p\vee q)'}]$. Then*

$$\mathscr{S}(\mathbb{R}^n) \subset AH^{s,\tau}_{p,q}(\mathbb{R}^n) \subset \mathscr{S}'(\mathbb{R}^n)$$

and $\mathscr{S}(\mathbb{R}^n)$ is dense in $AH^{s,\tau}_{p,q}(\mathbb{R}^n)$.

Proof. By similarity and $AH^{s,0}_{p,q}(\mathbb{R}^n) = A^s_{p,q}(\mathbb{R}^n)$, we only give the proof of the space $FH^{s,\tau}_{p,q}(\mathbb{R}^n)$ in the case when $\tau > 0$.

We first show that $FH^{s,\tau}_{p,q}(\mathbb{R}^n) \subset \mathscr{S}'(\mathbb{R}^n)$, namely, there exists an $M\in\mathbb{N}$ such that for all $f\in FH^{s,\tau}_{p,q}(\mathbb{R}^n)$ and $\phi\in\mathscr{S}(\mathbb{R}^n)$,

$$|\langle f,\phi\rangle| \lesssim \|f\|_{FH^{s,\tau}_{p,q}(\mathbb{R}^n)} \|\phi\|_{\mathscr{S}_{M+1}}.$$

Let Φ, φ, Ψ and ψ be as in (2.6). Let $\widetilde{\omega}$ be a nonnegative Borel measurable function satisfying (7.3) and (7.12). Then by Lemma 2.3, Lemma 2.4 and Remark 7.1(iii), we obtain

$$
\begin{aligned}
|\langle f, \phi\rangle| &\le \sum_{j=0}^{\infty} \int_{\mathbb{R}^n} |\varphi_j * \phi(x)||\psi_j * f(x)|\,dx \\
&\lesssim \|\phi\|_{\mathscr{S}_{M+1}} \sum_{j=0}^{\infty} 2^{-jM+jn\tau} \int_{\mathbb{R}^n} \frac{1}{(1+|x|)^{n+M}} |\psi_j * f(x)|[\widetilde{\omega}^j(x)]^{-1}\,dx \\
&\lesssim \|\phi\|_{\mathscr{S}_{M+1}} \sum_{j=0}^{\infty} 2^{-jM+jn\tau} \sum_{l=0}^{\infty} 2^{-l(n+M)} \int_{|x|<2^l} |\psi_j * f(x)|[\widetilde{\omega}^j(x)]^{-1}\,dx.
\end{aligned}
$$

Choosing $M > n\tau - s$, by Hölder's inequality, we have

$$
\begin{aligned}
|\langle f, \phi\rangle| &\lesssim \|\phi\|_{\mathscr{S}_{M+1}} \sum_{j=0}^{\infty} 2^{-jM+jn\tau-js} \sum_{l=0}^{\infty} 2^{-l(n+M)} 2^{ln(1-1/p)} \\
&\quad \times \left(\int_{|x|<2^l} 2^{jsp} |\psi_j * f(x)|^p [\widetilde{\omega}^j(x)]^{-p}\,dx\right)^{1/p} \\
&\lesssim \|f\|_{FH^{s,\tau}_{p,q}(\mathbb{R}^n)} \|\phi\|_{\mathscr{S}_{M+1}} \sum_{j=0}^{\infty} 2^{-jM+jn\tau-js} \sum_{l=0}^{\infty} 2^{-l(n+M)} 2^{ln(1-1/p)} \\
&\lesssim \|f\|_{FH^{s,\tau}_{p,q}(\mathbb{R}^n)} \|\phi\|_{\mathscr{S}_{M+1}}.
\end{aligned}
$$

Thus, $FH^{s,\tau}_{p,q}(\mathbb{R}^n) \subset \mathscr{S}'(\mathbb{R}^n)$.

Next we prove that $\mathscr{S}(\mathbb{R}^n) \subset FH^{s,\tau}_{p,q}(\mathbb{R}^n)$. For all $\phi \in \mathscr{S}(\mathbb{R}^n)$, by Lemma 2.4, we have

$$
\begin{aligned}
\|\phi\|_{FH^{s,\tau}_{p,q}(\mathbb{R}^n)} &\lesssim \|\phi\|_{\mathscr{S}_{M+1}} \inf_{\omega} \left\| \left\{ \sum_{k=0}^{\infty} \frac{2^{ksq} 2^{-kMq}}{(1+|\cdot|)^{nq+Mq}} [\omega^k]^{-q} \right\}^{1/q} \right\|_{L^p(\mathbb{R}^n)} \\
&\equiv \|\phi\|_{\mathscr{S}_{M+1}} \mathrm{J},
\end{aligned}
$$

where $M \in \mathbb{N}$ will be determined later.

Let ε be a positive real number such that $n\tau + \varepsilon > n\tau(p \vee q)'$. For all $l \in \mathbb{Z}_+$, set

$$
\omega_l(x,t) \equiv \left[\kappa 2^{-ln\tau(p\vee q)'} \min\left\{1, \left(\frac{2^l}{\sqrt{|x|^2+t^2}}\right)^{n\tau+\varepsilon}\right\}\right]^{1/(p\vee q)'},
$$

where the positive constant κ is same as in the proof of Lemma 7.1. Then, by an argument similar to that used in the proof of Lemma 7.1, we see that ω_l satisfies (7.3), and hence

$$
\begin{aligned}
\mathrm{J} &= \inf_{\omega}\Bigg\{\int_{|x|<1}\left[\sum_{k=0}^{\infty}\frac{2^{ksq}2^{-kMq}}{(1+|x|)^{nq+Mq}}[\omega^k(x)]^{-q}\right]^{p/q}dx \\
&\quad+\sum_{l=1}^{\infty}\int_{2^{l-1}\le|x|<2^l}\left[\sum_{k=0}^{\infty}\frac{2^{ksq}2^{-kMq}}{(1+|x|)^{nq+Mq}}[\omega^k(x)]^{-q}\right]^{p/q}dx\Bigg\}^{1/p} \\
&\lesssim \inf_{\omega}\left\{\int_{|x|<1}\left[\sum_{k=0}^{\infty}\frac{2^{ksq}2^{-kMq}}{(1+|x|)^{nq+Mq}}[\omega^k(x)]^{-q}\right]^{p/q}dx\right\}^{1/p} \\
&\quad+\sum_{l=1}^{\infty}2^{l\rho}\inf_{\omega}\left\{\int_{2^{l-1}\le|x|<2^l}\left[\sum_{k=0}^{\infty}\frac{2^{ksq}2^{-kMq}}{(1+|x|)^{nq+Mq}}[\omega^k(x)]^{-q}\right]^{p/q}dx\right\}^{1/p}.
\end{aligned}
$$

Notice that $\omega_l \sim 2^{-ln\tau}$ in $T(B(0,2^l))$. If we choose

$$M > \max\{s, n\tau+\rho-n(1-1/p)\},$$

then

$$
\begin{aligned}
\mathrm{J} &\lesssim \left\{\int_{|x|<1}\left[\sum_{k=0}^{\infty}\frac{2^{ksq}2^{-kMq}}{(1+|x|)^{nq+Mq}}\right]^{p/q}dx\right\}^{1/p} \\
&\quad+\sum_{l=1}^{\infty}2^{ln\tau+l\rho}\left\{\int_{2^{l-1}\le|x|<2^l}\left[\sum_{k=0}^{\infty}\frac{2^{ksq}2^{-kMq}}{(1+|x|)^{nq+Mq}}\right]^{p/q}dx\right\}^{1/p} \\
&\lesssim \sum_{l=0}^{\infty}2^{ln\tau+l\rho}2^{-l(n+M)}\left\{\int_{|x|<2^l}\left[\sum_{k=0}^{\infty}2^{ksq}2^{-kMq}\right]^{p/q}dx\right\}^{1/p} \\
&< \infty.
\end{aligned}
$$

We then obtain $\|\phi\|_{FH^{s,\tau}_{p,q}(\mathbb{R}^n)} \lesssim \|\phi\|_{\mathscr{S}_{M+1}}$, and hence $\mathscr{S}(\mathbb{R}^n) \subset FH^{s,\tau}_{p,q}(\mathbb{R}^n)$.

To show $\mathscr{S}(\mathbb{R}^n)$ is dense in $FH^{s,\tau}_{p,q}(\mathbb{R}^n)$, let $f \in FH^{s,\tau}_{p,q}(\mathbb{R}^n)$, Φ, φ, Ψ and ψ be as in (2.6). Then by Lemma 2.3,

$$f = \sum_{j=0}^{\infty}\widetilde{\psi}_j * \varphi_j * f$$

in $\mathscr{S}'(\mathbb{R}^n)$. For all $m, s \in \mathbb{N}$, set

$$f_{m,s} \equiv \sum_{0\le j\le m}\widetilde{\psi}_j * [(\varphi_j * f)\gamma_s],$$

where $\gamma_s \in C^\infty(\mathbb{R}^n)$ satisfying $0 \le \gamma_s \le 1$, $\operatorname{supp} \gamma_s \subset B(0, 2s)$ and $\gamma_s \equiv 1$ on $B(0, s)$. Then, by [134, p. 23, Theorem 3.13], $f_{m,s} \in \mathscr{S}(\mathbb{R}^n)$ and $f_{m,s} \to f$ in $\mathscr{S}'(\mathbb{R}^n)$ as $s, m \to \infty$.

To prove $f_{m,s} \to f$ in $FH_{p,q}^{s,\tau}(\mathbb{R}^n)$ as $m, s \to \infty$, assume $\widetilde{\omega}$ satisfies (7.3) and (7.12). Then, using Lemma 2.4 and the vector-valued inequality of Fefferman and Stein, similarly to the proof of Proposition 7.2, we see that

$$\begin{aligned}
&\left\| f - \sum_{0\le j\le m} \widetilde{\psi}_j * \varphi_j * f \right\|_{FH_{p,q}^{s,\tau}(\mathbb{R}^n)} \\
&= \inf_{\omega} \left\| \left\{ \sum_{k=0}^{\infty} 2^{ksq} \left[\sum_{j>m} \chi_{\{j\in\mathbb{Z}:\, k-1\le j\le k+1\}}(j) \right.\right.\right. \\
&\quad \left.\left.\left. \times \int_{\mathbb{R}^n} |\varphi_k * \widetilde{\psi}_j(\cdot - y)||\varphi_j * f(y)|\, dy \right]^q [\omega^k(\cdot)]^{-q} \right\}^{\frac{1}{q}} \right\|_{L^p(\mathbb{R}^n)} \\
&\lesssim \left\| \left\{ \sum_{j>m} \left[\mathrm{M}(2^{js}|\varphi_j * f|[\widetilde{\omega}^j]^{-1}) \right]^q \right\}^{\frac{1}{q}} \right\|_{L^p(\mathbb{R}^n)} \\
&\lesssim \left\| \left(\sum_{j>m} 2^{jsq} |\varphi_j * f|^q [\widetilde{\omega}^j]^{-1} \right)^{\frac{1}{q}} \right\|_{L^p(\mathbb{R}^n)} \\
&\to 0
\end{aligned}$$

as $m \to 0$. Thus, for any $\varepsilon > 0$, there exists an $m_\varepsilon \in \mathbb{N}$ such that for all $m \ge m_\varepsilon$,

$$\left\| f - \sum_{0\le j\le m} \widetilde{\psi}_j * \varphi_j * f \right\|_{FH_{p,q}^{s,\tau}(\mathbb{R}^n)} < \varepsilon/2^{\rho+1}.$$

Fix $m \in \mathbb{N}$. Repeating the argument in the proof of Proposition 7.2 again, by the Lebesgue dominated convergence theorem and $\lim_{s\to\infty} \gamma_s(x) = 1$ for all $x \in \mathbb{R}^n$, we see that if $s \to \infty$, then

$$\begin{aligned}
&\left\| \sum_{0\le j\le m} \widetilde{\psi}_j * [(\varphi_j * f)\gamma_s] - \sum_{0\le j\le m} \widetilde{\psi}_j * \varphi_j * f \right\|_{FH_{p,q}^{s,\tau}(\mathbb{R}^n)} \\
&\lesssim \left\| \left(\sum_{0\le j\le m} 2^{jsq} |\varphi_j * f|^q |\gamma_s - 1|^q [\widetilde{\omega}^j]^{-q} \right)^{\frac{1}{q}} \right\|_{L^p(\mathbb{R}^n)} \\
&\to 0.
\end{aligned}$$

Thus, there exists an $s_{m,\varepsilon} \in \mathbb{N}$ such that for all $s \geq s_{m,\varepsilon}$,

$$\left\| \sum_{0\le j\le m} \widetilde{\psi}_j * [(\varphi_j * f)\gamma_s] - \sum_{0\le j\le m} \widetilde{\psi}_j * \varphi_j * f \right\|_{FH^{s,\tau}_{p,q}(\mathbb{R}^n)} < \varepsilon/2^{\rho+1}.$$

Thus, for any $\varepsilon > 0$, choosing $m \geq m_\varepsilon$ and $s \geq s_{m,\varepsilon}$, we have

$$\begin{aligned}
&\|f - f_{m,s}\|_{FH^{s,\tau}_{p,q}(\mathbb{R}^n)} \\
&\leq 2^\rho \left\| f - \sum_{0\le j\le m} \widetilde{\psi}_j * \varphi_j * f \right\|_{FH^{s,\tau}_{p,q}(\mathbb{R}^n)} \\
&\quad + 2^\rho \left\| \sum_{0\le j\le m} \widetilde{\psi}_j * [(\varphi_j * f)\gamma_s] - \sum_{0\le j\le m} \widetilde{\psi}_j * \varphi_j * f \right\|_{FH^{s,\tau}_{p,q}(\mathbb{R}^n)} \\
&< \varepsilon,
\end{aligned}$$

which implies $\mathscr{S}(\mathbb{R}^n)$ is dense in $FH^{s,\tau}_{p,q}(\mathbb{R}^n)$, and hence completes the proof of Lemma 7.9. □

To establish the dual relation between $AH^{s,\tau}_{p,q}(\mathbb{R}^n)$ and $A^{s,\tau}_{p,q}(\mathbb{R}^n)$, we need the following result.

Lemma 7.10. *Let $s \in \mathbb{R}$, $p \in (1,\infty)$, $q \in [1,\infty)$ and $\tau \in [0, \frac{1}{(p\vee q)'}]$, Φ and φ satisfy, respectively,* (2.1) *and* (2.2). *Then π_φ is a bounded linear operator from $AT^{s,\tau}_{p,q}(\mathbb{R}^{n+1}_{\mathbb{Z}_+})$ to $AH^{s,\tau}_{p,q}(\mathbb{R}^n)$.*

Proof. Let ω be a nonnegative Borel measurable function satisfy (7.3) and

$$\left\| \left\{ \sum_{j=0}^\infty 2^{jsq} |F^j|^q [\omega^j]^{-q} \right\}^{1/q} \right\|_{L^p(\mathbb{R}^n)} \leq 2\|F\|_{FT^{s,\tau}_{p,q}(\mathbb{R}^{n+1}_{\mathbb{Z}_+})}.$$

By Lemma 2.4 and Remark 7.1(iii), similarly to the proof of Lemma 7.7, we obtain that (7.8) holds in $\mathscr{S}'(\mathbb{R}^n)$ for all $F \in AT^{s,\tau}_{p,q}(\mathbb{R}^{n+1}_{\mathbb{Z}_+})$.

Repeating the argument in the proof of Proposition 7.2 with $\varphi_j * f$ replaced by F^j, we obtain that for all $F \in AT^{s,\tau}_{p,q}(\mathbb{R}^{n+1}_{\mathbb{Z}_+})$,

$$\|\pi_\varphi(F)\|_{AH^{s,\tau}_{p,q}(\mathbb{R}^n)} \lesssim \|F\|_{AT^{s,\tau}_{p,q}(\mathbb{R}^{n+1}_{\mathbb{Z}_+})},$$

which implies that π_φ is a bounded linear operator from $AT^{s,\tau}_{p,q}(\mathbb{R}^{n+1}_{\mathbb{Z}_+})$ to $AH^{s,\tau}_{p,q}(\mathbb{R}^n)$, and hence completes the proof of Lemma 7.10. □

We have the following dual theorem. The proof is similar to that of its homogeneous counterpart in [164, 165].

Theorem 7.3. *Let p, q, s, τ be as in Definition 7.5. The dual space of $AH^{s,\tau}_{p,q}(\mathbb{R}^n)$ is $A^{-s,\tau}_{p',q'}(\mathbb{R}^n)$ in the following sense: if $g \in A^{-s,\tau}_{p',q'}(\mathbb{R}^n)$, then the linear functional*

$$L(f) = \int_{\mathbb{R}^n} f(x)g(x)\,dx, \tag{7.13}$$

defined initially for all $f \in \mathscr{S}(\mathbb{R}^n)$, has a bounded extension to $AH^{s,\tau}_{p,q}(\mathbb{R}^n)$.

Conversely, if L is a bounded linear functional on $AH^{s,\tau}_{p,q}(\mathbb{R}^n)$, then there exists $g \in A^{-s,\tau}_{p',q'}(\mathbb{R}^n)$ so that

$$\|g\|_{A^{-s,\tau}_{p',q'}(\mathbb{R}^n)} \lesssim \|L\|$$

and L can be written in the form (7.13) for all $f \in \mathscr{S}(\mathbb{R}^n)$.

Proof. We only consider $FH^{s,\tau}_{p,q}(\mathbb{R}^n)$. Let Φ, φ, Ψ and ψ be as in (2.6). For all $f \in \mathscr{S}(\mathbb{R}^n)$ and $g \in F^{-s,\tau}_{p',q'}(\mathbb{R}^n)$, applying the Calderón reproducing formula in Lemmas 2.3, 7.7 and 7.10, and Theorem 7.1, we obtain

$$\begin{aligned}
|L_g(f)| = |\langle f, g\rangle| &\equiv \left|\left\langle f, \sum_{k=0}^{\infty} \widetilde{\psi}_k * \varphi_k * g\right\rangle\right| \\
&= \left|\int_{\mathbb{R}^n} \sum_{k=0}^{\infty} (\psi_k * f)(x)(\varphi_k * g)(x)\,dx\right| \\
&\lesssim \|\rho_\psi(f)\|_{FT^{s,\tau}_{p,q}(\mathbb{R}^{n+1}_{\mathbb{Z}_+})} \|\rho_\varphi(g)\|_{FW^{-s,\tau}_{p',q'}(\mathbb{R}^{n+1}_{\mathbb{Z}_+})} \\
&\sim \|f\|_{FH^{s,\tau}_{p,q}(\mathbb{R}^n)} \|g\|_{F^{-s,\tau}_{p',q'}(\mathbb{R}^n)}.
\end{aligned}$$

Thus each $g \in F^{-s,\tau p'}_{p',q'}(\mathbb{R}^n)$ induces a bounded linear functional L_g on the space $\mathscr{S}(\mathbb{R}^n)$ with

$$\|L_g\| \lesssim \|g\|_{F^{-s,\tau}_{p',q'}(\mathbb{R}^n)}.$$

By Lemma 7.9, $\mathscr{S}(\mathbb{R}^n)$ is dense in $FH^{s,\tau}_{p,q}(\mathbb{R}^n)$, so that L_g can be extended to a bounded linear functional on $FH^{s,\tau}_{p,q}(\mathbb{R}^n)$ with

$$\|L_g\| \lesssim \|g\|_{F^{-s,\tau}_{p',q'}(\mathbb{R}^n)}.$$

Conversely, let L be a bounded linear functional on $FH^{s,\tau}_{p,q}(\mathbb{R}^n)$. Set $\widetilde{L} \equiv L \circ \pi_\psi$. Obviously, $\widetilde{L}$ is linear. By Lemma 7.10, we know that for all $F \in FT^{s,\tau}_{p,q}(\mathbb{R}^{n+1}_{\mathbb{Z}_+})$, then $\pi_\psi(F) \in FH^{s,\tau}_{p,q}(\mathbb{R}^n)$ and

$$|\widetilde{L}(F)| = |L(\pi_\psi(F))| \le \|L\| \|\pi_\psi(F)\|_{FH^{s,\tau}_{p,q}(\mathbb{R}^n)} \lesssim \|L\| \|F\|_{FT^{s,\tau}_{p,q}(\mathbb{R}^{n+1}_{\mathbb{Z}_+})}.$$

Thus, $\widetilde{L}$ becomes a bounded linear functional on $FT_{p,q}^{s,\tau}(\mathbb{R}_{\mathbb{Z}_+}^{n+1})$ with $\|\widetilde{L}\| \lesssim \|L\|$. By Theorem 7.1, there exists a function $G \in FW_{p',q'}^{-s,\tau}(\mathbb{R}_{\mathbb{Z}_+}^{n+1})$ with

$$\|G\|_{FW_{p',q'}^{-s,\tau}(\mathbb{R}_{\mathbb{Z}_+}^{n+1})} \lesssim \|L\|$$

such that for all $F \in FT_{p,q}^{s,\tau}(\mathbb{R}_{\mathbb{Z}_+}^{n+1})$,

$$\widetilde{L}(F) = \int_{\mathbb{R}^n} \sum_{k=0}^{\infty} F^k(x) G^k(x)\,dx.$$

Notice that by Lemma 2.3, the composite operator $\pi_\psi \rho_\varphi$ is the identity on $\mathscr{S}(\mathbb{R}^n)$. For all $f \in \mathscr{S}(\mathbb{R}^n)$, we have

$$\begin{aligned} L(f) &= L(\pi_\psi(\rho_\varphi(f))) \\ &= \widetilde{L}(\rho_\varphi(f)) \\ &= \int_{\mathbb{R}^n} \sum_{k=0}^{\infty} \varphi_k * f(y) G^k(y)\,dy \\ &= \int_{\mathbb{R}^n} f(x) \sum_{k=0}^{\infty} (\widetilde{\varphi}_k * G^k)(x)\,dx \\ &= \int_{\mathbb{R}^n} f(x) g(x)\,dx, \end{aligned}$$

where $g \equiv \pi_{\widetilde{\varphi}}(G)$. By Lemma 7.7, $g \in F_{p',q'}^{-s,\tau}(\mathbb{R}^n)$ and

$$\|g\|_{F_{p',q'}^{-s,\tau}(\mathbb{R}^n)} \lesssim \|G\|_{FW_{p',q'}^{-s,\tau}(\mathbb{R}_{\mathbb{Z}_+}^{n+1})} \lesssim \|L\|,$$

which completes the proof of Theorem 7.3. □

Remark 7.4. When $\tau = 0$, then Theorem 7.3 is the classical dual result of inhomogeneous Besov spaces and Triebel-Lizorkin spaces; see [145, Theorem 2.11.2].

It turns out that the spaces $AH_{p,q}^{s,\tau}(\mathbb{R}^n)$ satisfy many properties similar to those of $A_{p,q}^{s,\tau}(\mathbb{R}^n)$. We begin with their φ-transform characterizations. We define the corresponding sequence spaces to $AH_{p,q}^{s,\tau}(\mathbb{R}^n)$ as follows. Recall that $\mathscr{Q}$ denotes the set of all dyadic cubes in $\mathbb{R}^n$ and

$$\mathscr{Q}_j \equiv \left\{Q \in \mathscr{Q} : \, l(Q) = 2^{-j}\right\}$$

for $j \in \mathbb{Z}$.

Definition 7.6. Let $p \in (1,\infty)$ and $s \in \mathbb{R}$.

(i) If $q \in [1,\infty)$ and $\tau \in [0, \frac{1}{(p\vee q)'}]$, the *sequence space* $bH_{p,q}^{s,\tau}(\mathbb{R}^n)$ is then defined to be the set of all $t \equiv \{t_Q\}_{l(Q)\le 1} \subset \mathbb{C}$ such that

$$\|t\|_{bH^{s,\tau}_{p,q}(\mathbb{R}^n)} \equiv \inf_{\omega} \left\{ \sum_{j=0}^{\infty} 2^{jsq} \left\| \sum_{Q\in\mathscr{Q}_j} |t_Q|\widetilde{\chi}_Q[\omega(\cdot,2^{-j})]^{-1} \right\|_{L^p(\mathbb{R}^n)}^q \right\}^{\frac{1}{q}} < \infty,$$

where the infimum is taken over all nonnegative Borel measurable functions ω on $\mathbb{R}^{n+1}_+$ such that ω satisfies (7.3) and with the restriction that for any $j \in \mathbb{Z}_+$, $\omega(\cdot, 2^{-j})$ is allowed to vanish only where $\sum_{Q\in\mathscr{Q}_j} |t_Q|\widetilde{\chi}_Q$ vanishes.

(ii) If $q \in (1,\infty)$ and $\tau \in [0, \frac{1}{(p\vee q)'}]$, the *sequence space* $fH^{s,\tau}_{p,q}(\mathbb{R}^n)$ is then defined to be the set of all $t \equiv \{t_Q\}_{l(Q)\le 1} \subset \mathbb{C}$ such that

$$\|t\|_{fH^{s,\tau}_{p,q}(\mathbb{R}^n)} \equiv \inf_{\omega} \left\| \left\{ \sum_{j=0}^{\infty} 2^{jsq} \left(\sum_{Q\in\mathscr{Q}_j} |t_Q|\widetilde{\chi}_Q[\omega(\cdot,2^{-j})]^{-1} \right)^q \right\}^{\frac{1}{q}} \right\|_{L^p(\mathbb{R}^n)} < \infty,$$

where the infimum is taken over all nonnegative Borel measurable functions ω on $\mathbb{R}^{n+1}_+$ with the same restrictions as in (i).

We also use $aH^{s,\tau}_{p,q}(\mathbb{R}^n)$ to denote either $bH^{s,\tau}_{p,q}(\mathbb{R}^n)$ or $fH^{s,\tau}_{p,q}(\mathbb{R}^n)$. Similarly to the proof of (7.4), we see that $\|\cdot\|_{aH^{s,\tau}_{p,q}(\mathbb{R}^n)}$ is a quasi-norm, namely, there exists a nonnegative constant $\rho \in [0,1]$ such that for all $t_1, t_2 \in aH^{s,\tau}_{p,q}(\mathbb{R}^n)$,

$$\|t_1+t_2\|_{aH^{s,\tau}_{p,q}(\mathbb{R}^n)} \le 2^{\rho}\left(\|t_1\|_{aH^{s,\tau}_{p,q}(\mathbb{R}^n)} + \|t_2\|_{aH^{s,\tau}_{p,q}(\mathbb{R}^n)}\right). \tag{7.14}$$

The homogeneous counterpart of $aH^{s,\tau}_{p,q}(\mathbb{R}^n)$, denoted by $a\dot{H}^{s,\tau}_{p,q}(\mathbb{R}^n)$, was already introduced in [168]. Similarly to $a^{s,\tau}_{p,q}(\mathbb{R}^n)$, we define

$$V:\ aH^{s,\tau}_{p,q}(\mathbb{R}^n) \to a\dot{H}^{s,\tau}_{p,q}(\mathbb{R}^n)$$

by setting $(Vt)_Q \equiv t_Q$ if $l(Q) \le 1$, and $(Vt)_Q \equiv 0$ otherwise. Then V is an isometric embedding of $aH^{s,\tau}_{p,q}(\mathbb{R}^n)$ in $a\dot{H}^{s,\tau}_{p,q}(\mathbb{R}^n)$. Define

$$W: a\dot{H}^{s,\tau}_{p,q}(\mathbb{R}^n) \to aH^{s,\tau}_{p,q}(\mathbb{R}^n)$$

by setting $(Wt)_Q \equiv t_Q$ if $l(Q) \le 1$. Then W is continuous and $W \circ V$ is the identity on $aH^{s,\tau}_{p,q}(\mathbb{R}^n)$.

Next we show that the inverse φ-transform is well defined on $aH^{s,\tau}_{p,q}(\mathbb{R}^n)$.

Lemma 7.11. *Let $p \in (1,\infty)$, $q \in [1,\infty)$, $s \in \mathbb{R}$ and $\tau \in [0, \frac{1}{(p\vee q)'}]$. Then for all $t \in aH^{s,\tau}_{p,q}(\mathbb{R}^n)$,*

$$T_\psi t \equiv \sum_{l(Q)\le 1} t_Q \psi_Q$$

converges in $\mathscr{S}'(\mathbb{R}^n)$; moreover, $T_\psi : aH^{s,\tau}_{p,q}(\mathbb{R}^n) \to \mathscr{S}'(\mathbb{R}^n)$ is continuous.

Proof. By similarity, we only consider the space $bH^{s,\tau}_{p,q}(\mathbb{R}^n)$.

Let

$$t \equiv \{t_Q\}_{l(Q)\le 1} \in bH^{s,\tau}_{p,q}(\mathbb{R}^n).$$

We need to show that there exists an $M \in \mathbb{Z}_+$ such that for all $\phi \in \mathscr{S}(\mathbb{R}^n)$,

$$\sum_{l(Q)\le 1} |t_Q||\langle \psi_Q, \phi\rangle| \lesssim \|\phi\|_{\mathscr{S}_M}.$$

Choose a Borel function ω on $\mathbb{R}^{n+1}_+$ satisfying (7.3) as well as

$$\left\{\sum_{j=0}^{\infty} 2^{jsq} \left\| \sum_{Q\in\mathscr{Q}_j} |t_Q|\widetilde{\chi}_Q[\omega(\cdot,2^{-j})]^{-1} \right\|^q_{L^p(\mathbb{R}^n)}\right\}^{\frac{1}{q}} \le 2\|t\|_{bH^{s,\tau}_{p,q}(\mathbb{R}^n)}. \tag{7.15}$$

By Remark 7.1(iii), for all $(x,s) \in \mathbb{R}^{n+1}_+$, $\omega(x,s) \lesssim s^{-n\tau}$. Then for all $Q \in \mathscr{Q}_j$, by Hölder's inequality and (7.15), we have

$$|t_Q| \le |Q|^{-\tau-\frac{1}{p}}|t_Q|\left(\int_Q [\omega(x,2^{-j})]^{-p}\,dx\right)^{\frac{1}{p}} \lesssim |Q|^{\frac{s}{n}+\frac{1}{2}-\tau-\frac{1}{p}}\|t\|_{bH^{s,\tau}_{p,q}(\mathbb{R}^n)}. \tag{7.16}$$

Recall that as a special case of [24, Lemma 2.11], there exists a positive constant L_0 such that for all $j \in \mathbb{Z}_+$,

$$\sum_{Q\in\mathscr{Q}_j} (1+|x_Q|^n)^{-L_0} \lesssim 2^{nj}. \tag{7.17}$$

Furthermore, since $\phi \in \mathscr{S}(\mathbb{R}^n)$ and ψ satisfies either (2.1) or (2.2), a standard computation gives us that if

$$L > \max\{1/p+1/2-s/n-\tau, 1/p+3/2+s/n+\tau, L_0\},$$

then there exists an $M \in \mathbb{Z}_+$ such that for all $Q \in \mathscr{Q}_j$,

$$|\langle \psi_Q, \phi\rangle| \lesssim \|\phi\|_{\mathscr{S}_M}(1+|x_Q|^n)^{-L}2^{-jnL}; \tag{7.18}$$

see also [165, p. 10] and [24, (3.18)]. Using (7.16), (7.18) and (7.17), we conclude that

$$\begin{aligned}\sum_{l(Q)\le 1} |t_Q||\langle \psi_Q, \phi\rangle| &\lesssim \|t\|_{bH^{s,\tau}_{p,q}(\mathbb{R}^n)}\|\phi\|_{\mathscr{S}_M}\\ &\quad\times \sum_{l(Q)\le 1} |Q|^{\frac{s}{n}+\frac{1}{2}-\tau-\frac{1}{p}}(1+|x_Q|^n)^{-L}2^{-nLj_Q}\\ &\lesssim \|t\|_{bH^{s,\tau}_{p,q}(\mathbb{R}^n)}\|\phi\|_{\mathscr{S}_M},\end{aligned}$$

which completes the proof of Lemma 7.11. □

Now we present our main result of this section, which also implies that the spaces $AH^{s,\tau}_{p,q}(\mathbb{R}^n)$ are independent of the choices of φ and Φ.

Theorem 7.4. *Let $p \in (1,\infty)$, $q \in [1,\infty)$, $s \in \mathbb{R}$, $\tau \in [0, \frac{1}{(p\vee q)'}]$, φ and ψ be as in Definition 2.1. Then*

$$S_\varphi : AH^{s,\tau}_{p,q}(\mathbb{R}^n) \to aH^{s,\tau}_{p,q}(\mathbb{R}^n)$$

and

$$T_\psi : aH^{s,\tau}_{p,q}(\mathbb{R}^n) \to AH^{s,\tau}_{p,q}(\mathbb{R}^n)$$

are bounded; moreover, $T_\psi \circ S_\varphi$ is the identity on $AH^{s,\tau}_{p,q}(\mathbb{R}^n)$.

Theorem 7.4 is the φ-transform characterization of $AH^{s,\tau}_{p,q}(\mathbb{R}^n)$. We remark that the homogeneous counterpart of Theorem 7.4 was already obtained in [168, Theorem 2.1]. To prove Theorem 7.4, we need to establish the counterparts of Lemmas 2.6, 2.8 and 2.9 on $AH^{s,\tau}_{p,q}(\mathbb{R}^n)$.

Let γ be a fixed integer. Replacing φ_j by $\varphi_{j-\gamma}$ (φ_0 by $\Phi_{-\gamma}$) in Definition 7.5, we obtain a quasi-norm in $AH^{s,\tau}_{p,q}(\mathbb{R}^n)$, which is denoted by $\|f\|^*_{AH^{s,\tau}_{p,q}(\mathbb{R}^n)}$. As a counterpart of Lemma 2.6, we have the following conclusion. Its proof is similar to those of Lemma 2.6 and [64, Lemma 12.1].

Lemma 7.12. *Let s, p, q, τ be as in Theorem 7.4. The quasi-norms $\|f\|^*_{AH^{s,\tau}_{p,q}(\mathbb{R}^n)}$ and $\|f\|_{AH^{s,\tau}_{p,q}(\mathbb{R}^n)}$ are equivalent.*

Proof. By similarity, we only consider $FH^{s,\tau}_{p,q}(\mathbb{R}^n)$ and the case when $\gamma > 0$. First we prove

$$\|f\|_{FH^{s,\tau}_{p,q}(\mathbb{R}^n)} \lesssim \|f\|^*_{FH^{s,\tau}_{p,q}(\mathbb{R}^n)}.$$

It suffices to prove that

$$\mathrm{I} \equiv \inf_\omega \left\{ \int_{\mathbb{R}^n} |\Phi * f(x)|^p [\omega(x,1)]^{-p}\, dx \right\}^{1/p} \lesssim \|f\|^*_{FH^{s,\tau}_{p,q}(\mathbb{R}^n)}.$$

To this end, let ω satisfy (7.3) and

$$\left\{ \int_{\mathbb{R}^n} \left[\sum_{j=0}^{\infty} 2^{jsq} |\varphi_{j-\gamma} * f(x)|^q [\omega(x, 2^{-j})]^{-q} \right]^{q/p} dx \right\}^{1/p} \lesssim \|f\|^*_{FH^{s,\tau}_{p,q}(\mathbb{R}^n)}.$$

For each $m \in \mathbb{N}$ and $i \in \{-\gamma, -\gamma+1, \cdots, 1\}$, define

$$\omega_{m,i}(x,t) \equiv 2^{-mn(\lfloor (p\vee q)' \rfloor + 2)} \sup\{\omega(y,s) : |x-y| \le 2^m,\ s = 2^i t\}.$$

Similarly to the proof of Proposition 7.2, we see that there exists a positive constant $C(\gamma)$ such that $C(\gamma)\,\omega_{m,i}$ satisfies (7.3). Moreover,

$$[\omega_{m,i}(x,1)]^{-1} \omega(x-y, 2^{-i}) \lesssim 1$$

for all $|y| \le 2^m$.

Similarly to the proof of Lemma 2.6, we choose $\eta_i \in \mathscr{S}(\mathbb{R}^n)$, $i=-\gamma,\cdots,1$, such that

$$\Phi * f \equiv \eta_{-\gamma} * \Phi_{-\gamma} * f + \sum_{i=-\gamma+1}^{1} \eta_i * \varphi_i * f.$$

On the other hand, let χ_0 be the characteristic function on unit ball $B(0,1)$ and χ_0 be the characteristic function on $B(0,2^m)\setminus B(0,2^{m-1})$ for all $m\in\mathbb{N}$. Then by Minkowski's inequality and the fact that $\eta_i \in \mathscr{S}(\mathbb{R}^n)$, choosing

$$M > \rho + n(\lfloor (p\vee q)' \rfloor + 2),$$

we have

$$\begin{aligned}
\mathrm{I} &\lesssim \sum_{i=-\gamma}^{1} \inf_{\omega} \left\{ \int_{\mathbb{R}^n} |\eta_i * \varphi_i * f(x)|^p [\omega(x,1)]^{-p}\,dx \right\}^{1/p} \\
&\lesssim \sum_{i=-\gamma}^{1} \sum_{m=0}^{\infty} 2^{m\rho} \left\{ \int_{\mathbb{R}^n} |(\eta_i \chi_m) * \varphi_i * f(x)|^p [\omega_{m,i}(x,1)]^{-p}\,dx \right\}^{1/p} \\
&\lesssim \sum_{i=-\gamma}^{1} \sum_{m=0}^{\infty} 2^{m\rho} 2^{mn(\lfloor (p\vee q)' \rfloor + 2)} \\
&\quad \times \int_{\mathbb{R}^n} |\eta_i(y)| \chi_m(y) \left\{ \int_{\mathbb{R}^n} |\varphi_i * f(x-y)|^p [\omega(x-y,2^{-i})]^{-p}\,dx \right\}^{1/p} dy \\
&\lesssim \|f\|^*_{FH^{s,\tau}_{p,q}(\mathbb{R}^n)} \sum_{i=-\gamma}^{1} \sum_{m=0}^{\infty} 2^{m\rho} 2^{mn(\lfloor (p\vee q)' \rfloor + 2)} \int_{\mathbb{R}^n} \frac{\chi_m(y)}{(1+|y|)^{n+M}}\,dy \\
&\lesssim \|f\|^*_{FH^{s,\tau}_{p,q}(\mathbb{R}^n)},
\end{aligned}$$

which implies that

$$\|f\|_{FH^{s,\tau}_{p,q}(\mathbb{R}^n)} \lesssim \|f\|^*_{FH^{s,\tau}_{p,q}(\mathbb{R}^n)}.$$

The converse inequality follows from a similar argument. This finishes the proof of Lemma 7.12. □

Lemma 7.13. *Let s, p, q, τ be as in Theorem 7.4 and $\lambda \in (n,\infty)$ be sufficiently large. Then there exists a positive constant C such that for all $t \in aH^{s,\tau}_{p,q}(\mathbb{R}^n)$,*

$$\|t\|_{aH^{s,\tau}_{p,q}(\mathbb{R}^n)} \le \|t^*_{p\wedge q,\lambda}\|_{aH^{s,\tau}_{p,q}(\mathbb{R}^n)} \le C\|t\|_{aH^{s,\tau}_{p,q}(\mathbb{R}^n)},$$

where $t^*_{p\wedge q,\lambda}$ is as in (2.17).

Proof. The inequality

$$\|t\|_{aH^{s,\tau}_{p,q}(\mathbb{R}^n)} \le \|t^*_{p\wedge q,\lambda}\|_{aH^{s,\tau}_{p,q}(\mathbb{R}^n)}$$

being trivial, we only need to concentrate on

$$\|t^*_{p\wedge q,\lambda}\|_{aH^{s,\tau}_{p,q}(\mathbb{R}^n)} \lesssim \|t\|_{aH^{s,\tau}_{p,q}(\mathbb{R}^n)}.$$

Also, by similarity, we only consider the space $bH^{s,\tau}_{p,q}(\mathbb{R}^n)$.

Let

$$t \equiv \{t_Q\}_{Q\in\mathscr{Q}} \in bH^{s,\tau}_{p,q}(\mathbb{R}^n).$$

We choose a Borel function ω as in the proof of Lemma 7.11. For all cubes $Q \in \mathscr{Q}_j$ and $m \in \mathbb{N}$, we set

$$A_0(Q) \equiv \{P \in \mathscr{Q}_j : 2^j|x_P - x_Q| \le 1\}$$

and

$$A_m(Q) \equiv \{P \in \mathscr{Q}_j : 2^{m-1} < 2^j|x_P - x_Q| \le 2^m\}.$$

The triangle inequality that

$$|x-y| \le |x-x_Q| + |x_Q - x_P| + |x_P - y|$$

gives us that

$$|x-y| \le 3\sqrt{n}2^{m-j}$$

provided $x \in Q$, $y \in P$ and $P \in A_m(Q)$.

For all $m \in \mathbb{Z}_+$ and $(x,s) \in \mathbb{R}^{n+1}_+$, we set

$$\omega_m(x,s) \equiv 2^{-mn(\lfloor (p\vee q)'\rfloor+2)} \sup\{\omega(y,s) : y \in \mathbb{R}^n,\ |y-x| < \sqrt{n}2^{m+2}s\}.$$

By the argument in the proof of Lemma 7.1, we know that ω_m still satisfies (7.3) modulo multiplicative constants independent of m. Also it follows from the definition of ω_m that for all $x \in Q$ with $Q \in \mathscr{Q}_j$ and $y \in P$ with $P \in A_m(Q)$,

$$\omega(y,2^{-j}) \lesssim 2^{mn(\lfloor (p\vee q)'\rfloor+2)}\omega_m(x,2^{-j}).$$

Then for all $r \in (0,\infty)$ and $a \in (0,r)$, using this estimate and the monotonicity of $\ell^{a/r}$, we obtain that for all $x \in Q$,

$$\begin{aligned}
&\sum_{P\in A_m(Q)} \frac{|t_P|^r}{(1+2^j|x_Q-x_P|)^\lambda}[\omega_m(x,2^{-j})]^{-r}\\
&\lesssim 2^{-m\lambda}\left\{\sum_{P\in A_m(Q)} |t_P|^a[\omega_m(x,2^{-j})]^{-a}\right\}^{r/a}\\
&\lesssim 2^{-m\lambda+jnr/a}\left\{\int_{\mathbb{R}^n}\sum_{P\in A_m(Q)} |t_P|^a\chi_P(y)[\omega_m(x,2^{-j})]^{-a}\,dy\right\}^{r/a}
\end{aligned}$$

$$\lesssim 2^{-m\lambda+jnr/a+mnr(\lfloor(p\vee q)'\rfloor+2)}\left\{\int_{\mathbb{R}^n}\sum_{P\in A_m(Q)}|t_P|^a\chi_P(y)[\omega(y,2^{-j})]^{-a}\,dy\right\}^{r/a}$$

$$\lesssim 2^{-m\lambda+mnr/a+mnr(\lfloor(p\vee q)'\rfloor+2)}\left\{\mathrm{M}\left(\sum_{P\in A_m(Q)}|t_P|^a\chi_P[\omega(\cdot,2^{-j})]^{-a}\right)(x)\right\}^{r/a}.$$

Recall that M denotes the Hardy-Littlewood maximal operator on $\mathbb{R}^n$.

For all $m\in\mathbb{Z}_+$, set $t^{*,m}_{r,\lambda}\equiv\{(t^{*,m}_{r,\lambda})_Q\}_{Q\in\mathcal{Q}}$ with

$$(t^{*,m}_{r,\lambda})_Q\equiv\left(\sum_{P\in A_m(Q)}\frac{|t_P|^r}{(1+l(P)^{-1}|x_P-x_Q|)^\lambda}\right)^{\frac{1}{r}}.$$

In what follows, choose $a\in(0,p\wedge q)$ and

$$\lambda>n(p\wedge q)/a+n(p\wedge q)\{\lfloor(p\vee q)'\rfloor+2\}+(p\wedge q)\rho,$$

where ρ is a nonnegative constant as in (7.14). By (7.14), the previous pointwise estimate and the $L^{\frac{p}{a}}(\mathbb{R}^n)$-boundedness of M, we obtain

$$\begin{aligned}
&\|t^*_{p\wedge q,\lambda}\|_{bH^{s,\tau}_{p,q}(\mathbb{R}^n)}\\
&\lesssim\sum_{m=0}^\infty 2^{\rho m}\|t^{*,m}_{p\wedge q,\lambda}\|_{bH^{s,\tau}_{p,q}(\mathbb{R}^n)}\\
&\lesssim\sum_{m=0}^\infty 2^{\rho m}\left\{\sum_{j=0}^\infty 2^{jsq}\left[\int_{\mathbb{R}^n}\sum_{Q\in\mathcal{Q}_j}\left(\sum_{P\in A_m(Q)}\frac{|t_P|^{p\wedge q}}{(1+l(P)^{-1}|x_P-x_Q|)^\lambda}\right)^{\frac{p}{p\wedge q}}\right.\right.\\
&\qquad\qquad\left.\left.\times\frac{\widetilde{\chi}_Q(x)^p}{[\omega_m(x,2^{-j})]^p}\,dx\right]^{\frac{q}{p}}\right\}^{\frac{1}{q}}\\
&\lesssim\sum_{m=0}^\infty 2^{-\frac{m}{p\wedge q}\{-\lambda+n(p\wedge q)/a+n(p\wedge q)\{\lfloor(p\vee q)'\rfloor+2\}+(p\wedge q)\rho\}}\\
&\qquad\times\left[\sum_{j=0}^\infty 2^{jsq}\left\{\int_{\mathbb{R}^n}\left[\mathrm{M}\left(\sum_{P\in\mathcal{Q}_j}\frac{(|t_P|\widetilde{\chi}_P)^a}{[\omega(\cdot,2^{-j})]^a}\right)(x)\right]^{\frac{p}{a}}dx\right\}^{\frac{q}{p}}\right]^{\frac{1}{q}}\\
&\lesssim\|t\|_{bH^{s,\tau}_{p,q}(\mathbb{R}^n)},
\end{aligned}$$

which completes the proof of Lemma 7.13. □

With Lemma 2.6 replaced by Lemma 7.12, similarly to the proof of Lemma 2.9, we have the following conclusion.

Lemma 7.14. *Let s, p, q, τ be as in Theorem 7.4 and $\gamma \in \mathbb{Z}_+$ be sufficiently large. Then there exists a constant $C \in [1, \infty)$ such that for all $f \in AH^{s,\tau}_{p,q}(\mathbb{R}^n)$,*

$$C^{-1}\|\inf_\gamma(f)\|_{aH^{s,\tau}_{p,q}(\mathbb{R}^n)} \le \|f\|_{AH^{s,\tau}_{p,q}(\mathbb{R}^n)} \le \|\sup(f)\|_{aH^{s,\tau}_{p,q}(\mathbb{R}^n)} \le C\|\inf_\gamma(f)\|_{aH^{s,\tau}_{p,q}(\mathbb{R}^n)}.$$

With Lemmas 7.12, 7.13 and 7.14, the proof of Theorem 7.4 is similar to that of Theorem 2.1. We omit the details.

Applying Theorem 7.4, we have the following Sobolev-type embedding properties of $AH^{s,\tau}_{p,q}(\mathbb{R}^n)$, whose homogeneous counterparts were already established in [168, Proposition 2.2].

Proposition 7.3. *Let $1 < p_0 < p_1 < \infty$ and $-\infty < s_1 < s_0 < \infty$. Assume in addition that $s_0 - n/p_0 = s_1 - n/p_1$.*

(i) If $q \in [1, \infty)$ and

$$\tau \in \left[0, \min\left\{\frac{1}{(p_0 \vee q)'}, \frac{1}{(p_1 \vee q)'}\right\}\right]$$

such that $\tau(p_0 \vee q)' = \tau(p_1 \vee q)'$, then

$$BH^{s_0,\tau}_{p_0,q}(\mathbb{R}^n) \hookrightarrow BH^{s_1,\tau}_{p_1,q}(\mathbb{R}^n).$$

(ii) If $q, r \in (1, \infty)$ and

$$\tau \in \left[0, \min\left\{\frac{1}{(p_0 \vee r)'}, \frac{1}{(p_1 \vee q)'}\right\}\right]$$

such that $\tau(p_0 \vee r)' \le \tau(p_1 \vee q)'$, then

$$FH^{s_0,\tau}_{p_0,r}(\mathbb{R}^n) \hookrightarrow FH^{s_1,\tau}_{p_1,q}(\mathbb{R}^n).$$

Proof. By Theorem 7.4 and similarity, it suffices to consider the corresponding sequence spaces $fH^{s,\tau}_{p,q}(\mathbb{R}^n)$, that is, to prove that

$$\|t\|_{fH^{s_1,\tau}_{p_1,q}(\mathbb{R}^n)} \lesssim \|t\|_{fH^{s_0,\tau}_{p_0,r}(\mathbb{R}^n)}$$

for all $t \in fH^{s_0,\tau}_{p_0,r}(\mathbb{R}^n)$.

It was already proved in [168, Proposition 2.2] that

$$\|t\|_{f\dot{H}^{s_1,\tau}_{p_1,q}(\mathbb{R}^n)} \lesssim \|t\|_{f\dot{H}^{s_0,\tau}_{p_0,r}(\mathbb{R}^n)}$$

for all $t \in f\dot{H}^{s_0,\tau}_{p_0,r}(\mathbb{R}^n)$. Recall that V is an isometric embedding of $aH^{s,\tau}_{p,q}(\mathbb{R}^n)$ in $a\dot{H}^{s,\tau}_{p,q}(\mathbb{R}^n)$. Then for all $t \in fH^{s_0,\tau}_{p_0,r}(\mathbb{R}^n)$, we have

$$\|t\|_{fH^{s_1,\tau}_{p_1,q}(\mathbb{R}^n)} \sim \|Vt\|_{f\dot{H}^{s_1,\tau}_{p_1,q}(\mathbb{R}^n)} \lesssim \|Vt\|_{f\dot{H}^{s_0,\tau}_{p_0,r}(\mathbb{R}^n)} \sim \|t\|_{fH^{s_0,\tau}_{p_0,r}(\mathbb{R}^n)},$$

which completes the proof of Proposition 7.3. □

Remark 7.5. When $\tau = 0$, Proposition 7.3 recovers the corresponding results on Besov spaces and Triebel-Lizorkin spaces in [145, p. 129]. We also remark that the restriction that $\tau(p_0 \vee q)' = \tau(p_1 \vee q)'$ in Proposition 7.3(i) is necessary, and sharp in this sense; see [168, Proposition 2.3]. However, it is still unclear that if the restriction that $\tau(p_0 \vee r)' \le \tau(p_1 \vee q)'$ in Proposition 7.3(ii) is sharp.

Corresponding to Theorem 7.3, we also have the dual result for sequence spaces. The homogeneous counterpart of the following conclusion was obtained in [168, Proposition 2.1].

Proposition 7.4. *Let s, p, q, τ be as in Theorem 7.4. Then*

$$(aH^{s,\tau}_{p,q}(\mathbb{R}^n))^* = a^{-s,\tau}_{p',q'}(\mathbb{R}^n)$$

in the following sense: if $t \equiv \{t_Q\}_{l(Q)\le 1} \in a^{-s,\tau}_{p',q'}(\mathbb{R}^n)$, then the map

$$\lambda = \{\lambda_Q\}_{l(Q)\le 1} \mapsto \langle \lambda, t\rangle \equiv \sum_{l(Q)\le 1} \lambda_Q \overline{t_Q}$$

defines a continuous linear functional on $aH^{s,\tau}_{p,q}(\mathbb{R}^n)$ with operator norm no more than a constant multiple of $\|t\|_{a^{-s,\tau}_{p',q'}(\mathbb{R}^n)}$. Conversely, every $L \in (aH^{s,\tau}_{p,q}(\mathbb{R}^n))^*$ is of this form for a certain $t \in a^{-s,\tau}_{p',q'}(\mathbb{R}^n)$ and $\|t\|_{a^{-s,\tau}_{p',q'}(\mathbb{R}^n)}$ is no more than a constant multiple of the operator norm $\|L\|$.

Proof. We only consider the spaces $bH^{s,\tau}_{p,q}(\mathbb{R}^n)$ because the assertion for $fH^{s,\tau}_{p,q}(\mathbb{R}^n)$ can be proved similarly.

For $t \equiv \{t_Q\}_{l(Q)\le 1} \in b^{-s,\tau}_{p',q'}(\mathbb{R}^n)$ and $\lambda \equiv \{\lambda_Q\}_{l(Q)\le 1} \in bH^{s,\tau}_{p,q}(\mathbb{R}^n)$, let F and G be functions on $\mathbb{R}^{n+1}_{\mathbb{Z}_+}$ defined by setting, for all $x \in \mathbb{R}^n$ and $j \in \mathbb{Z}_+$,

$$F(x, 2^{-j}) \equiv \sum_{Q\in\mathcal{Q}_j} |\lambda_Q| \widetilde{\chi}_Q$$

and

$$G(x, 2^{-j}) \equiv \sum_{P\in\mathcal{Q}_j} |t_P| \widetilde{\chi}_P.$$

Since

$$\|F\|_{BT^{s,\tau}_{p,q}(\mathbb{R}^{n+1}_{\mathbb{Z}_+})} \sim \|\lambda\|_{bH^{s,\tau}_{p,q}(\mathbb{R}^n)}$$

and

$$\|G\|_{BW^{-s,\tau}_{p',q'}(\mathbb{R}^{n+1}_{\mathbb{Z}_+})} \sim \|t\|_{b^{-s,\tau}_{p',q'}(\mathbb{R}^n)},$$

by Theorem 7.1, we have

$$\begin{aligned}\left|\sum_{l(Q)\le 1}\lambda_Q\bar{t}_Q\right| &\le \sum_{j\in\mathbb{Z}_+}\int_{\mathbb{R}^n}\sum_{Q\in\mathcal{Q}_j}\sum_{P\in\mathcal{Q}_j}|\lambda_Q|\widetilde{\chi}_Q(x)|t_P|\widetilde{\chi}_P(x)\,dx\\ &=\sum_{j\in\mathbb{Z}_+}\int_{\mathbb{R}^n}F(x,2^{-j})G(x,2^{-j})\,dx\\ &\lesssim \|F\|_{BT^{s,\tau}_{p,q}(\mathbb{R}^{n+1}_{\mathbb{Z}_+})}\|G\|_{BW^{-s,\tau}_{p',q'}(\mathbb{R}^{n+1}_{\mathbb{Z}_+})}\\ &\sim \|\lambda\|_{bH^{s,\tau}_{p,q}(\mathbb{R}^n)}\|t\|_{b^{-s,\tau}_{p',q'}(\mathbb{R}^n)},\end{aligned}$$

which implies that $b^{-s,\tau}_{p',q'}(\mathbb{R}^n)\hookrightarrow(bH^{s,\tau}_{p,q}(\mathbb{R}^n))^*$.

Conversely, since the subspace consisting of all sequences with finite non-vanishing elements are dense in $bH^{s,\tau}_{p,q}(\mathbb{R}^n)$, we know that every $L\in(bH^{s,\tau}_{p,q}(\mathbb{R}^n))^*$ is of the form $\lambda\mapsto\sum_{l(Q)\le 1}\lambda_Q\bar{t}_Q$ for a certain $t\equiv\{t_Q\}_{l(Q)\le 1}\subset\mathbb{C}$. It remains to show that

$$\|t\|_{b^{-s,\tau}_{p',q'}(\mathbb{R}^n)}\lesssim\|L\|_{(bH^{s,\tau}_{p,q}(\mathbb{R}^n))^*}.$$

Fix $P\in\mathcal{Q}$ and $a\in\mathbb{R}$. For $j\ge(j_P\vee 0)$, let X_j be the set of all $Q\in\mathcal{Q}_j$ satisfying $Q\subset P$ and let μ be a measure on X_j such that the μ-measure of the "point" Q is $|Q|/|P|^{\tau a}$. Also, let ℓ^q_P denote the set of all $\{a_j\}_{j\ge(j_P\vee 0)}\subset\mathbb{C}$ with

$$\|\{a_j\}_{j\ge(j_P\vee 0)}\|_{\ell^q_P}\equiv\left(\sum_{j=(j_P\vee 0)}^{\infty}|a_j|^q\right)^{1/q}$$

and $\ell^q_P(\ell^p(X_j,d\mu))$ denote the set of all $\{a_{Q,j}\}_{Q\in\mathcal{Q}_j,Q\subset P,j\ge(j_P\vee 0)}\subset\mathbb{C}$ with

$$\begin{aligned}&\|\{a_{Q,j}\}_{Q\in\mathcal{Q}_j(\mathbb{R}^n),Q\subset P,j\ge(j_P\vee 0)}\|_{\ell^q_P(\ell^p(X_j,d\mu))}\\ &\equiv\left(\sum_{j=(j_P\vee 0)}^{\infty}\left[\sum_{Q\in\mathcal{Q}_j,Q\subset P}|a_{Q,j}|^p|Q||P|^{-\tau a}\right]^{\frac{q}{p}}\right)^{1/q}.\end{aligned}$$

It is well known that the dual space of $\ell^q_P(\ell^p(X_j,d\mu))$ is $\ell^{q'}_P(\ell^{p'}(X_j,d\mu))$; see, for example, [145, p. 177]. Via this observation and the already proved conclusion of this proposition, we see that

$$\begin{aligned}&\frac{1}{|P|^{\tau}}\left\{\sum_{j=(j_P\vee 0)}^{\infty}\left[\sum_{Q\in\mathcal{Q}_j,Q\subset P}\left(|Q|^{-\frac{s}{n}-\frac12}|t_Q|\right)^{p'}|Q|\right]^{\frac{q'}{p'}}\right\}^{\frac{1}{q'}}\\ &=\|\{|Q|^{-\frac{s}{n}-\frac12}|t_Q|\}_{Q\in\mathcal{Q}_j,Q\subset P,j\ge(j_P\vee 0)}\|_{\ell^{q'}_P(\ell^{p'}(X_j,d\mu))}\end{aligned}$$

$$= \sup_{\|\{\lambda_Q\}_{Q\in\mathcal{Q}_j, Q\subset P, j\ge(j_P\vee 0)}\|_{\ell^q_p(\ell^p(X_j,d\mu))}\le 1} \left| \sum_{j=(j_P\vee 0)}^{\infty} \sum_{Q\in\mathcal{Q}_j, Q\subset P} \lambda_Q |Q|^{-\frac{s}{n}-\frac{1}{2}} |t_Q| \frac{|Q|}{|P|^{\tau p'}} \right|$$

$$\le \sup_{\|\{\lambda_Q\}_{Q\in\mathcal{Q}_j, Q\subset P, j\ge(j_P\vee 0)}\|_{\ell^q_p(\ell^p(X_j,d\mu))}\le 1} \|L\|_{(bH^{s,\tau}_{p,q}(\mathbb{R}^n))^*}$$

$$\times \left\| \{\lambda_Q |Q|^{-\frac{s}{n}-\frac{1}{2}} |Q|/|P|^{\tau p'}\}_{Q\in\mathcal{Q}_j, Q\subset P, j\ge(j_P\vee 0)} \right\|_{bH^{s,\tau}_{p,q}(\mathbb{R}^n)}.$$

To finish the proof of this proposition, it suffices to show that

$$\left\| \{\lambda_Q |Q|^{-\frac{s}{n}-\frac{1}{2}} |Q|/|P|^{\tau p'}\}_{Q\in\mathcal{Q}_j, Q\subset P, j\ge(j_P\vee 0)} \right\|_{bH^{s,\tau}_{p,q}(\mathbb{R}^n)} \lesssim 1$$

for all sequences λ satisfying

$$\|\{\lambda_Q\}_{Q\in\mathcal{Q}_j, Q\subset P, j\ge(j_P\vee 0)}\|_{\ell^q_p(\ell^p(X_j,d\mu))} \le 1.$$

In fact, let $B \equiv B(c_P, \sqrt{n}l(P)) \subset \mathbb{R}^n$ and ω be as in the proof of Lemma 7.7 associated with B, then ω satisfies (7.3) and for all $x \in P$ and $j \ge (j_P \vee 0)$,

$$[\omega(x, 2^{-j})]^{-1} \sim [l(P)]^{n\tau}.$$

We then obtain that

$$\left\| \{\lambda_Q |Q|^{-\frac{s}{n}-\frac{1}{2}} |Q|/|P|^{\tau p'}\}_{Q\in\mathcal{Q}_j, Q\subset P, j\ge(j_P\vee 0)} \right\|_{bH^{s,\tau}_{p,q}(\mathbb{R}^n)}$$

$$\lesssim \left\{ \sum_{j=(j_P\vee 0)}^{\infty} 2^{jsq} \left[\sum_{Q\in\mathcal{Q}_j, Q\subset P} \left(|\lambda_Q| |Q|^{-\frac{s}{n}-1} \frac{|Q|}{|P|^{\tau p'}} \right)^p \int_Q [\omega(x, 2^{-j})]^{-p} \, dx \right]^{\frac{q}{p}} \right\}^{\frac{1}{q}}$$

$$\sim \left\{ \sum_{j=(j_P\vee 0)}^{\infty} \left[\sum_{Q\in\mathcal{Q}_j(\mathbb{R}^n), Q\subset P} |\lambda_Q|^p |Q|/|P|^{\tau p'} \right]^{\frac{q}{p}} \right\}^{\frac{1}{q}}$$

$$\sim \|\{\lambda_Q\}_{Q\in\mathcal{Q}_j, Q\subset P, j\ge(j_P\vee 0)}\|_{\ell^q_p(\ell^p(X_j,d\mu))}$$

$$\lesssim 1,$$

which completes the proof of Proposition 7.4. □

Remark 7.6. By Proposition 7.4 and the φ-transform characterizations of $A^{s,\tau}_{p,q}(\mathbb{R}^n)$ and $AH^{s,\tau}_{p,q}(\mathbb{R}^n)$, we also obtain the duality that $(AH^{s,\tau}_{p,q}(\mathbb{R}^n))^* = A^{-s,\tau}_{p,q}(\mathbb{R}^n)$. This gives another proof of this conclusion, which is different from that of Theorem 7.3.

Next we establish the smooth atomic and molecular decomposition characterizations of $AH^{s,\tau}_{p,q}(\mathbb{R}^n)$. We first obtain the boundedness on $aH^{s,\tau}_{p,q}(\mathbb{R}^n)$ of almost diagonal operators in Definition 3.1.

Theorem 7.5. *Let* $p \in (1,\infty)$, $q \in [1,\infty)$, $s \in \mathbb{R}$, $\varepsilon \in (0,\infty)$ *and* $\tau \in [0, \frac{1}{(p\vee q)'}]$. *Then all the* ε*-almost diagonal operators are bounded on* $aH^{s,\tau}_{p,q}(\mathbb{R}^n)$ *if* $\tau > 2n\tau$.

To prove this theorem, we need some technical lemmas established in [168]. For the reader's convenient, we give their proofs.

Lemma 7.15. *Let* $d \in (0,n]$ *and* Ω *be an open set in* $\mathbb{R}^n$ *such that* $\Omega = \cup_{j=1}^{\infty} B_j$, *where* $\{B_j\}_{j=1}^{\infty} \equiv \{B(X_j,R_j)\}_{j=1}^{\infty}$ *is a countable collection of balls. Define*

$$\Lambda_d^{(\infty)}(\Omega,\{B_j\}_{j=1}^{\infty})$$
$$\equiv \inf\left\{\sum_{k=1}^{\infty} r_k^d : \Omega \subset \bigcup_{k=1}^{\infty} B(x_k,r_k),\, B(x_k,r_k) \supset B_j \text{ if } B_j \cap B(x_k,r_k) \neq \emptyset\right\}.$$

Then there exists a positive constant C, *independent of* Ω, $\{B_j\}_{j=1}^{\infty}$ *and* d, *such that*

$$\Lambda_d^{(\infty)}(\Omega) \le \Lambda_d^{(\infty)}(\Omega,\{B_j\}_{j=1}^{\infty}) \le C(46)^d \Lambda_d^{(\infty)}(\Omega).$$

Proof. The first inequality is trivial. We only need to prove the second one. Without loss of generality, we may assume $\sup_{j\in\mathbb{N}} R_j < \infty$. By the well-known $(5r)$-covering lemma (see, for example, [53, Theorem 2.19]), there exists a subset J^* of $\mathbb{N}$ such that

$$\bigcup_{j=1}^{\infty}(3B_j) \subset \bigcup_{j\in J^*}(15B_j)$$

and $\chi_{j\in J^*}\chi_{(3B_j)} \le 1$. Furthermore, by its construction, if $B_{j'}$, $j' \in \mathbb{N}$, intersects B_j for some $j \in J^*$, we have that $(3B_{j'}) \subset (15B_j)$.

Let $\{B(x_k,r_k)\}_{k\in\mathbb{N}}$ be a collection of balls such that $\Omega \subset \cup_{k=1}^{\infty} B(x_k,r_k)$ and $\sum_{k=1}^{\infty} r_k^d \le 2\Lambda_d^{(\infty)}(\Omega)$. Set

$$K_1 \equiv \{k \in \mathbb{N} : \text{When } B(x_k,45r_k) \cap B_j \neq \emptyset \text{ for any } j \in \mathbb{N}, \text{ then } r_k \ge 135R_j\}$$

and

$$J_1 \equiv \{j \in \mathbb{N} : B_j \cap B(x_k,45r_k) \neq \emptyset \text{ for some } k \in K_1\}.$$

Also define $J_2 \equiv (\mathbb{N}\setminus J_1)$ and $K_2 \equiv (\mathbb{N}\setminus K_1)$. We remark that if $k \in K_2$, then there exists $j \in J_2$ such that $B_j \cap B(x_k,45r_k) \neq \emptyset$ and $135R_j > r_k$. Notice that

$$B_j \subset \Omega \subset \left(\bigcup_{k=1}^{\infty} B(x_k,r_k)\right).$$

Thus, for each $j \in J_2$, we have

$$B_j \subset \bigcup_{k\in K_2,\, B(x_k,r_k)\cap B_j \neq \emptyset} B(x_k,r_k),$$

and then, by $d \le n$ and the monotonicity of $\ell^{\frac{d}{n}}$, we see that

$$\begin{aligned}\sum_{k\in K_2} r_k^d &\sim \sum_{k\in K_2} |B(x_k,r_k)|^{\frac{d}{n}}\\ &\gtrsim \sum_{j\in J^*\cap J_2} \sum_{k\in K_2,\, B_j\cap B(x_k,45r_k)\neq\emptyset} |B(x_k,r_k)|^{\frac{d}{n}}\\ &\gtrsim \sum_{j\in J^*\cap J_2} \left(\sum_{k\in K_2,\, B_j\cap B(x_k,45r_k)\neq\emptyset} |B(x_k,r_k)|\right)^{\frac{d}{n}}\\ &\gtrsim \sum_{j\in J^*\cap J_2} R_j^d,\end{aligned}$$

which further yields that

$$\sum_{k\in K_1} r_k^d + \sum_{j\in J^*\cap J_2} R_j^d \lesssim \sum_{k\in K} r_k^d.$$

On the other hand, we have

$$\Omega \subset \bigcup_{j=1}^{\infty} B_j \subset \bigcup_{j\in J^*} (15B_j) = \left\{\bigcup_{j\in J^*\cap J_1} (15B_j)\right\} \bigcup \left\{\bigcup_{j\in J^*\cap J_2} (15B_j)\right\}$$

and hence

$$\Omega \subset \left\{\bigcup_{k\in K_1} B(x_k,46r_k)\right\} \bigcup \left\{\bigcup_{j\in J^*\cap J_2} (15B_j)\right\}.$$

Notice that for $k\in K_1$, $B(x_k,45r_k)$ meets B_j for some $j\in\mathbb{N}$, which gives us $r_k \ge 135R_j$ and further implies that $B(x_k,46r_k)\supset B_j$. Also, for $j\in J^*$ and $j'\in\mathbb{N}$, if $B_j\cap B_{j'}\neq\emptyset$, then $(15B_j)\supset B_{j'}$. As a result, we conclude that

$$\{B(x_k,46r_k)\}_{k\in K_1} \bigcup \{15B_j\}_{j\in J^*\cap J_2}$$

is the desired covering of Ω, and hence

$$\Lambda_d^{(\infty)}(\Omega,\{B_j\}_{j=1}^{\infty}) \le \sum_{k\in K_1} (46r_k)^d + \sum_{j\in J^*\cap J_2} (15R_j)^d \lesssim (46)^d \Lambda_d^{(\infty)}(\Omega),$$

which completes the proof of Lemma 7.15. □

Lemma 7.16. *Let $\beta\in[1,\infty)$, $\lambda\in(0,\infty)$ and ω be a nonnegative Borel measurable function on $\mathbb{R}_+^{n+1}$. Then there exists a positive constant C, independent of β, ω and λ, such that*

$$\Lambda_d^{(\infty)}(\{x\in\mathbb{R}^n : N_\beta\omega(x)>\lambda\}) \le C\beta^d \Lambda_d^{(\infty)}(\{x\in\mathbb{R}^n : N\omega(x)>\lambda\}),$$

where

$$N_\beta \omega(x) \equiv \sup_{|y-x|<\beta t} \omega(y,t).$$

Proof. Observe that

$$\{x \in \mathbb{R}^n : N\omega(x) > \lambda\} = \bigcup_{t\in(0,\infty)} \bigcup_{\substack{y\in\mathbb{R}^n \\ \omega(y,t)>\lambda}} B(y,t)$$

and that

$$\{x \in \mathbb{R}^n : N_\beta \omega(x) > \lambda\} = \bigcup_{t\in(0,\infty)} \bigcup_{\substack{y\in\mathbb{R}^n \\ \omega(y,t)>\lambda}} B(y,\beta t).$$

By the Linderöf covering lemma, there exists a countable subset $\{B_l\}_{l=0}^\infty$ of

$$\{B(y,t) : t \in (0,\infty),\ y \in \mathbb{R}^n \text{ satisfy } \omega(y,t) > \lambda\}$$

such that

$$\{x \in \mathbb{R}^n : N_\beta \omega(x) > \lambda\} = \left\{\bigcup_{l=0}^\infty (\beta B_l)\right\}$$

and

$$\{x \in \mathbb{R}^n : N\omega(x) > \lambda\} \supset \left(\bigcup_{l=0}^\infty B_l\right).$$

By Lemma 7.15, it suffices to prove that

$$\Lambda_d^{(\infty)}(\{x \in \mathbb{R}^n : N_\beta \omega(x) > \lambda\}, \{\beta B_l\}_{l=0}^\infty) \lesssim \beta^d \Lambda_d^{(\infty)}\left(\bigcup_{l=0}^\infty B_l, \{B_l\}_{l=0}^\infty\right).$$

Let $\{B_k^*\}_{k=0}^\infty$ be a ball covering of $\cup_{l\in\mathbb{N}} B_l$ such that

$$\sum_{k=0}^\infty r_{B_k^*}^d \le 2\Lambda_d^{(\infty)}(\cup_{l=0}^\infty B_l, \{B_l\}_{l=0}^\infty).$$

and that B_k^* engulfs B_l whenever they intersect, where $r_{B_k^*}$ denotes the radius of B_k^*. Therefore, βB_k^* engulfs βB_l whenever they intersect and

$$\{x \in \mathbb{R}^n : N_\beta \omega(x) > \lambda\} \subset \left\{\bigcup_{k=0}^\infty (\beta B_k^*)\right\}.$$

We then have

$$2\beta^d \Lambda_d^{(\infty)}\left(\bigcup_{l=0}^{\infty} B_l, \{B_l\}_{l=0}^{\infty}\right) \geq \sum_{l=0}^{\infty}(\beta r_{B_k^*})^d$$
$$\geq \Lambda_d^{(\infty)}\left(\{x\in\mathbb{R}^n : N_\beta\omega(x) > \lambda\}, \{\beta B_l\}_{l=0}^{\infty}\right),$$

which completes the proof of Lemma 7.16. □

Proof of Theorem 7.5. Without loss of generality, we may assume $s = 0$, since this case implies the general case. In fact, let $t \equiv \{t_Q\}_{l(Q)\leq 1} \in aH^{s,\tau}_{p,q}(\mathbb{R}^n)$ and A be a ε-almost diagonal operator associated with the matrix $\{a_{QP}\}_{l(Q),l(P)\leq 1}$ and $\varepsilon \in (0,\infty)$. If the conclusion holds for $s = 0$, let $\widetilde{t}_P \equiv l(P)^{-s}t_P$ and B be the operator associated with the matrix $\{b_{QP}\}_{l(Q),l(P)\leq 1}$, where $b_{QP} \equiv (l(P)/l(Q))^s a_{QP}$ for all $l(Q), l(P) \leq 1$. Then we have

$$\|At\|_{aH^{s,\tau}_{p,q}(\mathbb{R}^n)} = \|B\widetilde{t}\|_{aH^{0,\tau}_{p,q}(\mathbb{R}^n)} \lesssim \|\widetilde{t}\|_{aH^{0,\tau}_{p,q}(\mathbb{R}^n)} \sim \|t\|_{aH^{s,\tau}_{p,q}(\mathbb{R}^n)},$$

which deduces the desired conclusions.

By similarity, we only consider $fH^{s,\tau}_{p,q}(\mathbb{R}^n)$. By the Aoki theorem (see [8]), there exists a $\kappa \in (0,1]$ such that $\|\cdot\|^{\kappa}_{fH^{0,\tau}_{p,q}(\mathbb{R}^n)}$ becomes a norm in $fH^{0,\tau}_{p,q}(\mathbb{R}^n)$. Let $t \in f\dot{H}^{0,\tau}_{p,q}(\mathbb{R}^n)$. For $Q \in \mathscr{Q}$, we write $A \equiv A_0 + A_1$ with

$$(A_0 t)_Q \equiv \sum_{\{P\in\mathscr{Q}:\ l(Q)\leq l(P)\}} a_{QP}t_P$$

and

$$(A_1 t)_Q \equiv \sum_{\{P\in\mathscr{Q}:\ l(P)< l(Q)\}} a_{QP}t_P.$$

By Definition 3.1, we see that for $Q \in \mathscr{Q}$,

$$|(A_0 t)_Q| \lesssim \sum_{\{P\in\mathscr{Q}: l(Q)\leq l(P)\}} \left(\frac{l(Q)}{l(P)}\right)^{\frac{n+\varepsilon}{2}} \frac{|t_P|}{(1+l(P)^{-1}|x_Q - x_P|)^{n+\varepsilon}}.$$

Thus, we have

$$\|A_0 t\|_{fH^{0,\tau}_{p,q}(\mathbb{R}^n)} \lesssim \inf_{\omega}\left\|\left\{\sum_{j=0}^{\infty}\sum_{Q\in\mathscr{Q}_j} |Q|^{-\frac{q}{2}}\chi_Q \left[\sum_{i=-\infty}^{j}\sum_{P\in\mathscr{Q}_i} 2^{(i-j)\frac{n+\varepsilon}{2}} \right.\right.\right.$$
$$\left.\left.\left.\times \frac{|t_P|[\omega(\cdot,2^{-j})]^{-1}}{(1+2^i|x_Q - x_P|)^{n+\varepsilon}}\right]^q\right\}^{\frac{1}{q}}\right\|_{L^p(\mathbb{R}^n)}.$$

Let ω be a nonnegative Borel measurable function satisfying (7.3) and

$$\left\| \left\{ \sum_{j=0}^{\infty} \sum_{Q\in\mathcal{Q}_j} |t_Q|^q [\widetilde{\chi}_Q \omega(\cdot, 2^{-j})]^{-q} \right\}^{\frac{1}{q}} \right\|_{L^p(\mathbb{R}^n)} \lesssim \|t\|_{fH_{p,q}^{0,\tau}(\mathbb{R}^n)}.$$

Let

$$A_{0,i}(Q) \equiv \{P \in \mathcal{Q}_i : 2^i |x_P - x_Q| \le \sqrt{n}/2\}$$

and

$$A_{m,i}(Q) \equiv \{P \in \mathcal{Q}_i : 2^{m-1}\sqrt{n}/2 < 2^i |x_P - x_Q| \le 2^m \sqrt{n}/2\}$$

for all $i \in \mathbb{Z}$ and $m \in \mathbb{Z}_+$. Define

$$\omega_m(x,t) \equiv 2^{-mn\tau} \sup_{y \in B(x, \sqrt{n} 2^{m+1} t)} \omega(y,t)$$

for all $(x,t) \in \mathbb{R}_+^{n+1}$. Then $N\omega_m \lesssim 2^{-mn\tau} N_{\sqrt{n}2^{m+2}} \omega$ and

$$[\omega_m(x, 2^{-j})]^{-1} \omega(y, 2^{-i}) \lesssim 2^{mn\tau}$$

for $m \in \mathbb{Z}_+$, $x \in Q$ with $Q \in \mathcal{Q}_j$, $y \in P$ with $P \in A_{m,i}(Q)$ and $i \le j$. Moreover, using Lemma 7.16, we see that a constant multiple of ω_m also satisfies (7.3). Similarly to the proof of Lemma 2.4, we have that for all $x \in Q$,

$$\sum_{P \in A_{m,i}(Q)} \frac{|t_P| [\omega_m(x, 2^{-j})]^{-1}}{(1 + 2^i |x_Q - x_P|)^{n+\varepsilon}} \lesssim 2^{-m\varepsilon + mn\tau} \mathrm{M} \left(\sum_{P \in A_{m,i}(Q)} |t_P| \chi_P [\omega(\cdot, 2^{-i})]^{-1} \right)(x).$$

Thus, choosing $\varepsilon > n\tau$, by Fefferman-Stein's vector valued inequality, we obtain

$$\begin{aligned}
\|A_0 t\|_{fH_{p,q}^{0,\tau}(\mathbb{R}^n)}^{\kappa} &\lesssim \sum_{m=0}^{\infty} \inf_{\omega} \left\{ \left\| \left\{ \sum_{j=0}^{\infty} \sum_{Q\in\mathcal{Q}_j} |Q|^{-\frac{q}{2}} \chi_Q \left[\sum_{i=-\infty}^{j} \sum_{P \in A_{m,i}(Q)} 2^{(i-j)\frac{n+\varepsilon}{2}} \right.\right.\right.\right. \\
&\qquad \left.\left.\left.\left. \times \frac{|t_P| [\omega(\cdot, 2^{-j})]^{-1}}{(1 + 2^i |x_Q - x_P|)^{n+\varepsilon}} \right]^q \right\}^{\frac{1}{q}} \right\|_{L^p(\mathbb{R}^n)} \right\}^{\kappa} \\
&\lesssim \sum_{m=0}^{\infty} \left\| \left\{ \sum_{j=0}^{\infty} \sum_{Q\in\mathcal{Q}_j} |Q|^{-\frac{q}{2}} \chi_Q \left[\sum_{i=-\infty}^{j} \sum_{P \in A_{m,i}(Q)} 2^{(i-j)\frac{n+\varepsilon}{2}} \right.\right.\right. \\
&\qquad \left.\left.\left. \times \frac{|t_P| [\omega_m(\cdot, 2^{-j})]^{-1}}{(1 + 2^i |x_Q - x_P|)^{n+\varepsilon}} \right]^q \right\}^{\frac{1}{q}} \right\|_{L^p(\mathbb{R}^n)}^{\kappa}
\end{aligned}$$

$$\lesssim \sum_{m=0}^{\infty} 2^{m(n\tau-\varepsilon)\kappa} \left\| \left\{ \sum_{j=0}^{\infty} \sum_{Q\in\mathcal{Q}_j} \chi_Q \left[\sum_{i=-\infty}^{j} 2^{(i-j)\varepsilon/2} \right.\right.\right.$$
$$\left.\left.\left. \times \mathrm{M}\left(\sum_{P\in A_{m,i}(Q)} |t_P|\widetilde{\chi}_P[\omega(\cdot,2^{-i})]^{-1} \right) \right]^q \right\}^{\frac{1}{q}} \right\|_{L^p(\mathbb{R}^n)}^{\kappa}$$
$$\lesssim \|t\|_{fH_{p,q}^{0,\tau}(\mathbb{R}^n)}^{\kappa}.$$

The proof for $A_1 t$ is similar. Indeed, we have

$$|(A_1 t)_Q| \lesssim \sum_{\{P\in\mathcal{Q}:\, l(P)\le l(Q)\}} \left(\frac{l(P)}{l(Q)}\right)^{\frac{n+\varepsilon}{2}} \frac{|t_P|}{(1+l(Q)^{-1}|x_Q-x_P|)^{n+\varepsilon}}.$$

Thus,

$$\|A_1 t\|_{fH_{p,q}^{0,\tau}(\mathbb{R}^n)} \lesssim \inf_{\omega} \left\| \left\{ \sum_{j=0}^{\infty} \sum_{Q\in\mathcal{Q}_j} |Q|^{-\frac{q}{2}} \chi_Q \left[\sum_{l=0}^{\infty} \sum_{P\in\mathcal{Q}_{j+l}} 2^{-l\frac{n+\varepsilon}{2}} \right.\right.\right.$$
$$\left.\left.\left. \times \frac{|t_P|[\omega(\cdot,2^{-j})]^{-1}}{(1+2^j|x_Q-x_P|)^{n+\varepsilon}} \right]^q \right\}^{\frac{1}{q}} \right\|_{L^p(\mathbb{R}^n)}.$$

Let

$$\widetilde{A}_{0,j,l}(Q) \equiv \{P\in\mathcal{Q}_{j+l} : 2^j|x_P-x_Q| \le \sqrt{n}/2\}$$

and

$$\widetilde{A}_{m,j,l}(Q) \equiv \{P\in\mathcal{Q}_{j+l} : 2^{m-1}\sqrt{n}/2 < 2^j|x_P-x_Q| \le 2^m\sqrt{n}/2\}$$

for all $j\in\mathbb{Z}$ and $m,l\in\mathbb{Z}_+$. Set

$$\widetilde{\omega}_m(x,s) \equiv 2^{-(m+l)n\tau} \sup\{\omega(y,s) : y\in\mathbb{R}^n,\ |y-x| < \sqrt{n}2^{m+l+1}s\}$$

for all $m\in\mathbb{Z}_+$ and $(x,s)\in\mathbb{R}_+^{n+1}$. Similarly, we have that a constant multiple of $\widetilde{\omega}_m$ satisfies (7.3) and

$$[\widetilde{\omega}_m(x,2^{-j})]^{-1}\omega(y,2^{-j-l}) \lesssim 2^{(m+l)n\tau}$$

for $m,l\in\mathbb{Z}_+$, $x\in Q$ with $Q\in\mathcal{Q}_j$, $y\in P$ with $P\in\widetilde{A}_{m,j,l}(Q)$. Choosing $\varepsilon > 2n\tau$, similarly to the estimate of $\|A_0 t\|_{fH_{p,q}^{0,\tau}(\mathbb{R}^n)}$, we also have

$$\|A_1 t\|_{fH_{p,q}^{0,\tau}(\mathbb{R}^n)} \lesssim \|t\|_{fH_{p,q}^{0,\tau}(\mathbb{R}^n)},$$

which completes the proof of Theorem 7.5. □

From Theorem 7.5, we deduce the smooth atomic and molecular decomposition characterizations of $AH_{p,q}^{s,\tau}(\mathbb{R}^n)$. We begin with the smooth synthesis molecules, the smooth analysis molecules and the smooth atoms for $AH_{p,q}^{s,\tau}(\mathbb{R}^n)$ as follows.

Definition 7.7. Let $s \in \mathbb{R}$, $\tau \in [0,\infty)$, $p \in (1,\infty)$, $q \in [1,\infty)$, $s^* = s - \lfloor s \rfloor$ and

$$N \equiv \max(\lfloor -s+2n\tau \rfloor, -1).$$

Let Q be a dyadic cube with $l(Q) \le 1$.

(i) A function m_Q is said to be an *inhomogeneous smooth synthesis molecule for* $AH_{p,q}^{s,\tau}(\mathbb{R}^n)$ supported near Q if there exist a $\delta \in (\max\{s^*, (s+n\tau)^*\}, 1]$ and an $M \in (n+2n\tau, \infty)$ such that

$$\int_{\mathbb{R}^n} x^\gamma m_Q(x)\,dx = 0 \quad \text{if} \quad |\gamma| \le N \quad \text{and} \quad l(Q) < 1,$$

$$|m_Q(x)| \le (1+|x-x_Q|)^{-M} \quad \text{if} \quad l(Q) = 1,$$

$$|m_Q(x)| \le |Q|^{-1/2}(1+[l(Q)]^{-1}|x-x_Q|)^{-\max(M,M-s)} \quad \text{if} \quad l(Q) < 1,$$

$$|\partial^\gamma m_Q(x)| \le |Q|^{-1/2-|\gamma|/n}(1+[l(Q)]^{-1}|x-x_Q|)^{-M} \quad \text{if} \quad |\gamma| \le \lfloor s+3n\tau \rfloor,$$

and

$$|\partial^\gamma m_Q(x) - \partial^\gamma m_Q(y)| \le |Q|^{-1/2-|\gamma|/n-\delta/n}|x-y|^\delta \times \sup_{|z|\le|x-y|} (1+[l(Q)]^{-1}|x-z-x_Q|)^{-M}$$

if $|\gamma| = \lfloor s+3n\tau \rfloor$.

A collection $\{m_Q\}_{l(Q)\le 1}$ is called a family of inhomogeneous smooth synthesis molecules for $AH_{p,q}^{s,\tau}(\mathbb{R}^n)$, if each m_Q is an inhomogeneous smooth synthesis molecule for $AH_{p,q}^{s,\tau}(\mathbb{R}^n)$ supported near Q.

(ii) A function b_Q is said to be an *inhomogeneous smooth analysis molecule for* $AH_{p,q}^{s,\tau}(\mathbb{R}^n)$ supported near Q if there exist a $\rho \in ((-s)^*, 1]$ and an $M \in (n+2n\tau, \infty)$ such that

$$\int_{\mathbb{R}^n} x^\gamma b_Q(x)\,dx = 0 \quad \text{if} \quad |\gamma| \le \lfloor s+3n\tau \rfloor \quad \text{and} \quad l(Q) < 1,$$

$$|b_Q(x)| \le (1+|x-x_Q|)^{-M} \quad \text{if} \quad l(Q) = 1,$$

$$|b_Q(x)| \le |Q|^{-1/2}(1+[l(Q)]^{-1}|x-x_Q|)^{-\max(M,M+n+s+n\tau-J)} \quad \text{if} \quad l(Q) < 1,$$

$$|\partial^\gamma b_Q(x)| \le |Q|^{-1/2-|\gamma|/n}(1+[l(Q)]^{-1}|x-x_Q|)^{-M} \quad \text{if} \quad |\gamma| \le N,$$

and

$$|\partial^\gamma b_Q(x) - \partial^\gamma b_Q(y)| \le |Q|^{-1/2-|\gamma|/n-\rho/n}|x-y|^\rho \times \sup_{|z|\le|x-y|} (1+[l(Q)]^{-1}|x-z-x_Q|)^{-M} \quad \text{if} \quad |\gamma| = N.$$

A collection $\{b_Q\}_{l(Q)\le 1}$ is called a family of inhomogeneous smooth synthesis molecules for $AH^{s,\tau}_{p,q}(\mathbb{R}^n)$, if each b_Q is an inhomogeneous smooth analysis molecule for $AH^{s,\tau}_{p,q}(\mathbb{R}^n)$ supported near Q.

Definition 7.8. A function a_Q is called an *inhomogeneous smooth atom for* $AH^{s,\tau}_{p,q}(\mathbb{R}^n)$ supported near a dyadic cube Q with $l(Q)\le 1$ if

$$\operatorname{supp} a_Q \subset 3Q,$$

$$\int_{\mathbb{R}^n} x^\gamma a_Q(x)\,dx = 0 \quad \text{if} \quad |\gamma| \le \max\{\lfloor -s+2n\tau \rfloor, -1\} \quad \text{and} \quad l(Q)<1,$$

and

$$\|\partial^\gamma a_Q\|_{L^\infty(\mathbb{R}^n)} \le |Q|^{-1/2-|\gamma|/n} \quad \text{if} \quad |\gamma| \le \max\{\lfloor s+3n\tau+1 \rfloor, 0\}.$$

A collection $\{a_Q\}_{l(Q)\le 1}$ is called a family of inhomogeneous smooth atoms for $AH^{s,\tau}_{p,q}(\mathbb{R}^n)$, if each a_Q is an inhomogeneous smooth atom for $AH^{s,\tau}_{p,q}(\mathbb{R}^n)$ supported near Q.

We remark that the smooth molecules and atoms for $AH^{s,\tau}_{p,q}(\mathbb{R}^n)$ are also the smooth molecules and atoms for $A^{s,\tau}_{p,q}(\mathbb{R}^n)$ in Definitions 3.2 and 3.3.

Similarly to the proofs of Theorems 3.2 and 3.3, we have the following decomposition characterizations of $AH^{s,\tau}_{p,q}(\mathbb{R}^n)$.

Theorem 7.6. *Let* $p\in(1,\infty)$, $q\in[1,\infty)$, $s\in\mathbb{R}$ *and* $\tau\in[0,\frac{1}{(p\vee q)'}]$.

(i) If $\{m_Q\}_{l(Q)\le 1}$ *is a family of smooth synthesis molecules for* $AH^{s,\tau}_{p,q}(\mathbb{R}^n)$, *then there exists a positive constant* C *such that for all* $t\equiv\{t_Q\}_{l(Q)\le 1}\in aH^{s,\tau}_{p,q}(\mathbb{R}^n)$,

$$\left\| \sum_{l(Q)\le 1} t_Q m_Q \right\|_{AH^{s,\tau}_{p,q}(\mathbb{R}^n)} \le C\|t\|_{aH^{s,\tau}_{p,q}(\mathbb{R}^n)}.$$

(ii) If $\{b_Q\}_{l(Q)\le 1}$ *is a family of smooth analysis molecules for* $AH^{s,\tau}_{p,q}(\mathbb{R}^n)$, *then there exists a positive constant* C *such that for all* $f\in AH^{s,\tau}_{p,q}(\mathbb{R}^n)$,

$$\|\{\langle f,b_Q\rangle\}_{l(Q)\le 1}\|_{aH^{s,\tau}_{p,q}(\mathbb{R}^n)} \le C\|f\|_{AH^{s,\tau}_{p,q}(\mathbb{R}^n)}.$$

Theorem 7.7. *Let* s, p, q, τ *be as in Theorem 7.6. Then for each* $f\in AH^{s,\tau}_{p,q}(\mathbb{R}^n)$, *there exist a family* $\{a_Q\}_{l(Q)\le 1}$ *of smooth atoms for* $AH^{s,\tau}_{p,q}(\mathbb{R}^n)$, *a coefficient sequence* $t\equiv\{t_Q\}_{l(Q)\le 1}$, *and a positive constant* C *depending only on* p, q, s, τ *such that*

$$f = \sum_{l(Q)\le 1} t_Q a_Q$$

in $\mathscr{S}'(\mathbb{R}^n)$ *and*

$$\|t\|_{aH^{s,\tau}_{p,q}(\mathbb{R}^n)} \le C\|f\|_{AH^{s,\tau}_{p,q}(\mathbb{R}^n)}.$$

Conversely, there exists a positive constant C *depending only on* p,q,s,τ *such that for all families* $\{a_Q\}_{l(Q)\le 1}$ *of smooth atoms for* $AH^{s,\tau}_{p,q}(\mathbb{R}^n)$ *and*

$$t \equiv \{t_Q\}_{l(Q)\le 1} \in aH^{s,\tau}_{p,q}(\mathbb{R}^n),$$

$$\left\| \sum_{l(Q)\le 1} t_Q a_Q \right\|_{AH^{s,\tau}_{p,q}(\mathbb{R}^n)} \le C\|t\|_{aH^{s,\tau}_{p,q}(\mathbb{R}^n)}.$$

Based on these smooth atomic and molecular decomposition characterizations, similarly to the arguments in Chaps. 5 and 6, we obtain that the mapping properties of pseudo-differential operators in Theorem 5.1, lifting properties in Proposition 5.1, pointwise multiplier properties in Theorem 6.1 and diffeomorphism properties in Theorem 6.7 have counterparts for the spaces $AH^{s,\tau}_{p,q}(\mathbb{R}^n)$.

Theorem 7.8. *Let* $s,\mu \in \mathbb{R}$, $p \in (1,\infty)$, $q \in [1,\infty)$, $\tau \in [0, \frac{1}{(p\vee q)'}]$ *and* N *be as in Definition 7.7. Assume that* $a \in \mathscr{S}^{\mu}_{1,1}(\mathbb{R}^n)$ *and* $a(x,D)$ *be the corresponding pseudo-differential operator. If* $s > 2n\tau$, *then* $a(x,D)$ *is a bounded linear operator from* $AH^{s+\mu,\tau}_{p,q}(\mathbb{R}^n)$ *to* $AH^{s,\tau}_{p,q}(\mathbb{R}^n)$. *If* $s \le 2n\tau$ *and* (5.1) *holds, then* $a(x,D)$ *is a bounded linear operator from* $AH^{s+\mu,\tau}_{p,q}(\mathbb{R}^n)$ *to* $AH^{s,\tau}_{p,q}(\mathbb{R}^n)$.

Proposition 7.5. *Let* $s,\sigma \in \mathbb{R}$, $p \in (1,\infty)$, $q \in [1,\infty)$ *and* $\tau \in [0, \frac{1}{(p\vee q)'}]$. *Then the lifting operator* I_σ *maps* $AH^{s,\tau}_{p,q}(\mathbb{R}^n)$ *isomorphically onto* $AH^{s-\sigma,\tau}_{p,q}(\mathbb{R}^n)$.

Theorem 7.9. *Let* $s \in \mathbb{R}$, $p \in (1,\infty)$, $q \in [1,\infty)$ *and* $\tau \in [0, \frac{1}{(p\vee q)'}]$. *If* $m \in \mathbb{N}$ *is sufficiently large, then there exists a positive constant* $C(m)$ *such that for all* $g \in BC^m(\mathbb{R}^n)$ *and* $f \in AH^{s,\tau}_{p,q}(\mathbb{R}^n)$,

$$\|gf\|_{AH^{s,\tau}_{p,q}(\mathbb{R}^n)} \le C(m)\left(\sum_{|\alpha|\le m} \|\partial^\alpha g\|_{L^\infty(\mathbb{R}^n)}\right)\|f\|_{AH^{s,\tau}_{p,q}(\mathbb{R}^n)}.$$

Theorem 7.10. *Let* $m \in \mathbb{N}$, $s \in \mathbb{R}$, $p \in (1,\infty)$, $q \in [1,\infty)$, $\tau \in [0, \frac{1}{(p\vee q)'}]$ *and* ψ *be an m-diffeomorphism. If* $m \in \mathbb{N}$ *is sufficiently large, then* D_ψ *is an isomorphic mapping of* $AH^{s,\tau}_{p,q}(\mathbb{R}^n)$ *onto itself.*

Also, we establish the trace property for the space $AH^{s,\tau}_{p,q}(\mathbb{R}^n)$.

Theorem 7.11. *Let* $n \ge 2$, $p \in (1,\infty)$, $q \in [1,\infty)$, $s \in (\frac{1}{p} + 2n\tau, \infty)$ *and* $\tau \in [0, \frac{n-1}{n(p\vee q)'}]$. *Then there exists a surjective and continuous operator*

$$\mathrm{Tr}: f \in AH^{s,\tau}_{p,q}(\mathbb{R}^n) \mapsto \mathrm{Tr}(f) \in AH^{s-\frac{1}{p},\frac{n}{n-1}\tau}_{p,q}(\mathbb{R}^{n-1})$$

such that $\mathrm{Tr}(f)(x') = f(x',0)$ *for all* $x' \in \mathbb{R}^{n-1}$ *and smooth atoms* f *for* $AH^{s,\tau}_{p,q}(\mathbb{R}^n)$.

To prove this theorem, we need a technical lemma; see [168, Lemma 4.1].

Lemma 7.17. *Let $d \in (0,n]$ and Ω be an open set in $\mathbb{R}^n$. Define*

$$\Lambda_{d,*}^{(\infty)}(\Omega) \equiv \inf\left\{ \sum_{j=1}^{\infty} r_j^d : \Omega \subset \bigcup_{j=1}^{\infty} B(x_r, r_j),\, r_j > \frac{\operatorname{dist}(x_j, \partial\Omega)}{10000} \right\}.$$

Then $\Lambda_d^{(\infty)}(\Omega)$ and $\Lambda_{d,*}^{(\infty)}(\Omega)$ are equivalent for all Ω.

Proof. The inequality $\Lambda_d^{(\infty)}(\Omega) \le \Lambda_{d,*}^{(\infty)}(\Omega)$ is trivial from the definitions. To prove the converse, we choose a ball covering $\{B(x_j, r_j)\}_{j=1}^{\infty}$ of Ω such that

$$\sum_{j=1}^{\infty} r_j^d \le 2\Lambda_d^{(\infty)}(\Omega).$$

Let $\{B(X_j, R_j)\}_{j=1}^{\infty}$ be a Whitney covering of Ω satisfying

$$\Omega = \cup_{j=1}^{\infty} B(X_j, R_j), \quad R_j/1000 \le \operatorname{dist}(X_j, \partial\Omega) \le R_j/100$$

and $\sum_{j\in\mathbb{N}} \chi_{R_j} \le C_n$; see, for example, [68, Proposition 7.3.4]. Set

$$J_1 \equiv \left\{ j \in \mathbb{N} : (B(X_j, R_j) \cap B(x_k, r_k)) \ne \emptyset \text{ and } R_j \le 4r_k \text{ for some } k \in \mathbb{N} \right\}$$

and $J_2 \equiv (\mathbb{N} \setminus J_1)$. Notice that if $k \in \mathbb{N}$ satisfies

$$(B(X_j, R_j) \cap B(x_k, r_k)) \ne \emptyset$$

for some $j \in J_2$, then $B(x_k, r_k) \subset B(X_j, 2R_j)$, since $r_k < R_j/4$. With this in mind, we define

$$K_2 \equiv \left\{ k \in \mathbb{N} : (B(x_k, r_k) \cap B(X_j, R_j)) \ne \emptyset \ \text{ for some } j \in J_2 \right\},$$

and $K_1 \equiv (\mathbb{N} \setminus K_2)$. It is easy to see that

$$\bigcup_{k=1}^{\infty} B(x_k, r_k) \subset \left(\bigcup_{k\in K_1} B(x_k, r_k) \bigcup \bigcup_{j\in J_2} B(X_j, 2R_j) \right). \tag{7.19}$$

Furthermore, for each $k \in \mathbb{N}$, the cardinality of the set

$$\left\{ j \in J_2 : (B(x_k, r_k) \cap B(X_j, R_j)) \ne \emptyset \right\}$$

is bounded by a constant depending only on the dimension. Thus, we have

$$\begin{aligned}\sum_{k=1}^{\infty} r_k^d &= \sum_{k\in K_1} r_k^d + \sum_{k\in K_2} r_k^d \\ &\sim \sum_{k\in K_1} r_k^d + \sum_{j\in J_2}\left(\sum_{k\in K_2,\,(B(x_k,r_k)\cap B(X_j,R_j))\neq\emptyset} r_k^d\right) \\ &\sim \sum_{k\in K_1} r_k^d + \sum_{j\in J_2}\left(\sum_{k\in K_2,\,(B(x_k,r_k)\cap B(X_j,R_j))\neq\emptyset} |B(x_k,r_k)|^{\frac{d}{n}}\right).\end{aligned}$$

Notice that

$$B(X_j,R_j)\subset\Omega\subset\left(\bigcup_{k=1}^{\infty} B(x_k,r_k)\right).$$

Then for each $j\in J_2$, we have

$$B(X_j,R_j)\subset\left\{\bigcup_{k\in K_2,\,(B(x_k,r_k)\cap B(X_j,R_j))\neq\emptyset} B(x_k,r_k)\right\}$$

Since $d\in(0,n]$, by the monotonicity of $\ell^{\frac{d}{n}}$, we see that

$$\begin{aligned}&\left(\sum_{k\in K_2,\,(B(x_k,r_k)\cap B(X_j,R_j))\neq\emptyset} |B(x_k,r_k)|^{\frac{d}{n}}\right) \\ &\quad\geq\left(\sum_{k\in K_2,\,(B(x_k,r_k)\cap B(X_j,R_j))\neq\emptyset} |B(x_k,r_k)|\right)^{\frac{d}{n}} \\ &\quad\geq |B(X_j,R_j)|^{\frac{d}{n}}.\end{aligned}$$

As a consequence,

$$\sum_{k=0}^{\infty} r_k^d \gtrsim \sum_{k\in K_1} r_k^d + \sum_{j\in J_2} R_j^d,$$

which combined with (7.19) yields that

$$\Lambda_{d,*}^{(\infty)}(\Omega)\leq\sum_{k\in K_1} r_k^d+\sum_{j\in J_2}(2R_j)^d\lesssim\sum_{k\in K_1} r_k^d+\sum_{j\in J_2} R_j^d\lesssim\sum_{k=0}^{\infty} r_k^d\lesssim\Lambda_d^{(\infty)}(\Omega).$$

This finishes the proof of Lemma 7.17. □

Proof of Theorem 7.11. For similarity, we concentrate on $BH^{s,\tau}_{p,q}(\mathbb{R}^n)$. By Theorem 7.7, any $f \in BH^{s,\tau}_{p,q}(\mathbb{R}^n)$ admits a decomposition

$$f = \sum_{l(Q)\le 1} t_Q a_Q$$

in $\mathscr{S}'(\mathbb{R}^n)$, where each a_Q is a smooth atom for $BH^{s,\tau}_{p,q}(\mathbb{R}^n)$ and $t \equiv \{t_Q\}_{l(Q)\le 1} \subset \mathbb{C}$ satisfies

$$\|t\|_{bH^{s,\tau}_{p,q}(\mathbb{R}^n)} \lesssim \|f\|_{BH^{s,\tau}_{p,q}(\mathbb{R}^n)}.$$

Since $s > 1/p + 2n\tau$, there is no need to postulate any moment condition on a_Q. Define

$$\mathrm{Tr}(f) \equiv \sum_{l(Q)\le 1} t_Q a_Q(*',0) = \sum_{l(Q)\le 1} \frac{t_Q}{[l(Q)]^{\frac{1}{2}}}[l(Q)]^{\frac{1}{2}} a_Q(*',0).$$

By the support condition of atoms, the above summation can be re-written as

$$\mathrm{Tr}(f) \equiv \sum_{i=0}^{2} \sum_{\substack{Q'\in\mathscr{Q}(\mathbb{R}^{n-1})\\ l(Q')\le 1}} \frac{t_{Q'\times[(i-1)l(Q'),il(Q'))}}{[l(Q')]^{\frac{1}{2}}}[l(Q')]^{\frac{1}{2}} a_{Q'\times[(i-1)l(Q'),il(Q'))}(*',0). \quad (7.20)$$

We need to show that (7.20) converges in $\mathscr{S}'(\mathbb{R}^{n-1})$ and

$$\|\mathrm{Tr}(f)\|_{BH^{s-\frac{1}{p},\frac{n}{n-1}\tau}_{p,q}(\mathbb{R}^{n-1})} \lesssim \|f\|_{BH^{s,\tau}_{p,q}(\mathbb{R}^n)}.$$

By Theorem 7.7 again, we only need to prove that for each $Q' \in \mathscr{Q}(\mathbb{R}^{n-1})$ with $l(Q') \le 1$,

$$[l(Q')]^{\frac{1}{2}} a_{Q'\times[(i-1)l(Q'),il(Q'))}(*',0)$$

is a smooth atom for $BH^{s-\frac{1}{p},\frac{n}{n-1}\tau}_{p,q}(\mathbb{R}^{n-1})$ supported near Q' and for all $i \in \{0,1,2\}$,

$$\left\|\left\{[l(Q')]^{-\frac{1}{2}} t_{Q'\times[(i-1)l(Q'),il(Q'))}\right\}_{Q'\in\mathscr{Q}(\mathbb{R}^{n-1})}\right\|_{bH^{s-\frac{1}{p},\frac{n}{n-1}\tau}_{p,q}(\mathbb{R}^{n-1})} < \infty. \quad (7.21)$$

Indeed, it was already proved in Sect. 6.3 that

$$[l(Q')]^{\frac{1}{2}} a_{Q'\times[(i-1)l(Q'),il(Q'))}(*',0)$$

is a smooth atom for $BH^{s-\frac{1}{p},\frac{n}{n-1}\tau}_{p,q}(\mathbb{R}^{n-1})$. By similarity, we only prove (7.21) when $i = 1$. Let ω be a nonnegative function on $\mathbb{R}^{n+1}_+$ satisfying (7.3) and

$$\left\{\sum_{j=0}^{\infty}\left[\sum_{Q\in\mathscr{Q}_j(\mathbb{R}^n)} |Q|^{-(\frac{s}{n}+\frac{1}{2})p}|t_Q|^p \int_Q [\omega(x,2^{-j})]^{-p}\,dx\right]^{\frac{q}{p}}\right\}^{\frac{1}{q}} \lesssim \|t\|_{bH^{s,\tau}_{p,q}(\mathbb{R}^n)}.$$

For all $\lambda \in (0,\infty)$, set

$$E_\lambda \equiv \left\{x \in \mathbb{R}^n : [N\omega(x)]^{(p\vee q)'} > \lambda\right\}.$$

Then there exists a ball cover $\{B_m\}_m$ such that

$$\Lambda^{(\infty)}_{n\tau(p\vee q)'}(E_\lambda) \sim \sum_m r_{B_m}^{n\tau(p\vee q)'}.$$

Let $\widetilde{\Lambda}^{(\infty)}_{n\tau(p\vee q)'}$ be the $\{(n-1)\frac{n\tau}{n-1}(p\vee q)'\}$-Hausdorff capacity in $\mathbb{R}^{n-1}$ and define $\widetilde{\omega}$ on $\mathbb{R}^n_+$ by setting, for all $x' \in \mathbb{R}^{n-1}$ and $t \in (0,\infty)$,

$$\widetilde{\omega}(x',t) \equiv \widetilde{C} \sup_{|x_n|<t} \omega((x',x_n),t),$$

where $\widetilde{C}$ is a positive constant chosen so that $N\widetilde{\omega}(x') \le N\omega(x',0)$ for all $x' \in \mathbb{R}^{n-1}$. Therefore, if $[N\widetilde{\omega}(x')]^{(p\vee q)'} > \lambda$, then $[N\omega(x',0)]^{(p\vee q)'} > \lambda$, and hence $(x',0) \in B_m$ for some m, which further implies that

$$\widetilde{E}_\lambda \equiv \left\{x' \in \mathbb{R}^{n-1} : [N\widetilde{\omega}(x')]^{(p\vee q)'} > \lambda\right\} \subset \left(\bigcup_m B_m^*\right),$$

where B_m^* is the projection of B_m from $\mathbb{R}^n$ to $\mathbb{R}^{n-1}$. Thus, we obtain

$$\begin{aligned}\int_{\mathbb{R}^{n-1}} [N\widetilde{\omega}(x')]^{(p\vee q)'}\, d\widetilde{\Lambda}^{(\infty)}_{n\tau(p\vee q)'}(x') &= \int_0^\infty \widetilde{\Lambda}^{(\infty)}_{n\tau(p\vee q)'}(\widetilde{E}_\lambda)\,d\lambda\\ &\lesssim \int_0^\infty \Lambda^{(\infty)}_{n\tau(p\vee q)'}(E_\lambda)\,d\lambda\\ &\lesssim 1.\end{aligned}$$

Furthermore,

$$\begin{aligned}&\left\|\left\{[l(Q')]^{-\frac{1}{2}} t_{Q'\times[0,l(Q'))}\right\}_{Q'\in\mathcal{Q}(\mathbb{R}^{n-1}),\, l(Q')\le 1}\right\|_{bH^{s-\frac{1}{p},\frac{n}{n-1}\tau}_{p,q}(\mathbb{R}^{n-1})}\\ &\quad\lesssim \left\{\sum_{j=0}^\infty \left[\sum_{Q'\in\mathcal{Q}_j(\mathbb{R}^{n-1})} [l(Q')]^{-sp-\frac{np}{2}+1} |t_{Q'\times[0,l(Q'))}|^p \int_{Q'} [\widetilde{\omega}(x',2^{-j})]^{-p}\,dx'\right]^{\frac{q}{p}}\right\}^{\frac{1}{q}}\\ &\quad\lesssim \left\{\sum_{j=0}^\infty \left[\sum_{Q'\in\mathcal{Q}_j(\mathbb{R}^{n-1})} [l(Q')]^{-sp-\frac{np}{2}} |t_{Q'\times[0,l(Q'))}|^p \int_{Q} [\omega(x,2^{-j})]^{-p}\,dx\right]^{\frac{q}{p}}\right\}^{\frac{1}{q}}\\ &\quad\lesssim \|t\|_{bH^{s,\tau}_{p,q}(\mathbb{R}^n)},\end{aligned}$$

which implies that Tr is bounded from $BH^{s,\tau}_{p,q}(\mathbb{R}^n)$ to $BH^{s-\frac{1}{p},\frac{n}{n-1}\tau}_{p,q}(\mathbb{R}^{n-1})$.

Let us show that Tr is surjective. To this end, for any $f \in BH_{p,q}^{s-\frac{1}{p},\frac{n}{n-1}\tau}(\mathbb{R}^{n-1})$, by Theorem 7.7, there exist a sequence $\{a_{Q'}\}_{Q'\in\mathscr{Q}(\mathbb{R}^{n-1}),l(Q')\le 1}$ of smooth atoms for $BH_{p,q}^{s-\frac{1}{p},\frac{n}{n-1}\tau}(\mathbb{R}^{n-1})$ and coefficients $t \equiv \{t_{Q'}\}_{Q'\in\mathscr{Q}(\mathbb{R}^{n-1}),l(Q')\le 1}$ such that

$$f = \sum_{Q'\in\mathscr{Q}(\mathbb{R}^{n-1}),l(Q')\le 1} t_{Q'} a_{Q'}$$

in $\mathscr{S}'(\mathbb{R}^{n-1})$ and

$$\|t\|_{bH_{p,q}^{s-\frac{1}{p},\frac{n}{n-1}\tau}(\mathbb{R}^{n-1})} \lesssim \|f\|_{BH_{p,q}^{s-\frac{1}{p},\frac{n}{n-1}\tau}(\mathbb{R}^{n-1})}.$$

Let $\varphi \in C_c^\infty(\mathbb{R})$ with $\operatorname{supp}\varphi \subset (-\frac{1}{2},\frac{1}{2})$ and $\varphi(0)=1$. For all $Q' \in (\mathbb{R}^{n-1})$ and $x\in\mathbb{R}$, set $\varphi_{Q'}(x) \equiv \varphi(2^{-\log_2 l(Q')}x)$. Under this notation, we define

$$F \equiv \sum_{Q'\in\mathscr{Q}(\mathbb{R}^{n-1}),l(Q')\le 1} t_{Q'} a_{Q'} \otimes \varphi_{Q'}.$$

It is easy to check that for all $Q' \in \mathscr{Q}(\mathbb{R}^{n-1})$ with $l(Q') \le 1$, $[l(Q')]^{-\frac{1}{2}} a_{Q'} \otimes \varphi_{Q'}$ is a smooth atom for $BH_{p,q}^{s,\tau}(\mathbb{R}^n)$ supported near $Q' \times [0, l(Q'))$. Thus, to show $F \in BH_{p,q}^{s,\tau}(\mathbb{R}^n)$, by Theorem 7.7, it suffices to prove that

$$\left\|\{[l(Q')]^{\frac{1}{2}} t_{Q'}\}_{Q'\in\mathscr{Q}(\mathbb{R}^{n-1}),l(Q')\le 1}\right\|_{bH_{p,q}^{s,\tau}(\mathbb{R}^n)} \lesssim \|f\|_{BH_{p,q}^{s-\frac{1}{p},\frac{n}{n-1}\tau}(\mathbb{R}^{n-1})}.$$

Let $\widetilde{\omega}$ satisfy

$$\int_{\mathbb{R}^{n-1}} [N\widetilde{\omega}(x')]^{(p\vee q)'}\,d\widetilde{\Lambda}_{n\tau(p\vee q)'}^{(\infty)}(x') \le 1$$

such that

$$\left\{\sum_{j=0}^{\infty}\left[\sum_{Q'\in\mathscr{Q}_j(\mathbb{R}^{n-1})} |Q'|^{-(\frac{s-1/p}{n-1}+\frac{1}{2})p} |t_{Q'}|^p \int_{Q'} [\widetilde{\omega}(x',2^{-j})]^{-p}\,dx\right]^{\frac{q}{p}}\right\}^{\frac{1}{q}}$$

is equivalent to $\|t\|_{bH_{p,q}^{s-\frac{1}{p},\frac{n}{n-1}\tau}(\mathbb{R}^{n-1})}$. By Lemma 7.17, for each $\lambda \in (0,\infty)$, there exists a ball covering $\{B_m^*\}_m \equiv \{B(x_{B_m^*}, r_{B_m^*})\}_m$ of $\widetilde{E}_\lambda$ such that

$$\sum_m (r_{B_m^*})^{n\tau(p\vee q)'} \sim \widetilde{\Lambda}_{n\tau(p\vee q)'}^{(\infty)}(\widetilde{E}_\lambda)$$

and

$$r_{B_m^*} > \operatorname{dist}(x_{B_m^*}, \partial\widetilde{E}_\lambda)/10000$$

for all m. For all $x=(x',x_n)\in\mathbb{R}^n$ and $t\in(0,\infty)$, define

$$\omega(x,t)\equiv\widetilde{\omega}(x',t)\chi_{[0,t)}(x_n).$$

Notice that if $N\omega(x',x_n)>\lambda^{\frac{1}{(p\vee q)'}}$, then

$$\widetilde{\omega}(y',t)=\omega(y',y_n,t)>\lambda^{\frac{1}{(p\vee q)'}}$$

for some $|(y',y_n)-(x',x_n)|<t$ and $y_n\in[0,t)$. Then $N\widetilde{\omega}(y')>\lambda^{\frac{1}{(p\vee q)'}}$ and thus, $y'\in B_m^*$ for some m. Since, for all $z'\in B(y',t)$,

$$N\widetilde{\omega}(z')\geq\widetilde{\omega}(y',t)>\lambda^{\frac{1}{(p\vee q)'}},$$

we see that

$$B(y',t)\subset\widetilde{E}_\lambda\subset\left(\bigcup_m B_m^*\right),$$

and hence $t\leq 10000r_{B_m^*}$. Since $x_n\in[0,t)$, we have

$$(x',x_n)\in(20000B_m^*)\times[0,20000r_{B_m^*})$$

and

$$E_\lambda\subset\bigcup_m(20000B_m^*)\times[0,20000r_{B_m^*}),$$

which further implies that

$$\Lambda^{(\infty)}_{n\tau(p\vee q)'}(E_\lambda)\lesssim\sum_m(r_{B_m^*})^{n\tau(p\vee q)'}\lesssim\widetilde{\Lambda}^{(\infty)}_{n\tau(p\vee q)'}(\widetilde{E}_\lambda)$$

and

$$\begin{aligned}\int_{\mathbb{R}^n}[N\omega(x',x_n)]^{(p\vee q)'}\,d\Lambda^{(\infty)}_{n\tau(p\vee q)'}(x)&=\int_0^\infty\Lambda^{(\infty)}_{n\tau(p\vee q)'}(E_\lambda)\,d\lambda\\&\lesssim\int_0^\infty\widetilde{\Lambda}^{(\infty)}_{n\tau(p\vee q)'}(\widetilde{E}_\lambda)\,d\lambda\\&\lesssim\int_{\mathbb{R}^{n-1}}[N\widetilde{\omega}(x')]^{(p\vee q)'}\,d\widetilde{\Lambda}^{(\infty)}_{n\tau(p\vee q)'}(x')\\&\lesssim 1.\end{aligned}$$

Therefore, we have

$$\begin{aligned}&\left\|\{[l(Q')]^{\frac{1}{2}}t_{Q'}\}_{Q'\in\mathcal{Q}(\mathbb{R}^{n-1}),l(Q')\leq1}\right\|_{bH^{s,\tau}_{p,q}(\mathbb{R}^n)}\\&\quad\lesssim\left\{\sum_{j=0}^\infty\left[\sum_{Q'\in\mathcal{Q}_j(\mathbb{R}^{n-1})}[l(Q')]^{-(\frac{s}{n}+\frac{1}{2})pn+\frac{p}{2}}|t_{Q'}|^p\int_{Q'\times[0,l(Q'))}[\omega(x,2^{-j})]^{-p}\,dx\right]^{\frac{q}{p}}\right\}^{\frac{1}{q}}\end{aligned}$$

$$\lesssim \left\{ \sum_{j=0}^{\infty} \left[\sum_{Q' \in \mathscr{Q}_j(\mathbb{R}^{n-1})} |Q'|^{-(\frac{s-1/p}{n-1}+\frac{1}{2})p} |t_{Q'}|^p \int_{Q'} [\widetilde{\omega}(x', 2^{-j})]^{-p}\, dx' \right]^{\frac{q}{p}} \right\}^{\frac{1}{q}}$$
$$\lesssim \|t\|_{bH_{p,q}^{s-\frac{1}{p},\frac{n}{n-1}\tau}(\mathbb{R}^{n-1})}$$
$$\lesssim \|f\|_{BH_{p,q}^{s-\frac{1}{p},\frac{n}{n-1}\tau}(\mathbb{R}^{n-1})},$$

which implies that $F \in BH_{p,q}^{s,\tau}(\mathbb{R}^n)$ and

$$\|F\|_{BH_{p,q}^{s,\tau}(\mathbb{R}^n)} \lesssim \|f\|_{BH_{p,q}^{s-\frac{1}{p},\frac{n}{n-1}\tau}(\mathbb{R}^{n-1})}.$$

Furthermore, the definition of F implies $\mathrm{Tr}(F) = f$, which completes the proof of Theorem 7.11. □

7.3 A (vmo, h^1)-Type Duality Result

When $p = q \in (1,\infty)$, applying the atomic decomposition of $AT_{p,p}^{s,\tau}(\mathbb{R}^n)$ in Proposition 7.1, we also find a predual space of $AH_{p,p}^{s,\tau}(\mathbb{R}^n)$. In what follows, we denote by ${}_0A_{p,p}^{s,\tau}(\mathbb{R}^n)$ *the closure of* $C_c^\infty(\mathbb{R}^n)$ *in* $A_{p,p}^{s,\tau}(\mathbb{R}^n)$. Recall that

$$C_c^\infty(\mathbb{R}^n) \subset \mathscr{S}(\mathbb{R}^n) \subset A_{p,p}^{s,\tau}(\mathbb{R}^n);$$

see Proposition 2.3.

Theorem 7.12. *Let* $s \in \mathbb{R}$, $p \in (1,\infty)$ *and* $\tau \in [0, \frac{1}{p}]$. *Then the dual space of* ${}_0A_{p,p}^{s,\tau}(\mathbb{R}^n)$ *is* $AH_{p',p'}^{-s,\tau}(\mathbb{R}^n)$ *in the following sense: if* $f \in AH_{p',p'}^{-s,\tau}(\mathbb{R}^n)$, *then the linear map*

$$\nu \mapsto \int_{\mathbb{R}^n} f(x)\nu(x)\, dx, \tag{7.22}$$

defined initially for all $\nu \in C_c^\infty(\mathbb{R}^n)$, *has a bounded extension to* ${}_0A_{p,p}^{s,\tau}(\mathbb{R}^n)$ *with the operator norm no more than a constant multiple of* $\|f\|_{AH_{p',p'}^{-s,\tau}(\mathbb{R}^n)}$.

Conversely, if $L \in ({}_0A_{p,p}^{s,\tau}(\mathbb{R}^n))^*$, *then there exists an* $f \in AH_{p',p'}^{-s,\tau}(\mathbb{R}^n)$ *with* $\|f\|_{AH_{p',p'}^{-s,\tau}(\mathbb{R}^n)}$ *no more than a constant multiple of* $\|L\|$ *such that* L *has the form* (7.22) *for all* $\nu \in C_c^\infty(\mathbb{R}^n)$.

We remark that Theorem 7.12 generalizes the classical result that

$$(\mathrm{cmo}\,(\mathbb{R}^n))^* = h^1(\mathbb{R}^n)$$

by taking $s=0$, $p=2$ and $\tau=1/2$, where $\mathrm{cmo}(\mathbb{R}^n)$ is the local $\mathrm{CMO}(\mathbb{R}^n)$ space and $h^1(\mathbb{R}^n)$ is the local Hardy space; see, for example, [43]. The homogeneous counterpart of Theorem 7.12 was already established in [166].

To prove Theorem 7.12, we need several functional analysis results and some technical conclusions. We first obtain the corresponding result for sequence spaces. Let ${}_0a^{s,\tau}_{p,p}(\mathbb{R}^n)$ be the subspace of $a^{s,\tau}_{p,p}(\mathbb{R}^n)$ consisting of all sequences with finite non-vanishing elements.

Proposition 7.6. *Let $s\in\mathbb{R}$, $p\in(1,\infty)$ and $\tau\in[0,\frac{1}{p}]$. Then*

$$({}_0a^{s,\tau}_{p,p}(\mathbb{R}^n))^* = aH^{-s,\tau}_{p',p'}(\mathbb{R}^n)$$

in the following sense: for each $t=\{t_Q\}_{l(Q)\le1}\in aH^{-s,\tau}_{p',p'}(\mathbb{R}^n)$, the map

$$\lambda\equiv\{\lambda_Q\}_{l(Q)\le1}\mapsto\langle\lambda,t\rangle\equiv\sum_{l(Q)\le1}\lambda_Q\overline{t_Q} \tag{7.23}$$

induces a continuous linear functional on ${}_0a^{s,\tau}_{p,p}(\mathbb{R}^n)$ with the operator norm no more than a constant multiple of $\|t\|_{aH^{-s,\tau}_{p',p'}(\mathbb{R}^n)}$. Conversely, every $L\in({}_0a^{s,\tau}_{p,p}(\mathbb{R}^n))^*$ is of the form (7.23) for a certain $t\in aH^{-s,\tau}_{p',p'}(\mathbb{R}^n)$ and $\|t\|_{aH^{-s,\tau}_{p',p'}(\mathbb{R}^n)}$ is no more than a constant multiple of $\|L\|$.

Proof. Since Proposition 7.6 when $\tau=0$ is just the classic result on Triebel-Lizorkin spaces, we only need consider the case that $\tau>0$. By Proposition 7.4 and the definition of ${}_0a^{s,\tau}_{p,p}(\mathbb{R}^n)$, we have that

$${}_0a^{s,\tau}_{p}(\mathbb{R}^n)\subset a^{s,\tau}_{p}(\mathbb{R}^n)=(aH^{-s,\tau}_{p',p'}(\mathbb{R}^n))^*,$$

which implies that

$$aH^{-s,\tau}_{p',p'}(\mathbb{R}^n)\subset(aH^{-s,\tau}_{p',p'}(\mathbb{R}^n))^{**}\subset({}_0a^{s,\tau}_{p,p}(\mathbb{R}^n))^*.$$

To show

$$({}_0a^{s,\tau}_{p,p}(\mathbb{R}^n))^*\subset aH^{-s,\tau}_{p',p'}(\mathbb{R}^n),$$

we first claim that if this is true when $s=0$, then it is also true for all $s\in\mathbb{R}$. In fact, for all $u\in\mathbb{R}$, define an operator T_u by setting, for all sequences $t\equiv\{t_Q\}_{l(Q)\le1}\subset\mathbb{C}$ and dyadic cubes Q satisfying $l(Q)\le1$, $(T_ut)_Q\equiv|Q|^{-\frac{u}{n}}t_Q$. Then T_u is an isometric isomorphism from $a^{s,\tau}_{p,p}(\mathbb{R}^n)$ to $a^{s+u,\tau}_{p,p}(\mathbb{R}^n)$ and from $aH^{s,\tau}_{p,p}(\mathbb{R}^n)$ to $aH^{s+u,\tau}_{p,p}(\mathbb{R}^n)$. If $L\in({}_0a^{s,\tau}_{p,p}(\mathbb{R})^*$, then $L\circ T_s\in({}_0a^{0,\tau}_{p,p}(\mathbb{R}^n))^*$ and hence there exists a sequence

$$\lambda\equiv\{\lambda_Q\}_{l(Q)\le1}\in aH^{0,\tau}_{p',p'}(\mathbb{R}^n)$$

such that

$$L \circ T_s(t) = \sum_{l(Q) \le 1} t_Q \overline{\lambda_Q}$$

for all $t \in {}_0a_p^{0,\tau}(\mathbb{R}^n)$. Since $T_s \circ T_{-s}$ is the identity on ${}_0a_{p,p}^{s,\tau}(\mathbb{R}^n)$ and T_{-s} is an isometric isomorphism from ${}_0a_{p,p}^{s,\tau}(\mathbb{R}^n)$ onto ${}_0a_{p,p}^{0,\tau}(\mathbb{R}^n)$, then

$$L(t) = L \circ T_s \circ T_{-s}(t) = \sum_{l(Q) \le 1} (T_{-s}t)_Q \overline{\lambda_Q} = \sum_{l(Q) \le 1} t_Q \overline{(T_{-s}\lambda)_Q}$$

for all $t \in {}_0a_{p,p}^{s,\tau}(\mathbb{R}^n)$. Since $\lambda \in aH_{p',p'}^{0,\tau}(\mathbb{R}^n)$, we see that $T_{-s}\lambda \in aH_{p',p'}^{-s,\tau}(\mathbb{R}^n)$ and

$$\|T_{-s}\lambda\|_{aH_{p',p'}^{-s,\tau}(\mathbb{R}^n)} = \|\lambda\|_{aH_{p',p'}^{0,\tau}(\mathbb{R}^n)}.$$

Thus, the above claim is true.

Next we prove that

$$({}_0a_{p,p}^{0,\tau}(\mathbb{R}^n))^* \subset aH_{p',p'}^{0,\tau}(\mathbb{R}^n).$$

Notice that ${}_0a_{p,p}^{0,\tau}(\mathbb{R}^n)$ consists of all sequences in $a_{p,p}^{0,\tau}(\mathbb{R}^n)$ with finite non-vanishing elements. We know that every $L \in ({}_0a_{p,p}^{0,\tau}(\mathbb{R}^n))^*$ is of the form

$$\lambda \mapsto \sum_{l(Q) \le 1} \lambda_Q \overline{t_Q}$$

for a certain $t \equiv \{t_Q\}_{l(Q) \le 1} \subset \mathbb{C}$. In fact, for any $m \in \mathbb{N}$, let ${}_0^m a_{p,p}^{0,\tau}(\mathbb{R}^n)$ denote the set of all sequences $\lambda \equiv \{\lambda_Q\}_{l(Q) \le 1} \in a_{p,p}^{0,\tau}(\mathbb{R}^n)$, where $\lambda_Q = 0$ if $Q \cap [-2^m, 2^m)^n = \emptyset$ or $l(Q) < 2^{-m}$. Then $L \in ({}_0^m a_{p,p}^{0,\tau}(\mathbb{R}^n))^*$. It is easy to see that each linear functional in $({}_0^m a_{p,p}^{0,\tau}(\mathbb{R}^n))^*$ has the form (7.23). Thus, there exists $t_m \equiv \{(t_m)_Q\}_{l(Q) \le 1}$, where $(t_m)_Q = 0$ if $Q \cap (-2^m, 2^m]^n = \emptyset$ or $l(Q) < 2^{-m}$, such that $L(\lambda)$ for all $\lambda \in {}_0^m a_{p,p}^{0,\tau}(\mathbb{R}^n)$ has the form (7.23) with t replaced by t_m. By this construction, we are easy to see that $(t_{m+1})_Q = (t_m)_Q$ if $Q \subset [-2^m, 2^m)^n$ and $2^{-m} \le l(Q) \le 1$. Thus, if let $t_Q \equiv (t_m)_Q$ when $Q \subset [-2^m, 2^m)^n$ and $2^{-m} \le l(Q) \le 1$, then $t \equiv \{t_Q\}_{l(Q) \le 1}$ is the desired sequence. To complete the proof of Proposition 7.6, we need to show that

$$\|t\|_{aH_{p',p'}^{0,\tau}(\mathbb{R}^n)} \lesssim \|L\|_{({}_0a_{p,p}^{0,\tau}(\mathbb{R}^n))^*}.$$

To this end, for all $m \in \mathbb{N}$, define χ_m by setting $\chi_m(Q) \equiv 1$ if $Q \subset [-2^m, 2^m)^n$ and $2^{-m} \le l(Q) \le 1$, $\chi_m(Q) \equiv 0$ otherwise. Then for all $\lambda \equiv \{\lambda_Q\}_{l(Q) \le 1} \in a_{p,p}^{0,\tau}(\mathbb{R}^n)$ with $\|\lambda\|_{a_{p,p}^{0,\tau}(\mathbb{R}^n)} \le 1$, we have

$$\lambda_m \equiv \{\lambda_Q \chi_m(Q)\}_{l(Q) \le 1} \in {}_0a_{p,p}^{0,\tau}(\mathbb{R}^n)$$

and $\|\lambda_m\|_{a^{0,\tau}_{p,p}(\mathbb{R}^n)} \le 1$. Thus, using Fatou's lemma yields

$$\begin{aligned}\sum_{l(Q)\le 1} |\lambda_Q||t_Q| &\le \lim_{m\to\infty} \sum_{l(Q)\le 1} |\lambda_Q|\chi_m(Q)|t_Q| \\ &= \lim_{m\to\infty} \sum_{l(Q)\le 1} \frac{|\lambda_Q|t_Q}{|t_Q|}\chi_m(Q)\overline{t_Q} \\ &\le \lim_{m\to\infty} \|L\|_{({}_0a^{0,\tau}_{p,p}(\mathbb{R}^n))^*}\|\lambda_m\|_{a^{0,\tau}_{p,p}(\mathbb{R}^n)} \\ &\le \|L\|_{({}_0a^{0,\tau}_{p,p}(\mathbb{R}^n))^*}. \end{aligned} \tag{7.24}$$

Notice that for all $m\in\mathbb{N}$,

$$t_m \equiv \{t_Q\chi_m(Q)\}_{l(Q)\le 1} \in aH^{0,\tau}_{p',p'}(\mathbb{R}^n).$$

For each m, we define function $F^{(m)}$ by setting, for all $x\in\mathbb{R}^n$ and $j\in\mathbb{Z}_+$,

$$F^{(m)}(x,2^{-j}) \equiv \sum_{Q\in\mathcal{Q}_j} |t_Q|\chi_m(Q)\widetilde{\chi}_Q(x).$$

Then

$$F^{(m)} \in AT^{0,\tau}_{p',p'}(\mathbb{R}^{n+1}_{\mathbb{Z}_+}) \quad\text{and}\quad \|F^{(m)}\|_{AT^{0,\tau}_{p',p'}(\mathbb{R}^{n+1}_{\mathbb{Z}_+})} \sim \|t_m\|_{aH^{0,\tau}_{p',p'}(\mathbb{R}^n)}.$$

Applying Theorem 7.2, we see that

$$\begin{aligned}\|F^{(m)}\|_{AT^{0,\tau}_{p',p'}(\mathbb{R}^{n+1}_{\mathbb{Z}_+})} &\lesssim \sup\left\{\left|\int_{\mathbb{R}^n}\sum_{j=0}^{\infty} F^{(m)}(x,2^{-j})G(x,2^{-j})\,dx\right|\right\} \\ &\lesssim \sup\left\{\left|\sum_{j=0}^{\infty}\sum_{Q\in\mathcal{Q}_j} |t_Q|\chi_m(Q)|Q|^{-1/2}\int_Q G(x,2^{-j})\,dx\right|\right\},\end{aligned}$$

where the supremum is taken over all functions $G\in AW^{0,\tau}_{p,p}(\mathbb{R}^{n+1}_{\mathbb{Z}_+})$ with compact support satisfying $\|G\|_{AW^{0,\tau}_{p,p}(\mathbb{R}^{n+1}_{\mathbb{Z}_+})} \le 1$. If we set

$$\lambda_Q \equiv |Q|^{-1/2}\int_Q G(x,2^{-j})\,dx$$

and $\lambda \equiv \{\lambda_Q\}_{l(Q)\le 1}$, then using Hölder's inequality, we obtain

$$\|\lambda\|_{a^{0,\tau}_{p,p}(\mathbb{R}^n)} \lesssim \|G\|_{AW^{0,\tau}_{p,p}(\mathbb{R}^{n+1}_{\mathbb{Z}_+})} \lesssim 1,$$

and hence

$$\begin{aligned}\|t_m\|_{aH^{0,\tau}_{p',p'}(\mathbb{R}^n)} &\sim \|F^{(m)}\|_{AT^{0,\tau}_{p',p'}(\mathbb{R}^{n+1}_{\mathbb{Z}_+})}\\ &\lesssim \sup\left\{\sum_{l(Q)\le 1}|\lambda_Q||t_Q| : \lambda \in a^{0,\tau}_{p,p}(\mathbb{R}^n),\ \|\lambda\|_{a^{0,\tau}_{p,p}(\mathbb{R}^n)} \le 1\right\},\end{aligned}$$

which together with (7.24) yields

$$\|t_m\|_{aH^{0,\tau}_{p',p'}(\mathbb{R}^n)} \sim \|F^{(m)}\|_{AT^{0,\tau}_{p',p'}(\mathbb{R}^{n+1}_{\mathbb{Z}_+})} \lesssim \|L\|_{({}_0a^{0,\tau}_{p,p}(\mathbb{R}^n))^*}.$$

To show $t \in aH^{0,\tau}_{p',p'}(\mathbb{R}^n)$, let F be the function on $\mathbb{R}^{n+1}_{\mathbb{Z}_+}$ defined by setting, for all $x \in \mathbb{R}^n$ and $j \in \mathbb{Z}_+$,

$$F(x, 2^{-j}) \equiv \sum_{Q\in\mathcal{Q}_j} |t_Q|\widetilde{\chi}_Q(x).$$

Notice that

$$\|t\|_{aH^{0,\tau}_{p',p'}(\mathbb{R}^n)} \sim \|F\|_{AT^{0,\tau}_{p',p'}(\mathbb{R}^{n+1}_{\mathbb{Z}_+})}.$$

It suffices to prove that $F \in AT^{0,\tau}_{p',p'}(\mathbb{R}^{n+1}_{\mathbb{Z}_+})$.

Recall that

$$\|F^{(m)}\|_{AT^{0,\tau}_{p',p'}(\mathbb{R}^{n+1}_{\mathbb{Z}_+})} \lesssim \|L\|_{({}_0a^{0,\tau}_{p,p}(\mathbb{R}^n))^*}.$$

By Lemma 7.6, there exist a subsequence $\{F^{(m_i)}\}_{i\in\mathbb{N}}$ and $\widetilde{F} \in AT^{0,\tau}_{p,p}(\mathbb{R}^{n+1}_{\mathbb{Z}_+})$ such that for all $G \in AW^{0,\tau}_{p',p'}(\mathbb{R}^{n+1}_{\mathbb{Z}_+})$ with compact support,

$$\langle F^{(m_i)}, G\rangle \to \langle \widetilde{F}, G\rangle$$

as $i \to \infty$ and its quasi-norm

$$\|\widetilde{F}\|_{AT^{0,\tau}_{p,p}(\mathbb{R}^{n+1}_{\mathbb{Z}_+})} \lesssim \|L\|_{({}_0a^{0,\tau}_{p,p}(\mathbb{R}^n))^*},$$

which together with the uniqueness of the weak limit and the fact that $F^{(m)} \to F$ pointwise as $m \to \infty$ yields that $F = \widetilde{F}$ in $AT^{0,\tau}_{p,p}(\mathbb{R}^{n+1}_{\mathbb{Z}_+})$ and

$$\|F\|_{AT^{0,\tau}_{p,p}(\mathbb{R}^{n+1}_{\mathbb{Z}_+})} \lesssim \|L\|_{({}_0a^{0,\tau}_{p,p}(\mathbb{R}^n))^*}.$$

This finishes the proof of Proposition 7.6. □

Let $\widetilde{{}_0A^{s,\tau}_{p,p}(\mathbb{R}^n)}$ denote the *closure of* $\mathcal{S}(\mathbb{R}^n)$ *in* $A^{s,\tau}_{p,p}(\mathbb{R}^n)$. As an immediate consequence of Proposition 7.6 and the φ-transform characterizations of $AH^{s,\tau}_{p,q}(\mathbb{R}^n)$ and $A^{s,\tau}_{p,q}(\mathbb{R}^n)$, we have the following theorem, which generalizes the classical results on Besov spaces and Triebel-Lizorkin spaces when $p = q$; see, for example, [145, p. 180].

Theorem 7.13. *Let $s \in \mathbb{R}$, $p \in (1,\infty)$ and $\tau \in [0, \frac{1}{p}]$. Then the dual space of $\widetilde{{}_0A^{s,\tau}_{p,p}(\mathbb{R}^n)}$ is $AH^{-s,\tau}_{p',p'}(\mathbb{R}^n)$ in the following sense: If $f \in AH^{-s,\tau}_{p',p'}(\mathbb{R}^n)$, then the linear map defined as in (7.22) for all $v \in \mathcal{S}(\mathbb{R}^n)$, has a bounded extension to $\widetilde{{}_0A^{s,\tau}_{p,p}(\mathbb{R}^n)}$ with operator norm no more than a constant multiple of $\|f\|_{AH^{-s,\tau}_{p',p'}(\mathbb{R}^n)}$.*

Conversely, if $L \in (\widetilde{{}_0A^{s,\tau}_{p,p}(\mathbb{R}^n)})^*$, then there exists an $f \in AH^{-s,\tau}_{p',p'}(\mathbb{R}^n)$ with $\|f\|_{AH^{-s,\tau}_{p',p'}(\mathbb{R}^n)}$ no more than a constant multiple of $\|L\|$ such that L has the form (7.22) for all $v \in \mathcal{S}(\mathbb{R}^n)$.

Proof. Since the case that $\tau = 0$ is known (see [145, p. 180]), we only need consider the case that $\tau > 0$. By Theorem 7.3 and the definition of $\widetilde{{}_0A^{s,\tau}_{p,p}(\mathbb{R}^n)}$, we have that

$$\widetilde{{}_0A^{s,\tau}_{p,p}(\mathbb{R}^n)} \subset A^{s,\tau}_{p,p}(\mathbb{R}^n) = (AH^{-s,\tau}_{p',p'}(\mathbb{R}^n))^*,$$

which implies that

$$AH^{-s,\tau}_{p',p'}(\mathbb{R}^n) \subset (AH^{-s,\tau}_{p',p'}(\mathbb{R}^n))^{**} \subset (\widetilde{{}_0A^{s,\tau}_{p,p}(\mathbb{R}^n)})^*.$$

To show

$$(\widetilde{{}_0A^{s,\tau}_{p,p}(\mathbb{R}^n)})^* \subset AH^{-s,\tau}_{p',p'}(\mathbb{R}^n),$$

let Φ and φ satisfy, respectively, (2.1) and (2.2) such that (2.6) holds with Ψ and ψ replaced, respectively, by Φ and φ. If $L \in (\widetilde{{}_0A^{s,\tau}_{p,p}(\mathbb{R}^n)})^*$, then applying Theorem 7.4, we see that

$$\widetilde{L} \equiv L \circ T_\varphi \in ({}_0a^{s,\tau}_{p,p}(\mathbb{R}^n))^*.$$

By Proposition 7.6, there exists a $\lambda \equiv \{\lambda_Q\}_{l(Q) \le 1} \in aH^{-s,\tau}_{p',p'}(\mathbb{R}^n)$ such that

$$\widetilde{L}(t) = \sum_{l(Q) \le 1} t_Q \overline{\lambda_Q}$$

for all $t \equiv \{t_Q\}_{l(Q) \le 1} \in {}_0a^{s,\tau}_{p,p}(\mathbb{R}^n)$ and

$$\|\lambda\|_{aH^{-s,\tau}_{p',p'}(\mathbb{R}^n)} \lesssim \|\widetilde{L}\|_{({}_0a^{s,\tau}_{p,p}(\mathbb{R}^n))^*} \lesssim \|L\|_{(\widetilde{{}_0A^{s,\tau}_{p,p}(\mathbb{R}^n)})^*}.$$

Notice that

$$\widetilde{L} \circ S_\varphi = L \circ T_\varphi \circ S_\varphi = L.$$

Thus, for all $f \in \mathcal{S}(\mathbb{R}^n)$, if letting

$$g \equiv T_\varphi(\lambda) \equiv \sum_{l(Q) \le 1} \lambda_Q \varphi_Q,$$

then

$$L(f) = \widetilde{L} \circ S_\varphi(f) = \sum_{l(Q)\le 1} (S_\varphi f)_Q \overline{\lambda_Q} = \langle f, g\rangle.$$

Furthermore, by Theorem 7.4 again, we have

$$\|g\|_{AH^{-s,\tau}_{p',p'}(\mathbb{R}^n)} \lesssim \|\lambda\|_{aH^{-s,\tau}_{p',p'}(\mathbb{R}^n)} \lesssim \|L\|_{(\widetilde{_0A^{s,\tau}_{p,p}(\mathbb{R}^n)})^*}.$$

This finishes the proof of Theorem 7.13. □

Now we are ready to prove Theorem 7.12.

Proof of Theorem 7.12. By Proposition 2.3, we see that

$$C_c^\infty(\mathbb{R}^n) \subset \mathscr{S}(\mathbb{R}^n) \subset A^{s,\tau}_{p,p}(\mathbb{R}^n),$$

and hence

$$_0A^{s,\tau}_{p,p}(\mathbb{R}^n) \subset \widetilde{_0A^{s,\tau}_{p,p}(\mathbb{R}^n)}.$$

Therefore, to obtain Theorem 7.12, by Theorem 7.13, it suffices to prove that

$$\widetilde{_0A^{s,\tau}_{p,p}(\mathbb{R}^n)} \subset {_0A^{s,\tau}_{p,p}(\mathbb{R}^n)}.$$

Let $f \in \widetilde{_0A^{s,\tau}_{p,p}(\mathbb{R}^n)}$ and $\varepsilon > 0$. By the definition of $\widetilde{_0A^{s,\tau}_{p,p}(\mathbb{R}^n)}$, there exists a function $g \in \mathscr{S}(\mathbb{R}^n)$ such that

$$\|f-g\|_{A^{s,\tau}_{p,p}(\mathbb{R}^n)} < \varepsilon/2.$$

Thus, to complete the proof, it suffices to find a function $h \in C_c^\infty(\mathbb{R}^n)$ such that

$$\|g-h\|_{A^{s,\tau}_{p,p}(\mathbb{R}^n)} < \varepsilon/2.$$

By the proof of Proposition 2.3, we know that for all $\varphi \in \mathscr{S}(\mathbb{R}^n)$,

$$\|\varphi\|_{A^{s,\tau}_{p,p}(\mathbb{R}^n)} \le C\|\varphi\|_{\mathscr{S}_M}$$

when

$$M > \max\{0, s+n\tau, n(1/p-1)\}+1.$$

On the other hand, since $g \in \mathscr{S}(\mathbb{R}^n)$, for each fixed $M \in \mathbb{N}$, there exists a function $h \in C_c^\infty(\mathbb{R}^n)$ such that $\|g-h\|_{\mathscr{S}_M} < \varepsilon/(2C)$. Thus, we have

$$\|g-h\|_{A^{s,\tau}_{p,p}(\mathbb{R}^n)} \le C\|g-h\|_{\mathscr{S}_M} < \varepsilon/2,$$

which completes the proof of Theorem 7.12. □

The proof of Theorem 7.12 implies that ${}_0\widetilde{A^{s,\tau}_{p,p}(\mathbb{R}^n)} = {}_0A^{s,\tau}_{p,p}(\mathbb{R}^n)$.

We have the following interesting remark.

Remark 7.7.

(i) We first claim that when $\tau > 0$, the dual property in Theorem 7.12 is not possible to be correct for ${}_0B^{s,\tau}_{p,q}(\mathbb{R}^n)$ and $BH^{-s,\tau}_{p',q'}(\mathbb{R}^n)$ with $p \in (1,\infty)$, $q \in [1,\infty)$ and $q > p$, which is quite different from the case that $\tau = 0$. Recall that when $\tau = 0$, $p \in (1,\infty)$ and $q \in [1,\infty)$,

$${}_0B^{s,\tau}_{p,q}(\mathbb{R}^n) = B^{s,\tau}_{p,q}(\mathbb{R}^n) \quad \text{and} \quad (\mathrm{B}^{s,\tau}_{p,q}(\mathbb{R}^n))^* = \mathrm{B}^{-s,\tau}_{p',q'}(\mathbb{R}^n);$$

see [145, p. 244].

To show the claim, by Remark 7.5 (see also [168, Propositions 2.2(i) and 2.3(i)]), we know that if $1 < p_0 < p_1 < \infty$, $-\infty < s_1 < s_0 < \infty$, $q \in [1,\infty)$ and

$$\tau \in \left[0, \min\left\{\frac{1}{(p_0 \vee q)'}, \frac{1}{(p_q \vee q)'}\right\}\right]$$

such that $s_0 - n/p_0 = s_1 - n/p_1$, then

$$BH^{s_0,\tau}_{p_0,q}(\mathbb{R}^n) \subset BH^{s_1,\tau}_{p_1,q}(\mathbb{R}^n) \Longleftrightarrow \tau(p_0 \vee q)' = \tau(p_1 \vee q)'.$$

When $\tau > 0$, the sufficient and necessary condition that $\tau(p_0 \vee q)' = \tau(p_1 \vee q)'$ is equivalent to that $q \geq p_1$. If we assume that Theorem 7.12 is correct for ${}_0B^{s,\tau}_{p,q}(\mathbb{R}^n)$ and $BH^{-s,\tau}_{p',q'}(\mathbb{R}^n)$ with $\tau > 0$ and certain $1 < p < q < \infty$, then by this assumption together with an argument by duality and the embedding

$$B^{s,\tau}_{p,q}(\mathbb{R}^n) \subset B^{s-n/p+n/q,\tau}_{q,q}(\mathbb{R}^n)$$

in Corollary 2.2, we see that

$$BH^{-s+n/p-n/q,\tau}_{q',q'}(\mathbb{R}^n) \subset BH^{s,\tau}_{p',q'}(\mathbb{R}^n),$$

which is not true since $q' < p'$. Thus, the claim is true.

From the above claim, it follows that if $\tau > 0$ and $p \neq q$, only when $1 \leq q < p < \infty$, the conclusion of Theorem 7.12 may be true for the spaces ${}_0B^{s,\tau}_{p,q}(\mathbb{R}^n)$ and $BH^{-s,\tau}_{p',q'}(\mathbb{R}^n)$, which is unclear so far to us; see also Remark 7.2.

(ii) Similarly, we claim that when $\tau > 0$, the dual property in Theorem 7.12 is not possible to be correct for all ${}_0F^{s,\tau}_{p,q}(\mathbb{R}^n)$ and $FH^{-s,\tau}_{p',q'}(\mathbb{R}^n)$ with $p, q \in (1,\infty)$ and $q > p$. In fact, by Remark 7.5, we know that the embedding

$$FH^{s_0,\tau}_{p_0,r}(\mathbb{R}^n) \subset FH^{s_1,\tau}_{p_1,q}(\mathbb{R}^n)$$

is true only when

$$\tau(p_0 \vee r)' \leq \tau(p_1 \vee q)' + \tau(1/p_0 - 1/p_1)(p_0 \vee r)'(p_1 \vee q)'.$$

If we assume that

$$({}_0F^{s,\tau}_{p,q}(\mathbb{R}^n))^* = FH^{-s,\tau}_{p',q'}(\mathbb{R}^n)$$

for all $s \in \mathbb{R}$, $\tau > 0$ and $1 < p < q < \infty$, then by the embedding

$$F^{s,\tau}_{p,q}(\mathbb{R}^n) \subset F^{s-n/p+n/q,\tau}_{q,r}(\mathbb{R}^n)$$

in Corollary 2.2 with $r > q$ together with an argument by duality, we have

$$FH^{-s+n/p-n/q,\tau}_{q',r'}(\mathbb{R}^n) \subset FH^{-s,\tau}_{p',q'}(\mathbb{R}^n),$$

which is not possible by the above conclusion. Thus, the claim is also true. It is also unclear that when $p \neq q$, for which range of $p, q \in (1,\infty)$, the conclusion of Theorem 7.12 is true.

By Theorem 7.12 and Corollary 2.2, a dual argument yields that following conclusion, which improves Proposition 7.3(ii) in the case that $p = q$.

Proposition 7.7. *Let $s_0, s_1 \in \mathbb{R}$, $p_0, p_1 \in (1,\infty)$ and $\tau \in [0, \frac{1}{p'}]$ such that $p_0 < p_1$ and $s_0 - n/p_0 = s_1 - n/p_1$. Then*

$$FH^{s_0,\tau}_{p_0,p_0}(\mathbb{R}^n) \subset FH^{s_1,\tau}_{p_1,p_1}(\mathbb{R}^n).$$

Also, from Proposition 7.6, Theorem 3.1 and a dual argument, we deduce the following result.

Proposition 7.8. *Let $\varepsilon \in (0,\infty)$, $s \in \mathbb{R}$, $p \in (1,\infty)$ and $\tau \in [0, 1/p']$. Then all the ε-almost diagonal operators are bounded on $aH^{s,\tau}_{p,p}(\mathbb{R}^n)$.*

We remark that Proposition 7.8 improves Theorem 7.5 in the case that $p = q$, since in Theorem 7.5, we need an additional condition that $\varepsilon > 2n\tau$.

From Proposition 7.8 and the arguments in Sect. 7.2, we deduce that when $p = q$, the smooth atomic and molecular decomposition characterizations of $AH^{s,\tau}_{p,q}(\mathbb{R}^n)$ in Theorems 7.6 and 7.7 can be improved. Precisely, in the case that $p = q$, via replacing the conditions $N \equiv \max\{\lfloor s + 2n\tau \rfloor, -1\}$, $M \in (n + 2n\tau, \infty)$ and $|\gamma| \le \lfloor s + 3n\tau \rfloor$ in Definition 7.7, respectively, by $N \equiv \max\{\lfloor s \rfloor, -1\}$, $M \in (n, \infty)$ and $|\gamma| \le \lfloor s + n\tau \rfloor$, we obtain a class of "weaker" molecules. Proposition 7.8 and the arguments in Sect. 7.2 then yield that Theorem 7.6 is still true for these new molecules in the case that $p = q$. Also, via replacing the conditions

$$|\gamma| \le \max\{\lfloor -s + 2n\tau \rfloor, -1\} \quad \text{and} \quad |\gamma| \le \max\{\lfloor s + 3n\tau + 1 \rfloor, -1\}$$

in Definition 7.8, respectively, by

$$|\gamma| \le \max\{\lfloor -s \rfloor, -1\} \quad \text{and} \quad |\gamma| \le \max\{\lfloor s + n\tau + 1 \rfloor, -1\},$$

we obtain a class of "weaker" atoms. Proposition 7.8 and the arguments in Sect. 7.2 then yield that in the case that $p = q$, Theorem 7.7 is still true for these new atoms in the case that $p = q$.

Via these improvements, in the case that $p = q$, the conditions $s > 2n\tau$ or $s \leq 2n\tau$ in Theorem 7.8 can be replaced, respectively, by $s > 0$ or $s \leq 0$, and the condition $s \in (1/p + 2n\tau, \infty)$ in Theorem 7.11 can be replaced by $s \in (1/p, \infty)$. We omit the details.

7.4 Real Interpolation

In this section we are concerned with the interpolation properties of the spaces $AH_{p,q}^{s,\tau}(\mathbb{R}^n)$. Nowadays interpolation theory is a well established tool in various branches of mathematics, but in particular in the theory of partial differential equations.

To establish the real interpolation properties of $AH_{p,q}^{s,\tau}(\mathbb{R}^n)$, we need some preparations (see, for example, [145, pp. 62–63]). Let $\mathscr{H}$ be a linear complex Hausdorff space and A_0 and A_1 be complex quasi-Banach spaces such that $A_0, A_1 \subset \mathscr{H}$. Let $A_0 + A_1$ be the set of all elements $a \in \mathscr{H}$ such that a can be represented as $a = a_0 + a_1$ with $a_0 \in A_0$ and $a_1 \in A_1$. As usual, for $t \in (0, \infty)$ and $a \in A_0 + A_1$, Peetre's celebrated K-functional is defined by

$$K(t, a; A_0, A_1) \equiv \inf(\|a_0\|_{A_0} + t\|a_1\|_{A_1}),$$

where the infimum is taken over all representations of a of the above form.

Let $\theta \in (0,1)$ and $q \in (0, \infty]$. The interpolation space $(A_0, A_1)_{\theta,q}$ is defined to be the set of all $a \in A_0 + A_1$ such that $\|a\|_{(A_0,A_1)_{\theta,q}} < \infty$, where

$$\|a\|_{(A_0,A_1)_{\theta,q}} \equiv \left(\int_0^\infty [t^{-\theta} K(t, a; A_0, A_1)]^q \frac{dt}{t} \right)^{1/q}$$

with suitable modifications when $q = \infty$.

Lemma 7.9 is the basis for the real interpolation theory of $AH_{p,q}^{s,\tau}(\mathbb{R}^n)$, which shows that $\mathscr{S}'(\mathbb{R}^n)$ can be identified as the Hausdorff space $\mathscr{H}$ mentioned above. The following result partially generalizes [145, Theorem 2.4.2].

Theorem 7.14. *Let $\theta \in (0,1)$, q_0, q_1, $q \in [1, \infty)$, $p \in (1, \infty)$, $\tau \in [0, 1/p']$ and s_0, $s_1 \in (0, \infty)$ satisfy $s_0 \neq s_1$, $s = (1-\theta)s_0 + \theta s_1$ and*

$$\tau(p \vee q)' = \tau(p \vee q_0)' = \tau(p \vee q_1)'.$$

Then

$$(AH_{p,q_0}^{s_0,\tau}(\mathbb{R}^n), AH_{p,q_1}^{s_1,\tau}(\mathbb{R}^n))_{\theta,q} = BH_{p,q}^{s,\tau}(\mathbb{R}^n).$$

Proof. When $\tau = 0$, Theorem 7.14 is just the classic result obtained in [145, Theorem 2.4.2]. We only consider the case when $\tau > 0$, under which the restriction that

$$\tau(p \vee q)' = \tau(p \vee q_0)' = \tau(p \vee q_1)'$$

implies that $p \geq \max\{q_0, q_1, q\}$. Let $q_2 \in [q_0 \vee q_1, p]$. We first show that

$$(BH^{s_0,\tau}_{p,q_2}(\mathbb{R}^n), BH^{s_1,\tau}_{p,q_2}(\mathbb{R}^n))_{\theta,q} \subset BH^{s,\tau}_{p,q}(\mathbb{R}^n). \tag{7.25}$$

Without loss of generality, we may assume that $s_0 > s_1$. Notice that

$$(0,\infty) = \bigcup_{k=-\infty}^{\infty} [2^{(k-1)(s_0-s_1)}, 2^{k(s_0-s_1)}).$$

Then

$$\begin{aligned}
&\int_0^\infty t^{-\theta q}[K(t,f;BH^{s_0,\tau}_{p,q_2}(\mathbb{R}^n), BH^{s_1,\tau}_{p,q_2}(\mathbb{R}^n))]^q \frac{dt}{t} \\
&\quad \gtrsim \sum_{k=0}^{\infty} 2^{-\theta q k(s_0-s_1)}[K(2^{k(s_0-s_1)},f;BH^{s_0,\tau}_{p,q_2}(\mathbb{R}^n), BH^{s_1,\tau}_{p,q_2}(\mathbb{R}^n))]^q.
\end{aligned}$$

Write $f \equiv f_0 + f_1$ with $f_0 \in BH^{s_0,\tau}_{p,q_2}(\mathbb{R}^n)$ and $f_1 \in BH^{s_1,\tau}_{p,q_2}(\mathbb{R}^n)$ such that

$$\begin{aligned}
&\|f_0\|_{BH^{s_0,\tau}_{p,q_2}(\mathbb{R}^n))} + 2^{k(s_0-s_1)}\|f_1\|_{BH^{s_1,\tau}_{p,q_2}(\mathbb{R}^n))} \\
&\quad \leq 2K(2^{k(s_0-s_1)},f;BH^{s_0,\tau}_{p,q_2}(\mathbb{R}^n), BH^{s_1,\tau}_{p,q_2}(\mathbb{R}^n)).
\end{aligned} \tag{7.26}$$

There exist ω_0, ω_1 satisfying (7.3) with q replaced by q_2 such that for $i = 0, 1$,

$$\sum_{k=0}^{\infty} 2^{ks_i q_i} \left\| \varphi_k * f_i[\omega_i(\cdot,2^{-k})]^{-1} \right\|^{q_i}_{L^p(\mathbb{R}^n)} \lesssim \|f_i\|^{q_i}_{BH^{s_i,\tau}_{p,q_2}(\mathbb{R}^n))}. \tag{7.27}$$

Set $\omega \equiv (\omega_0 + \omega_1)/2$. Then (7.27) remains true if we replace ω_i by ω, which together with (7.26) further yields that

$$\begin{aligned}
&2^{ks_0}\left\|\varphi_k * f[\omega(\cdot,2^{-k})]^{-1}\right\|_{L^p(\mathbb{R}^n)} \\
&\quad \leq 2^{ks_0}\left\|\varphi_k * f_0[\omega(\cdot,2^{-k})]^{-1}\right\|_{L^p(\mathbb{R}^n)} + 2^{k(s_0-s_1)}2^{ks_1}\left\|\varphi_k * f_1[\omega(\cdot,2^{-k})]^{-1}\right\|_{L^p(\mathbb{R}^n)} \\
&\quad \leq \|f_0\|_{BH^{s_0,\tau}_{p,q_2}(\mathbb{R}^n)} + 2^{k(s_0-s_1)}\|f_1\|_{BH^{s_1,\tau}_{p,q_2}(\mathbb{R}^n)} \\
&\quad \lesssim K(2^{k(s_0-s_1)},f;BH^{s_0,\tau}_{p,q_2}(\mathbb{R}^n), BH^{s_1,\tau}_{p,q_2}(\mathbb{R}^n)).
\end{aligned}$$

Notice that ω also satisfies (7.3). We then have

$$\begin{aligned}\|f\|^q_{BH^{s,\tau}_{p,q}(\mathbb{R}^n)} &\lesssim \sum_{k=0}^{\infty} 2^{ksq} \left\| \varphi_k * f[\omega(\cdot,2^{-k})]^{-1} \right\|^q_{L^p(\mathbb{R}^n)} \\ &\lesssim \sum_{k=0}^{\infty} 2^{k(s-s_0)q} \left[K(2^{k(s_0-s_1)}, f; BH^{s_0,\tau}_{p,q_2}(\mathbb{R}^n), BH^{s_1,\tau}_{p,q_2}(\mathbb{R}^n)) \right]^q \\ &\lesssim \int_0^{\infty} t^{-\theta q} [K(t,f;BH^{s_0,\tau}_{p,q_2}(\mathbb{R}^n), BH^{s_1,\tau}_{p,q_2}(\mathbb{R}^n))]^q \frac{dt}{t}.\end{aligned}$$

This implies that (7.25) holds.

Let $r \in [1, \min\{q_0, q_1\}]$. Next we prove that

$$BH^{s,\tau}_{p,q}(\mathbb{R}^n) \subset (BH^{s_0,\tau}_{p,r}(\mathbb{R}^n), BH^{s_1,\tau}_{p,r}(\mathbb{R}^n))_{\theta,q}. \tag{7.28}$$

Since $s > s_1$, applying Hölder's inequality concludes that

$$BH^{s,\tau}_{p,q}(\mathbb{R}^n) \subset BH^{s_1,\tau}_{p,r}(\mathbb{R}^n),$$

which further implies that

$$K(t,f;BH^{s_0,\tau}_{p,r}(\mathbb{R}^n), BH^{s_1,\tau}_{p,r}(\mathbb{R}^n)) \lesssim t\|f\|_{BH^{s_1,\tau}_{p,r}(\mathbb{R}^n)} \lesssim t\|f\|_{BH^{s,\tau}_{p,q}(\mathbb{R}^n)}.$$

Thus,

$$\int_0^1 t^{-\theta q}[K(t,f;BH^{s_0,\tau}_{p,r}(\mathbb{R}^n), BH^{s_1,\tau}_{p,r}(\mathbb{R}^n))]^q \frac{dt}{t} \lesssim \|f\|^q_{BH^{s,\tau}_{p,q}(\mathbb{R}^n)}.$$

It remains to estimate

$$\mathrm{I} \equiv \int_1^{\infty} t^{-\theta q}[K(t,f;BH^{s_0,\tau}_{p,r}(\mathbb{R}^n), BH^{s_1,\tau}_{p,r}(\mathbb{R}^n))]^q \frac{dt}{t}.$$

Similarly, we have

$$\mathrm{I} \lesssim \sum_{k=0}^{\infty} 2^{-\theta qk(s_0-s_1)}[K(2^{k(s_0-s_1)}, f; BH^{s_0,\tau}_{p,r}(\mathbb{R}^n), BH^{s_1,\tau}_{p,r}(\mathbb{R}^n))]^q.$$

Let Φ and φ satisfy, respectively, (2.1) and (2.2). Assume further that (2.6) holds with Ψ and ψ replaced, respectively, by Φ and φ. Then by the Calderón reproducing formula in Lemma 2.3, we can write $f \equiv f_0 + f_1$ with

$$f_0 \equiv \sum_{j=0}^{k} \varphi_j * f \quad \text{and} \quad f_1 \equiv \sum_{j=k+1}^{\infty} \varphi_j * f,$$

where when $j=0$, φ_0 is replaced by Φ. Then

$$\mathrm{I}\lesssim\sum_{k=0}^{\infty}2^{kq(s-s_0)}\|f_0\|^q_{BH^{s_0,\tau}_{p,r}(\mathbb{R}^n)}+\sum_{k=0}^{\infty}2^{kqs-kqs_0}2^{kq(s_0-s_1)}\|f_1\|^q_{BH^{s_1,\tau}_{p,r}(\mathbb{R}^n)}\equiv\mathrm{I}_1+\mathrm{I}_2.$$

Notice that $\varphi_m*\varphi_j\equiv 0$ if $|m-j|>1$. For I_1, we have

$$\begin{aligned}\mathrm{I}_1&=\sum_{k=0}^{\infty}2^{kq(s-s_0)}\inf_{\omega}\left\{\sum_{m=0}^{\infty}2^{mrs_0}\left[\int_{\mathbb{R}^n}|\varphi_m*f_0|^p[\omega(x,2^{-m})]^{-p}\,dx\right]^{r/p}\right\}^{q/r}\\
&\le\inf_{\omega}\sum_{k=0}^{\infty}2^{kq(s-s_0)}\left\{\sum_{m=0}^{\infty}2^{mrs_0}\left[\int_{\mathbb{R}^n}|\varphi_m*f_0|^p[\omega(x,2^{-m})]^{-p}\,dx\right]^{r/p}\right\}^{q/r}\\
&\le\inf_{\omega}\sum_{k=0}^{\infty}2^{kq(s-s_0)}\left\{\sum_{j=0}^{k}\sum_{m=(j-1)\vee 0}^{j+1}2^{mrs_0}\right.\\
&\qquad\times\left.\left[\int_{\mathbb{R}^n}|\varphi_m*\varphi_j*\varphi_j*f|^p[\omega(x,2^{-m})]^{-p}\,dx\right]^{r/p}\right\}^{q/r}.\end{aligned}$$

Let $t_0\in(s,s_0)$. By Hölder's inequality with $r/q+r/\sigma=1$, we further have

$$\begin{aligned}\mathrm{I}_1&\lesssim\inf_{\omega}\sum_{k=0}^{\infty}2^{kq(s-s_0)}\left(\sum_{j=0}^{k}\sum_{m=(j-1)\vee 0}^{j+1}2^{m\sigma(s_0-t_0)}\right)^{q/\sigma}\sum_{j=0}^{k}\sum_{m=(j-1)\vee 0}^{j+1}2^{mqt_0}\\
&\quad\times\left[\int_{\mathbb{R}^n}|\varphi_m*\varphi_j*\varphi_j*f|^p[\omega(x,2^{-m})]^{-p}\,dx\right]^{q/p}\\
&\lesssim\inf_{\omega}\sum_{k=0}^{\infty}2^{kq(s-t_0)}\sum_{j=0}^{k}\sum_{m=(j-1)\vee 0}^{j+1}2^{mqt_0}\\
&\quad\times\left[\int_{\mathbb{R}^n}|\varphi_m*\varphi_j*\varphi_j*f|^p[\omega(x,2^{-m})]^{-p}\,dx\right]^{q/p}\\
&\lesssim\inf_{\omega}\sum_{j=0}^{\infty}2^{jq(s-t_0)}\sum_{m=(j-1)\vee 0}^{j+1}2^{mqt_0}\left[\int_{\mathbb{R}^n}|\varphi_m*\varphi_j*\varphi_j*f|^p[\omega(x,2^{-m})]^{-p}\,dx\right]^{q/p}.\end{aligned}$$

Similarly to the proof of [164, Propostion 5.1], we obtain that the last line of the above inequalities can be dominated by $\|f\|_{BH^{s,\tau}_{p,q}(\mathbb{R}^n)}$, and hence $\mathrm{I}_1\lesssim\|f\|_{BH^{s,\tau}_{p,q}(\mathbb{R}^n)}$. The proof of the estimate that $\mathrm{I}_2\lesssim\|f\|_{BH^{s,\tau}_{p,q}(\mathbb{R}^n)}$ is similar. In fact, for I_2, we also have

$$\begin{aligned}\mathrm{I}_2&\le\inf_{\omega}\sum_{k=0}^{\infty}2^{kq(s-s_1)}\left\{\sum_{j=k+1}^{\infty}\sum_{m=(j-1)\vee 0}^{j+1}2^{mrs_1}\right.\\
&\qquad\times\left.\left[\int_{\mathbb{R}^n}|\varphi_m*\varphi_j*\varphi_j*f|^p[\omega(x,2^{-m})]^{-p}\,dx\right]^{r/p}\right\}^{q/r}.\end{aligned}$$

Let $t_1 \in (s_1, s)$. By Hölder's inequality with $r/q + r/\sigma = 1$ again, we obtain that

$$\begin{aligned}
\mathrm{I}_2 &\lesssim \inf_{\omega} \sum_{k=0}^{\infty} 2^{kq(s-s_1)} \left(\sum_{j=k+1}^{\infty} \sum_{m=(j-1)\vee 0}^{j+1} 2^{m\sigma(s_1-t_1)} \right)^{q/\sigma} \sum_{j=0}^{k} \sum_{m=(j-1)\vee 0}^{j+1} 2^{mqt_1} \\
&\quad \times \left[\int_{\mathbb{R}^n} |\varphi_m * \varphi_j * \varphi_j * f|^p [\omega(x, 2^{-m})]^{-p} \, dx \right]^{q/p} \\
&\lesssim \inf_{\omega} \sum_{j=0}^{\infty} 2^{jq(s-t_1)} \sum_{m=(j-1)\vee 0}^{j+1} 2^{mqt_1} \left[\int_{\mathbb{R}^n} |\varphi_m * \varphi_j * \varphi_j * f|^p [\omega(x, 2^{-m})]^{-p} \, dx \right]^{q/p}.
\end{aligned}$$

Similarly to the estimate of I_1, we have $\mathrm{I}_2 \lesssim \|f\|_{BH^{s,\tau}_{p,q}(\mathbb{R}^n)}$, which further yields (7.28).

By (7.25), (7.28), [145, Remark 2.4.1/4] and the trivial embedding that for $i = 0, 1$,

$$BH^{s_i,\tau}_{p,r}(\mathbb{R}^n) \subset BH^{s_i,\tau}_{p,q_i}(\mathbb{R}^n) \subset BH^{s_i,\tau}_{p,q_2}(\mathbb{R}^n),$$

we see that

$$\begin{aligned}
BH^{s,\tau}_{p,q}(\mathbb{R}^n) &\subset (BH^{s_0,\tau}_{p,r}(\mathbb{R}^n), BH^{s_1,\tau}_{p,r}(\mathbb{R}^n))_{\theta,q} \\
&\subset (BH^{s_0,\tau}_{p,q_0}(\mathbb{R}^n), BH^{s_1,\tau}_{p,q_1}(\mathbb{R}^n))_{\theta,q} \\
&\subset (BH^{s_0,\tau}_{p,q_2}(\mathbb{R}^n), BH^{s_1,\tau}_{p,q_2}(\mathbb{R}^n))_{\theta,q} \\
&\subset BH^{s,\tau}_{p,q}(\mathbb{R}^n).
\end{aligned}$$

This proves Theorem 7.14 for Besov-Hausdorff spaces.

The interpolation conclusion for Triebel-Lizorkin-Hausdorff spaces follows form that for Besov-Hausdorff spaces and the trivial embedding that

$$BH^{s,\tau}_{p,\min\{p,q\}}(\mathbb{R}^n) \subset FH^{s,\tau}_{p,q}(\mathbb{R}^n) \subset BH^{s,\tau}_{p,\max\{p,q\}}(\mathbb{R}^n),$$

which completes the proof of Theorem 7.14. □

Recall that when $\tau = 0$, the Besov-Hausdorff and Triebel-Lizorkin-Hausdorff spaces are just, respectively, Besov spaces and Triebel-Lizorkin spaces. Then Theorem 7.14 partially generalizes the classical real interpolation conclusions in [145, Theorem 2.4.2].

Chapter 8
Homogeneous Spaces

In this chapter we deal with the homogeneous counterpart of $A^{s,\tau}_{p,q}(\mathbb{R}^n)$. The homogeneous Besov-type spaces $\dot{B}^{s,\tau}_{p,q}(\mathbb{R}^n)$ and Triebel-Lizorkin-type spaces $\dot{F}^{s,\tau}_{p,q}(\mathbb{R}^n)$ were introduced and investigated in [127, 164–167].

8.1 The Definition and Some Preliminaries

To recall definitions of $\dot{B}^{s,\tau}_{p,q}(\mathbb{R}^n)$ and $\dot{F}^{s,\tau}_{p,q}(\mathbb{R}^n)$ in [164, 165], we need some notation. Following Triebel's [145], we set

$$\mathscr{S}_\infty(\mathbb{R}^n) \equiv \left\{\varphi \in \mathscr{S}(\mathbb{R}^n) : \int_{\mathbb{R}^n} \varphi(x)x^\gamma\,dx = 0 \text{ for all multi-indices } \gamma \in \mathbb{Z}^n_+\right\}$$

and use $\mathscr{S}'_\infty(\mathbb{R}^n)$ to denote the *topological dual* of $\mathscr{S}_\infty(\mathbb{R}^n)$, namely, the set of all continuous linear functionals on $\mathscr{S}_\infty(\mathbb{R}^n)$ endowed with weak $*$-topology. Let $\varphi \in \mathscr{S}(\mathbb{R}^n)$ such that

$$\operatorname{supp}\widehat{\varphi} \subset \{\xi \in \mathbb{R}^n : 1/2 \le |\xi| \le 2\}, \quad |\widehat{\varphi}(\xi)| \ge C > 0 \text{ if } 3/5 \le |\xi| \le 5/3. \tag{8.1}$$

Then $\varphi \in \mathscr{S}_\infty(\mathbb{R}^n)$. Moreover, it is well known that there exists a function $\psi \in \mathscr{S}(\mathbb{R}^n)$ satisfying (8.1) such that

$$\sum_{j\in\mathbb{Z}} \overline{\widehat{\varphi}(2^j\xi)}\widehat{\psi}(2^j\xi) = 1$$

for all $\xi \in \mathbb{R}^n \setminus \{0\}$; see [65, Lemma (6.9)].

Let $\mathscr{P}(\mathbb{R}^n)$ denote the set of all polynomials on $\mathbb{R}^n$. We endow $\mathscr{S}'(\mathbb{R}^n)/\mathscr{P}(\mathbb{R}^n)$ with the quotient topology (namely, O is open in $\mathscr{S}'(\mathbb{R}^n)/\mathscr{P}(\mathbb{R}^n)$ if and only if $\pi^{-1}(O)$ is open in $\mathscr{S}'(\mathbb{R}^n)$), where π is the quotient map form $\mathscr{S}'(\mathbb{R}^n)$ to $\mathscr{S}'(\mathbb{R}^n)/\mathscr{P}(\mathbb{R}^n)$. The following assertion is well known. For completeness, we give its proof.

D. Yang et al., *Morrey and Campanato Meet Besov, Lizorkin and Triebel*, Lecture Notes in Mathematics 2005, DOI 10.1007/978-3-642-14606-0_8,

Proposition 8.1. *$\mathscr{S}'(\mathbb{R}^n)/\mathscr{P}(\mathbb{R}^n)$ is identified with $\mathscr{S}'_\infty(\mathbb{R}^n)$ as topological spaces, namely, there exists a homeomorphism T mapping $\mathscr{S}'(\mathbb{R}^n)/\mathscr{P}(\mathbb{R}^n)$ onto $\mathscr{S}'_\infty(\mathbb{R}^n)$.*

Proof. Define the map

$$T : \mathscr{S}'(\mathbb{R}^n)/\mathscr{P}(\mathbb{R}^n) \to \mathscr{S}'_\infty(\mathbb{R}^n)$$

by setting

$$T([f]) \equiv f|_{\mathscr{S}_\infty(\mathbb{R}^n)} \quad \text{for all} \quad [f] = f + \mathscr{P}(\mathbb{R}^n) \in \mathscr{S}'(\mathbb{R}^n)/\mathscr{P}(\mathbb{R}^n).$$

Notice that for any $f \in \mathscr{S}'(\mathbb{R}^n)$ and $P_1, P_2 \in \mathscr{P}(\mathbb{R}^n)$,

$$(f+P_1)|_{\mathscr{S}_\infty(\mathbb{R}^n)} = (f+P_2)|_{\mathscr{S}_\infty(\mathbb{R}^n)}.$$

Then T is well defined. To show that T is a homeomorphism, we need to prove that T is injective, surjective, continuous and its inverse T^{-1} is also continuous.

Step 1. T is injective. It suffices to show that if $T([f]) = 0$ in $\mathscr{S}'_\infty(\mathbb{R}^n)$, then $[f] = [0]$ in $\mathscr{S}'(\mathbb{R}^n)/\mathscr{P}(\mathbb{R}^n)$, equivalently, if $f \in \mathscr{S}'(\mathbb{R}^n)$ satisfying $f|_{\mathscr{S}_\infty(\mathbb{R}^n)} = 0$, then $f \in \mathscr{P}(\mathbb{R}^n)$. In fact, since $f|_{\mathscr{S}_\infty(\mathbb{R}^n)} = 0$, we know that for all $\varphi \in \mathscr{S}_\infty(\mathbb{R}^n)$, $\langle f, \varphi\rangle = 0$, and hence $\langle \widehat{f}, \widehat{\varphi}\rangle = 0$. We then claim that $\operatorname{supp} \widehat{f} \subset \{0\}$. To see this, by [67, p. 12, Definition 1.4.1],

$$\operatorname{supp} \widehat{f} = \mathbb{R}^n \setminus \{x \in \mathbb{R}^n : \widehat{f} = 0 \quad \text{on a neighborhood of } x\}.$$

If $\operatorname{supp} \widehat{f} \not\subset \{0\}$, we can find $x_0 \in \mathbb{R}^n \setminus \{0\}$ such that for any $\varepsilon > 0$, there exists a $\varphi \in \mathscr{S}(\mathbb{R}^n)$ satisfying $\operatorname{supp} \widehat{\varphi} \subset B(x_0, \varepsilon)$ and $\langle \widehat{f}, \widehat{\varphi}\rangle \neq 0$. Since $x_0 \neq 0$, if ε is sufficiently small, then $\widehat{\varphi} \equiv 0$ in a neighborhood of 0. Thus, for all α, $\partial^\alpha \widehat{\varphi}(0) = 0$, namely, $\varphi \in \mathscr{S}_\infty(\mathbb{R}^n)$. Then $\langle \widehat{f}, \widehat{\varphi}\rangle \neq 0$ contradicts $f|_{\mathscr{S}_\infty(\mathbb{R}^n)} = 0$. This finishes the proof of the above claim.

By [67, p. 36, Theorem 3.2.1], there exists $N \in \mathbb{Z}_+$ such that

$$\widehat{f} = \sum_{|\alpha| \le N} C_\alpha \partial^\alpha \delta,$$

where $C_\alpha \in \mathbb{C}$ and δ is dirac function. This observation together with $\widehat{x^\alpha} = C_\alpha \partial^\alpha \delta$ yields that $f \in \mathscr{P}(\mathbb{R}^n)$ and then, T is injective.

Step 2. T is surjective. Notice that $\mathscr{S}(\mathbb{R}^n)$ is a locally convex space. Then by [116, p. 61, Theorem 3.6], for each $f \in \mathscr{S}'_\infty(\mathbb{R}^n)$, there exists a $\widetilde{f} \in \mathscr{S}'(\mathbb{R}^n)$ such that $\widetilde{f}|_{\mathscr{S}_\infty(\mathbb{R}^n)} = f$. Thus, $T([\widetilde{f}]) = f$.

Step 3. T is continuous. It suffices to show that for all open sets $V \subset \mathscr{S}'_\infty(\mathbb{R}^n)$, $T^{-1}(V)$ is open in $\mathscr{S}'(\mathbb{R}^n)/\mathscr{P}(\mathbb{R}^n)$. Let

$$i : \mathscr{S}'(\mathbb{R}^n) \to \mathscr{S}'_\infty(\mathbb{R}^n)$$

be the map defined as $i(f) \equiv f|_{\mathscr{S}_\infty(\mathbb{R}^n)}$ for all $f \in \mathscr{S}'(\mathbb{R}^n)$. Then i is a continuous, surjective and closed map. Since i is continuous, for an open set $V \subset \mathscr{S}'_\infty(\mathbb{R}^n)$, $i^{-1}(V)$ is open in $\mathscr{S}'(\mathbb{R}^n)$. Then

$$\pi^{-1}(\pi \circ i^{-1}(V)) = i^{-1}(V) + \mathscr{P}(\mathbb{R}^n) = \cup_{P \in \mathscr{P}(\mathbb{R}^n)}(i^{-1}(V) + P)$$

is open in $\mathscr{S}'(\mathbb{R}^n)$. By the definition of quotient topology, $\pi \circ i^{-1}(V)$ is open in $\mathscr{S}'(\mathbb{R}^n)/\mathscr{P}(\mathbb{R}^n)$. Thus, $T^{-1}(V) = \pi \circ i^{-1}(V)$ is open in $\mathscr{S}'(\mathbb{R}^n)/\mathscr{P}(\mathbb{R}^n)$ and T is continuous.

Step 4. T^{-1} is continuous from $\mathscr{S}'_\infty(\mathbb{R}^n)$ to $\mathscr{S}'(\mathbb{R}^n)/\mathscr{P}(\mathbb{R}^n)$. It suffices to show that for all open sets $V \subset \mathscr{S}'(\mathbb{R}^n)/\mathscr{P}(\mathbb{R}^n)$, $T(V)$ is open in $\mathscr{S}'_\infty(\mathbb{R}^n)$.

We first claim that for all sets $O \subset \mathscr{S}'_\infty(\mathbb{R}^n)$, if $i^{-1}(O)$ is open in $\mathscr{S}'(\mathbb{R}^n)$, then O is open in $\mathscr{S}'_\infty(\mathbb{R}^n)$. Observe that $i^{-1}(O^c) = [i^{-1}(O)]^c$. In fact, for all $x \in i^{-1}(O^c)$, $i(x) \in O^c$. If $x \notin [i^{-1}(O)]^c$, then $x \in i^{-1}(O)$. Thus, $i(x) \in O$, which contradicts $i(x) \in O^c$. Thus, $i^{-1}(O^c) \subset [i^{-1}(O)]^c$. On the other hand, if $x \in [i^{-1}(O)]^c$, then $x \notin i^{-1}(O)$, hence $i(x) \notin O$. Thus, $x \in i^{-1}(O^c)$ and $[i^{-1}(O)]^c \subset i^{-1}(O^c)$. This observation implies that $i^{-1}(O)$ is open in $\mathscr{S}'(\mathbb{R}^n)$ if and only if $i^{-1}(O^c)$ is closed in $\mathscr{S}'(\mathbb{R}^n)$. Then the above claim follows from the fact that i is a closed map.

Since $T^{-1} \circ i = \pi$ is continuous, for all open sets $V \subset \mathscr{S}'(\mathbb{R}^n)/\mathscr{P}(\mathbb{R}^n)$,

$$i^{-1}(T(V)) = i^{-1} \circ T(V) = (T^{-1} \circ i)^{-1}(V) = \pi^{-1}(V)$$

is open in $\mathscr{S}'(\mathbb{R}^n)$, which together with the above claim implies that $T(V)$ is open in $\mathscr{S}'_\infty(\mathbb{R}^n)$.

Combining Steps 1 through 4, we obtain that T is a homeomorphism, which completes the proof of Proposition 8.1. $\square$

Following Triebel's [145], we use the distribution space $\mathscr{S}'_\infty(\mathbb{R}^n)$ in the following Definition 8.1.

Definition 8.1. Let $s \in \mathbb{R}$, $\tau \in [0,\infty)$, $q \in (0,\infty]$ and $\varphi \in \mathscr{S}(\mathbb{R}^n)$ satisfy (8.1).

(i) Let $p \in (0,\infty]$. The *Besov-type space* $\dot{B}^{s,\tau}_{p,q}(\mathbb{R}^n)$ is defined to be the set of all $f \in \mathscr{S}'_\infty(\mathbb{R}^n)$ such that $\|f\|_{\dot{B}^{s,\tau}_{p,q}(\mathbb{R}^n)} < \infty$, where

$$\|f\|_{\dot{B}^{s,\tau}_{p,q}(\mathbb{R}^n)} \equiv \sup_{P \in \mathscr{Q}} \frac{1}{|P|^\tau} \left\{ \sum_{j=j_P}^{\infty} \left[\int_P (2^{js}|\varphi_j * f(x)|)^p \, dx \right]^{q/p} \right\}^{1/q}$$

with suitable modifications made when $p = \infty$ or $q = \infty$.

(ii) Let $p \in (0,\infty)$. The *Triebel-Lizorkin-type space* $\dot{F}^{s,\tau}_{p,q}(\mathbb{R}^n)$ is defined to be the set of all $f \in \mathscr{S}'_\infty(\mathbb{R}^n)$ such that $\|f\|_{\dot{F}^{s,\tau}_{p,q}(\mathbb{R}^n)} < \infty$, where

$$\|f\|_{\dot{F}^{s,\tau}_{p,q}(\mathbb{R}^n)} \equiv \sup_{P \in \mathscr{Q}} \frac{1}{|P|^\tau} \left\{ \int_P \left[\sum_{j=j_P}^{\infty} (2^{js}|\varphi_j * f(x)|)^q \, dx \right]^{p/q} \right\}^{1/p}$$

with suitable modification made when $q = \infty$.

Remark 8.1. These spaces are called homogeneous because of the following fact: There exists a positive constant C such that for all $\lambda \in (0,\infty)$ and $f \in \dot{B}^{s,\tau}_{p,q}(\mathbb{R}^n)$ or $f \in \dot{F}^{s,\tau}_{p,q}(\mathbb{R}^n)$,

$$\|f(\lambda\cdot)\|_{\dot{B}^{s,\tau}_{p,q}(\mathbb{R}^n)} \le C\lambda^{s-n/p+n\tau}\|f\|_{\dot{B}^{s,\tau}_{p,q}(\mathbb{R}^n)}$$

and

$$\|f(\lambda\cdot)\|_{\dot{F}^{s,\tau}_{p,q}(\mathbb{R}^n)} \le C\lambda^{s-n/p+n\tau}\|f\|_{\dot{F}^{s,\tau}_{p,q}(\mathbb{R}^n)}.$$

Let $\dot{A}^{s,\tau}_{p,q}(\mathbb{R}^n)$ denote either $\dot{B}^{s,\tau}_{p,q}(\mathbb{R}^n)$ or $\dot{F}^{s,\tau}_{p,q}(\mathbb{R}^n)$. It was proved in [165, Corollary 3.1] that the spaces $\dot{A}^{s,\tau}_{p,q}(\mathbb{R}^n)$ are independent of the choices of φ. Furthermore,

$$\mathscr{S}_\infty(\mathbb{R}^n) \subset \dot{A}^{s,\tau}_{p,q}(\mathbb{R}^n) \subset \mathscr{S}'_\infty(\mathbb{R}^n);$$

see [165, Propositions 3.1(ix) and 3.4]. These spaces unify and generalize the classical homogeneous Besov spaces, Triebel-Lizorkin spaces, Q spaces and Morrey spaces; see [127, 164, 165]. An important tool to study $\dot{A}^{s,\tau}_{p,q}(\mathbb{R}^n)$ is the following Calderón reproducing formula; see [62, Lemma 2.1] and [164, Lemma 2.1].

Lemma 8.1. *Let* φ, $\psi \in \mathscr{S}(\mathbb{R}^n)$ *satisfying* (8.1) *such that*

$$\sum_{j\in\mathbb{Z}} \overline{\widehat{\varphi}(2^j\xi)}\widehat{\psi}(2^j\xi) = 1$$

for all $\xi \in \mathbb{R}^n \setminus \{0\}$. *Then for any* $f \in \mathscr{S}_\infty(\mathbb{R}^n)$,

$$f = \sum_{j\in\mathbb{Z}} \psi_j * \widetilde{\varphi}_j * f = \sum_{j\in\mathbb{Z}} 2^{-jn} \sum_{k\in\mathbb{Z}^n} \widetilde{\varphi}_j * f(2^{-j}k)\,\psi_j(\cdot - 2^{-j}k) = \sum_{j\in\mathbb{Z}} \sum_{l(Q)=2^{-j}} \langle f, \varphi_Q\rangle\,\psi_Q$$

in $\mathscr{S}_\infty(\mathbb{R}^n)$. *Moreover, for any* $f \in \mathscr{S}'_\infty(\mathbb{R}^n)$, *the above equalities also hold in* $\mathscr{S}'_\infty(\mathbb{R}^n)$.

The corresponding sequence spaces were introduced in [165, Definition 3.1].

Definition 8.2. Let $s \in \mathbb{R}$, $\tau \in [0,\infty)$ and $q \in (0,\infty]$.

(i) Let $p \in (0,\infty]$. The *sequence space* $\dot{b}^{s,\tau}_{p,q}(\mathbb{R}^n)$ is defined to be the set of all sequences $t \equiv \{t_Q\}_{Q\in\mathscr{Q}} \subset \mathbb{C}$ such that $\|t\|_{\dot{b}^{s,\tau}_{p,q}(\mathbb{R}^n)} < \infty$, where

$$\|t\|_{\dot{b}^{s,\tau}_{p,q}(\mathbb{R}^n)} \equiv \sup_{P\in\mathscr{Q}} \frac{1}{|P|^\tau} \left\{ \sum_{j=j_P}^{\infty} 2^{jsq} \left[\int_P \left(\sum_{l(Q)=2^{-j}} |t_Q \widetilde{\chi}_Q(x)| \right)^p dx \right]^{q/p} \right\}^{1/q}.$$

(ii) Let $p \in (0,\infty)$. The *sequence space* $\dot{f}^{s,\tau}_{p,q}(\mathbb{R}^n)$ is defined to be the set of all sequences $t \equiv \{t_Q\}_{Q\in\mathscr{Q}} \subset \mathbb{C}$ such that $\|t\|_{\dot{f}^{s,\tau}_{p,q}(\mathbb{R}^n)} < \infty$, where

$$\|t\|_{\dot{f}^{s,\tau}_{p,q}(\mathbb{R}^n)} \equiv \sup_{P\in\mathscr{Q}} \frac{1}{|P|^\tau} \left\{ \int_P \left[\sum_{Q\subset P} \left(|Q|^{-s/n} |t_Q \widetilde{\chi}_Q(x)| \right)^q \right]^{p/q} dx \right\}^{1/p}.$$

Similarly to Remark 2.4, from [37, Proposition 2.2] again, we deduce the discretization of $\dot{f}^{s,\tau}_{p,q}(\mathbb{R}^n)$ in the following Remark 8.2.

Remark 8.2. Let $\tau \in [0,\infty)$, $s \in \mathbb{R}$, $p \in (0,\infty)$ and $q \in (0,\infty]$. If $p \geq q$, then there exists a positive constant C, depending only on p and q, such that for all $t \in \dot{f}^{s,\tau}_{p,q}(\mathbb{R}^n)$,

$$\begin{aligned} C^{-1}\|t\|_{\dot{f}^{s,\tau}_{p,q}(\mathbb{R}^n)} &\leq \sup_{P\in\mathcal{Q}} \frac{1}{|P|^\tau} \left\{ \sum_{j=j_P}^{\infty} \sum_{\substack{l(Q)=2^{-j}\\ Q\subset P}} (|Q|^{-s/n-1/2+1/q}|t_Q|)^q \right. \\ &\qquad \left. \times \left[\frac{1}{|Q|} \sum_{\substack{R\in\mathcal{Q}\\ R\subset Q}} (|R|^{-s/n-1/2+1/q}|t_R|)^q \right]^{p/q-1} \right\}^{1/p} \\ &\leq C\|t\|_{\dot{f}^{s,\tau}_{p,q}(\mathbb{R}^n)}. \end{aligned}$$

The spaces $\dot{A}^{s,\tau}_{p,q}(\mathbb{R}^n)$ also have the following φ-transform characterization; see [164, Theorem 3.1].

Theorem 8.1. *Let $s \in \mathbb{R}$, $\tau \in [0,\infty)$, $p,q \in (0,\infty]$, φ and ψ be as in Lemma 8.1. Then*

$$S_\varphi : \dot{A}^{s,\tau}_{p,q}(\mathbb{R}^n) \to \dot{a}^{s,\tau}_{p,q}(\mathbb{R}^n)$$

and

$$T_\psi : \dot{a}^{s,\tau}_{p,q}(\mathbb{R}^n) \to \dot{A}^{s,\tau}_{p,q}(\mathbb{R}^n)$$

are bounded; moreover, $T_\psi \circ S_\varphi$ is the identity on $\dot{A}^{s,\tau}_{p,q}(\mathbb{R}^n)$.

We remark that $\dot{B}^{s,\tau}_{p,q}(\mathbb{R}^n)$ and $\dot{F}^{s,\tau}_{p,q}(\mathbb{R}^n)$ have some similar properties to $B^{s,\tau}_{p,q}(\mathbb{R}^n)$ and $F^{s,\tau}_{p,q}(\mathbb{R}^n)$ such as Sobolev-type embedding properties, smooth atomic and molecular decomposition characterizations, boundedness of pseudo-differential operators with homogeneous symbols and trace theorems. These properties have been studied in [127, 164, 165]. Also, the maximal function and local mean characterizations of $\dot{B}^{s,\tau}_{p,q}(\mathbb{R}^n)$ and $\dot{F}^{s,\tau}_{p,q}(\mathbb{R}^n)$ were already obtained in [167].

However, similarly to homogeneous Besov spaces $\dot{B}^{s}_{p,q}(\mathbb{R}^n)$ and Triebel-Lizorkin spaces $\dot{F}^{s}_{p,q}(\mathbb{R}^n)$ (see [145, p. 238]), some of the most striking features of the spaces $B^{s,\tau}_{p,q}(\mathbb{R}^n)$ and $F^{s,\tau}_{p,q}(\mathbb{R}^n)$ have no counterparts, such as the pointwise multipliers theorem and the diffeomorphism property. Thus, we cannot expect to find counterparts of Theorems 6.1 and 6.7.

8.2 The Wavelet Characterization of $\dot{B}^{s,\tau}_{p,q}(\mathbb{R}^n)$ and $\dot{F}^{s,\tau}_{p,q}(\mathbb{R}^n)$

Let $\dot{a}^{s,\tau}_{p,q}(\mathbb{R}^n)$ denote either $\dot{b}^{s,\tau}_{p,q}(\mathbb{R}^n)$ or $\dot{f}^{s,\tau}_{p,q}(\mathbb{R}^n)$. We now focus on the wavelet characterization of $\dot{A}^{s,\tau}_{p,q}(\mathbb{R}^n)$. Differently from those in Chap. 4, what we deal with below are so called "wavelets with two humps". These wavelet basis have no compact support of their own, and their Fourier transforms have compact support; see, for example, [99, Sect. 6.11] or [65, Sect. 7].

We begin with the one-dimensional case. Let $\varphi \in \mathscr{S}(\mathbb{R})$ be as in [65, Theorem, (7.11)]. That is, φ is a real valued function such that

$$\operatorname{supp} \widehat{\varphi} \subset \left[-\frac{8}{3}\pi, -\frac{2}{3}\pi\right] \bigcup \left[\frac{2}{3}\pi, \frac{8}{3}\pi\right] \tag{8.2}$$

and the collection

$$\{\varphi_{jk} : j, k \in \mathbb{Z}\} \equiv \left\{2^{j/2}\varphi(2^j \cdot -k) : j, k \in \mathbb{Z}\right\} \tag{8.3}$$

is an orthonormal basis of $L^2(\mathbb{R})$. We call φ the *mother function* of the *wavelet basis* $\{\varphi_{jk} : j, k \in \mathbb{Z}\}$.

Define

$$\widetilde{S}_\varphi f \equiv \{\langle f, \varphi_Q\rangle\}_{Q\in\mathscr{Q}} \equiv \{f(\varphi_Q)\}_{Q\in\mathscr{Q}}$$

for $f \in \mathscr{S}'_\infty(\mathbb{R})$, and let

$$\widetilde{T}_\varphi t \equiv \sum_{Q\in\mathscr{Q}} t_Q\varphi_Q$$

when $t = \{t_Q\}_{Q\in\mathscr{Q}}$. To obtain the wavelet characterization of $\dot{A}^{s,\tau}_{p,q}(\mathbb{R})$, we need to prove that the coefficient sequence of a wavelet expansion of an $f \in \dot{A}^{s,\tau}_{p,q}(\mathbb{R})$ is an element of $\dot{a}^{s,\tau}_{p,q}(\mathbb{R})$.

Theorem 8.2. *Let $s \in \mathbb{R}$, $p, q \in (0,\infty]$ and τ be as in Lemma 3.1. The operator $\widetilde{S}_\varphi$ is bounded from $\dot{A}^{s,\tau}_{p,q}(\mathbb{R})$ to $\dot{a}^{s,\tau}_{p,q}(\mathbb{R})$ and $\widetilde{T}_\varphi$ is bounded from $\dot{a}^{s,\tau}_{p,q}(\mathbb{R})$ to $\dot{A}^{s,\tau}_{p,q}(\mathbb{R})$. Furthermore, $\widetilde{T}_\varphi \circ \widetilde{S}_\varphi$ and $\widetilde{S}_\varphi \circ \widetilde{T}_\varphi$ are, respectively, the identities on $\dot{A}^{s,\tau}_{p,q}(\mathbb{R})$ and $\dot{a}^{s,\tau}_{p,q}(\mathbb{R})$.*

The proof of Theorem 8.2 is similar to that for [65, Theorem (7.20)]. For the reader's convenience, we give the details.

Proof of Theorem 8.2. By (8.3), $\{\varphi_Q\}_{Q\in\mathscr{Q}}$ is an orthonormal basis of $L^2(\mathbb{R})$. Thus,

$$f = \sum_{Q\in\mathscr{Q}} \langle f, \varphi_Q\rangle\varphi_Q$$

holds in $L^2(\mathbb{R})$. It was further proved in [65, p. 71] that the identity

$$f = \sum_{Q\in\mathscr{Q}} \langle f, \varphi_Q\rangle\varphi_Q$$

also holds in $\mathscr{S}_\infty(\mathbb{R})$ and $\mathscr{S}'_\infty(\mathbb{R})$, which implies that $\widetilde{T}_\varphi \circ \widetilde{S}_\varphi$ is identity transformation.

Recall that $\varphi \in \mathscr{S}(\mathbb{R})$ and satisfies (8.2). It is easy to check that each φ_Q is a smooth synthesis molecule for $\dot{A}^{s,\tau}_{p,q}(\mathbb{R})$ in [165, Definition 4.2] (up to a constant factor that is independent of Q). Then applying the homogeneous smooth molecular decomposition characterization of $\dot{A}^{s,\tau}_{p,q}(\mathbb{R})$ obtained in [165, Theorem 4.2], we know that for all $t \in \dot{a}^{s,\tau}_{p,q}(\mathbb{R})$,

$$\left\| \sum_{Q\in\mathscr{Q}} t_Q \varphi_Q \right\|_{\dot{A}^{s,\tau}_{p,q}(\mathbb{R})} \lesssim \|t\|_{\dot{a}^{s,\tau}_{p,q}(\mathbb{R})},$$

which further implies that $\widetilde{T}_\varphi$ is bounded from $\dot{a}^{s,\tau}_{p,q}(\mathbb{R})$ to $\dot{A}^{s,\tau}_{p,q}(\mathbb{R})$. By (8.2) again, for all $t \in \dot{a}^{s,\tau}_{p,q}(\mathbb{R})$ and $P \in \mathscr{Q}$,

$$\left\langle \sum_{Q\in\mathscr{Q}} t_Q \varphi_Q, t_P \right\rangle = t_P,$$

which shows that $\widetilde{S}_\varphi \circ \widetilde{T}_\varphi$ is the identity on $\dot{a}^{s,\tau}_{p,q}(\mathbb{R})$.

To obtain the boundedness of $\widetilde{S}_\varphi$, let $\psi \in \mathscr{S}(\mathbb{R})$ such that

$$\operatorname{supp} \widehat{\psi} \subset [-2,-1/2] \cup [1/2,2]$$

and for all $\xi \in \mathbb{R}\setminus\{0\}$,

$$\sum_{k\in\mathbb{Z}} |\widehat{\psi}(2^k\xi)|^2 = 1.$$

Let $f \in \dot{A}^{s,\tau}_{p,q}(\mathbb{R})$. Then by the Calderón reproducing formula in [165, Lemma 2.1],

$$f = \sum_{Q\in\mathscr{Q}} \langle f, \psi_Q \rangle \psi_Q$$

holds in $\mathscr{S}'_\infty(\mathbb{R})$. Moreover, by the φ-transform characterization of $\dot{A}^{s,\tau}_{p,q}(\mathbb{R})$ in [165, Theorem 3.1], we further have

$$\|\{\langle f, \psi_Q \rangle\}_{Q\in\mathscr{Q}}\|_{\dot{a}^{s,\tau}_{p,q}(\mathbb{R})} \lesssim \|f\|_{\dot{A}^{s,\tau}_{p,q}(\mathbb{R})}.$$

By [165, Lemma 2.1] again, we obtain that

$$\varphi_Q = \sum_{P\in\mathscr{Q}} \langle \varphi_Q, \psi_P \rangle \psi_P$$

in $\mathscr{S}_\infty(\mathbb{R}^n)$. Thus,

$$\langle f, \varphi_Q \rangle = \sum_{P\in\mathscr{Q}} \overline{\langle \varphi_Q, \psi_P \rangle} \langle f, \psi_P \rangle \equiv \sum_{P\in\mathscr{Q}} a_{QP} \langle f, \psi_P \rangle,$$

where $a_{QP} \equiv \overline{\langle \varphi_Q, \psi_P \rangle}$. Since φ_Q is a constant multiple of a homogeneous smooth synthesis molecule in [165, Definition 4.2], using [165, Corollary 4.1], we know that the matrix operator $A \equiv \{a_{QP}\}_{Q,P\in\mathcal{Q}}$ is ε_1-almost diagonal on $\dot{a}^{s,\tau}_{p,q}(\mathbb{R})$ as in [165, Definition 4.1]. Then [165, Theorem 4.1] tells us

$$\|\{\langle f,\varphi_Q\rangle\}_{Q\in\mathcal{Q}}\|_{\dot{a}^{s,\tau}_{p,q}(\mathbb{R})} \lesssim \|\{\langle f,\psi_Q\rangle\}_{Q\in\mathcal{Q}}\|_{\dot{a}^{s,\tau}_{p,q}(\mathbb{R})} \lesssim \|f\|_{\dot{A}^{s,\tau}_{p,q}(\mathbb{R})},$$

which yields that $\widetilde{S}_\varphi$ is bounded from $\dot{A}^{s,\tau}_{p,q}(\mathbb{R})$ to $\dot{a}^{s,\tau}_{p,q}(\mathbb{R})$, and then, completes the proof of Theorem 8.2. □

For the n-dimensional case, the well-known *tensor product ansatz* yields a wavelet basis

$$\left\{2^{jn/2}\varphi^i(2^jx-k):\ j\in\mathbb{Z}, k\in\mathbb{Z}^n, i\in\{1,\cdots,2^n-1\}\right\}.$$

The 2^n-1 functions φ^i belong to the Schwartz class $\mathcal{S}(\mathbb{R}^n)$ and the Fourier transforms $\widehat{\varphi^i}$ of φ^i vanish in a neighborhood of 0 and have compact support; moreover,

$$\{2^{jn/2}\varphi^i(2^jx-k):\ j\in\mathbb{Z}, k\in\mathbb{Z}^n, i\in\{1,\cdots,2^n-1\}\}$$

yields an orthonormal basis of $L^2(\mathbb{R}^n)$; see, [99, p. 168] or [65, p. 73]. We remark that Theorem 8.2 still holds in this case.

Remark 8.3. Theorem 8.2 generalizes the corresponding results on homogeneous Besov and Triebel-Lizorkin spaces established in [65, Sect. 7] by taking $\tau=0$.

Next we establish the difference characterization and the wavelet characterization of $\dot{A}^{s,\tau}_{p,q}(\mathbb{R}^n)$ in the sense of Chap. 4, namely, wavelets with compact supports. We need some preparations. Recall that $L^{\overline{p}}_{\mathrm{loc}}(\mathbb{R}^n)$ consists of all $\overline{p}$-locally integrable functions and $\overline{p}\equiv\max\{p,1\}$.

Proposition 8.2. *Let $p,q\in(0,\infty]$, $\tau\in[0,\infty)$ and $s\in(0,\infty)$. Then $\dot{A}^{s,\tau}_{p,q}(\mathbb{R}^n)\subset L^{\overline{p}}_{\mathrm{loc}}(\mathbb{R}^n)$ in the sense of $\mathcal{S}'_\infty(\mathbb{R}^n)$.*

Proof. Notice that Proposition 2.1(i) and (iii) are also correct for $\dot{A}^{s,\tau}_{p,q}(\mathbb{R}^n)$. It suffices to consider $\dot{B}^{s,\tau}_{p,\infty}(\mathbb{R}^n)$. Let $f\in\dot{B}^{s,\tau}_{p,\infty}(\mathbb{R}^n)$. We need to prove that there exists a function g such that $f=g$ in $\mathcal{S}'_\infty(\mathbb{R}^n)$ and

$$\int_P|g(x)|^{\overline{p}}\,dx<\infty$$

for all $P\equiv[-2^m,2^m)^n$ and $m\in\mathbb{N}$.

Let $L\in\mathbb{N}$ be sufficiently large and

$$\mathrm{I}(x)\equiv\sum_{j=-\infty}^{j_P-1}2^{-jn}\sum_{k\in\mathbb{Z}^n}\widetilde{\varphi}_j*f(2^{-j}k)\left[\psi_j(x-2^{-j}k)-\sum_{|\gamma|\le L}\frac{(\partial^\gamma\psi_j)(-2^{-j}k)}{\gamma!}(-x)^\gamma\right]$$

and

$$\mathrm{II}(x) \equiv \sum_{j=j_P}^{\infty} \psi_j * \widetilde{\varphi}_j * f(x).$$

By Lemma 8.1, for any $\phi \in \mathscr{S}_\infty(\mathbb{R}^n)$, we have

$$\langle f, \phi \rangle = \langle \mathrm{I}, \phi \rangle + \langle \mathrm{II}, \phi \rangle,$$

and hence $f \equiv \mathrm{I} + \mathrm{II}$ in $\mathscr{S}'_\infty(\mathbb{R}^n)$.

We first estimate $\int_P |\mathrm{II}(x)|^{\overline{p}}\,dx$. If $p \in (1,\infty]$, by Minkowski's inequality and $s > 0$, we see that

$$\left(\int_P |\mathrm{II}(x)|^p\,dx\right)^{1/p} \le \sum_{j=j_P}^{\infty} \left(\int_P |\psi_j * \widetilde{\varphi}_j * f(x)|^p\,dx\right)^{1/p} \lesssim 2^{-j_P s}|P|^\tau \|f\|_{\dot{B}^{s,\tau}_{p,\infty}(\mathbb{R}^n)}.$$

If $p \in (0,1)$, since

$$\mathrm{II} \equiv \sum_{j=j_P}^{\infty} 2^{-jn} \sum_{k\in\mathbb{Z}^n} \widetilde{\varphi}_j * f(2^{-j}k)\psi_j(\cdot - 2^{-j}k)$$

in $\mathscr{S}'_\infty(\mathbb{R}^n)$, by (2.11), for all $x \in P$, we have

$$|\mathrm{II}(x)| \lesssim \left[\sum_{j=j_P}^{\infty} \sum_{k\in\mathbb{Z}^n} |\widetilde{\varphi}_j * f(2^{-j}k)|^p \frac{1}{(1+|2^j x - k|)^{(n+\delta)p}}\right]^{1/p},$$

where $\delta \in (0,\infty)$ will be determined later. Decomposing

$$\sum_{k\in\mathbb{Z}^n} \equiv \sum_{\substack{k\in\mathbb{Z}^n\\ |x-2^{-j}k|\le l(P)}} + \sum_{i=1}^{\infty} \sum_{\substack{k\in\mathbb{Z}^n\\ 2^{i-1}l(P)<|x-2^{-j}k|\le 2^i l(P)}}$$

and noticing that

$$\widetilde{\varphi}_j * f(2^{-j}k) = 2^{jn/2}\langle f, \varphi_Q\rangle,$$

then by Theorem 8.1, we see that

$$\begin{aligned}
|\mathrm{II}(x)| &\lesssim [l(P)]^{-(n+\delta)} \left[\sum_{j=j_P}^{\infty} \sum_{i=0}^{\infty} \sum_{\substack{k\in\mathbb{Z}^n\\ |x-2^{-j}k|\le 2^i l(P)}} |\widetilde{\varphi}_j * f(2^{-j}k)|^p 2^{-(i+j)(n+\delta)p}\right]^{\frac{1}{p}} \\
&\lesssim [l(P)]^{-(n+\delta)} \|f\|_{\dot{B}^{s,\tau}_{p,\infty}(\mathbb{R}^n)} \left[\sum_{j=j_P}^{\infty} \sum_{i=0}^{\infty} |2^i P|^{\tau p} 2^{-jsp} 2^{jn} 2^{-(i+j)(n+\delta)p}\right]^{\frac{1}{p}} \\
&\lesssim 2^{-j_P s}|P|^{\tau - 1/p} \|f\|_{\dot{B}^{s,\tau}_{p,\infty}(\mathbb{R}^n)},
\end{aligned}$$

where we choose $\delta > \max\{n\tau - n, n/p - n - s\}$. Thus,

$$\int_P |\mathrm{II}(x)|\,dx \lesssim 2^{-j_P s}|P|^{\tau-1/p+1}\|f\|_{\dot{B}^{s,\tau}_{p,\infty}(\mathbb{R}^n)}.$$

Next we estimate $\int_P |\mathrm{I}(x)|^{\overline{p}}\,dx$. By the mean value theorem, there exists $\theta \in [0,1]$ such that

$$\begin{aligned}|\mathrm{I}(x)| &\le \sum_{j=-\infty}^{j_P-1} 2^{-jn} \sum_{k\in\mathbb{Z}^n} |\widetilde{\varphi}_j * f(2^{-j}k)| \sup_{|\gamma|=L+1} |x|^{L+1}|(\partial^\gamma \psi_j)(\theta x - 2^{-j}k)| \\ &\lesssim \sum_{j=-\infty}^{j_P-1}\sum_{k\in\mathbb{Z}^n} |\widetilde{\varphi}_j * f(2^{-j}k)|2^{j(L+1)}|x|^{L+1}(1+|2^j\theta x - k|)^{-(n+\delta)}.\end{aligned}$$

Noticing that

$$|\widetilde{\varphi}_j * f(2^{-j}k)| \le 2^{-js-jn\tau+jn/p}\|f\|_{\dot{B}^{s,\tau}_{p,\infty}(\mathbb{R}^n)},$$

we then have

$$\begin{aligned}|\mathrm{I}(x)| &\lesssim \sum_{j=-\infty}^{j_P-1}\sum_{k\in\mathbb{Z}^n} 2^{-js-jn\tau+jn/p}2^{j(L+1)}|x|^{L+1}(1+|2^j\theta x - k|)^{-(n+\delta)}\|f\|_{\dot{B}^{s,\tau}_{p,\infty}(\mathbb{R}^n)} \\ &\lesssim 2^{-j_P s + j_P(L+1)}|x|^{L+1}|P|^{\tau-1/p}\|f\|_{\dot{B}^{s,\tau}_{p,\infty}(\mathbb{R}^n)},\end{aligned}$$

where we choose $\delta > 0$ and $L > s + n\tau - n/p - 1$. Thus,

$$\left(\int_P |\mathrm{I}(x)|^{\overline{p}}\,dx\right)^{1/\overline{p}} \lesssim 2^{-j_P s}|P|^{\tau}|P|^{1/\overline{p}-1/p}\|f\|_{\dot{B}^{s,\tau}_{p,\infty}(\mathbb{R}^n)},$$

which completes the proof of Proposition 8.2. □

By Proposition 8.2, in what follows, when $f \in \dot{A}^{s,\tau}_{p,q}(\mathbb{R}^n)$, we also use f to denote its representative in $L^{\overline{p}}_{\mathrm{loc}}(\mathbb{R}^n)$.

Theorem 8.3. *Let $s \in (0,\infty)$, $p \in (0,\infty)$, $q \in (0,\infty]$, $\tau \in [0,\infty)$ and $N_1, M \in \mathbb{N}$ such that $M \le N_1$, $s < \{M \wedge (M + n(1/p - \tau))\}$ and $N_1 \ge \lfloor s + n\tau \rfloor$. Then for all $f \in \dot{A}^{s,\tau}_{p,q}(\mathbb{R}^n)$,*

$$C^{-1}\|f\|_{\dot{A}^{s,\tau}_{p,q}(\mathbb{R}^n)} \le \|f\|_{\overset{\blacktriangle}{A}{}^{s,\tau}_{p,q}(\mathbb{R}^n)} \equiv \sum_{i=1}^{2^n-1}\left\|\{\langle f, \psi_{i,j,k}\rangle\}_{j\in\mathbb{Z},k\in\mathbb{Z}^n}\right\|_{\dot{a}^{s,\tau}_{p,q}(\mathbb{R}^n)} \le C\|f\|_{\dot{A}^{s,\tau}_{p,q}(\mathbb{R}^n)},$$

where $\psi_{i,j,k}$ are wavelets in Sect. 4.2.

We give the proof of Theorem 8.3 in the next section.

8.3 The Characterization by Differences of $\dot{B}^{s,\tau}_{p,q}(\mathbb{R}^n)$ and $\dot{F}^{s,\tau}_{p,q}(\mathbb{R}^n)$

Let a_t be as in (4.12). For all $f \in L^1_{\mathrm{loc}}(\mathbb{R}^n)$, set

$$\|f\|^{\clubsuit}_{\dot{B}^{s,\tau}_{p,q}(\mathbb{R}^n)} \equiv \sup_{P\in\mathcal{Q}} \frac{1}{|P|^\tau} \left\{ \int_0^{2l(P)} t^{-sq} \left(\int_P [a_t(x)]^p \, dx \right)^{q/p} \frac{dt}{t} \right\}^{1/q}$$

and

$$\|f\|^{\clubsuit}_{\dot{F}^{s,\tau}_{p,q}(\mathbb{R}^n)} \equiv \sup_{P\in\mathcal{Q}} \frac{1}{|P|^\tau} \left\{ \int_P \left(\int_0^{2l(P))} t^{-sq} [a_t(x)]^q \frac{dt}{t} \right)^{p/q} dx \right\}^{1/p}.$$

We have the following difference characterizations for $\dot{B}^{s,\tau}_{p,q}(\mathbb{R}^n)$ and $\dot{F}^{s,\tau}_{p,q}(\mathbb{R}^n)$.

Theorem 8.4. *Let s, p, q, τ, N_1 and M be as in Theorem 8.3. Then $\|f\|_{\dot{A}^{s,\tau}_{p,q}(\mathbb{R}^n)}$ is equivalent to $\|f\|^{\clubsuit}_{\dot{A}^{s,\tau}_{p,q}(\mathbb{R}^n)}$ for all $f \in \dot{A}^{s,\tau}_{p,q}(\mathbb{R}^n)$.*

To prove Theorems 8.3 and 8.4, let $L \in (0,1/2]$ and $\widetilde{\Theta}(\mathbb{R}^n, L)$ be the collection of all Schwartz functions φ satisfying

$$\operatorname{supp} \widehat{\varphi} \subset \{\xi \in \mathbb{R}^n : L \le |\xi| \le 2\}$$

and

$$\sum_{j\in\mathbb{Z}} \widehat{\varphi}(2^{-j}\xi) = 1$$

for all $\xi \in \mathbb{R}^n \setminus \{0\}$. Similarly to the proof of Lemma 4.1, we obtain the following conclusion.

Lemma 8.2. *The space $\dot{A}^{s,\tau}_{p,q}(\mathbb{R}^n)$ is independent of the choices of $L \in (0,1/2]$ and $\varphi \in \widetilde{\Theta}(\mathbb{R}^n, L)$.*

We need to construct a representation of $\varphi_j * f$ by an integral mean of differences of f. Let $\psi \in \mathcal{S}_\infty(\mathbb{R}^n)$ such that

$$\operatorname{supp} \widehat{\psi} \subset \{\xi \in \mathbb{R}^n : 1/2 \le |\xi| \le 2\} \quad \text{and} \quad \sum_{j\in\mathbb{Z}} \widehat{\psi}(2^{-j}\xi) = 1$$

for all $\xi \in \mathbb{R}^n \setminus \{0\}$. Define φ by setting, for all $\xi \in \mathbb{R}^n$,

$$\widehat{\varphi}(\xi) \equiv (-1)^{M+1} \sum_{i=0}^{M} (-1)^i \binom{M}{i} \widehat{\psi}((M-i)\xi). \tag{8.4}$$

It is easy to check that $\varphi \in \widetilde{\Theta}(\mathbb{R}^n, 1/(2M))$. Furthermore, for all locally integrable functions f,

$$\varphi_j * f(x) = (-1)^{M+1} \int_{\mathbb{R}^n} \left(\Delta^M_{-2^{-j}y} f(x)\right) \psi(y)\, dy. \tag{8.5}$$

Via these constructions, we have the following result.

Lemma 8.3. *Let $p, q \in (0,\infty]$, $\tau \in [0,\infty)$ and $s \in (0,\infty)$. There exists a positive constant C such that for all $f \in \dot{A}^{s,\tau}_{p,q}(\mathbb{R}^n)$,*

$$\|f\|_{\dot{A}^{s,\tau}_{p,q}(\mathbb{R}^n)} \le C\|f\|^{\clubsuit}_{\dot{A}^{s,\tau}_{p,q}(\mathbb{R}^n)}.$$

Proof. By Proposition 8.2, we know that each $f \in \dot{A}^{s,\tau}_{p,q}(\mathbb{R}^n)$ is a locally integrable function in the sense of $\mathscr{S}'_\infty(\mathbb{R}^n)$. Let φ be as in (8.4). Then (8.5) holds for all $f \in \dot{A}^{s,\tau}_{p,q}(\mathbb{R}^n)$, which together with Lemma 8.2 yields that

$$\|f\|_{\dot{B}^{s,\tau}_{p,q}(\mathbb{R}^n)} \lesssim \sup_{P\in\mathscr{Q}} \frac{1}{|P|^\tau} \left\{ \sum_{j=j_P}^{\infty} 2^{jsq} \left[\int_P \left(\int_{\mathbb{R}^n} |\Delta^M_{-2^{-j}y} f(x)||\psi(y)|\, dy \right)^p dx \right]^{\frac{q}{p}} \right\}^{\frac{1}{q}}$$

and

$$\|f\|_{\dot{F}^{s,\tau}_{p,q}(\mathbb{R}^n)} \lesssim \sup_{P\in\mathscr{Q}} \frac{1}{|P|^\tau} \left\{ \int_P \left[\sum_{j=j_P}^{\infty} 2^{jsq} \left(\int_{\mathbb{R}^n} |\Delta^M_{-2^{-j}y} f(x)||\psi(y)|\, dy \right)^q \right]^{\frac{p}{q}} dx \right\}^{\frac{1}{p}}.$$

Then a modification of the proof of Lemma 4.3 gives us the desired inequalities. □

Next we show that $\|f\|^{\clubsuit}_{\dot{A}^{s,\tau}_{p,q}(\mathbb{R}^n)} \lesssim \|f\|^{\blacktriangle}_{\dot{A}^{s,\tau}_{p,q}(\mathbb{R}^n)}$.

Lemma 8.4. *Let $s \in (0,\infty)$, $p \in (0,\infty)$, $q \in (0,\infty]$, $\tau \in [0,\infty)$ and $M \in \mathbb{N}$ such that $M \le N_1$ and $s < \{M \wedge (M + n(1/p - \tau))\}$. Then there exists a positive constant C such that for all $f \in \dot{A}^{s,\tau}_{p,q}(\mathbb{R}^n)$,*

$$\|f\|^{\clubsuit}_{\dot{A}^{s,\tau}_{p,q}(\mathbb{R}^n)} \le C\|f\|^{\blacktriangle}_{\dot{A}^{s,\tau}_{p,q}(\mathbb{R}^n)}.$$

Proof. By similarity, we only consider the spaces $\dot{F}^{s,\tau}_{p,q}(\mathbb{R}^n)$. Let $f \in \dot{F}^{s,\tau}_{p,q}(\mathbb{R}^n)$. By Proposition 8.2 and [156, Theorem 8.4], similarly to the argument in Sect. 4.3, we see that

$$f = \sum_{i=1}^{2^n-1} \sum_{j\in\mathbb{Z}} \sum_{k\in\mathbb{Z}^n} a_{i,j,k} \psi_{i,j,k}$$

holds in $L^p_{\mathrm{loc}}(\mathbb{R}^n)$ when $p \in (0,\infty)$, where $a_{i,j,k} \equiv \langle f, \psi_{i,j,k}\rangle$. Therefore, for all $P \in \mathscr{Q}$,

$$\begin{aligned}
&\frac{1}{|P|^\tau}\left\{\int_P\left(\int_0^{2l(P)} t^{-sq}[a_t(x)]^q\frac{dt}{t}\right)^{p/q}dx\right\}^{1/p}\\
&\lesssim \frac{1}{|P|^\tau}\left\{\int_P\left(\sum_{m=j_P-1}^{\infty} 2^{msq}\left[2^{mn}\int_{2^{-m-2}\le|h|<2^{-m}}|\Delta_h^M f(x)|\,dx\right]^q\right)^{p/q}dx\right\}^{1/p}\\
&\lesssim \frac{1}{|P|^\tau}\left\{\int_P\left(\sum_{m=j_P-1}^{\infty} 2^{msq}\left[2^{mn}\sum_{i=1}^{2^n-1}\sum_{j=m+1}^{\infty}\sum_{k\in\mathscr{I}_{P,j}}|a_{i,j,k}|\right.\right.\right.\\
&\quad\left.\left.\left.\times\int_{2^{-m-2}\le|h|<2^{-m}}|\Delta_h^M \psi_{i,j,k}(x)|\,dx\right]^q\right)^{p/q}dx\right\}^{1/p}+\frac{1}{|P|^\tau}\left\{\cdots\sum_{j=-\infty}^{m}\cdots\right\}^{1/p}\\
&\equiv \mathrm{I}+\mathrm{II}.
\end{aligned}$$

It suffices to show that both I and II are dominated by $\|f\|_{\dot{F}^{s,\tau}_{p,q}(\mathbb{R}^n)}^{\blacktriangle}$.

For I, by the support condition of ψ_i (see (4.20)), we see that for all $|h| < 2^{-m}$ and $l \in \{0,\cdots,M\}$,

$$\operatorname{supp}\,\psi_i(2^j(\cdot+(M-l)h)-k) \subset 2^{-j}(k+[-N_3,N_3]^n)+2^{-m}M(-1,1)^n \equiv E_{M,m,j}.$$

Thus,

$$\begin{aligned}
\mathrm{I} \lesssim \sum_{l=0}^{M}\frac{1}{|P|^\tau}&\left\{\int_P\left(\sum_{m=j_P-1}^{\infty} 2^{msq}\left[2^{mn}\sum_{i=1}^{2^n-1}\sum_{j=m+1}^{\infty} 2^{jn/2}\sum_{k\in\mathscr{I}_{P,j}}|a_{i,j,k}|\chi_{E_{M,m,j}}(x)\right.\right.\right.\\
&\left.\left.\left.\times\int_{2^{-m-2}\le|h|<2^{-m}}|\psi_i(2^j(x+(M-l)h)-k)|\,dh\right]^q\right)^{p/q}dx\right\}^{1/p}.
\end{aligned}$$

If $q \in (0,1]$, since $\psi_i \in C^{N_1}(\mathbb{R}^n)$ and has compact support, by (2.11), we have

$$\mathrm{I} \lesssim \frac{1}{|P|^\tau}\left\{\int_P\left(\sum_{m=j_P-1}^{\infty} 2^{msq}\sum_{i=1}^{2^n-1}\sum_{j=m+1}^{\infty} 2^{jnq/2}\sum_{k\in\mathscr{I}_{P,j}}|a_{i,j,k}|^q\chi_{E_{M,m,j}}(x)\right)^{p/q}dx\right\}^{1/p}.$$

Since $s > 0$, for any fix $x \in P$,

$$\begin{aligned}
&\sum_{m=j_P-1}^{\infty} 2^{msq}\sum_{j=m+1}^{\infty} 2^{jnq/2}\sum_{k\in\mathscr{I}_{P,j}}|a_{i,j,k}|^q\chi_{E_{M,m,j}}(x)\\
&= 2^{-sq}\sum_{j=j_P}^{\infty}\sum_{k\in\mathscr{I}_{P,j}} 2^{j(s+n/2)q}|a_{i,j,k}|^q\chi_{E_{M,j-1,j}}(x),
\end{aligned}$$

which together with the fact that $E_{M,j-1,j}$ is covered by a union of finite number dyadic cubes in $\mathcal{Q}_j$ implies that $\mathrm{I} \lesssim \|f\|_{\dot{F}^{s,\tau}_{p,q}(\mathbb{R}^n)}^{\blacktriangle}$. If $q \in (1,\infty]$, letting $\varepsilon \in (0,s)$, by Hölder's inequality, we see that

$$\begin{aligned}
\mathrm{I} &\lesssim \sum_{l=0}^{M} \frac{1}{|P|^\tau} \left\{ \int_P \left(\sum_{m=j_P-1}^{\infty} 2^{msq} \left[\sum_{i=1}^{2^n-1} \sum_{j=m+1}^{\infty} 2^{jn/2} \right.\right.\right. \\
&\qquad \left.\left.\left. \times \left(\sum_{k\in\mathcal{I}_{P,j}} |a_{i,j,k}|^q \chi_{E_{M,m,j}}(x) \right)^{1/q} \right]^q \right)^{p/q} dx \right\}^{1/p} \\
&\lesssim \sum_{l=0}^{M} \frac{1}{|P|^\tau} \left\{ \int_P \left(\sum_{m=j_P-1}^{\infty} 2^{m(s-\varepsilon)q} \sum_{i=1}^{2^n-1} \sum_{j=m+1}^{\infty} 2^{jnq/2+j\varepsilon q} \right.\right. \\
&\qquad \left.\left. \times \sum_{k\in\mathcal{I}_{P,j}} |a_{i,j,k}|^q \chi_{E_{M,m,j}}(x) \right)^{p/q} dx \right\}^{1/p} \\
&\lesssim \|f\|_{\dot{F}^{s,\tau}_{p,q}(\mathbb{R}^n)}^{\blacktriangle}.
\end{aligned}$$

The estimate of II is similar to that of Lemma 4.8. In fact, by (4.28), we have

$$\begin{aligned}
\mathrm{II} &\lesssim \frac{1}{|P|^\tau} \left\{ \int_P \left(\sum_{m=j_P-1}^{\infty} 2^{msq} \left[2^{mn} \int_{2^{-m-2}\le|h|<2^{-m}} \sum_{i=1}^{2^n-1} \sum_{j=-\infty}^{m} 2^{jn/2} \right.\right.\right. \\
&\qquad \left.\left.\left. \times \sum_{k\in\mathcal{I}_{P,j}} \frac{|a_{i,j,k}|\chi_{E_{M,m,j}}(x)|h|^M 2^{jM}}{(1+|2^j(x+\theta_1 h+\cdots+\theta_M h)-k|)^{n+\delta}} \, dh \right]^q \right)^{p/q} dx \right\}^{1/p},
\end{aligned}$$

where $\delta \in (0,\infty)$ can be sufficiently large. Since $j \le m$ and $|h| < 2^{-m}$, we see that

$$|2^j(x+\theta_1 h+\cdots+\theta_M h)-k| \ge |2^j x-k| - M2^j|h| \ge |2^j x-k| - M,$$

which further deduces that

$$\begin{aligned}
\mathrm{II} &\lesssim \frac{1}{|P|^\tau} \left\{ \int_P \left(\sum_{m=j_P-1}^{\infty} 2^{msq} \left[\sum_{i=1}^{2^n-1} \sum_{j=-\infty}^{m} 2^{jn/2} 2^{(j-m)M} \right.\right.\right. \\
&\qquad \left.\left.\left. \times \sum_{k\in\mathcal{I}_{P,j}} \frac{|a_{i,j,k}|\chi_{E_{M,m,j}}(x)}{(1+|2^j x-k|)^{n+\delta}} \right]^q \right)^{p/q} dx \right\}^{1/p}
\end{aligned}$$

$$\lesssim \frac{1}{|P|^\tau}\left\{\int_P\left(\sum_{m=j_P-1}^{\infty}2^{msq}\left[\sum_{j=-\infty}^{m}2^{-js}2^{(j-m)M}2^{jn/p}\right.\right.\right.$$
$$\left.\left.\left.\times\left(\sum_{i=1}^{2^n-1}\sum_{k\in\mathscr{I}_{P,j}}2^{jsp}2^{jnp/2}2^{-jn}|a_{i,j,k}|^p\right)^{1/p}\chi_{E_{M,m,j}}(x)\right]^q\right)^{p/q}dx\right\}^{1/p},$$

where the last inequality follows from (2.11) when $p\le 1$ or Hölder's inequality when $p>1$. Since $j\le m$, a finite union of dyadic cubes in $\mathscr{Q}_j$ covers $E_{M,m,j}$. Then

$$\left(\sum_{i=1}^{2^n-1}\sum_{k\in\mathscr{I}_{P,j}}2^{jsp}2^{jnp/2}2^{-jn}|a_{i,j,k}|^p\right)^{1/p}\lesssim 2^{-(j\wedge j_P)n\tau}\|f\|_{\dot{F}^{s,\tau}_{p,q}(\mathbb{R}^n)}^{\blacktriangle}$$

and hence, by $s<\{M\wedge M+n(1/p-\tau)\}$, we have

$$\mathrm{II}\lesssim \frac{1}{|P|^\tau}\left\{\int_P\left(\sum_{m=j_P-1}^{\infty}2^{msq}\left[\sum_{j=-\infty}^{m}2^{-js}2^{(j-m)M}2^{jn/p}2^{-(j\wedge j_P)n\tau}\right.\right.\right.$$
$$\left.\left.\left.\times\chi_{E_{M,m,j}}(x)\right]^q\right)^{p/q}dx\right\}^{1/p}\|f\|_{\dot{F}^{s,\tau}_{p,q}(\mathbb{R}^n)}^{\blacktriangle}$$
$$\lesssim \|f\|_{\dot{F}^{s,\tau}_{p,q}(\mathbb{R}^n)}^{\blacktriangle},$$

which completes the proof of Lemma 8.4. □

By Lemmas 8.3 and 8.4, to prove Theorems 8.3 and 8.4, we only need to prove that

$$\|f\|_{\dot{A}^{s,\tau}_{p,q}(\mathbb{R}^n)}^{\blacktriangle}\lesssim\|f\|_{\dot{A}^{s,\tau}_{p,q}(\mathbb{R}^n)}.$$

In fact, letting $N_1\ge\lfloor s+n\tau\rfloor$, by the properties of wavelets (see [99, p. 108]), we know that each $\psi_{i,j,k}$ is a constant multiple of a homogeneous smooth analysis molecule for $\dot{A}^{s,\tau}_{p,q}(\mathbb{R}^n)$ introduced in [165, Definition 4.2]. From the smooth molecular decomposition characterization of $\dot{A}^{s,\tau}_{p,q}(\mathbb{R}^n)$ in [165, Theorem 4.2], we deduce that

$$\|\{\langle f,\psi_{i,j,k}\rangle\}_{j\in\mathbb{Z},k\in\mathbb{Z}^n}\|_{\dot{a}^{s,\tau}_{p,q}(\mathbb{R}^n)}\lesssim\|f\|_{\dot{A}^{s,\tau}_{p,q}(\mathbb{R}^n)},$$

which further implies that

$$\|f\|_{\dot{A}^{s,\tau}_{p,q}(\mathbb{R}^n)}^{\blacktriangle}\lesssim\|f\|_{\dot{A}^{s,\tau}_{p,q}(\mathbb{R}^n)}$$

and then yields Theorems 8.3 and 8.4.

Remark 8.4. For the wavelet characterizations on $\dot{B}^s_{p,q}(\mathbb{R}^n)$ and $\dot{F}^s_{p,q}(\mathbb{R}^n)$, we refer to, e. g., [65, 99]. For the difference characterization on $\dot{B}^s_{p,q}(\mathbb{R}^n)$ with $p, q \in [1,\infty]$, we refer to [9, p. 147, Theorem 6.3.1] (see also [145, Sect. 5.2.3]); while for the difference characterization on $\dot{F}^s_{p,q}(\mathbb{R}^n)$ with $p, q \in [1,\infty]$, see [22].

Corresponding to Sect. 4.4, we now establish the oscillation characterization of $\dot{A}^{s,\tau}_{p,q}(\mathbb{R}^n)$. For all $f \in L^1_{\rm loc}(\mathbb{R}^n)$, define

$$\|f\|^{\sharp}_{\dot{B}^{s,\tau}_{p,q}(\mathbb{R}^n)} \equiv \sup_{P\in\mathcal{Q}} \frac{1}{|P|^\tau} \left\{ \int_0^{2l(P)} t^{-sq} \left(\int_P \left[\operatorname{osc}_1^{M-1} f(x,Mt)\right]^p dx \right)^{q/p} \frac{dt}{t} \right\}^{1/q}$$

and

$$\|f\|^{\sharp}_{\dot{F}^{s,\tau}_{p,q}(\mathbb{R}^n)} \equiv \sup_{P\in\mathcal{Q}} \frac{1}{|P|^\tau} \left\{ \int_P \left(\int_0^{2l(P)} t^{-sq} \left[\operatorname{osc}_1^{M-1} f(x,Mt)\right]^q \frac{dt}{t} \right)^{p/q} dx \right\}^{1/p}.$$

Then we have the following conclusion.

Theorem 8.5. *Let s, p, q, τ, N_1 and M be as in Theorem 8.3. If $f \in \dot{A}^{s,\tau}_{p,q}(\mathbb{R}^n)$, then there exists a function $g \in L^{\overline{p}}_{\rm loc}(\mathbb{R}^n)$ such that $f = g$ in $\mathcal{S}'_\infty(\mathbb{R}^n)$ and*

$$\|g\|^{\sharp}_{\dot{A}^{s,\tau}_{p,q}(\mathbb{R}^n)} \le C\|f\|_{\dot{A}^{s,\tau}_{p,q}(\mathbb{R}^n)};$$

if $g \in L^{\overline{p}}_{\rm loc}(\mathbb{R}^n) \cap \mathcal{S}'_\infty(\mathbb{R}^n)$ satisfies that $\|g\|^{\sharp}_{\dot{A}^{s,\tau}_{p,q}(\mathbb{R}^n)} < \infty$, then $g \in \dot{A}^{s,\tau}_{p,q}(\mathbb{R}^n)$ and

$$\|g\|_{\dot{A}^{s,\tau}_{p,q}(\mathbb{R}^n)} \le C\|g\|^{\sharp}_{\dot{A}^{s,\tau}_{p,q}(\mathbb{R}^n)},$$

where C is a positive constant independent of f and g.

Proof. We only consider Triebel-Lizorkin spaces. If $g \in L^{\overline{p}}_{\rm loc}(\mathbb{R}^n) \cap \mathcal{S}'_\infty(\mathbb{R}^n)$ satisfies that $\|g\|^{\sharp}_{\dot{A}^{s,\tau}_{p,q}(\mathbb{R}^n)} < \infty$, similarly to the proof of Lemma 4.10, we obtain that for all $g \in L^{\overline{p}}_{\rm loc}(\mathbb{R}^n) \cap \mathcal{S}'_\infty(\mathbb{R}^n)$,

$$\|g\|^{\clubsuit}_{\dot{A}^{s,\tau}_{p,q}(\mathbb{R}^n)} \lesssim \|g\|^{\sharp}_{\dot{A}^{s,\tau}_{p,q}(\mathbb{R}^n)},$$

which together with Theorem 8.4 yields that $g \in \dot{A}^{s,\tau}_{p,q}(\mathbb{R}^n)$ and

$$\|g\|_{\dot{A}^{s,\tau}_{p,q}(\mathbb{R}^n)} \lesssim \|g\|^{\sharp}_{\dot{A}^{s,\tau}_{p,q}(\mathbb{R}^n)}.$$

On the other hand, for $f \in \dot{A}^{s,\tau}_{p,q}(\mathbb{R}^n)$, by Proposition 8.2, there exists $g \in L^{\overline{p}}_{\mathrm{loc}}(\mathbb{R}^n)$ such that $f = g$ in $\mathscr{S}'_\infty(\mathbb{R}^n)$. It remains to show that

$$\|g\|^\sharp_{\dot{A}^{s,\tau}_{p,q}(\mathbb{R}^n)} \lesssim \|g\|_{\dot{A}^{s,\tau}_{p,q}(\mathbb{R}^n)}.$$

To this end, we need to estimate

$$\mathrm{I}_R \equiv \frac{1}{|R|^\tau}\left\{\int_R \left(\int_0^{2l(R)} t^{-sq}\left[\mathrm{osc}_1^{M-1} g(x,Mt)\right]^q \frac{dt}{t}\right)^{\frac{p}{q}} dx\right\}^{\frac{1}{p}},$$

where R is dyadic cube.

It is easy to see that

$$\left\{\int_R \left[\int_0^{2l(R)} t^{-sq}\left(\inf_{P\in\mathscr{P}_{M-1}(\mathbb{R}^n)} t^{-n}\int_{B(x,Mt)} |g(y)-P(y)|\,dy\right)^q \frac{dt}{t}\right]^{\frac{p}{q}} dx\right\}^{\frac{1}{p}}$$
$$\lesssim \left\{\int_R \left[\sum_{k=j_R-1}^{\infty} 2^{ksq}\left(\inf_{P\in\mathscr{P}_{M-1}(\mathbb{R}^n)} 2^{kn}\int_{B(x,M2^{-k})} |g(y)-P(y)|\,dy\right)^q\right]^{\frac{p}{q}} dx\right\}^{\frac{1}{p}}.$$

Recall that

$$g = \sum_{i=1}^{2^n-1}\sum_{j\in\mathbb{Z}}\sum_{m\in\mathbb{Z}^n} a_{i,j,m}\psi_{i,j,m}$$

holds in $L^p_{\mathrm{loc}}(\mathbb{R}^n)$ when $p \in (0,\infty)$, where $a_{i,j,m} \equiv \langle g, \psi_{i,j,m}\rangle$ (see the proof of Lemma 8.4). Then for any $\varepsilon > 0$, there exists an $N \equiv N(x,k,R) \in \mathbb{N}$ such that $N > j_R$ and

$$\int_{B(x,M2^{-k})}\left|\sum_{i=1}^{2^n-1}\sum_{|j|>N}\sum_{m\in\mathbb{Z}^n} a_{i,j,m}\psi_{i,j,m}(y)\right| dy < \varepsilon 2^{-k(M+n)}2^{-j_R(s+n\tau-n/p-M)}.$$

Thus, by the support condition of $\psi_{i,j,m}$ and $s<M$, we see that $\psi_{i,j,m}(y) = 0$ if $y \notin MR$ and

$$\mathrm{I}_R \lesssim \frac{1}{|R|^\tau}\left\{\int_R \left[\sum_{k=j_R-1}^{\infty} 2^{ksq}\left(\inf_{P\in\mathscr{P}_{M-1}(\mathbb{R}^n)} 2^{kn}\int_{B(x,M2^{-k})}\right.\right.\right.$$
$$\left.\left.\left.\times\left|\sum_{i=1}^{2^n-1}\sum_{|j|\le N}\sum_{m\in I_{MR,j}} a_{i,j,m}\psi_{i,j,m}(y)-P(y)\right| dy\right)^q\right]^{\frac{p}{q}} dx\right\}^{\frac{1}{p}} + \varepsilon,$$

where $I_{MR,j}$ is the collection of all $m \in \mathbb{Z}^n$ such that $|\operatorname{supp} \psi_{i,j,m} \cap (MR)| > 0$ for some $i \in \{1, \cdots, 2^{n-1}\}$. Similarly to the proof of Lemma 4.10, by (4.20) and employing the Taylor remainders of order M of $\phi_{i,j,m}$, we have

$$\begin{aligned}
\mathrm{I}_R &\lesssim \frac{1}{|R|^\tau} \left\{ \int_R \left[\sum_{k=j_R-1}^{\infty} 2^{ksq} \left(2^{kn} \int_{B(x,M2^{-k})} \sum_{i=1}^{2^n-1} \sum_{|j|\le N} \right.\right.\right. \\
&\quad \left.\left.\left. \times \sum_{m\in I_{MR,j}} |a_{i,j,m}| 2^{jn/2} \frac{2^{jM}|x-y|^M \chi_{2^{-j}([-N_3,N_3]^n+m)}(x)}{(1+|2^j x-m|)^{n+\delta}} \, dy \right)^q \right]^{\frac{p}{q}} dx \right\}^{\frac{1}{p}} + \varepsilon \\
&\lesssim \frac{1}{|R|^\tau} \left\{ \int_R \left[\sum_{k=j_R-1}^{\infty} 2^{ksq} \left(\sum_{i=1}^{2^n-1} \sum_{j=-N}^{j_R} \right.\right.\right. \\
&\quad \left.\left.\left. \times \sum_{m\in I_{MR,j}} |a_{i,j,m}| 2^{jn/2} 2^{jM} 2^{-kM} \chi_{2^{-j}([-N_3,N_3]^n+m)}(x) \right)^q \right]^{\frac{p}{q}} dx \right\}^{\frac{1}{p}} \\
&\quad + \frac{1}{|R|^\tau} \left\{ \int_R \left[\sum_{k=j_R-1}^{\infty} 2^{ksq} \left(\sum_{i=1}^{2^n-1} \sum_{j=j_R}^{N} \right.\right.\right. \\
&\quad \left.\left.\left. \times \sum_{m\in I_{MR,j}} |a_{i,j,m}| 2^{jn/2} \frac{2^{jM} 2^{-kM} \chi_{2^{-j}([-N_3,N_3]^n+m)}(x)}{(1+|2^j x-m|)^{n+\delta}} \right)^q \right]^{\frac{p}{q}} dx \right\}^{\frac{1}{p}} + \varepsilon,
\end{aligned}$$

where $\delta \in (0,\infty)$ can be sufficient large. Similarly to the estimates of I and II in the proof of Lemma 8.4, we obtain that $\mathrm{I}_R \lesssim \|g\|_{\dot{F}^{s,\tau}_{p,q}(\mathbb{R}^n)}^{\blacktriangle} + \varepsilon$. By Theorem 8.3 and the arbitrariness of ε, we further have

$$\mathrm{I}_R \lesssim \|g\|_{\dot{F}^{s,\tau}_{p,q}(\mathbb{R}^n)} \sim \|f\|_{\dot{F}^{s,\tau}_{p,q}(\mathbb{R}^n)},$$

which implies that

$$\|g\|_{\dot{A}^{s,\tau}_{p,q}(\mathbb{R}^n)}^{\sharp} \lesssim \|f\|_{\dot{A}^{s,\tau}_{p,q}(\mathbb{R}^n)}$$

and then completes the proof of Theorem 8.5. □

8.4 Homogeneous Besov-Hausdorff and Triebel-Lizorkin-Hausdorff Spaces

Similarly to Chap. 7, we also determine the predual spaces of $\dot{A}^{s,\tau}_{p,q}(\mathbb{R}^n)$, which are called homogeneous Besov-Hausdorff space or Triebel-Lizorkin-Hausdorff space.

Definition 8.3. Let φ be as in Definition 8.1, $s \in \mathbb{R}$, $p \in (1,\infty)$, $q \in [1,\infty)$ and $\tau \in [0, \frac{1}{(p\vee q)'}]$. Then the space $A\dot{H}^{s,\tau}_{p,q}(\mathbb{R}^n)$ is defined to be the set of all $f \in \mathcal{S}'_\infty(\mathbb{R}^n)$ such that $\|f\|_{A\dot{H}^{s,\tau}_{p,q}(\mathbb{R}^n)}$, where when $A\dot{H}^{s,\tau}_{p,q}(\mathbb{R}^n) = B\dot{H}^{s,\tau}_{p,q}(\mathbb{R}^n)$,

$$\|f\|_{B\dot{H}^{s,\tau}_{p,q}(\mathbb{R}^n)} \equiv \inf_{\omega} \left\{ \sum_{j\in\mathbb{Z}} 2^{jsq} \|\varphi_j * f[\omega(\cdot,2^{-j})]^{-1}\|^q_{L^p(\mathbb{R}^n)} \right\}^{\frac{1}{q}} < \infty$$

and when $A\dot{H}^{s,\tau}_{p,q}(\mathbb{R}^n) = F\dot{H}^{s,\tau}_{p,q}(\mathbb{R}^n)$ $(q \neq 1)$,

$$\|f\|_{F\dot{H}^{s,\tau}_{p,q}(\mathbb{R}^n)} \equiv \inf_{\omega} \left\| \left\{ \sum_{j\in\mathbb{Z}} 2^{jsq} |\varphi_j * f[\omega(\cdot,2^{-j})]^{-1}|^q \right\}^{\frac{1}{q}} \right\|_{L^p(\mathbb{R}^n)} < \infty,$$

and the function ω runs over all nonnegative Borel measurable functions on $\mathbb{R}^{n+1}_+$ satisfying (7.3) and with the restriction that for any $j \in \mathbb{Z}$, $\omega(\cdot,2^{-j})$ is allowed to vanish only where $\varphi_j * f$ vanishes.

The spaces $B\dot{H}^{s,\tau}_{p,q}(\mathbb{R}^n)$ and $F\dot{H}^{s,\tau}_{p,q}(\mathbb{R}^n)$ were original introduced, respectively, in [164, Sect. 5] and [165, Sect. 6] and proved therein to be the predual spaces of $\dot{B}^{-s,\tau}_{p',q'}(\mathbb{R}^n)$ and $\dot{F}^{-s,\tau}_{p',q'}(\mathbb{R}^n)$. We also recall that the spaces $B\dot{H}^{s,\tau}_{p,q}(\mathbb{R}^n)$ and $F\dot{H}^{s,\tau}_{p,q}(\mathbb{R}^n)$ cover the Hardy-Hausdorff space $HH^1_{-\alpha}(\mathbb{R}^n)$, which was introduced in [44] and proved therein to be the predual of the space $Q_\alpha(\mathbb{R}^n)$; see [164, Remark 5.1]. We remark that all results in Chap. 7 have counterparts for homogeneous case. Indeed, in [168], we obtained the φ-transform characterization, Sobolev-type embedding properties, smooth atomic and molecular decomposition characterizations for $B\dot{H}^{s,\tau}_{p,q}(\mathbb{R}^n)$ and $F\dot{H}^{s,\tau}_{p,q}(\mathbb{R}^n)$, which further deduced the trace theorem, the boundedness of pseudo-differential operators with homogeneous symbols and the lifting property of $B\dot{H}^{s,\tau}_{p,q}(\mathbb{R}^n)$ and $F\dot{H}^{s,\tau}_{p,q}(\mathbb{R}^n)$. The dual properties in Sect. 7.3 have homogeneous counterparts (see [166]) and the real interpolation properties in Sect. 7.4 are true for the homogeneous spaces $B\dot{H}^{s,\tau}_{p,q}(\mathbb{R}^n)$ and $F\dot{H}^{s,\tau}_{p,q}(\mathbb{R}^n)$ by a similar proof. We also point out that in [167], the maximal function and local mean characterizations of $B\dot{H}^{s,\tau}_{p,q}(\mathbb{R}^n)$ and $F\dot{H}^{s,\tau}_{p,q}(\mathbb{R}^n)$ were established.

References

1. D. R. Adams, A note on Choquet integrals with respect to Hausdorff capacity, In: Function spaces and applications (Lund, 1986), 115–124, Lecture Notes in Math. **1302**, Springer, Berlin, 1988.
2. D. R. Adams, Choquet integrals in potential theory, Publ. Mat. **42** (1998), 3–66.
3. H. Amann, Anisotropic Function Spaces and Maximal Regularity for Parabolic Problems. Part 1: Function Spaces, J. Nečas Center for Math. Modeling, Lect. Notes **6**, matfyzpress, Prague, 2009.
4. T. I. Amanov, Spaces of Differentiable Functions with Dominating Mixed Derivatives, Nauka Kaz. SSR, Alma-Ata, 1976.
5. J.-M. Angeletti, S. Mazet and P. Tchamitchian, Analysis of second order elliptic operators without boundary conditions and with VMO or Hölderian coefficients. In: Multiscale wavelet methods for PDE's, Academic Press, Boston, 1997, 495–539.
6. J. Appell and P. P. Zabrejko, Nonlinear Superposition Operators, Cambridge University Press, Cambridge, 1990.
7. R. Aulaskari, J. Xiao and R. H. Zhao, On subspaces and subsets of $BMOA$ and UBC, Analysis **15** (1995), 101–121.
8. T. Aoki, Locally bounded linear topological space, Proc. Imp. Acad. Tokyo **18** (1942), 588–594.
9. J. Bergh and J. Löfström, Interpolation Spaces, Springer, Berlin, 1976.
10. S. N. Bernštein, On properties of homogeneous functional classes, Dokl. Acad. Nauk SSSR **57** (1947), 111–114.
11. O. V. Besov, On a family of function spaces. Embedding theorems and extensions, Dokl. Acad. Nauk SSSR **126** (1959), 1163–1165.
12. O. V. Besov, On a family of function spaces in connection with embeddings and extensions, Trudy Mat. Inst. Steklov **60** (1961), 42–81.
13. O. V. Besov, V. P. Il'in and S. M. Nikol'skij, Integral Representations of Functions and Imbedding Theorems. Vol. I+II, V. H. Winston & Sons, Washington, D.C., 1978, 1979, Transalated from the Russian. Scripta Series in Mathematics, Edited by Mitchell H. Taibleson.
14. O. V. Besov, V. P. Il'in and S. M. Nikol'skij, Integralnye predstavleniya funktsii i teoremy vlozheniya, Second edition, Fizmatlit "Nauka", Moscow, 1996.
15. B. Bojarski, Sharp maximal operators of fractional order and Sobolev imbedding inequalities, Bull. Polish Acad. Sci. Math. **33** (1985), 7–16.
16. G. Bourdaud, L^p estimates for certain nonregular pseudodifferential operators, Comm. Partial Differential Equations **7** (1982), 1023–1033.
17. G. Bourdaud, The functional calculus in Sobolev spaces. In: Function spaces, differential operators and nonlinear analysis, Teubner-Texte Math. **133**, Teubner, Stuttgart, Leipzig, 1993, 127–142.
18. G. Bourdaud, Ondelettes et espaces de Besov, Rev. Mat. Iberoamericana **11** (1995), 477–512.
19. G. Bourdaud and Y. Meyer, Fonctions qui operent sur les espaces de Sobolev, J. Funct. Anal. **97** (1991), 351–360.

20. G. Bourdaud, M. Moussai and W. Sickel, An optimal symbolic calculus on Besov algebras, Ann. I. H. Poincaré-AN **23** (2006), 949–956.
21. G. Bourdaud, M. Moussai and W. Sickel, Towards sharp superposition theorems in Besov and Lizorkin-Triebel spaces, Nonlinear Anal. **68** (2008), 2889–2912.
22. G. Bourdaud, M. Moussai and W. Sickel, Composition operators on Lizorkin-Triebel spaces, J. Funct. Anal. **259** (2010), 1098–1128.
23. M. Bownik, Atomic and molecular decompositions of anisotropic Besov spaces, Math. Z. **250** (2005), 539–571.
24. M. Bownik, Anisotropic Triebel-Lizorkin spaces with doubling measures, J. Geom. Anal. **17** (2007), 387–424.
25. M. Bownik and K.-P. Ho, Atomic and molecular decompositions of anisotropic Triebel-Lizorkin spaces, Trans. Amer. Math. Soc. **358** (2006), 1469–1510.
26. Yu. A. Brudnyi, Spaces defined by means of local approximation, Trudy Moskov. Mat. Obsc. **24** (1971), 69–132.
27. Yu. A. Brudnyi, Whitney's inequality for quasi-Banach spaces. In: Function spaces and their applications to differential equations (ed. V. Maslennikova), Izd. Ross. Univ. Druzhby Narodov, Moscow, 1992, 20–27.
28. V. I. Burenkov and M. L. Gol'dman, On the extensions of functions of L_p, Trudy Math. Inst. Steklov **150** (1979), 31–51. (English transl. 1981, no. **4**, 33–53.)
29. V. I. Burenkov and V. S. Guliev, Necessary and sufficient conditions for boundedness of the maximal operator in local Morrey-type spaces, Studia Math. **163** (2004), 157–176.
30. V. I. Burenkov, H. V. Guliev and V. S. Guliev, Necessary and sufficient conditions for boundedness of the fractional maximal operator in local Morrey-type spaces, Dokl. Acad. Nauk **409** (2006), 443–447.
31. V. I. Burenkov and V. S. Guliev, Necessary and sufficient conditions for boundedness of the Riesz potential in local Morrey-type spaces, Potential Anal. **30** (2009), 211–249.
32. A. P. Calderón, An atomic decomposition of distributions in parabolic H^p spaces, Adv. Math. **25** (1977), 216–225.
33. S. Campanato, Proprieta di inclusione per spaci di Morrey, Richerche Mat. **12** (1963), 67–86.
34. S. Campanato, Proprieta di hölderianita di alcune classi di funzioni, Ann. Scuola Norm. Sup. Pisa **17** (1963), 175–188.
35. S. Campanato, Proprieta di una famiglia di spazi funzionali, Ann. Scuola Norm. Sup. Pisa **18** (1964), 137–160.
36. S. Campanato, Teoremi di interpolazione per transformazioni che applicano L^p in $C^{k,\alpha}$, Ann. Scuola Norm. Sup. Pisa **18** (1964), 345–360.
37. C. Cascante, J. M. Ortega and I. E. Verbitsky, Nonlinear potentials and two weight trace inequalities for general dyadic and radial kernels, Indiana Univ. Math. J. **53** (2004), 845–882.
38. S.-Y. A. Chang and R. Fefferman, A continuous version of duality of H^1 and BMO on the bidisc, Ann. of Math. (2) **112** (1980), 179–201.
39. F. Chiarenza and M. Frasca, Morrey spaces and Hardy-Littlewood maximal function, Rend. Math. **7** (1987), 273–279.
40. M. Christ, The extension problem for certain function spaces involving fractional order s of differentiability, Ark. Mat. **22** (1984), 63–81.
41. R. Coifman and R. Rochberg, Representation theorems for holomorphic and harmonic functions in L^p, Astérisque **77** (1980), 11–66.
42. R. Coifman and G. Weiss, Extensions of Hardy spaces and their use in analysis, Bull. Amer. Math. Soc. **83** (1977), 569–645.
43. G. Dafni, Local *VMO* and weak convergence in h^1, Canad. Math. Bull. **45** (2002), 46–59.
44. G. Dafni and J. Xiao, Some new tent spaces and duality theorem for fractional Carleson measures and $Q_\alpha(\mathbb{R}^n)$, J. Funct. Anal. **208** (2004), 377–422.
45. G. Dafni and J. Xiao, The dyadic structure and atomic decomposition of Q spaces in several real variables, Tohoku Math. J. **57** (2005), 119–145.
46. R. A. DeVore and R. C. Sharpley, Maximal functions measuring smoothness, Mem. Amer. Math. Soc. **49** (1984), no. 293, viii+115 pp.

47. L. Diening, P. Hastö and S. Roudenko, Function spaces of variable smoothness and integrability, J. Funct. Anal. **256** (2009), 1731–1768.
48. N. Dunford and J. Schwartz, Linear Operators. I, Interscience, New York and London, 1964 (First edition, 1958).
49. A. El Baraka, An embedding theorem for Campanato spaces, Electron. J. Differential Equations **66** (2002), 1–17.
50. A. El Baraka, Function spaces of BMO and Campanato type, Proceedings of the 2002 Fez Conference on Partial Differential Equations, 109–115 (electronic), Electron. J. Differ. Equ. Conf. **9**, Southwest Texas State Univ., San Marcos, TX, 2002.
51. A. El Baraka, Littlewood-Paley characterization for Campanato spaces, J. Funct. Spaces Appl. **4** (2006), 193–220.
52. D. Drihem, Some embeddings and equivalent norms of the $\mathscr{L}^{\lambda}_{p,q}$ spaces, Funct. Appro. Comment. Math. **41** (2009), 15–40.
53. J. Duoandikoetxea, Fourier Analysis, Graduate Studies in Mathematics **29**, American Mathematical Society, Providence, R. I., 2001.
54. M. Essén, S. Janson, L. Peng and J. Xiao, Q spaces of several real variables, Indiana Univ. Math. J. **49** (2000), 575–615.
55. M. Essén and J. Xiao, Some results on Q_p spaces, $0 < p < 1$, J. Reine Angew. Math. **485** (1997), 173–195.
56. W. Farkas, J. Johnsen and W. Sickel, Traces of Besov-Lizorkin–Triebel spaces – a complete collection of all limiting cases, Math. Bohemica **125** (2000), 1–37.
57. W. Farkas and H.-G. Leopold, Characterisations of function spaces of generalised smoothness, Ann. Mat. Pura Appl. **185** (2006), 1–62.
58. C. Fefferman and E. M. Stein, Some maximal inequalities, Amer. J. Math. **93** (1971), 107–115.
59. J. Franke, On the spaces $F^s_{p,q}$ of Triebel-Lizorkin type: pointwise multipliers and spaces on domains, Math. Nachr. **125** (1986), 29–68.
60. J. Franke and T. Runst, Regular elliptic boundary value problems in Besov-Triebel-Lizorkin spaces, Math. Nachr. **174** (1995), 113–149.
61. M. Frazier, Y.-S. Han, B. Jawerth and G. Weiss, The $T1$ theorem for Triebel-Lizorkin spaces, In: Harmonic analysis and partial differential equations (El Escorial, 1987), 168–181, Lecture Notes in Math. **1384**, Springer, Berlin, 1989.
62. M. Frazier and B. Jawerth, Decomposition of Besov spaces, Indiana Univ. Math. J. **34** (1985), 777–799.
63. M. Frazier and B. Jawerth, The ϕ-transform and applications to distribution spaces, In: Function spaces and applications (Lund, 1986), 223–246, Lecture Notes in Math. **1302**, Springer, Berlin, 1988.
64. M. Frazier and B. Jawerth, A discrete transform and decompositions of distribution spaces, J. Funct. Anal. **93** (1990), 34–171.
65. M. Frazier, B. Jawerth and G. Weiss, Littlewood-Paley Theory and The Study of Function Spaces, CBMS Regional Conference Series in Mathematics **79**, Published for the Conference Board of the Mathematical Sciences, Washington, DC; by the American Mathematical Society, Providence, RI, 1991.
66. M. Frazier, R. Torres and G. Weiss, The boundedness of Calderón-Zygmund operators on the spaces $\dot{F}^{\alpha,q}_p$, Rev. Mat. Iberoamericana **4** (1988), 41–72.
67. F. G. Friedlander and M. Joshi, Introduction to the Theory of Distributions, Cambridge University Press, Cambridge, 1998.
68. L. Grafakos, Modern Fourier Analysis, Second edition, Graduate Texts in Math. **250**, Springer, New York, 2008.
69. L. Grafakos and R. H. Torres, Pseudodifferential operators with homogeneous symbols, Michigan Math. J. **46** (1999), 261–269.
70. L. I. Hedberg and Y. V. Netrusov, An axiomatic approach to function spaces, spectral synthesis, and Luzin approximation, Mem. Amer. Math. Soc. **188** (2007), no. 882, vi+97 pp.
71. J. Heinonen, Lectures on Analysis on Metric Spaces, Springer-Verlag, New York, 2001.

72. E. Hernández and G. Weiss, A First Course on Wavelets, Studies in Advanced Mathematics, CRC Press, Boca Raton, FL, 1996.
73. L. Hörmander, Pseudodifferential operators of type 1,1, Comm. Partial Differential Equations **13** (1988), 1085–1111.
74. L. Hörmander, Continuity of pseduodifferential operators of type 1,1, Comm. Partial Differential Equations **14** (1989), 231–243.
75. S. Janson, On functions with conditions on the mean oszillation, Ark. Mat. **4** (1976), 189–196.
76. S. Janson, On the space Q_p and its dyadic counterpart, In: C. Kiselman, A. Vretblad (Eds.), Complex Analysis and Differential Equations, Proceeding, Marcus Wallenberg Symposium in Honor of Matts Essén, Uppsala, 1997; Acta Univ. Upsaliensis **64** (1999), 194–205.
77. B. Jawerth, The trace of Sobolev and Besov spaces if $0 < p < 1$, Studia Math. **62** (1978), 65–71.
78. H. Jia and H. Wang, Decomposition of Hardy-Morrey spaces, J. Math. Anal. Appl. **354** (2009), 99–110.
79. F. John and L. Nirenberg, On functions of bounded mean oszillation, Comm. Pure Appl. Math. **14** (1961), 415–426.
80. J. Johnsen, Pointwise multiplication of Besov and Triebel-Lizorkin spaces, Math. Nachr. **175** (1995), 85–133.
81. J. Johnsen and W. Sickel, On the trace problem for Lizorkin–Triebel spaces with mixed norms, Math. Nachr. **281** (2008), 1–28.
82. J.-P. Kahane and P.-G. Lemarie-Rieuseut, Fourier Series and Wavelets, Gordon and Breach Publishers, 1995.
83. G. A. Kalyabin, Characterizations of function spaces of Besov-Lizorkin-Triebel type, Dokl. Acad. Nauk. SSSR **236** (1977), 1056–1059.
84. G. A. Kalyabin, Multiplier conditions of function spaces of Besov and Lizorkin-Triebel type, Dokl. Acad. Nauk SSSR **251** (1980), 25–26.
85. G. A. Kalyabin, The description of functions of classes of Besov-Lizorkin-Triebel type, Trudy Mat. Inst. Steklov **156** (1980), 82–109.
86. G. A. Kalyabin, Criteria of the multiplication property and the embeddings in C of spaces of Besov-Lizorkin-Triebel type, Mat. Zametki **30** (1981), 517–526.
87. H. Koch and W. Sickel, Pointwise multipliers of Besov spaces of smoothness zero and spaces of continuous functions, Rev. Mat. Iberoamericana **18** (2002), 587–626.
88. H. Kozono and M. Yamazaki, Semilinear heat equations and the Navier-Stokes equation with distributions in new function spaces as initial data, Comm. Partial Differential Equations **19** (1994), 959–1014.
89. A. Kufner, O. John and S. Fučik, Function Spaces, Academia, Prague, 1977.
90. P. Li and Z. Zhai, Well-posedness and regularity of generalized Navier-Stokes equations in some critical Q-spaces, J. Funct. Anal. (to appear).
91. P. I. Lizorkin, Operators connected with fractional derivatives and classes of differentiable functions, Trudy Mat. Inst. Steklov **117** (1972), 212–243.
92. P. I. Lizorkin, Properties of functions of the spaces $\Lambda^r_{p\theta}$, Trudy Mat. Inst. Steklov **131** (1974), 158–181.
93. J. Marschall, Some remarks on Triebel spaces, Studia Math. **87** (1987), 79–92.
94. J. Marschall, Nonregular pseudo-differential operators, Z. Anal. Anwendungen **1** (1996), 109–148.
95. V. G. Maz'ya and T. O. Shaposnikova, Theory of Multipliers in Spaces of Differentiable Functions, Pitman, Boston, 1985.
96. V. G. Maz'ya and T. O. Shaposnikova, Theory of Sobolev Multipliers, Springer, Berlin, 2009.
97. A. Mazzucato, Function space theory and applications to non-linear PDE, Trans. Amer. Math. Soc. **355** (2003), 1297–1369.
98. Y. Meyer, Remarques sur un théorème de J.M. Bony, Suppl. Rend. Circ. Mat. Palermo, Serie II, **1** (1981), 1–20.
99. Y. Meyer, Wavelets and Operators, Cambridge University Press, Cambridge, 1992.
100. Y. Meyer and R. Coifman, Wavelets. Calderon-Zygmund and Multilinear Operators, Cambridge University Press, Cambridge, 1997.

101. G. N. Meyers, Mean oscillation over cubes and Hölder continuity, Proc. Amer. Math. Soc. **15** (1964), 717–721.
102. C. B. Morrey, On the solutions of quasi linear elliptic partial differential equations, Trans. Amer. Math. Soc. **43** (1938), 126–166.
103. A. M. Najafov, On some properties of the functions from Sobolev-Morrey type spaces, Cent. Eur. J. Math. **3** (2005), 496–507.
104. A. M. Najafov, Some properties of functions from the intersection of Besov-Morrey type spaces with dominant mixed derivatives, Proc. A. Razmadze Math. Inst. **139** (2005), 71–82.
105. A. M. Najafov, Embedding theorems in the Sobolev-Morrey-type spaces $S_{p,a,\kappa,\tau}W(G)$ with dominating mixed derivatives, Siberian Math. J. **47** (2006), 505–516.
106. Y. V. Netrusov, Theorems on traces and multipliers for functions in Lizorkin-Triebel spaces, Zapsiki Nauchnykh Seminarov POMI **200** (1992), 132–138.
107. M. V. Nevskii, Approximation of functions in Orlicz classes, In: Studies in the theory of functions of several real variables (ed. Yu. A. Brudnyi), Yaroslavl. Gos. Univ., Yaroslavl', 1984, 83–101.
108. S. M. Nikol'skij, Inequalities for entire analytic functions of finite order and their application to the theory of differentiable functions of several variables, Trudy Mat. Inst. Steklov **38** (1951), 244–278.
109. S. M. Nikol'skij, Approximation of Functions of Several Variables and Imbedding Theorems, Springer-Verlag, Berlin, 1975.
110. Y. Peetre, On the theory of $\mathscr{L}_{p,\lambda}$ spaces, J. Funct. Anal. **4** (1969), 71–87.
111. Y. Peetre, Remarques sur les espaces de Besov. Le cas $0 < p < 1$, C. R. Acad. Sci. Paris Sér. A-B **277** (1973), 947–950.
112. Y. Peetre, The trace of Besov spaces – a limiting case, Technical Report, Lund 1975.
113. Y. Peetre, On spaces of Triebel-Lizorkin type, Ark. Mat. **13** (1975), 123–130.
114. Y. Peetre, New Thougths on Besov Spaces, Duke University Mathematics Series, Duke University Press, Durham, 1976.
115. F. Ricci and M. Taibleson, Boundary values of harmonic functions in mixed norm spaces and their atomic structure, Ann. Scuola Norm. Sup. Pisa Cl. Sci. **10** (1983), 1–54.
116. W. Rudin, Functional Analysis. Second edition, McGraw-Hill, Inc., New York, 1991.
117. T. Runst, Pseudo-differential operators of the "exotic" class $L^0_{1,1}$ in spaces of Besov and Triebel-Lizorkin type, Annals Global Analysis Geometry **3** (1985), 13–28.
118. T. Runst, Mapping properties of non-linear operators in spaces of Triebel-Lizorkin and Besov type, Anal. Math. **12** (1986), 313–346.
119. T. Runst and W. Sickel, Sobolev Spaces of Fractional Order, Nemytskij Operators, and Nonlinear Partial Differential Equations, de Gruyter Series in Nonlinear Analysis and Applications **3**, Walter de Gruyter, Berlin, 1996.
120. V. S. Rychkov, Littlewood-Paley theory and function spaces with $A_p^{\ell oc}$ weights, Math. Nachr. **224** (2001), 145–180.
121. V. S. Rychkov, On restrictions and extensions of the Besov and Triebel-Lizorkin spaces with respect to Lipschitz domains, J. London Math. Soc. **60** (1999), 237–257.
122. D. Sarason, Functions of vanishing mean oscillation, Trans. Amer. Math. Soc. **207** (1975), 391–405.
123. Y. Sawano, Wavelet characterization of Besov-Morrey and Triebel-Lizorkin-Morrey spaces, Funct. Approx. Comment. Math. **38** (2008), 93–107.
124. Y. Sawano, Besov-Morrey spaces and Triebel-Lizorkin-Morrey spaces on domains, Math. Nachr. **283** (2010), 1–32.
125. Y. Sawano, Brézis-Gallouët-Wainger type inequality for Besov-Morrey spaces, Studia. Math. **196** (2010), 91–101.
126. Y. Sawano and H. Tanaka, Decompositions of Besov-Morrey spaces and Triebel-Lizorkin-Morrey spaces, Math. Z. **257** (2007), 871–905.
127. Y. Sawano, D. Yang and W. Yuan, New applications of Besov-type and Triebel-Lizorkin-type spaces, J. Math. Anal. Appl. **363** (2010), 73–85.

128. H.-J. Schmeisser, Recent developments in the theory of function spaces with dominating mixed smoothness, Proc. Conf. NAFSA-8, (ed. J. Rakosnik), Inst. Math. Acad. Sci. Prague, 2007, 145–204.
129. H.-J. Schmeisser and H. Triebel, Topics in Fourier analysis and function spaces, Akademische Verlagsgesellschaft Geest & Portig K.-G., Leipzig, 1987, (Also John Wiley, 1987).
130. A. Seeger, A note on Triebel-Lizorkin spaces, Banach Center Publ. **22**, Warsaw, PWN Polish Sci. Publ. 1989, 391–400.
131. W. Sickel, On pointwise multipliers for $F^s_{p,q}(\mathbb{R}^n)$ in case $\sigma_{p,q} < s < n/p$, Annali Mat. Pura Appl. **176** (1999), 209–250.
132. W. Sickel, Pointwise multipliers for Lizorkin-Triebel spaces, Operator theory: Advances and Appl. **110** (1999), 295–321.
133. W. Sickel and H. Triebel, Hölder inequalities and sharp embeddings in function spaces of $B^s_{p,q}$ and $F^s_{p,q}$ type, Z. Anal. Anwendungen **14** (1995), 105–140.
134. E. M. Stein and G. Weiss, Introduction to Fourier Analysis on Euclidean Spaces, Princeton University Press, Princeton, N. J., 1971.
135. R. S. Strichartz, Multipliers on fractional Sobolev spaces, J. Math. Mech. **16** (1967), 1031–1060.
136. M. H. Taibleson, On the theory of Lipschitz spaces of distributions on Euclidean n-space. I. Principal properties, J. Math. Mech. **13** (1964), 407–479.
137. M. H. Taibleson, On the theory of Lipschitz spaces of distributions on Euclidean n-space. II. Translation invariant operators, duality, and interpolation, J. Math. Mech. **14** (1965), 821–839.
138. M. H. Taibleson, On the theory of Lipschitz spaces of distributions on Euclidean n-space. III. Smoothness and integrability of Fourier tansforms, smoothness of convolution kernels, J. Math. Mech. **15** (1966), 973–981.
139. L. Tang and J. Xu, Some properties of Morrey type Besov-Triebel spaces, Math. Nachr. **278** (2005), 904–917.
140. R. H. Torres, Continuity properties of pseudodifferential operators of type 1, 1, Comm. Partial Differential Equations **15** (1990), 1313–1328.
141. R. H. Torres, Boundedness results for operators with singular kernels on distribution spaces, Mem. Amer. Math. Soc. **90** (1991), no. 442, viii+172 pp.
142. H. Triebel, Spaces of distributions of Besov type on Euclidean n-space. Duality, interpolation, Ark. Mat. **11** (1973), 13–64.
143. H. Triebel, Multiplication properties of the spaces $B^s_{p,q}$ and $F^s_{p,q}$. Quasi-Banach Algebras of functions, Ann. Mat. Pura Appl. (4) **113** (1977), 33–42.
144. H. Triebel, Spaces of Besov-Sobolev-Hardy Type, Teubner-Texte Math. **9**, Teubner, Leipzig, 1978.
145. H. Triebel, Theory of Function Spaces, Birkhäuser Verlag, Basel, 1983.
146. H. Triebel, Theory of Function Spaces II, Birkhäuser Verlag, Basel, 1992.
147. H. Triebel, Fractals and Spectra, Birkhäuser Verlag, Basel, 1997.
148. H. Triebel, Theory of Function Spaces III, Birkhäuser Verlag, Basel, 2006.
149. H. Triebel, Function Spaces and Wavelets on Domains, EMS Publishing House, Zürich, 2008.
150. A. Uchiyama, A constructive proof of the Fefferman-Stein decomposition of BMO$(\mathbb{R}^n)$, Acta Math. **148** (1982), 215–241.
151. J. Vybíral, Function spaces with dominating mixed smoothness, Diss. Math. **436** (2006), 73 pp.
152. J. Vybíral, Sobolev and Jawerth embeddings for spaces with variable smoothness and integrability, Ann. Acad. Fenn. Math. **34** (2009), 529–544.
153. H. Wallin, New and old function spaces, In: Function spaces and applications (Lund, 1986), 97–114, Lecture Notes in Math. **1302**, Springer, Berlin, 1988.
154. H. Wang, Decomposition for Morrey type Besov-Triebel spaces, Math. Nachr. **282** (2009), 774–787.
155. J. M. Wilson, On the atomic decomposition for Hardy spaces, Pacific J. Math. **116** (1985), 201–207.
156. P. Wojtasczyk, A Mathematical Introduction to Wavelets, Cambridge University Press, Cambridge, 1997.

157. Z. Wu and C. Xie, Decomposition theorems for Q_p spaces, Ark. Mat. **40** (2002), 383–401.
158. Z. Wu and C. Xie, Q spaces and Morrey spaces, J. Funct. Anal. **201** (2003), 282–297.
159. J. Xiao, Holomorphic Q Classes, Lecture Notes in Math. **1767**, Springer, Berlin, 2001.
160. J. Xiao, Geometric Q_p Functions, Birkhäuser Verlag, Basel, 2006.
161. J. Xiao, Homothetic variant of fractional Sobolev space with application to Navier-Stokes system, Dyn. Partial Differ. Equ. **4** (2007), 227–245.
162. J. Xiao, The Q_p Carleson measure problem, Adv. Math. **217** (2008), 2075–2088.
163. D. Yang and W. Yuan, A note on dyadic Hausdorff capacities, Bull. Sci. Math. **132** (2008), 500–509.
164. D. Yang and W. Yuan, A new class of function spaces connecting Triebel-Lizorkin spaces and Q spaces, J. Funct. Anal. **255** (2008), 2760–2809.
165. D. Yang and W. Yuan, New Besov-type spaces and Triebel-Lizorkin-type spaces including Q spaces, Math. Z. **265** (2010), 451–480.
166. D. Yang and W. Yuan, Dual properties of Triebel-Lizorkin-Type spaces and their applications, Z. Anal. Anwendungen (to appear).
167. D. Yang and W. Yuan, Characterizations of Besov-type and Triebel-Lizorkin-type spaces via maximal functions and local means, Nonlinear Anal. (to appear).
168. W. Yuan, Y. Sawano and D. Yang, Decompositions of Besov-Hausdorff and Triebel-Lizorkin-Hausdorff spaces and their applications, J. Math. Anal. Appl. **369** (2010), 736–757.
169. H. Yue and G. Dafni, A John-Nirenberg type inequality for $Q_\alpha(\mathbb{R}^n)$, J. Math. Anal. Appl. **351** (2009), 428–439.
170. A. Zygmund, Smooth functions, Duke Math. J. **12** (1945), 47–76.

Index

Lecture Notes in Mathematics

For information about earlier volumes
please contact your bookseller or Springer
LNM Online archive: springerlink.com

Vol. 1817: E. Koelink, W. Van Assche (Eds.), Orthogonal Polynomials and Special Functions. Leuven 2002 (2003)
Vol. 1818: M. Bildhauer, Convex Variational Problems with Linear, nearly Linear and/or Anisotropic Growth Conditions (2003)
Vol. 1819: D. Masser, Yu. V. Nesterenko, H. P. Schlickewei, W. M. Schmidt, M. Waldschmidt, Diophantine Approximation. Cetraro, Italy 2000. Editors: F. Amoroso, U. Zannier (2003)
Vol. 1820: F. Hiai, H. Kosaki, Means of Hilbert Space Operators (2003)
Vol. 1821: S. Teufel, Adiabatic Perturbation Theory in Quantum Dynamics (2003)
Vol. 1822: S.-N. Chow, R. Conti, R. Johnson, J. Mallet-Paret, R. Nussbaum, Dynamical Systems. Cetraro, Italy 2000. Editors: J. W. Macki, P. Zecca (2003)
Vol. 1823: A. M. Anile, W. Allegretto, C. Ringhofer, Mathematical Problems in Semiconductor Physics. Cetraro, Italy 1998. Editor: A. M. Anile (2003)
Vol. 1824: J. A. Navarro González, J. B. Sancho de Salas, $\mathscr{C}^\infty$ – Differentiable Spaces (2003)
Vol. 1825: J. H. Bramble, A. Cohen, W. Dahmen, Multiscale Problems and Methods in Numerical Simulations, Martina Franca, Italy 2001. Editor: C. Canuto (2003)
Vol. 1826: K. Dohmen, Improved Bonferroni Inequalities via Abstract Tubes. Inequalities and Identities of Inclusion-Exclusion Type. VIII, 113 p, 2003.
Vol. 1827: K. M. Pilgrim, Combinations of Complex Dynamical Systems. IX, 118 p, 2003.
Vol. 1828: D. J. Green, Grbner Bases and the Computation of Group Cohomology. XII, 138 p, 2003.
Vol. 1829: E. Altman, B. Gaujal, A. Hordijk, Discrete-Event Control of Stochastic Networks: Multimodularity and Regularity. XIV, 313 p, 2003.
Vol. 1830: M. I. Gil', Operator Functions and Localization of Spectra. XIV, 256 p, 2003.
Vol. 1831: A. Connes, J. Cuntz, E. Guentner, N. Higson, J. E. Kaminker, Noncommutative Geometry, Martina Franca, Italy 2002. Editors: S. Doplicher, L. Longo (2004)
Vol. 1832: J. Azéma, M. Émery, M. Ledoux, M. Yor (Eds.), Séminaire de Probabilités XXXVII (2003)
Vol. 1833: D.-Q. Jiang, M. Qian, M.-P. Qian, Mathematical Theory of Nonequilibrium Steady States. On the Frontier of Probability and Dynamical Systems. IX, 280 p, 2004.
Vol. 1834: Yo. Yomdin, G. Comte, Tame Geometry with Application in Smooth Analysis. VIII, 186 p, 2004.
Vol. 1835: O.T. Izhboldin, B. Kahn, N.A. Karpenko, A. Vishik, Geometric Methods in the Algebraic Theory of Quadratic Forms. Summer School, Lens, 2000. Editor: J.-P. Tignol (2004)
Vol. 1836: C. Năstăsescu, F. Van Oystaeyen, Methods of Graded Rings. XIII, 304 p, 2004.
Vol. 1837: S. Tavaré, O. Zeitouni, Lectures on Probability Theory and Statistics. Ecole d'Eté de Probabilités de Saint-Flour XXXI-2001. Editor: J. Picard (2004)
Vol. 1838: A.J. Ganesh, N.W. O'Connell, D.J. Wischik, Big Queues. XII, 254 p, 2004.
Vol. 1839: R. Gohm, Noncommutative Stationary Processes. VIII, 170 p, 2004.
Vol. 1840: B. Tsirelson, W. Werner, Lectures on Probability Theory and Statistics. Ecole d'Eté de Probabilités de Saint-Flour XXXII-2002. Editor: J. Picard (2004)
Vol. 1841: W. Reichel, Uniqueness Theorems for Variational Problems by the Method of Transformation Groups (2004)
Vol. 1842: T. Johnsen, A. L. Knutsen, K_3 Projective Models in Scrolls (2004)
Vol. 1843: B. Jefferies, Spectral Properties of Noncommuting Operators (2004)
Vol. 1844: K.F. Siburg, The Principle of Least Action in Geometry and Dynamics (2004)
Vol. 1845: Min Ho Lee, Mixed Automorphic Forms, Torus Bundles, and Jacobi Forms (2004)
Vol. 1846: H. Ammari, H. Kang, Reconstruction of Small Inhomogeneities from Boundary Measurements (2004)
Vol. 1847: T.R. Bielecki, T. Bjrk, M. Jeanblanc, M. Rutkowski, J.A. Scheinkman, W. Xiong, Paris-Princeton Lectures on Mathematical Finance 2003 (2004)
Vol. 1848: M. Abate, J. E. Fornaess, X. Huang, J. P. Rosay, A. Tumanov, Real Methods in Complex and CR Geometry, Martina Franca, Italy 2002. Editors: D. Zaitsev, G. Zampieri (2004)
Vol. 1849: Martin L. Brown, Heegner Modules and Elliptic Curves (2004)
Vol. 1850: V. D. Milman, G. Schechtman (Eds.), Geometric Aspects of Functional Analysis. Israel Seminar 2002-2003 (2004)
Vol. 1851: O. Catoni, Statistical Learning Theory and Stochastic Optimization (2004)
Vol. 1852: A.S. Kechris, B.D. Miller, Topics in Orbit Equivalence (2004)
Vol. 1853: Ch. Favre, M. Jonsson, The Valuative Tree (2004)
Vol. 1854: O. Saeki, Topology of Singular Fibers of Differential Maps (2004)
Vol. 1855: G. Da Prato, P.C. Kunstmann, I. Lasiecka, A. Lunardi, R. Schnaubelt, L. Weis, Functional Analytic Methods for Evolution Equations. Editors: M. Iannelli, R. Nagel, S. Piazzera (2004)
Vol. 1856: K. Back, T.R. Bielecki, C. Hipp, S. Peng, W. Schachermayer, Stochastic Methods in Finance, Bressanone/Brixen, Italy, 2003. Editors: M. Fritelli, W. Runggaldier (2004)
Vol. 1857: M. Émery, M. Ledoux, M. Yor (Eds.), Séminaire de Probabilités XXXVIII (2005)
Vol. 1858: A.S. Cherny, H.-J. Engelbert, Singular Stochastic Differential Equations (2005)
Vol. 1859: E. Letellier, Fourier Transforms of Invariant Functions on Finite Reductive Lie Algebras (2005)
Vol. 1860: A. Borisyuk, G.B. Ermentrout, A. Friedman, D. Terman, Tutorials in Mathematical Biosciences I. Mathematical Neurosciences (2005)

Vol. 1861: G. Benettin, J. Henrard, S. Kuksin, Hamiltonian Dynamics – Theory and Applications, Cetraro, Italy, 1999. Editor: A. Giorgilli (2005)
Vol. 1862: B. Helffer, F. Nier, Hypoelliptic Estimates and Spectral Theory for Fokker-Planck Operators and Witten Laplacians (2005)
Vol. 1863: H. Führ, Abstract Harmonic Analysis of Continuous Wavelet Transforms (2005)
Vol. 1864: K. Efstathiou, Metamorphoses of Hamiltonian Systems with Symmetries (2005)
Vol. 1865: D. Applebaum, B.V. R. Bhat, J. Kustermans, J. M. Lindsay, Quantum Independent Increment Processes I. From Classical Probability to Quantum Stochastic Calculus. Editors: M. Schürmann, U. Franz (2005)
Vol. 1866: O.E. Barndorff-Nielsen, U. Franz, R. Gohm, B. Kümmerer, S. Thorbjønsen, Quantum Independent Increment Processes II. Structure of Quantum Lévy Processes, Classical Probability, and Physics. Editors: M. Schürmann, U. Franz, (2005)
Vol. 1867: J. Sneyd (Ed.), Tutorials in Mathematical Biosciences II. Mathematical Modeling of Calcium Dynamics and Signal Transduction. (2005)
Vol. 1868: J. Jorgenson, S. Lang, $Pos_n(R)$ and Eisenstein Series. (2005)
Vol. 1869: A. Dembo, T. Funaki, Lectures on Probability Theory and Statistics. Ecole d'Eté de Probabilités de Saint-Flour XXXIII-2003. Editor: J. Picard (2005)
Vol. 1870: V.I. Gurariy, W. Lusky, Geometry of Mntz Spaces and Related Questions. (2005)
Vol. 1871: P. Constantin, G. Gallavotti, A.V. Kazhikhov, Y. Meyer, S. Ukai, Mathematical Foundation of Turbulent Viscous Flows, Martina Franca, Italy, 2003. Editors: M. Cannone, T. Miyakawa (2006)
Vol. 1872: A. Friedman (Ed.), Tutorials in Mathematical Biosciences III. Cell Cycle, Proliferation, and Cancer (2006)
Vol. 1873: R. Mansuy, M. Yor, Random Times and Enlargements of Filtrations in a Brownian Setting (2006)
Vol. 1874: M. Yor, M. Émery (Eds.), In Memoriam Paul-Andr Meyer - Sminaire de Probabilits XXXIX (2006)
Vol. 1875: J. Pitman, Combinatorial Stochastic Processes. Ecole d'Et de Probabilits de Saint-Flour XXXII-2002. Editor: J. Picard (2006)
Vol. 1876: H. Herrlich, Axiom of Choice (2006)
Vol. 1877: J. Steuding, Value Distributions of L-Functions (2007)
Vol. 1878: R. Cerf, The Wulff Crystal in Ising and Percolation Models, Ecole d'Et de Probabilités de Saint-Flour XXXIV-2004. Editor: Jean Picard (2006)
Vol. 1879: G. Slade, The Lace Expansion and its Applications, Ecole d'Et de Probabilits de Saint-Flour XXXIV-2004. Editor: Jean Picard (2006)
Vol. 1880: S. Attal, A. Joye, C.-A. Pillet, Open Quantum Systems I, The Hamiltonian Approach (2006)
Vol. 1881: S. Attal, A. Joye, C.-A. Pillet, Open Quantum Systems II, The Markovian Approach (2006)
Vol. 1882: S. Attal, A. Joye, C.-A. Pillet, Open Quantum Systems III, Recent Developments (2006)
Vol. 1883: W. Van Assche, F. Marcellàn (Eds.), Orthogonal Polynomials and Special Functions, Computation and Application (2006)
Vol. 1884: N. Hayashi, E.I. Kaikina, P.I. Naumkin, I.A. Shishmarev, Asymptotics for Dissipative Nonlinear Equations (2006)
Vol. 1885: A. Telcs, The Art of Random Walks (2006)
Vol. 1886: S. Takamura, Splitting Deformations of Degenerations of Complex Curves (2006)
Vol. 1887: K. Habermann, L. Habermann, Introduction to Symplectic Dirac Operators (2006)
Vol. 1888: J. van der Hoeven, Transseries and Real Differential Algebra (2006)
Vol. 1889: G. Osipenko, Dynamical Systems, Graphs, and Algorithms (2006)
Vol. 1890: M. Bunge, J. Funk, Singular Coverings of Toposes (2006)
Vol. 1891: J.B. Friedlander, D.R. Heath-Brown, H. Iwaniec, J. Kaczorowski, Analytic Number Theory, Cetraro, Italy, 2002. Editors: A. Perelli, C. Viola (2006)
Vol. 1892: A. Baddeley, I. Bárány, R. Schneider, W. Weil, Stochastic Geometry, Martina Franca, Italy, 2004. Editor: W. Weil (2007)
Vol. 1893: H. Hanßmann, Local and Semi-Local Bifurcations in Hamiltonian Dynamical Systems, Results and Examples (2007)
Vol. 1894: C.W. Groetsch, Stable Approximate Evaluation of Unbounded Operators (2007)
Vol. 1895: L. Molnár, Selected Preserver Problems on Algebraic Structures of Linear Operators and on Function Spaces (2007)
Vol. 1896: P. Massart, Concentration Inequalities and Model Selection, Ecole d'Été de Probabilités de Saint-Flour XXXIII-2003. Editor: J. Picard (2007)
Vol. 1897: R. Doney, Fluctuation Theory for Lévy Processes, Ecole d'Été de Probabilités de Saint-Flour XXXV-2005. Editor: J. Picard (2007)
Vol. 1898: H.R. Beyer, Beyond Partial Differential Equations, On linear and Quasi-Linear Abstract Hyperbolic Evolution Equations (2007)
Vol. 1899: Séminaire de Probabilités XL. Editors: C. Donati-Martin, M. Émery, A. Rouault, C. Stricker (2007)
Vol. 1900: E. Bolthausen, A. Bovier (Eds.), Spin Glasses (2007)
Vol. 1901: O. Wittenberg, Intersections de deux quadriques et pinceaux de courbes de genre 1, Intersections of Two Quadrics and Pencils of Curves of Genus 1 (2007)
Vol. 1902: A. Isaev, Lectures on the Automorphism Groups of Kobayashi-Hyperbolic Manifolds (2007)
Vol. 1903: G. Kresin, V. Maz'ya, Sharp Real-Part Theorems (2007)
Vol. 1904: P. Giesl, Construction of Global Lyapunov Functions Using Radial Basis Functions (2007)
Vol. 1905: C. Prévôt, M. Röckner, A Concise Course on Stochastic Partial Differential Equations (2007)
Vol. 1906: T. Schuster, The Method of Approximate Inverse: Theory and Applications (2007)
Vol. 1907: M. Rasmussen, Attractivity and Bifurcation for Nonautonomous Dynamical Systems (2007)
Vol. 1908: T.J. Lyons, M. Caruana, T. Lévy, Differential Equations Driven by Rough Paths, Ecole d'Été de Probabilités de Saint-Flour XXXIV-2004 (2007)
Vol. 1909: H. Akiyoshi, M. Sakuma, M. Wada, Y. Yamashita, Punctured Torus Groups and 2-Bridge Knot Groups (I) (2007)
Vol. 1910: V.D. Milman, G. Schechtman (Eds.), Geometric Aspects of Functional Analysis. Israel Seminar 2004-2005 (2007)
Vol. 1911: A. Bressan, D. Serre, M. Williams, K. Zumbrun, Hyperbolic Systems of Balance Laws. Cetraro, Italy 2003. Editor: P. Marcati (2007)

Vol. 1912: V. Berinde, Iterative Approximation of Fixed Points (2007)
Vol. 1913: J.E. Marsden, G. Misiołek, J.-P. Ortega, M. Perlmutter, T.S. Ratiu, Hamiltonian Reduction by Stages (2007)
Vol. 1914: G. Kutyniok, Affine Density in Wavelet Analysis (2007)
Vol. 1915: T. Bıyıkoğlu, J. Leydold, P.F. Stadler, Laplacian Eigenvectors of Graphs. Perron-Frobenius and Faber-Krahn Type Theorems (2007)
Vol. 1916: C. Villani, F. Rezakhanlou, Entropy Methods for the Boltzmann Equation. Editors: F. Golse, S. Olla (2008)
Vol. 1917: I. Veselić, Existence and Regularity Properties of the Integrated Density of States of Random Schrdinger (2008)
Vol. 1918: B. Roberts, R. Schmidt, Local Newforms for GSp(4) (2007)
Vol. 1919: R.A. Carmona, I. Ekeland, A. Kohatsu-Higa, J.-M. Lasry, P.-L. Lions, H. Pham, E. Taflin, Paris-Princeton Lectures on Mathematical Finance 2004. Editors: R.A. Carmona, E. inlar, I. Ekeland, E. Jouini, J.A. Scheinkman, N. Touzi (2007)
Vol. 1920: S.N. Evans, Probability and Real Trees. Ecole d'Été de Probabilités de Saint-Flour XXXV-2005 (2008)
Vol. 1921: J.P. Tian, Evolution Algebras and their Applications (2008)
Vol. 1922: A. Friedman (Ed.), Tutorials in Mathematical BioSciences IV. Evolution and Ecology (2008)
Vol. 1923: J.P.N. Bishwal, Parameter Estimation in Stochastic Differential Equations (2008)
Vol. 1924: M. Wilson, Littlewood-Paley Theory and Exponential-Square Integrability (2008)
Vol. 1925: M. du Sautoy, L. Woodward, Zeta Functions of Groups and Rings (2008)
Vol. 1926: L. Barreira, V. Claudia, Stability of Nonautonomous Differential Equations (2008)
Vol. 1927: L. Ambrosio, L. Caffarelli, M.G. Crandall, L.C. Evans, N. Fusco, Calculus of Variations and Non-Linear Partial Differential Equations. Cetraro, Italy 2005. Editors: B. Dacorogna, P. Marcellini (2008)
Vol. 1928: J. Jonsson, Simplicial Complexes of Graphs (2008)
Vol. 1929: Y. Mishura, Stochastic Calculus for Fractional Brownian Motion and Related Processes (2008)
Vol. 1930: J.M. Urbano, The Method of Intrinsic Scaling. A Systematic Approach to Regularity for Degenerate and Singular PDEs (2008)
Vol. 1931: M. Cowling, E. Frenkel, M. Kashiwara, A. Valette, D.A. Vogan, Jr., N.R. Wallach, Representation Theory and Complex Analysis. Venice, Italy 2004. Editors: E.C. Tarabusi, A. D'Agnolo, M. Picardello (2008)
Vol. 1932: A.A. Agrachev, A.S. Morse, E.D. Sontag, H.J. Sussmann, V.I. Utkin, Nonlinear and Optimal Control Theory. Cetraro, Italy 2004. Editors: P. Nistri, G. Stefani (2008)
Vol. 1933: M. Petkovic, Point Estimation of Root Finding Methods (2008)
Vol. 1934: C. Donati-Martin, M. Émery, A. Rouault, C. Stricker (Eds.), Séminaire de Probabilités XLI (2008)
Vol. 1935: A. Unterberger, Alternative Pseudodifferential Analysis (2008)
Vol. 1936: P. Magal, S. Ruan (Eds.), Structured Population Models in Biology and Epidemiology (2008)
Vol. 1937: G. Capriz, P. Giovine, P.M. Mariano (Eds.), Mathematical Models of Granular Matter (2008)
Vol. 1938: D. Auroux, F. Catanese, M. Manetti, P. Seidel, B. Siebert, I. Smith, G. Tian, Symplectic 4-Manifolds and Algebraic Surfaces. Cetraro, Italy 2003. Editors: F. Catanese, G. Tian (2008)
Vol. 1939: D. Boffi, F. Brezzi, L. Demkowicz, R.G. Durán, R.S. Falk, M. Fortin, Mixed Finite Elements, Compatibility Conditions, and Applications. Cetraro, Italy 2006. Editors: D. Boffi, L. Gastaldi (2008)
Vol. 1940: J. Banasiak, V. Capasso, M.A.J. Chaplain, M. Lachowicz, J. Miękisz, Multiscale Problems in the Life Sciences. From Microscopic to Macroscopic. Będlewo, Poland 2006. Editors: V. Capasso, M. Lachowicz (2008)
Vol. 1941: S.M.J. Haran, Arithmetical Investigations. Representation Theory, Orthogonal Polynomials, and Quantum Interpolations (2008)
Vol. 1942: S. Albeverio, F. Flandoli, Y.G. Sinai, SPDE in Hydrodynamic. Recent Progress and Prospects. Cetraro, Italy 2005. Editors: G. Da Prato, M. Rckner (2008)
Vol. 1943: L.L. Bonilla (Ed.), Inverse Problems and Imaging. Martina Franca, Italy 2002 (2008)
Vol. 1944: A. Di Bartolo, G. Falcone, P. Plaumann, K. Strambach, Algebraic Groups and Lie Groups with Few Factors (2008)
Vol. 1945: F. Brauer, P. van den Driessche, J. Wu (Eds.), Mathematical Epidemiology (2008)
Vol. 1946: G. Allaire, A. Arnold, P. Degond, T.Y. Hou, Quantum Transport. Modelling, Analysis and Asymptotics. Cetraro, Italy 2006. Editors: N.B. Abdallah, G. Frosali (2008)
Vol. 1947: D. Abramovich, M. Mariño, M. Thaddeus, R. Vakil, Enumerative Invariants in Algebraic Geometry and String Theory. Cetraro, Italy 2005. Editors: K. Behrend, M. Manetti (2008)
Vol. 1948: F. Cao, J-L. Lisani, J-M. Morel, P. Mus, F. Sur, A Theory of Shape Identification (2008)
Vol. 1949: H.G. Feichtinger, B. Helffer, M.P. Lamoureux, N. Lerner, J. Toft, Pseudo-Differential Operators. Quantization and Signals. Cetraro, Italy 2006. Editors: L. Rodino, M.W. Wong (2008)
Vol. 1950: M. Bramson, Stability of Queueing Networks, Ecole d'Eté de Probabilits de Saint-Flour XXXVI-2006 (2008)
Vol. 1951: A. Moltó, J. Orihuela, S. Troyanski, M. Valdivia, A Non Linear Transfer Technique for Renorming (2009)
Vol. 1952: R. Mikhailov, I.B.S. Passi, Lower Central and Dimension Series of Groups (2009)
Vol. 1953: K. Arwini, C.T.J. Dodson, Information Geometry (2008)
Vol. 1954: P. Biane, L. Bouten, F. Cipriani, N. Konno, N. Privault, Q. Xu, Quantum Potential Theory. Editors: U. Franz, M. Schuermann (2008)
Vol. 1955: M. Bernot, V. Caselles, J.-M. Morel, Optimal Transportation Networks (2008)
Vol. 1956: C.H. Chu, Matrix Convolution Operators on Groups (2008)
Vol. 1957: A. Guionnet, On Random Matrices: Macroscopic Asymptotics, Ecole d'Eté de Probabilits de Saint-Flour XXXVI-2006 (2009)
Vol. 1958: M.C. Olsson, Compactifying Moduli Spaces for Abelian Varieties (2008)
Vol. 1959: Y. Nakkajima, A. Shiho, Weight Filtrations on Log Crystalline Cohomologies of Families of Open Smooth Varieties (2008)
Vol. 1960: J. Lipman, M. Hashimoto, Foundations of Grothendieck Duality for Diagrams of Schemes (2009)

Vol. 1961: G. Buttazzo, A. Pratelli, S. Solimini, E. Stepanov, Optimal Urban Networks via Mass Transportation (2009)
Vol. 1962: R. Dalang, D. Khoshnevisan, C. Mueller, D. Nualart, Y. Xiao, A Minicourse on Stochastic Partial Differential Equations (2009)
Vol. 1963: W. Siegert, Local Lyapunov Exponents (2009)
Vol. 1964: W. Roth, Operator-valued Measures and Integrals for Cone-valued Functions and Integrals for Cone-valued Functions (2009)
Vol. 1965: C. Chidume, Geometric Properties of Banach Spaces and Nonlinear Iterations (2009)
Vol. 1966: D. Deng, Y. Han, Harmonic Analysis on Spaces of Homogeneous Type (2009)
Vol. 1967: B. Fresse, Modules over Operads and Functors (2009)
Vol. 1968: R. Weissauer, Endoscopy for GSP(4) and the Cohomology of Siegel Modular Threefolds (2009)
Vol. 1969: B. Roynette, M. Yor, Penalising Brownian Paths (2009)
Vol. 1970: M. Biskup, A. Bovier, F. den Hollander, D. Ioffe, F. Martinelli, K. Netočný, F. Toninelli, Methods of Contemporary Mathematical Statistical Physics. Editor: R. Kotecký (2009)
Vol. 1971: L. Saint-Raymond, Hydrodynamic Limits of the Boltzmann Equation (2009)
Vol. 1972: T. Mochizuki, Donaldson Type Invariants for Algebraic Surfaces (2009)
Vol. 1973: M.A. Berger, L.H. Kauffmann, B. Khesin, H.K. Moffatt, R.L. Ricca, De W. Sumners, Lectures on Topological Fluid Mechanics. Cetraro, Italy 2001. Editor: R.L. Ricca (2009)
Vol. 1974: F. den Hollander, Random Polymers: École d'Été de Probabilités de Saint-Flour XXXVII – 2007 (2009)
Vol. 1975: J.C. Rohde, Cyclic Coverings, Calabi-Yau Manifolds and Complex Multiplication (2009)
Vol. 1976: N. Ginoux, The Dirac Spectrum (2009)
Vol. 1977: M.J. Gursky, E. Lanconelli, A. Malchiodi, G. Tarantello, X.-J. Wang, P.C. Yang, Geometric Analysis and PDEs. Cetraro, Italy 2001. Editors: A. Ambrosetti, S.-Y.A. Chang, A. Malchiodi (2009)
Vol. 1978: M. Qian, J.-S. Xie, S. Zhu, Smooth Ergodic Theory for Endomorphisms (2009)
Vol. 1979: C. Donati-Martin, M. Émery, A. Rouault, C. Stricker (Eds.), Śeminaire de Probablitiés XLII (2009)
Vol. 1980: P. Graczyk, A. Stos (Eds.), Potential Analysis of Stable Processes and its Extensions (2009)
Vol. 1981: M. Chlouveraki, Blocks and Families for Cyclotomic Hecke Algebras (2009)
Vol. 1982: N. Privault, Stochastic Analysis in Discrete and Continuous Settings. With Normal Martingales (2009)
Vol. 1983: H. Ammari (Ed.), Mathematical Modeling in Biomedical Imaging I. Electrical and Ultrasound Tomographies, Anomaly Detection, and Brain Imaging (2009)
Vol. 1984: V. Caselles, P. Monasse, Geometric Description of Images as Topographic Maps (2010)
Vol. 1985: T. Linß, Layer-Adapted Meshes for Reaction-Convection-Diffusion Problems (2010)
Vol. 1986: J.-P. Antoine, C. Trapani, Partial Inner Product Spaces. Theory and Applications (2009)
Vol. 1987: J.-P. Brasselet, J. Seade, T. Suwa, Vector Fields on Singular Varieties (2010)
Vol. 1988: M. Broué, Introduction to Complex Reflection Groups and Their Braid Groups (2010)
Vol. 1989: I.M. Bomze, V. Demyanov, Nonlinear Optimization. Cetraro, Italy 2007. Editors: G. di Pillo, F. Schoen (2010)
Vol. 1990: S. Bouc, Biset Functors for Finite Groups (2010)
Vol. 1991: F. Gazzola, H.-C. Grunau, G. Sweers, Polyharmonic Boundary Value Problems (2010)
Vol. 1992: A. Parmeggiani, Spectral Theory of Non-Commutative Harmonic Oscillators: An Introduction (2010)
Vol. 1993: P. Dodos, Banach Spaces and Descriptive Set Theory: Selected Topics (2010)
Vol. 1994: A. Baricz, Generalized Bessel Functions of the First Kind (2010)
Vol. 1995: A.Y. Khapalov, Controllability of Partial Differential Equations Governed by Multiplicative Controls (2010)
Vol. 1996: T. Lorenz, Mutational Analysis. A Joint Framework for Cauchy Problems *In* and *Beyond* Vector Spaces (2010)
Vol. 1997: M. Banagl, Intersection Spaces, Spatial Homology Truncation, and String Theory (2010)
Vol. 1998: M. Abate, E. Bedford, M. Brunella, T.-C. Dinh, D. Schleicher, N. Sibony, Holomorphic Dynamical Systems. Cetraro, Italy 2008. Editors: G. Gentili, J. Guenot, G. Patrizio (2010)
Vol. 1999: H. Schoutens, The Use of Ultraproducts in Commutative Algebra (2010)
Vol. 2000: H. Yserentant, Regularity and Approximability of Electronic Wave Functions (2010)
Vol. 2001: O.E. Barndorff-Nielson, J. Bertoin, J. Jacod, C. Kl"uppelberg (Eds.), Lévy Matters I (2010)
Vol. 2002: C. Pötzsche, Geometric Theory of Discrete Nonautonomous Dynamical Systems (2010)
Vol. 2003: A. Cousin, S. Crépey, O. Guéant, D. Hobson, M. Jeanblanc, J.-M. Lasry, J.-P. Laurent, P.-L. Lions, P. Tankov, Paris-Princeton Lectures on Mathematical Finance 2010. Editors: R.A. Carmona, E. Cinlar, I. Ekeland, E. Jouini, J.A. Scheinkman, N. Touzi (2010)
Vol. 2004: K. Diethelm, The Analysis of Fractional Differential Equations (2010)
Vol. 2005: W. Yuan, W. Sickel, D. Yang, Morrey and Campanato Meet Besov, Lizorkin and Triebel (2010)

Recent Reprints and New Editions

Vol. 1702: J. Ma, J. Yong, Forward-Backward Stochastic Differential Equations and their Applications. 1999 – Corr. 3rd printing (2007)
Vol. 830: J.A. Green, Polynomial Representations of GL_n, with an Appendix on Schensted Correspondence and Littelmann Paths by K. Erdmann, J.A. Green and M. Schoker 1980 – 2nd corr. and augmented edition (2007)
Vol. 1693: S. Simons, From Hahn-Banach to Monotonicity (Minimax and Monotonicity 1998) – 2nd exp. edition (2008)
Vol. 470: R.E. Bowen, Equilibrium States and the Ergodic Theory of Anosov Diffeomorphisms. With a preface by D. Ruelle. Edited by J.-R. Chazottes. 1975 – 2nd rev. edition (2008)
Vol. 523: S.A. Albeverio, R.J. Høegh-Krohn, S. Mazzucchi, Mathematical Theory of Feynman Path Integral. 1976 – 2nd corr. and enlarged edition (2008)
Vol. 1764: A. Cannas da Silva, Lectures on Symplectic Geometry 2001 – Corr. 2nd printing (2008)

LECTURE NOTES IN MATHEMATICS **Springer**

Edited by J.-M. Morel, F. Takens, B. Teissier, P.K. Maini

Editorial Policy (for the publication of monographs)

1. Lecture Notes aim to report new developments in all areas of mathematics and their applications - quickly, informally and at a high level. Mathematical texts analysing new developments in modelling and numerical simulation are welcome.
   Monograph manuscripts should be reasonably self-contained and rounded off. Thus they may, and often will, present not only results of the author but also related work by other people. They may be based on specialised lecture courses. Furthermore, the manuscripts should provide sufficient motivation, examples and applications. This clearly distinguishes Lecture Notes from journal articles or technical reports which normally are very concise. Articles intended for a journal but too long to be accepted by most journals, usually do not have this "lecture notes" character. For similar reasons it is unusual for doctoral theses to be accepted for the Lecture Notes series, though habilitation theses may be appropriate.

2. Manuscripts should be submitted either to Springer's mathematics editorial in Heidelberg, or to one of the series editors. In general, manuscripts will be sent out to 2 external referees for evaluation. If a decision cannot yet be reached on the basis of the first 2 reports, further referees may be contacted: The author will be informed of this. A final decision to publish can be made only on the basis of the complete manuscript, however a refereeing process leading to a preliminary decision can be based on a pre-final or incomplete manuscript. The strict minimum amount of material that will be considered should include a detailed outline describing the planned contents of each chapter, a bibliography and several sample chapters.
   Authors should be aware that incomplete or insufficiently close to final manuscripts almost always result in longer refereeing times and nevertheless unclear referees' recommendations, making further refereeing of a final draft necessary.
   Authors should also be aware that parallel submission of their manuscript to another publisher while under consideration for LNM will in general lead to immediate rejection.

3. Manuscripts should in general be submitted in English. Final manuscripts should contain at least 100 pages of mathematical text and should always include

   - a table of contents;
   - an informative introduction, with adequate motivation and perhaps some historical remarks: it should be accessible to a reader not intimately familiar with the topic treated;
   - a subject index: as a rule this is genuinely helpful for the reader.

   For evaluation purposes, manuscripts may be submitted in print or electronic form, in the latter case preferably as pdf- or zipped ps-files. Lecture Notes volumes are, as a rule, printed digitally from the authors' files. To ensure best results, authors are asked to use the LaTeX2e style files available from Springer's web-server at:

   ftp://ftp.springer.de/pub/tex/latex/svmonot1/ (for monographs).
   ftp://ftp.springer.de/pub/tex/latex/svmultt1/ (for summer schools/tutorials).

Additional technical instructions, if necessary, are available on request from: lnm@springer.com.

4. Careful preparation of the manuscripts will help keep production time short besides ensuring satisfactory appearance of the finished book in print and online. After acceptance of the manuscript authors will be asked to prepare the final LaTeX source files (and also the corresponding dvi-, pdf- or zipped ps-file) together with the final printout made from these files. The LaTeX source files are essential for producing the full-text online version of the book (see www.springerlink.com/content/110312 for the existing online volumes of LNM).
   The actual production of a Lecture Notes volume takes approximately 12 weeks.

5. Authors receive a total of 50 free copies of their volume, but no royalties. They are entitled to a discount of 33.3% on the price of Springer books purchased for their personal use, if ordering directly from Springer.

6. Commitment to publish is made by letter of intent rather than by signing a formal contract. Springer-Verlag secures the copyright for each volume. Authors are free to reuse material contained in their LNM volumes in later publications: a brief written (or e-mail) request for formal permission is sufficient.

Addresses:

Professor J.-M. Morel, CMLA,
École Normale Supérieure de Cachan,
61 Avenue du Président Wilson, 94235 Cachan Cedex, France
E-mail: Jean-Michel.Morel@cmla.ens-cachan.fr

Professor F. Takens, Mathematisch Instituut,
Rijksuniversiteit Groningen, Postbus 800,
9700 AV Groningen, The Netherlands
E-mail: F.Takens@math.rug.nl

Professor B. Teissier, Institut Mathématique de Jussieu,
UMR 7586 du CNRS, Équipe "Géométrie et Dynamique",
175 rue du Chevaleret
75013 Paris, France
E-mail: teissier@math.jussieu.fr

For the "Mathematical Biosciences Subseries" of LNM:

Professor P.K. Maini, Center for Mathematical Biology,
Mathematical Institute, 24-29 St Giles,
Oxford OX1 3LP, UK
E-mail: maini@maths.ox.ac.uk

Springer, Mathematics Editorial I, Tiergartenstr. 17
69121 Heidelberg, Germany,
Tel.: +49 (6221) 487-8259
Fax: +49 (6221) 4876-8259
E-mail: lnm@springer.com